D1222117

*Origin and Evolution
of Gymnosperms*

Origin and Evolution
of Gymnosperms

EDITED BY

CHARLES B. BECK

ROBERT MANNING
STROZIER LIBRARY

JUL 26 1989

Tallahassee, Florida

Columbia University Press
New York

Sci
QE
975
O75
1988

ROBERT MANNING
STROZIER LIBRARY

JUL 26 1989

Tallahassee, Florida

Library of Congress Cataloging-in-Publication Data
Origin and evolution of gymnosperms / edited by Charles B. Beck.
p. cm.
Includes bibliographies and index.
ISBN 0-231-06358-X
1. Gymnosperms—Origin. 2. Gymnosperms—Evolution.
3. Paleobotany. I. Beck, Charles B.
QK494.O75 1988
561'.5—dc19   88-14758
CIP

Columbia University Press
New York   Guildford, Surrey

Copyright © 1988 Columbia University Press
All rights reserved
Printed in the United States of America

Casebound editions of Columbia University Press books are Smyth-sewn
and are printed on permanent and durable acid-free paper

*Dedicated to the memory of Sergei V. Meyen,
friend, distinguished colleague, original thinker*

Contents

Contributors

Charles B. Beck is Professor of Botany and Director of the Museum of Paleontology at the University of Michigan, Ann Arbor. Since his discovery in 1960 of the organic connection between *Archaeopteris* and *Callixylon,* his primary research interests have been the progymnosperms and early seed plants of the Paleozoic, and their phylogenetic relationships. He is currently president of the International Organization of Paleobotany.

Johanna A. Clement-Westerhof is an adjunct collaborator in the Laboratory of Palaeobotany and Palynology of the State University of Utrecht, the Netherlands. Her studies of the Permian conifer families Walchiaceae and Majoniaceae have provided new hypotheses of conifer phylogeny. She has a special interest in conifer cuticles and, in the future, plans to utilize cuticle characters in phylogenetic analyses.

Peter R. Crane is Associate Curator of Paleobotany at the Field Museum of Natural History and Lecturer in the Committee on Evolutionary Biology at the University of Chicago. His research focuses on the biology, ecology, and systematics of fossil flowering plants, and one of his recent interests has been the application of cladistic techniques to resolving relationships among angiosperms and other groups of seed plants. He is a recipient of the

Bicentenary Medal of the Linnean Society of London, and is a former coeditor of the journal *Paleobiology*.

Jean Galtier is Directeur de Recherche at the CNRS (National Scientific Research Center), Laboratoire de Paléobotanique, Université des Sciences et Techniques at Montpellier, France. His research on Carboniferous floras of Europe concerns, in particular, permineralized plants of both the Lower and Upper Carboniferous. It is directed toward a better understanding of the early diversification and evolution of pteridophytes and early seed plants. He is a vice-president of the International Organization of Paleobotany.

Sergei V. Meyen, who died in 1987, was Head of the Laboratory of Paleofloristics of the Geological Institute of the USSR Academy of Sciences. He was a world authority on the Paleozoic floras of Angaraland and had made major contributions to our knowledge of Paleozoic and Mesozoic phytogeography and the phylogeny of gymnosperms. At the time of his death he was a vice-president of the International Organization of Paleobotany.

Charles N. Miller, Jr. is Professor of Botany at the University of Montana at Missoula. His research on structurally preserved conifer cones from the Mesozoic and Cenozoic of the United States led to an interest in the evolution of the Coniferophyta. His present research involves a detailed analysis of the phylogeny of the Pinaceae, and he plans to conduct similar studies on other conifer families in the future.

Gar W. Rothwell is Professor of Botany at Ohio University, Athens. His detailed studies of a diversity of Paleozoic seed plants have led to hypotheses concerning the origin and evolution of major gymnosperm taxa. He is an authority on the Callistophytales and the Cordaitales, and is currently studying plants from the Lower Carboniferous of Great Britain.

Stephen E. Scheckler is Associate Professor of Biology and Associate Director of the Center for Systematics Collections at Virginia Polytechnic Institute and State University at Blacksburg. His research on Devonian and Carboniferous plants is directed toward a clearer understanding of the biology of progymnosperms and early gymnosperms, the systematics of tree lycopods, and the paleoecology of the plant communities of this time. He is a past chairman of the Paleobotanical Section of the Botanical Society of America.

Thomas N. Taylor is Professor of Botany at The Ohio State University and a member of the Byrd Polar Research Center in Columbus. His research interests include fossil fungi; ultrastructure of pollen, spores, and cuticle; Antarctic fossil floras; reproductive biology in early seed plants; origin and evolution of land plants; and plant/animal interactions in the fossil record. He is a recipient of the Merit Award given for achievement in the plant sciences by the Botanical Society of America, editor of the *Plant Science Bulletin*, and a vice-president of the International Organization of Paleobotany.

Joan Watson is Lecturer in Palaeobotany in the Department of Geology, University of Manchester, England. She started a major revision of the English Wealden flora while a student of T. M. Harris and has since widened her studies to include the German Wealden, U. S. Potomac, and other Lower Cretaceous floras. Her particular interest in the fossil gymnosperms involves establishing a well-authenticated type specimen for each species, on which is based a modern redescription and diagnosis for publication in a series of monographic reference works.

David C. Wight, a recent postdoctoral fellow in the Department of Botany at Ohio University, Athens, is currently a member of the Edison Biotechnology Center there. His interests in the origin and evolution of progymnosperms, utilizing, in particular, characteristics of stelar anatomy, developed during his doctoral studies at the University of Michigan.

Preface

During the past two decades significant investigations of broad evolutionary patterns and phylogenetic relationships among seed plants have been undertaken both by neobotanists and by paleobotanists. Whereas most of these have dealt with angiosperms, gymnosperms have been largely neglected. This book, which considers gymnosperms and their immediate ancestors, the progymnosperms, is an effort to correct this imbalance. It consists of contributions from an illustrious group of authorities who together present the current state of the science. Each paper is a synthesis of recent data and thought in a particular subarea, and most papers also include original ideas and analyses reflecting recent research of the contributors.

This book is timely for several reasons. First, some of Rudolf Florin's time-honored interpretations of early conifer morphology have been contested recently. Second, there have been important discoveries of very primitive seeds from the late Devonian, some with preserved anatomy. Third, anatomically preserved remains of primitive conifer cones and compression specimens of conifer cones with well-preserved cuticles have been discovered in the recent past. These late Pennsylvanian and Permian fossils are providing new information or more accurate information than that from the compression specimens studied by earlier workers and, consequently, are yielding new interpretations. Fourth, material on pteridosperms, cycads, and cordaites demonstrating the organic connection of several organs or branch orders, or new morphological characteristics, has been studied recently in several laboratories. Fifth, gymnosperms of the Angara and Gondwana floras are being studied by several paleobotanists. Finally, new methods of phylogenetic and morphometric analysis are now being employed by many paleobotanists that are leading, in some cases, to new and different interpretations while, in others, corroborating earlier ideas and conclusions.

Topics and contributors have been chosen to emphasize evolutionary patterns and phylogenetic relationships rather than merely descriptions of morphology and anatomy, although the latter necessarily comprise important parts of most contributions. Several papers consist largely of detailed phylogenetic analyses.

The objective of paleobotanical research on gymnosperms is ultimately to understand their origin as well as the pattern and pathways of their evolution. This book, while presenting a summary of some of what is known about these subjects, and significant new information and analyses, also clearly demonstrates how little we really know and how much is yet to be learned. We hope it will stimulate more research that will contribute toward the achievement of our ultimate goals.

I wish to acknowledge the contributions to this book of Dr. William E. Stein, Jr., who has assisted me in many important ways throughout the planning of the project and the preparation of the manuscript for publication. He has been of inestimable service in facilitating the computer-assisted duplication of individual manuscripts, and, more important, he has served as a discriminating and highly knowledgeable reviewer of several manuscripts. Many thanks, Bill.

Origin and Evolution
of Gymnosperms

1

Progymnosperms

CHARLES B. BECK
DAVID C. WIGHT

In 1957, Beck erected the genus *Tetraxylopteris* and considered it to be "a pteridosperm precursor." Whereas the original conception of progymnosperms can be traced to this work, the group, Progymnospermopsida, was not formally established until three years later upon discovery that *Archaeopteris* and *Callixylon* represented parts of the same plant (Beck 1960a, 1960b).

Initially, three orders, Aneurophytales, Protopityales, and Pityales, were recognized as progymnosperms on the basis of their gymnospermous anatomy and free-sporing, pteridophytic reproduction. All produced secondary xylem with characteristics identical to those of gymnosperms, and all produced a phloem-like layer peripheral to the secondary xylem. The groups thus possessed a bifacial cambium.

As currently conceived, the progymnoserms include the three original orders, but the name Pityales was dropped in favor of Archaeopteridales when it was determined (Long 1963) that *Pitus (Pitys)* was probably a pteridosperm.

The progymnosperms, which extend from Middle Devonian (early Eifelian) to Lower Mississippian (Tournaisian), may be characterized as follows: Habit shrubby to arborescent; branching pseudomonopodial—i.e., no axillary buds produced; ultimate appendages (leaf homologues) and leaves small, dichotomously branched (or dissected?) or laminate with dichotomous venation, borne in three-dimensional or planate frond-like lateral branch systems. Vascular system a ribbed protostele, a medullated protostele, or a eustele from which leaf traces diverge radially from the ends or ribs of from axial bundles. Order of maturation of primary xylem mesarch in the shoot system. Secondary growth by a bifacial cambium producing secondary xylem and phloem; secondary xylem compact, consisting of tracheids and rays; tracheids bearing circular bordered pits. Reproduction pteridophytic (free-sporing), homosporous, or heterosporous; sporangia elliptical (fusiform); dehiscence by a longitudinal slit; sporangia borne along adaxial to lateral surfaces of pinnate branches of a dichotomous system or on adaxial surfaces of sporophylls.

Since 1960, the Progymnospermopsida has gained widespread acceptance as the most likely group from which seed plants evolved. Our understanding of the group has grown immeasurably through the work of many paleobotanists, and it continues today to be an important group in analyses of seed plant phylogeny (see e.g., Beck 1981; Rothwell 1982; Meyen 1984; Crane 1985a; Doyle and Donoghue 1986). Many questions remain to be answered, however. In this chapter we shall consider some of these, as well as some new insights and information that have appeared since the last comprehensive reviews of the group (Bonamo 1975; Beck 1976).

TAXONOMIC CONCEPTS IN PROGYMNOSPERMS

Taxonomic concepts are based upon and limited by the nature and amount of information that is known about specimens included in the taxon. Difficulties in the evaluation and comparison of taxa arise when the concepts (i.e., the information base) associated with taxonomic names vary widely and thus are only partially or not at all comparable. In paleobotany, the information available varies not only with the rigor and nature of the analysis but with the mode, quality, and extent of specimen preservation as well. These related problems of incomplete information and lack of comparability of evidence have resulted in disagreements both about the delimitation of the class Progymnospermopsida and about the delimitation of taxa within the class (e.g., Bonamo 1975, 1977; Beck 1976; Stein 1982b; Gensel 1984;

Gensel and Andrews 1984; Wight 1985). We shall summarize our opinions on these issues below.

The class Progymnospermopsida was initially established for plants of pteridophytic reproduction that produced gymnospermous secondary vascular tissues (Beck 1960b). Since that time, several workers (Bonamo 1975; Gensel and Andrews 1984) have maintained that the presence of these two characters is necessary for assignment to the class and thus have declined to include taxa for which such evidence is lacking. Although we understand and respect their desire for caution, we disagree with this point of view. Stein (1982b) and Wight (1985), following positions taken earlier by Scheckler and Banks (1971a, 1971b), Matten (1976), Beck (1976), and Stein and Beck (1983), have argued for a more inclusive circumscription of the group. They suggest that membership in the Progymnospermopsida should not necessarily be tied to the demonstrated presence of any specific subset of morphological characteristics found within the group (i.e., gymnospermous secondary vascular tissues and pteridophytic reproduction). Stein's (1982b) argument, in particular, emphasized the difference in relative value between primitive and derived characters for assigning taxa to more inclusive taxonomic groups and inferring evolutionary relationships. He argued that gymnospermous secondary vascular tissues, a derived (i.e., advanced) character thought to have evolved only once in the history of vascular plants (and, therefore, present only in members of the seed plant lineage), is of much greater significance in determining membership in the Progymnospermopsida than pteridophytic reproduction, a primitive character with respect to vascular plants as a whole (see Stein 1982b:619). Although not explicitly stated, this line of argumentation is implicit in the decisions of Scheckler and Banks (1971a, 1971b), Matten (1973), and Beck (1976) to include taxa for which reproductive structures are unknown within the Aneurophytales.

Stein (1982b) and Wight (1985) suggested that it is the demonstration of characters or sets of characters that are unique to the Progymnospermopsida, or to one of the orders within the group, that is critical to resolving questions of membership. Work on progymnosperms since 1960 has provided a broadened morphological basis for comparison between the class and potential members. We believe, therefore, that restricting membership to the demonstrated presence of the two original characters, gymnospermous secondary vascular tissues and the pteridophytic reproduction, ignores the progress that has been made in the last 25 years and utilizes only two out of many potentially useful characters. Within aneurophytes, the precise morphology of the primary vascular system and the morphology of the fertile organs are good examples. There are no data which suggest that another, unrelated group of plants existed that had similar morphological characteristics. Thus, although we agree with Bonamo (1975) and Gensel and Andrews (1984) that demonstration of gymnospermous secondary vascular tissues and pteridophytic reproduction is sufficient to establish membership in the Progymnospermopsida, we also believe that they are not necessary for such an assignment (Scheckler and Banks 1971a, 1971b; Matten 1973; Beck 1976; Stein 1982b; Stein and Beck 1983; Wight 1985). We suggest that each case of potential membership be considered on its own merits, and that all available information—anatomical, morphological, and stratigraphic—be considered.

For example, Stein (1982b) included *Reimannia* in the class, although it exhibited evidence of neither pteridophytic reproduction nor secondary growth. He justified this assignment to the Aneurophytales on the basis of a group of other characters he considered unique to aneurophytes, and on the stratigraphic position of *Reimannia*, and noted that the absence of evidence of secondary growth in a fragment of relatively small diameter does not preclude the possibility of secondary growth in the plant from which it came (e.g., Bonamo 1977; Dannenhoffer and Bonamo 1984).

Similarly, although Bonamo (1975) and Gensel and Andrews (1984) excluded *Siderella*, *Actinopodium*, and *Svalbardia* from the Archaeopteridales, we follow Beck (1976) and include them. We accept *Siderella* as an archaeopteridalean even though there is evidence of neither secondary growth nor pteridophytic reproduction because the pattern of its primary vascular tissues, including the form and pattern of divergence of traces to lateral appendages, is characteristic of that of *Archaeopteris*, the best-known genus in the order. *Actinopodium*, for which there is no information on reproductive structures, also has a primary vascular system apparently identical to that of *Archaeopteris*. We also accept *Svalbardia* as an archaeopterid even though it is known solely from external morphology—not because we know that its mode of reproduction was pteridophytic but because of the distinctive arrangement of sporangia of a certain shape and size borne adaxially on sporophylls in strobiloid shoots, characteristics found only among archaeopterids. Thus in all three cases above, we have evidence in the taxa in question of characters that are considered to be unique to Archaeopteridales, and it is upon these characters that we base our ordinal assignments. Similar cases of taxonomic identification on the basis of single or small groups of derived characters are common among members of the extant flora as well. For example, one would not hesitate to identify *Quercus* with confidence on the basis of a single acorn, or *Sequoia* on the basis of a fragment of secondary wood (by its distinctive ray tracheids).

In addition to recognizing the importance of derived versus primitive characters, Stein (1982b), as had Bonamo (1975), recognized the need for emphasizing what is known about each included taxon. In our discussion below of genera of Aneurophytales, we shall follow Stein's suggestion of grouping taxa into descriptive categories that emphasize only the nature of the evidence collected to date for each taxon and the type of information required to recognize taxa in the fossil record (see Stein 1982b:619–620). Our use of these categories is not meant to suggest that the taxa included in any of them are fundamentally different from those included in other categories, or that any taxon should be restricted to particular kinds of information. We consider all of the taxa to represent biologically meaningful entities that are distinct at some level in the taxonomic hierarchy. These purely artificial groupings of genera are for convenience of discussion only, and serve to emphasize where gaps in our understanding lie. New information can be added to taxa in any of the categories; the ultimate goal is that all taxa become recognizable and comparable on the basis of both external morphology and internal anatomy.

ANEUROPHYTALES

Kräusel and Weyland (1941) proposed the order Aneurophytales for taxa they considered ancestral to ferns and pteridosperms, and included *Aneurophyton* and *Rhacophyton* in the group. Some paleobotanists (Read 1937; Andrews 1940; Arnold 1947) considered *Aneurophyton* to be unrelated to ferns, however, and emphasized the similarities in stelar morphology and secondary vascular tissues of *Aneurophyton* and some early seed plants. The order Aneurophytales was revived by Beck in 1957 to include both *Aneurophyton* and a newly described plant, *Tetraxylopteris* (Beck 1957), which had many characters in common with *Aneurophyton*. Beck thought that these taxa represented intermediates between psilophytaleans and pteridosperms. With the discovery that *Archaeopteris* and *Callixylon* represent different parts of the same plant and the resulting establishment of the Progymnospermopsida, he included Aneurophytales as the most primitive order in the class (Beck 1960a, 1960b).

Work on aneurophytes done prior to the mid-1970s was summarized in reviews by Bonamo (1975) and Beck (1976). Since these reviews, another genus, *Reimannia* (Arnold 1935), has been added to the order (Stein 1982b) and significant new information has been presented on *Aneurophyton* (Serlin and Banks 1978; Schweitzer and Matten 1982), *Rellimia (Protopteridium)* (Schweitzer 1974; Mustafa 1975; Bonamo 1977; Schweitzer and Matten 1982), and *Triloboxylon* (Matten 1974; Scheckler 1975; Mustafa 1978; Stein and Beck 1983; Wight 1985). Additional specimens of three-ribbed aneurophytes have been described from the Middle Devonian of the southeastern United States (Wight 1985), including documentation of secondary phloem (Wight and Beck 1984), and an analysis of development in several aneurophytalean taxa has been presented (Scheckler 1976). These recent studies have provided a more complete understanding both of individual taxa and of the range of variation within the order.

Diagnoses and/or general descriptions of Aneurophytales have been presented elsewhere (e.g., Bonamo and Banks 1967; Scheckler and Banks 1971b; Bonamo 1975; Beck 1976). In this paper, instead of a similar general description of the order, we prefer to concentrate on characteristics that are, either by themselves or in combination with others, unique to aneurophytes. We believe this is useful because it focuses attention on the features that distinguish aneurophytes from other contemporaneous groups of plants, emphasizes the cohesiveness of aneurophytes as a group, and provides useful criteria with which to determine membership in the order.

Aneurophytes can be characterized by (1) a deeply ribbed, protostelic primary vascular system (except in ultimate appendages, where, in transverse view, it is circular in outline); (2) mesarch order of primary xylem maturation with protoxylem strands at the tips and along the midplanes of ribs and in the center of the protostele; (3) a bifacial vascular cambium that produced secondary vascular tissues (xylem and phloem) similar to those found in gymnosperms; (4) circular bordered pitting on metaxylem and secondary xylem tracheid walls; (5) groups of fibers and/or sclereids in the cortex; (6) branching (other than in ultimate appendages) primarily anisodichotomous, producing three-dimensional branching systems that are strongly pseudomonopodial with lateral axes arranged in a helical or decussate pattern; (7) fertile appendages based on a dichotomous and then pinnate branching pattern with sporangia borne on the ultimate pinnate divisions; (8) ultimate appendages isodichotomously divided and differentiated from other axis orders (stems).

We include the following taxa in the order: *Aneurophyton germanicum* Kräusel and Weyland (1923); *Cairoa lamanekii* Matten (1973); *Proteokalon petryi* Scheckler and Banks (1971b); *Reimannia aldenense*

Arnold (1935); *Rellimia thomsonii* (Dawson) Leclercq and Bonamo (1973); *Tetraxylopteris schmidtii* Beck (1957); *Triloboxylon ashlandicum* Matten and Banks (1966). Additional members of the group include *Triloboxylon arnoldii* Matten (1974), which will be transferred to a new genus as a result of some recent work (Wight 1985) and two unpublished taxa (Wight 1985). In the sections that follow, we shall include brief summaries of all aneurophyte taxa but shall emphasize advances made since the mid-1970s.

GENERA RECOGNIZED FROM EITHER INTERNAL ANATOMY OR EXTERNAL MORPHOLOGY

Tetraxylopteris *Tetraxylopteris* is the most completely known aneurophyte. Studies by Beck (1957), Bonamo and Banks (1967), and Scheckler and Banks (1971a) have resulted in an understanding of branching patterns, fertile organ morphology and position in the branching system, and have correlated information on primary vascular architecture. Distinctive features of *Tetraxylopteris* include (1) a four-ribbed primary vascular system in all branch orders except ultimate appendages (figures 1.1A, B, D); (2) a decussate branching pattern (figures 1.1B; 1.2A); (3) a fertile organ that is twice dichotomous proximally and three times pinnately branched distally, bearing sporangia on the ultimate subdivisions (figure 1.2); and (4) spores equivalent to *Rhabdosporites langii* of the sporae dispersae. Because of this combination of characteristics, *Tetraxylopteris* can be recognized in the fossil record on the basis of either external morphology or internal anatomy.

Most specimens of *Tetraxylopteris* are of Frasnian age and are from the northeastern United States (Beck 1957; Bonamo and Banks 1967; Scheckler and Banks 1971a). Recently however, Mustafa (1975) has described specimens he assigns to the taxon from the Middle Devonian (Upper Eifelian) of Germany. If *Sphenoxylon eupunctatum* (Thomas 1935; Read 1937) is accepted as a preservational form of *Tetraxylopteris* (Matten and Banks 1967; Scheckler and Banks 1971a), then the range of *Tetraxylopteris* can be extended downward into the Middle Devonian (Givetian) in North America as well.

The most recent significant work on *Tetraxylopteris* has been the addition of information on the anatomy of reproductive structures presented by Scheckler (1982). He described elliptical to V-shaped traces that depart the tips of ribs of the primary vascular column of the branch order bearing fertile appendages. These traces undergo two dichotomous divisions resulting in four terete strands that form

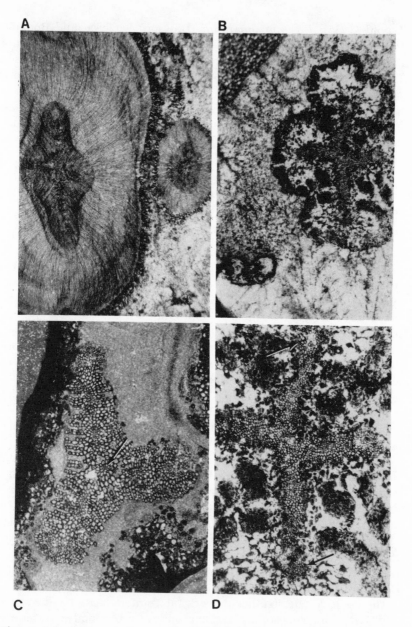

Figure 1.1. A Transverse section of first order axis of *Tetraxylopteris schmidtii* and one of a pair of oppositely arranged second order branches (other not preserved at this level). CUPC Type No. 145-31R. × 6.5. (Used by permission of S. E. Scheckler.) **B** Transverse section of fourth

the vascular supply to the four D2 (Bonamo and Banks 1967) axes of the dichotomizing fertile organ (figure 1.2). Scheckler (1982) also described sporangial morphology for *Tetraxylopteris* and pointed out similarities with rhyniophytes, trimerophytes, other progymnosperms, and the megasporangia of some early seed plants.

One apparent point of controversy remains about the morphology of *Tetraxylopteris*. Beck (1957) described *Tetraxylopteris* "stems" that bore "sterile fronds" in a helical pattern. Scheckler and Banks (1971a) interpreted the helical appearance of branches on the first order axes (stems of Beck) as due to crowding and suggested that they were actually borne in a decussate pattern, as are all other branch orders in *Tetraxylopteris*.

Tetraxylopteris, because of its decussate branching pattern, four-ribbed stelar morphology, and complex fertile organ, is the most distinctive and well-defined taxon within the Aneurophytales. The branching pattern and stelar morphology are characters shared by *Proteokalon* (see below), the taxon we consider to be most closely related. The fertile organ, although based on the dichotomous and then pinnate branching pattern common to all known aneurophytalean fertile organs, is larger and more complex than any other. It is twice dichotomously divided at the base, with each of the four resulting branches thrice pinnate (figure 1.2). Sporangia are borne on the ultimate divisions and number between 2,800 and 4,700 (Bonamo and Banks 1967). Fertile organs apparently occupy the position of penultimate branches, thus replacing the last three orders of branching (penultimate and ultimate branches and ultimate appendages) in the branching system (Beck 1957; Bonamo and Banks 1967). The characteristics cited above have been considered derived character states

---

order (ultimate) branch of *Tetraxylopteris schmidtii* showing four-ribbed primary xylem column, and outer sclerenchymatous cortex. Note pair of incipient traces to ultimate appendages, one at upper and one at lower rib of primary xylem column. CUPC Type No. 146-14R. × 12.5. (Used by permission of S. E. Scheckler.) **C** Transverse section of penultimate axis of *Cairoa lamanekii*. Note three-ribbed protostele with protoxylem strands at center (arrow), along mid-planes and at tips of ribs. Secondary xylem surrounds primary xylem. SIPC No. CQ74-2B. × 30. (Used by permission of W. E. Stein.) **D** Four-ribbed primary xylem column from figure 1.1B at higher magnification. Incipient traces to ultimate appendages are visible at tips of upper and lower ribs (arrows). CUPC Type No. 146-14R. × 25. (Used by permission of S. E. Scheckler.)

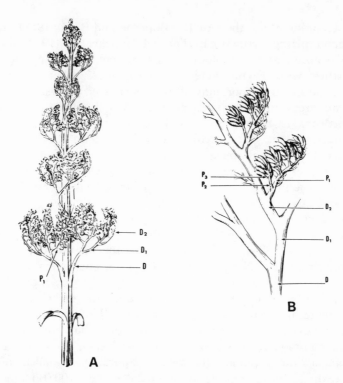

Figure 1.2. A Reconstruction of fertile branching system of *Tetraxylopteris schmidtii*. Sporangial complexes consist of a stalk (D) that dichotomizes twice forming four major branches (D₂). Further branching is pinnate, and sporangia are borne terminally on the third order pinnate branches. Sporangial complexes are borne in a decussate arrangement on the main axis. (From Bonamo and Banks 1967, by permission.) **B** Part of a sporangial complex of *Tetraxylopteris schmidtii* showing stalk (D), axes resulting from dichotomous divisions (D₁, D₂), first (P₁), second (P₂), and third (P₂) order pinnate branches, and terminal sporangia. (From Bonamo and Banks 1967, by permission.)

within the Aneurophytales (Scheckler and Banks 1971b; Bonamo 1975; Beck 1976), based primarily on the complexity of the pattern, the known stratigraphic range of *Tetraxylopteris* at the time, and the generalized occurrence of the helical and three-ribbed states within the group. Because of the presence of these putatively derived states, *Tetraxylopteris* has been considered the most advanced taxon in the

Aneurophytales. Although comparisons with the presumed outgroup, the Trimerophytina, support the derived nature of the states listed above, recent stratigraphic data (Mustafa 1975) suggest that *Tetraxylopteris* first appeared as early as any other aneurophyte with the possible exception of *Rellimia*. This evidence opens the possibility that aneurophytalean taxa represent a rapid diversification subsequent to the initial evolution of the Bauplan, and suggests straight-line or gradual evolution of one form into another (i.e., anagenesis), as has been suggested previously (Scheckler and Banks 1971a, 1971b; Beck 1976).

GENERA RECOGNIZED FROM INTERNAL ANATOMY

Cairoa *Cairoa* is known from a single permineralized specimen collected from the Middle Devonian (Givetian) of eastern New York (Matten 1973). In 1973, Matten believed *Cairoa* was clearly distinct from other aneurophytes, even those with three-ribbed primary vascular systems (at that time, only *Aneurophyton* and *Triloboxylon*) because of a difference in symmetry of the primary vascular system in penultimate and ultimate branches (figure 1.3A). This is still the primary character that serves to distinguish the taxon from other three-ribbed taxa (although see below). In many other features, *Cairoa* is similar to both *Reimannia aldenense* and *Triloboxylon ashlandicum*. All of these taxa have deeply three-ribbed primary vascular systems in penultimate axes (figures 1.1C; 1.3A; 1.4A; 1.5A) and similar histological features of the inner cortex or phloem. Both *Cairoa* and *Reimannia* also produce a subopposite pair of ultimate appendage traces from the ultimate branch trace in association with its departure from the vascular system of the penultimate axis (figures 1.3A; 1.4B), a characteristic that occurs in other aneurophytalean taxa as well (Wight 1985). However, *Cairoa* differs from *Reimannia* in that it (1) is smaller at nearly all equivalent levels; (2) has secondary xylem (figures 1.1C; 1.3A); (3) has alternating bundles of parenchyma and sclerenchyma in the outer cortex, and lacks a sclerenchymatous hypodermis; and (4) has traces to ultimate axes that are four-angled proximally (figure 1.3A) and that retain this shape in the ultimate axes. Of these four characters, we regard numbers (3) and (4) as most important in distinguishing between the two taxa. *Cairoa* differs from *Triloboxylon ashlandicum* in diameter of the primary vascular system, but the primary distinction is in the unique asymmetric nature of ultimate branch trace production in *Triloboxylon ashlandicum* (figures 1.5A, B, op-

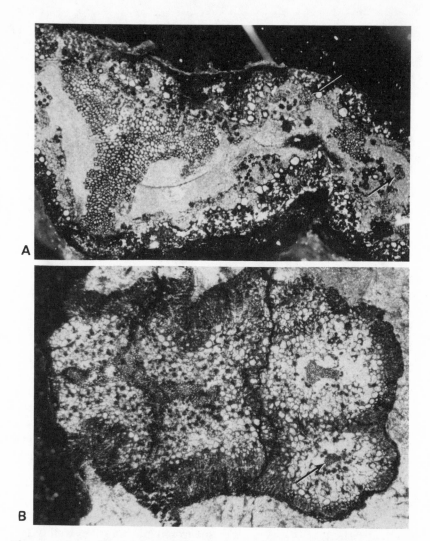

Figure 1.3. A Transverse section of penultimate branch of *Cairoa lamanekii* with attached ultimate axis. Note diamond-shaped trace to ultimate branch (right) and small, terete traces to pair of ultimate appendages (arrows) associated with ultimate branch trace. SIPC No. CQ74-10. × 30. (Used by permission of W. E. Stein.) **B** Transverse section of penultimate branch of *Proteokalon petryi* with one of a pair of opposite ultimate axes attached at right (other not preserved at this level). Note skewed four-ribbed primary vascular column in penultimate axis and three-ribbed primary vascular column in ultimate axis. One of a pair of opposite traces to ultimate appendages (arrow) is visible below ultimate branch trace (other not preserved at this level). CUPC Type No. 140-22. × 18.5. (Used by permission of S. E. Scheckler.)

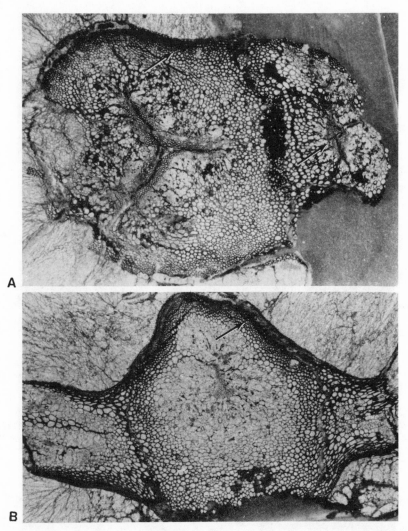

Figure 1.4. A Transverse section of penultimate branch of *Reimannia aldenense* with attached ultimate branch. Note three-ribbed primary vascular column of penultimate axis, sclerenchymatous hypodermal layer (arrow, top), and diamond-shaped trace to ultimate branch (arrow, right). UMMP No. 16231-2. × 10. (From Stein 1982b, by permission.) **B** Transverse section of ultimate branch of *Reimannia aldenense* distal to level shown in figure 1.4A. Note three-ribbed primary vascular column, sclerenchymatous hypodermal layer (arrow) and pair of ultimate appendages (right and left) still attached to ultimate branch. UMMP No. 16231-1A. × 8.5. (From Stein 1982b, by permission.)

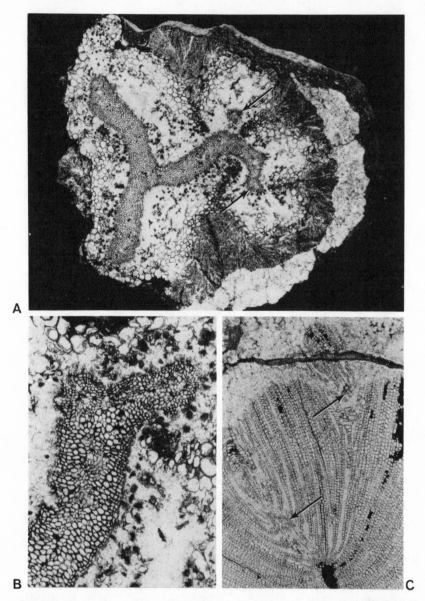

Figure 1.5. A Transverse section of penultimate axis of *Triloboxylon ashlandicum*. Note outer cortical sclerenchyma, three-ribbed primary vascular column, and extreme asymmetry of tip of rib extending to right. Asymmetry is result of presence of incipient trace to ultimate axis (extending to lower right). One ultimate appendage trace is visi-

posite). The distinctiveness of *Cairoa* with respect to other three-ribbed aneurophytalean taxa *(Aneurophyton* and *Rellimia)* will be considered further below.

On the basis of a change in symmetry of the primary vascular system from the branch order to the next, Matten (1973) suggested that *Cairoa* and *Proteokalon* represent a distinct line of evolution within the Aneurophytales and, therefore, should be recognized as a separate family within the order. Although Matten did not formally create the new taxonomic category, Barnard and Long (1975) subdivided the members of the Aneurophytales into the Proteokalonaceae and the Aneurophytaceae based on the differences in symmetry noted by Matten (1973) earlier. We disagree with this grouping of aneurophytes for several reasons. First, although both of these taxa exhibit changes in symmetry of the primary vascular system, that of *Proteokalon* changes from four-ribbed to three-ribbed (figure 1.3B), whereas that of *Cairoa* changes from three-ribbed to four-ribbed (figure 1.3A). We suggest, therefore, that this is a similarity only in a very broadly defined sense, and not a homology that indicates a close evolutionary relationship between the two taxa. Second, *Proteokalon* seems to be most closely related to *Tetraxylopteris*, and *Cairoa* to the three-ribbed taxa, *Reimannia, T. ashlandicum,* and perhaps *Rellimia* (see below). Third, Stein (1982b) doubted that the four-ribbed configuration was retained in more distal regions of the ultimate axes of *Cairoa*. Since both *Reimannia* (figure 1.4A; Stein, 1982b) and an unpublished taxon (Wight 1985) have traces to ultimate branches that proximally are four-angled like those of *Cairoa*, yet distally become distinctly three-ribbed, we wonder whether this feature will prove to be consistent when more distal regions of ultimate branches or new specimens of *Cairoa* are discovered.

---

ble above ultimate axis trace, and an incipient ultimate appendage trace is still attached to lower edge of ultimate axis trace (arrows). CUPC Type No. 151-21. × 14.5. **B** Tip of primary vascular column rib bearing incipient trace shown in figure 1.5A but at higher magnification and more proximal level in axis. Note pronounced asymmetry, even at this level. Extension of rib to left will become ultimate appendage trace; larger extension to right will become three-ribbed ultimate axis trace. CUPC Type No. 151-10. × 42.3. **C** Transverse section of part of main axis of *Triloboxylon arnoldii* showing pair of traces (arrows) distal to level of their separation from a rib of the primary vascular column (bottom). UMMP No. 65131-10B. × 19. (From Stein and Beck 1983, by permission.)

Proteokalon *Proteokalon* (Scheckler and Banks 1971b) was described from the Upper Devonian (Frasnian) of New York and is known from specimens preserved almost exclusively as permineralizations. Although reproductive structures are unknown, in most other respects *Proteokalon* closely resembles *Tetraxylopteris*, and it is to this taxon that we believe *Proteokalon* is most closely related. The most significant differences between the two taxa are (1) the "skewed" (i.e., not radially symmetrical; Scheckler and Banks 1971b) nature of the four-ribbed protostele (figure 1.3B) in penultimate axes (first order branches of Scheckler and Banks 1971b) of *Proteokalon* versus the cruciform shape in *Tetraxylopteris* (figures 1.1A, B, D); and (2) the change to a three-ribbed protostele (figure 1.3B) in ultimate axes (second order branches of Scheckler and Banks 1971b) of *Proteokalon* versus the maintenance of a four-ribbed protostele in all axis orders of *Tetraxylopteris* (figures 1.1A, B, D). Other reported differences between the two taxa include the presence of abundant xylem parenchyma, occasional dichotomous branching of axes, and planated ultimate appendages in *Proteokalon* (Scheckler and Banks 1971b).

The change in morphology of the primary vascular column from penultimate to ultimate branches (first and second order of Scheckler and Banks 1971b) in *Proteokalon* is correlated with a change in the arrangement of lateral appendages. The three-ribbed protostele in ultimate axes is T-shaped, with the "stem" of the T oriented abaxially (figure 1.3B). Ultimate appendage traces diverge from the tips of each of the ribs of the T. Production of a subopposite pair, one from each end of the top of the T, alternates between nodes with the production of a trace from the abaxial rib (stem) of the T. This results in an ultimate branching system that Scheckler and Banks (1971b) described as bilaterally symmetrical versus the radial symmetry present in penultimate axes and throughout the branching system of *Tetraxylopteris* as well as all other aneurophytes. They considered this the most advanced branching pattern in aneurophytes and suggested that it represents an early stage in the evolution of a frond-like leaf.

Although there seem to be distinct differences between *Proteokalon* and *Tetraxylopteris*, these taxa are more similar to each other than either is to any other aneurophyte. Since the only known specimens of *Proteokalon* were found in a locality from which *Tetraxylopteris* also has been collected (Scheckler and Banks 1971a, 1971b), it is possible that the differences between the two represent variation within a single species. Ranges of morphologic variation in taxa of extinct plants are not well established, however, so such a decision would be arbitrary. A more defensible position might be to regard the differences as variation within a single genus. Discovery of the reproductive organs of *Proteokalon* might provide a test of this proposed relationship.

Reimannia *Reimannia aldenense* (Arnold 1935) was based on permineralized specimens from the Middle Devonian (Givetian) of western New York. Arnold included *Reimannia* in his Iridopteridineae, a group defined primarily on the basis of protostelic primary vascular systems with "peripheral loops" and thought by him to be intermediate between "psilophytes" and coenopterid ferns (Arnold 1940, 1947). Other workers have discussed *Reimannia* and its evolutionary relationships and assigned new specimens to the genus (Read and Campbell 1939; Hoskins and Cross 1951; Beck 1960c; Banks 1966; Wilcox 1967; Matten and Banks 1969; Leclercq 1970; Matten 1973; Scheckler 1974). Stein (1981, 1982a, 1982b) reinvestigated the three members of Arnold's suborder and redefined it, excluding *Reimannia* (Stein 1982b) and leaving *Arachnoxylon* (Stein 1981) and *Iridopteris* (Stein 1982a) in the group.

As a result of Stein's (1982b) investigation, *Reimannia* can now be characterized by (1) three orders of branching in organic connection (figure 1.4); (2) a three-ribbed primary vascular system with protoxylem strands located at the tips and along the midplanes of the ribs in penultimate and ultimate axes (figure 1.4; first and second order axes of Stein 1982b); (3) traces to third order axes (ultimate appendages) circular in outline; (4) change in the configuration of the primary xylem in the second order axes from broadly elliptical or diamond-shaped proximally to distinctly three-ribbed distally (figure 1.4); and (5) lack of a "peripheral loop" associated with the protoxylem. On the basis of this new information, Stein concluded that *Reimannia* was a progymnosperm, and transferred it to the Aneurophytales. Although *Reimannia* lacks evidence of secondary growth, we agree with Stein (1982b) that the plant represents an aneurophytalean progymnosperm. Similarities with other aneurophytes in stelar morphology, branching pattern, and histology preclude its classification in any other contemporaneous group (see Gensel and Andrews 1984 for an alternative viewpoint). Stein (1982b) excluded all specimens except Arnold's (1940) type from the taxon.

Distinctive features of *Reimannia* include (1) a deeply three-ribbed protostele in penultimate and ultimate axes; (2) traces of ultimate axes that are elliptical to diamond-shaped proximally and that become three-ribbed distally; (3) a wide, homogeneous cortex composed of parenchyma (figure 1.4); and (4) a thick-walled, sclerenchymatous hypodermal layer (figure 1.4).

Reimannia is similar in several respects to both *Cairoa lamanekii* (Matten 1973, 1975) and *Triloboxylon ashlandicum* (Matten and Banks 1966; Scheckler and Banks 1971a). Differences in size, secondary vascular tissue development, cortical histology, and trace morphology, however, serve to separate the three taxa (Matten 1973; Stein 1982b; Wight 1985).

Triloboxylon *Triloboxylon ashlandicum. T. ashlandicum* was initially described from permineralized vegetative specimens collected from the Upper Devonian (Frasnian) of New York (Matten and Banks 1966; Scheckler and Banks 1971a). Scheckler (1975) described a possible fertile specimen (CUPC Type No. 152) that he assigned to *T. ashlandicum* on the basis of similarities to vegetative specimens in stelar morphology and trace departure. The specimen is a weathered impression, with the penultimate axis permineralized in part of the region that bears the putative fertile appendages. Anatomically, it is characterized by a three-ribbed primary vascular system surrounded by extensive secondary xylem; divergence of pairs of traces, elliptical in outline in transverse section, from the tips of successive ribs of the primary xylem column; clusters of sclereids in the inner cortex; and fiber-sclereid bundles in the outer cortex. The putative reproductive structures are similar to those described for other aneurophytes (Bonamo and Banks 1967; Leclerq and Bonamo 1971; Bonamo 1977; Serlin and Banks 1978; Schweitzer and Matten 1982) in that they are based on a branching pattern that is proximally dichotomous and distally pinnate. They dichotomize twice and bear sporangia, either singly or in pairs, on lateral, pinnately arranged stalks attached to the distal segments (figure 1.6A). Fertile organs are borne between regions of vegetative lateral branches on the penultimate axis (figure 1.6A). This differs from other aneurophytes that apparently bear fertile appendages only in distal regions of the branching system (Beck 1957; Bonamo and Banks 1967; Leclercq and Bonamo 1971; Bonamo 1977; Schweitzer and Matten 1982). Because of the poor preservation of this specimen and the absence of spores, however, its identification as a fertile organ remains somewhat problematical and awaits further corroboration.

Additional work on the genus *Triloboxylon* and morphologically similar specimens has supported the recognition of *T. ashlandicum* as a distinct taxon within the Aneurophytales (Wight 1985). It can be recognized primarily by the presence of a deeply three-ribbed pri-

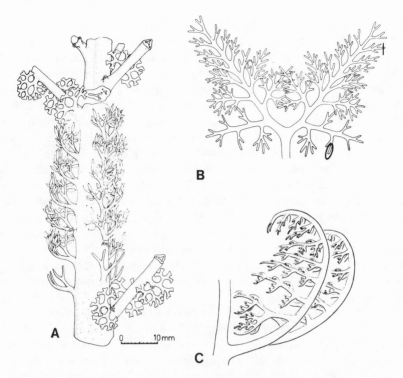

Figure 1.6. A Reconstruction of part of branching system of *Trilobox-ylon ashlandicum* (sensu Scheckler 1975) showing position of fertile appendages. Note two basal dichotomous divisions and pinnate arrangement of sporangia on ultimate subdivisions. (From Scheckler 1975, by permission.) **B** Diagram of fertile organ of *Rellimia* flattened to show basal dichotomy and pinnate branching pattern distally. Sporangia are borne terminally on the third order pinnate branches but are not shown. (From Schweitzer and Matten 1982, by permission.) **C** Diagram of fertile organ of *Rellimia* as thought to appear in life (cf. figure 1.16). (From Leclercq and Bonamo 1971, by permission.)

mary vascular column in penultimate and ultimate axes (figure 1.5A), the distinct asymmetric pattern of ultimate branch trace departure in penultimate axes (figures 1.5A, B); the lack of inner cortical sclerenchyma (figures 1.5A, B), the presence of fiber-sclereid bundles in the outer cortex (figure 1.5A), and the production of single traces from the tips of the ribs of the protostele in ultimate axes (Wight 1985). The fertile specimen assigned to the taxon by Scheckler (1975; CUPC Type No. 152) lacks this combination of essential features. Consequently, it has been removed from *T. ashlandicum* and reassigned to *T. arnoldii* (Wight 1985) because of a close correspondence in diagnostic features between it and *T. arnoldii*, as circumscribed in the recent work of Stein and Beck (1983; but see below for the taxonomic disposition of *T. arnoldii*). (See Wight [1985] for a complete list of specimens included in *T. ashlandicum* and *T. arnoldii*.) Thus *T. ashlandicum* is recognizable at this time only on the basis of internal anatomy. In addition to the features listed above, Matten and Banks (1966), Scheckler and Banks (1971a), and Scheckler (1975) have used the presence of a median band of flattened metaxylem tracheids as a distinctive character of *T. ashlandicum*. This no longer seems to be a useful character because of its intermittent occurrence in *Triloboxylon* as well as its presence in both *Reimannia* (Stein 1982b) and *Rellimia* (Bonamo 1977). Ultimate appendages in the taxon have been reported as planate (Scheckler and Banks 1971a). However, Wight (1985) showed that the vascular tissue in ultimate appendages divides at approximately right angles in successive dichotomous divisions, suggesting a three-dimensional appendage.

An additional specimen from the Middle Devonian of Germany was assigned to the taxon by Mustafa (1978). Although photographs (Mustafa 1978: plate 14, figures 11–16) suggest the asymmetric trace departure characteristic of *T. ashlandicum*, given the limited information available we believe this specimen could just as easily represent *Rellimia* or some other aneurophyte with a three-ribbed primary column. As a result, we exclude it from *T. ashlandicum*.

Whereas the asymmetric pattern of ultimate branch trace departure has been considered the primary distinguishing feature of *T. ashlandicum*, asymmetry of trace departure is also characteristic of other aneurophytalean taxa. Scheckler and Banks (1971a) figured and Scheckler (1975) first explicitly used this feature as diagnostic of *T. ashlandicum*. Scheckler (1975), however, considered the asymmetry itself, regardless of the kinds of traces produced, as the important characteristic. As a result, he placed in *T. ashlandicum* a specimen (CUPC Type No. 152) that lacked evidence of ultimate branch trace departure but showed asymmetry at the tip of the primary xylem rib in association with the departure of a pair of traces, each circular to elliptical in transverse outline. Work since that time (Stein and Beck 1983; Wight 1985) has demonstrated that asymmetry at the tip of a primary xylem rib is common when two such traces, each separating at a slightly different level, are produced as a pair from a single rib of the xylem column. Thus Wight (1985) suggested that the type of asymmetry present in CUPC Type No. 152 is unimportant taxonomically and fundamentally different from the consistent asymmetric pattern of ultimate branch trace departure that he documented in *T. ashlandicum.* (See above for taxonomic status of CUPC Type No. 152.)

Triloboxylon ashlandicum is similar to both *Cairoa lamenekii* and *Reimannia aldenense,* and to an unpublished taxon from the Middle Devonian of Virginia (Wight 1985). The characteristic mode of ultimate branch trace departure (figure 1.5A, B) separates *T. ashlandicum* from both *Cairoa* and *Reimannia.* In addition, *Cairoa* differs in the shape of the primary vascular system in ultimate branches and *Reimannia* differs in its lack of fiber-sclereid bundles in the outer cortex. Bonamo (1977) has noted similarities between *T. ashlandicum* and *Rellimia.* At present, comparisons between the taxa are difficult (see below).

Triloboxylon arnoldii. *T. arnoldii* was originally described from permineralized specimens from the Middle Devonian (Givetian) of western New York (Arnold 1940). Although reinvestigated by Scheckler and Banks (1971a), the taxon remained relatively poorly understood until the work of Stein and Beck (1983). It is now known to be characterized by a three-ribbed vascular system (figure 1.7) that produces pairs of small circular to elliptical traces from successive ribs (figure 1.5C). The specimens thus far described have extensive secondary vascular tissues (figure 1.5C; 1.7) as well as apparent extensive secondary cortical development. Cortical tissue is divided into an inner cortex consisting of parenchyma, occasional fibers, and clusters of sclereids and an outer cortex consisting of fiber-sclereid bundles separated by parenchyma (figure 1.7). There is also an extensive development of simple periderm. Specimens assigned to the taxon by Scheckler and Banks (1971a; CUPC Type No. 153) and Matten (1974; SIPC CQ 20) were excluded by Stein and Beck (1983) because they lacked some or all of the above characters.

Wight (1985) placed Scheckler's (1975; CUPC Type No. 152) fertile specimen in this taxon because of the close correspondence in all features between it and *T. arnoldii,* including the presence of paired traces, extensive secondary vascular tissue, and a heterogeneous inner cortex, characters considered important in the recognition of *T. arnoldii* (Stein and Beck 1983; Wight 1985). Although this changes the concept of the taxon somewhat, we believe that because the fertile nature of this specimen is in need of further corroboration, the taxon is still recognizable in the fossil record primarily on the basis of

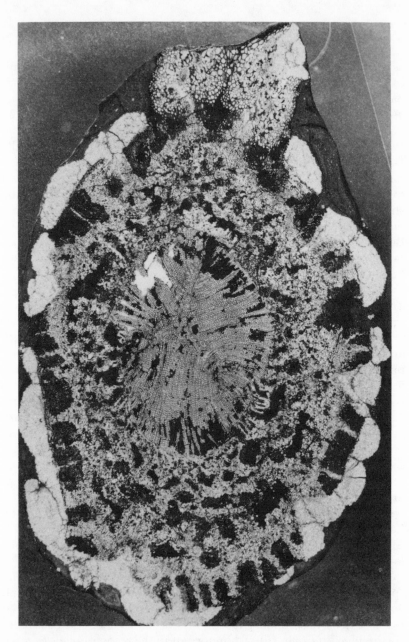

Figure 1.7. Transverse section of main axis of *Triloboxylon arnoldii* with attached lateral appendages (top). Note three-ribbed primary vascular column, extensive secondary xylem, clusters of sclereids in inner cortex, and fiber-sclereid bundles separated by parenchyma in outer cortex. UMMP No. 23848-6C. × 11. (From Stein and Beck 1983, by permission.)

internal anatomy. Inclusion of this specimen in *T. arnoldii* extends the stratigraphic range of the taxon upward into the Upper Devonian (Lower Frasnian).

Although it was originally described by Arnold (1940) as a species of *Aneurophyton*, Scheckler and Banks (1971a) transferred the taxon to *Triloboxylon* because it shared more characteristics with *T. ashlandicum* than with *A. germanicum*. This was a reasonable taxonomic decision, given the lack of information about Arnold's material and the lack of knowledge of other clearly defined taxa of aneurophytes with three-ribbed primary vascular systems. The proliferation of three-ribbed aneurophytes since that time (Matten 1973, 1975; Bonamo 1977; Stein 1982b; Wight 1985) and the new information derived from Arnold's specimens (Stein and Beck 1983), however, suggest that a reevaluation of the placement of this taxon in *Triloboxylon* is in order (Stein and Beck 1983). The degree of morphological difference between *T. arnoldii* and *T. ashlandicum* indicates that there is no reason to consider the taxon represented by Arnold's original material to be more closely related to *T. ashlandicum* than either is to any other three-ribbed aneurophytalean. Thus inclusion of both taxa in *Triloboxylon* is unjustified and Arnold's material should be placed in a new genus (Wight 1985).

At present, *T. arnoldii* is distinct from other aneurophytalean taxa. Stein and Beck (1983), however, investigated the possibility that the taxon might represent a later stage of growth of some other three-ribbed aneurophytalean taxon. Although their analysis was not conclusive, they suggested a possible developmental relationship between *T. arnoldii* and *T. ashlandicum*.

GENERA RECOGNIZED FROM EXTERNAL MORPHOLOGY

Aneurophyton *Aneurophyton germanicum* (Kräusel and Weyland 1923) was described from the Middle Devonian of Germany by Kräusel and Weyland (1923, 1926, 1929, 1935). Leclercq (1940) added some anatomical information, and other specimens have been assigned to the genus from time to time (Arnold 1940; Stockmans, 1948, 1968; Termier and Termier 1950; Streel 1964). Recent workers (Scheckler and Banks 1971a; Serlin and Banks 1978; Schweitzer and Matten 1982), however, consider *Aneurophyton germanicum* the only valid species in the genus. This species is known from the Upper Eifelian to Lower Frasnian and from localities in North America, Belgium, and Germany.

Descriptions of *Aneurophyton* in the literature might suggest that the taxon is recognizable in the fossil record on the basis of both internal anatomy and external morphology (e.g., Serlin and Banks 1978; Schweitzer and Matten 1982). Internally, *Aneurophyton* is thought to be characterized by a three-ribbed protostele (figure 1.8A) with four protoxylem strands in all axis orders (Leclercq 1940; Bonamo 1977; Serlin and Banks 1978; Schweitzer and Matten 1982). If true, this would serve to distinguish *Aneurophyton* anatomically from all other aneurophytalean taxa. Recent work on the genus, however, has failed to document this feature conclusively. Although some transverse sections with apparently four protoxylem strands have been figured (e.g., Serlin and Banks 1978; Schweitzer and Matten 1982), the material is generally either poorly preserved, of limited extent, or both. As a result, we regard the available data as equivocal. We would like to see well-preserved material that clearly shows not only the maintenance of four protoxylem strands throughout the length of a single axis order but also the continuity of this morphology from one axis order to the next.

Furthermore, for several reasons, we consider it unlikely that a constant number of protoxylem strands is maintained in different axis orders in *Aneurophyton*. First, it is well known that in other aneurophytalean taxa, protoxylem strand number is correlated with the diameter of the primary xylem column, which itself decreases as one examines successively higher order axes (see e.g., Scheckler 1976; Stein 1982b; Wight 1985). Because the diameter of the primary xylem column varies from penultimate to ultimate axes in *Aneurophyton*, we would expect protoxylem number to vary as well. Second, in at least some aneurophytes, protoxylem strand number and position are not constant features even within a single axis order (Beck 1957; Wight and Beck 1982; Wight 1985, 1987). Protoxylem architecture is directly related to the pattern of trace departure (Beck 1957; Wight and Beck 1982; Wight 1985, 1986, in press), and, as a result, the precise configuration at any level depends on its position relative to the insertion of lateral appendages. If *Aneurophyton* has a similar protoxylem archi-

Figure 1.8. A Transverse section of *Aneurophyton germanicum* showing three-ribbed protostele surrounded by secondary xylem. Note presence of centrally located protoxylem strand (a); protoxylem strand in upper right rib that appears to have separated from the central strand at a slightly more proximal level (b); and protoxylem strand in lower right rib approximately half way to tip of rib. Additional strands are present at the tips of ribs. This arrangement of protoxylem strands

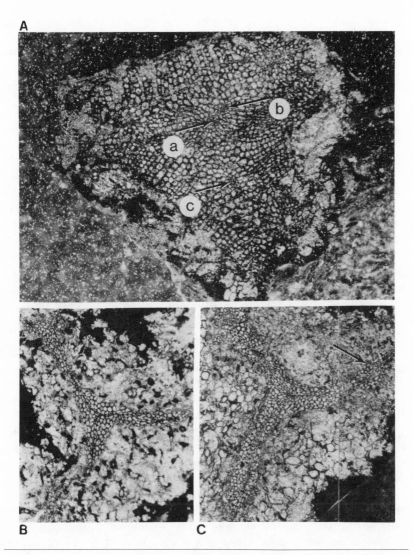

would be expected if a nonsympodial protoxylem architecture were present (see text for further explanation). CUPC 241.1-32. × 42.5. **B** Transverse section of an n + 3 branch of *Rellimia* borne by the n + 2 branch shown in figure 1.20. Note three-ribbed primary xylem column. This axis order bears the fertile appendages. 335.56.b.VIII.x.c. × 41.5. (From Bonamo 1977, by permission.) **C** Transverse section of n + 2 axis of *Rellimia*. Note three-ribbed primary vascular column and trace to n + 3 branch to right. 335.56.b.VIII.x.e. × 32.5. (From Bonamo 1977, by permission.)

tecture, then the number of protoxylem strands would not be constant either within or between axis orders. Published work on *Aneurophyton* provides some evidence in support of this position (figure 1.8A).

Recent work on *Aneurophyton* (Serlin and Banks 1978; Schweitzer and Matten 1982) also has not provided the detailed information on trace morphology and departure critical to distinguishing between taxa on the basis of anatomical characters. Because of this, and the uncertainty associated with stelar morphology discussed above, we would find it difficult to assign a specimen to *Aneurophyton* on the basis of anatomical information alone. Therefore, we suggest that the taxon is recognizable from external morphology only, primarily on the basis of its distinctive fertile organs. Although based on a proximally dichotomous and distally pinnate branching pattern like other aneurophytalean reproductive organs, *Aneurophyton* has the smallest and simplest ones in the order (Kräusel and Weyland 1923; Serlin and Banks 1978; Schweitzer and Matten 1982). The fertile organs are approximately 10 mm in length and bear approximately 20 sporangia, borne in distal regions of branching systems on axes that are primarily fertile. Fertile organs apparently occur in the position of ultimate appendages in the branching system of *Aneurophyton* (Serlin and Banks 1978; Schweitzer and Matten 1982). Sporangia are elongate with blunt tips and contain spores assignable to *Aneurospora goensis* (Streel 1964). These spores are distinct from those of other aneurophytes (*Rhabdosporites langii* for *Tetraxylopteris* [Bonamo and Banks 1967] and *Rellimia* [Leclercq and Bonamo 1971; Bonamo 1977]).

Aneurophyton has been considered to be the most primitive member of the Aneurophytales (e.g., Serlin and Banks 1978) because of the simplicity of its fertile organ and anatomical structure compared to other members of the order. Although this may be correct, *Rellimia*, *Tetraxylopteris*, and perhaps *Triloboxylon ashlandicum* all appear in the fossil record at approximately the same time as *Aneurophyton*. We believe that simplicity alone may not be a sufficient basis upon which to assess the relative primitiveness of character states or taxa. Furthermore, stating that *Aneurophyton* is most primitive may imply an ancestor-descendant relationship between it and other members of the order, a question that no one has addressed in detail.

Eospermatopteris (Goldring 1924), a genus of casts of the bases of large trees, is found in association with *Aneurophyton* in both New York and Germany (Goldring 1924; Kräusel and Weyland 1935; Serlin and Banks 1978). These casts have been considered to be the basal portions of *Aneurophyton* and are the basis for reconstructions that show *Aneurophyton* as a tree (Goldring 1924). Although the concept of

Aneurophyton as a tree appears throughout the literature on aneurophytalean progymnosperms (e.g., Bonamo 1975; Beck 1976), there is no definitive evidence to support such a reconstruction (Scheckler 1976; Schweitzer and Matten 1982). The taxonomic status of *Eospermatopteris* is unclear and thus it is excluded by us from the Progymnospermopsida.

Rellimia Dawson (1871a) provided the initial description of specimens now included within *Rellimia (Protopteridium)*. Since that time, there has been much confusion about the name that should be associated with the taxon, the number of species involved, the precise morphological nature of the specimens, and their taxonomic and evolutionary relationships (see, e.g., Dawson 1871a, 1871b, 1878, 1882, 1888; Carruthers 1873; Krejči 1880, 1881; Stur 1881; Piedboeuf 1887; Kidston 1903; Potonié and Bernard 1904; Arber 1921; Lang 1925, 1926; Kräusel and Weyland 1932, 1933, 1938; Halle 1936; Leclercq 1940; Høeg 1942; Chirkova-Zalesskaja 1957; Obrhel 1959–1961, 1966, 1968a, 1968b; Stockmans 1968; Iurina 1969). We believe that all workers today agree that there is only one species represented by the various specimens, but disagreement still exists about the proper name for the taxon. We have chosen to accept the arguments of Leclercq and Bonamo (see Leclercq and Bonamo 1971, 1973; Bonamo 1977, 1983), and use *Rellimia* as the name for the taxon (see Matten and Schweitzer 1982; Schweitzer and Matten 1982; Takhtajan and Zhilin 1976 for dissenting viewpoints).

Rellimia has been collected from northeastern North America (Bonamo 1977), Belgium (Leclercq 1940; Stockmans 1968; Leclercq and Bonamo 1971), Scotland (Dawson 1871a; Carruthers 1873; Lang 1925, 1926), Germany (Krejči 1880, 1881; Piedboeuf 1887; Kräusel and Weyland 1932, 1933, 1938; Obrhel 1959–1961; Mustafa 1975; Schweitzer and Matten 1982), Spitzbergen (Høeg 1942), and the Soviet Union (Chirkova-Zalesskaja 1957). It is of limited range in the geological column, having been collected only from the Middle Devonian (Lower Eifelian to Upper Givetian).

Although *Rellimia* had been known for a century, it was not until the work of Leclercq and Bonamo (1971) that the morphology of the plant was clearly understood (see Leclercq and Bonamo 1971 and Schweitzer and Matten 1982 for detailed records of morphologic and taxonomic interpretations of specimens). More recent work (Bonamo 1977; Schweitzer and Matten 1982; Dannenhoffer and Bonamo 1984) has confirmed many of the observations made by Leclercq and Bonamo (1971) and has added significant new information as well.

Specimens of *Rellimia* from New York (Bonamo 1977) showed clearly for the first time that the plant has a deeply three-ribbed protostelic primary vascular system in both penultimate and ultimate axes (figures 1.8B, C; n + 2 and n + 3 of Bonamo). Primary xylem maturation was mesarch, with protoxylem strands in positions typical of aneurophytalean progymnosperms (figures 1.8B, C). Bonamo (1977) also reported a continuous medial strand of protoxylem in each lobe of the protostele (figures 1.8B, C). Although similar tissue in *Triloboxylon* was also interpreted as protoxylem (Matten and Banks 1966), we agree with Scheckler and Banks (1971b) and Stein (1982b) that it is mostly likely metaxylem. Bonamo briefly discussed trace departure in the specimens, but the pattern was inferred on the basis of a single transverse section in each axis order. Consequently, we believe that the information requires further documentation. A preliminary description of secondary xylem in these New York specimens was presented by Dannenhoffer and Bonamo (1984). In all features, it appears to be typical of that of aneurophytalean progymnosperms (i.e., elongate tracheids with circular bordered pits on all walls; narrow, tall rays). They also report growth rings in the secondary xylem.

Schweitzer and Matten (1982) provided some additional anatomical information in a description of specimens from the Middle Devonian (Upper Lower Eifelian and Givetian) of Germany. They figure two sections from a permineralized segment of an apparently ultimate axis in a fertile branching system (= n + 3? of Leclercq and Bonamo 1971 and Bonamo 1977). The specimen is poorly preserved but has a shallowly three-ribbed primary vascular column approximately 0.5 mm "wide." According to Schweitzer and Matten, the protostele contains three protoxylem strands, one in each rib of the primary xylem column (Schweitzer and Matten 1982: text figures 33b, c), but they are difficult to see in the photographs. This stelar morphology appears to be quite different from that in an apparently equivalent axis order (n + 3) in *Rellimia* from New York (figure 1.8B; see also Bonamo 1977: figures 19, 20). Schweitzer and Matten (1982) recognized the apparent differences between the New York and German specimens and suggested similarities between their material and specimens from the Middle Devonian (Eifelian) of Germany that Mustafa (1975) described and called *Protopteridium* (i.e., *Rellimia*). They suggested that these differences in stelar morphology may represent species-level differences within the genus. Our comparison of the *Rellimia* of Schweitzer and Matten (1982: text figures 33b, c) with the anatomical preparations of Mustafa (1975: table 14, figures 5, 6; table 15, figures 1–5; table 16, figure 1) reveals no striking similarities. Furthermore, several authors (Bonamo 1977; Serlin and Banks 1978; Stein 1982b; Stein and Beck 1983), as well as Schweitzer and Matten (1982:86–87) themselves, have questioned the validity of Mustafa's (1975) assignment of these specimens to *Rellimia*. Although Mustafa's specimens almost certainly represent aneurophytalean progymnosperms, we concur with the authors cited above that taxonomic placement in *Rellimia* is uncertain at best. Therefore, we feel that suggestions of specific differences within *Rellimia* based on comparisons with these specimens are unsupported. However, given the apparent differences in anatomy and morphology (see below) between the German material described by Schweitzer and Matten (1982), Kräusel and Weyland's (1938) material, and the specimens from Belgium and New York (Leclercq and Bonamo 1971; Bonamo 1977), we believe it is possible that two taxa are represented, both with fertile structures characteristic of *Rellimia* (Leclercq and Bonamo 1971) and differentiated on the basis of anatomical features and branching patterns. This possibility should be investigated further.

Despite the anatomical information presented in the studies on *Rellimia* cited above, we still have only a limited understanding of the three-dimensional morphology of the primary vascular system, trace departure, and trace morphology, characters that are important in distinguishing among aneurophytalean taxa on the basis of internal anatomy (Stein and Beck 1983; Wight 1985). As a result, we believe that *Rellimia* is recognizable in the fossil record at present only on the basis of external morphology, primarily its distinctive fertile organ (Leclercq and Bonamo 1971; Bonamo 1977; Schweitzer and Matten 1982). The fertile organs are intermediate in size and complexity between those of *Aneurophyton* and *Tetraxylopteris*. They are borne helically in the branching system and have a single dichotomy near the base (figures 1.6B, C). The two resulting first order pinnae bear three to four orders of alternate pinnate subdivisions that ultimately bear sporangia (figures 1.6B, C). Approximately 400 sporangia are present in a single fertile organ.

New information on the external morphology of *Rellimia* was presented by Schweitzer and Matten (1982). In this study they corroborated the overall vegetative and reproductive morphology described earlier by Leclercq and Bonamo (1971) and Bonamo (1977), although in some cases their reconstructions seem to be at odds with prior descriptions (e.g., the arrangement and morphology of fertile organs depicted in text figure 28). Of special interest is their description of several specimens that bear dichotomizing ultimate appendages on axes other than those of the ultimate order. Prior to this discovery, ultimate appendages ("leaves") were thought to be confined to the ultimate axis order in *Rellimia* (Leclercq and Bonamo 1971) as well as in all other aneurophytalean progymnosperms (Scheckler 1975, 1976). If Schweitzer and Matten (1982) are correct, then analyses of aneurophytes that use the presence or absence of ultimate appendages as a criterion for locating particular axes in a hypothetical branching sequence, such as the developmental analysis of Scheckler (1976), may be subject to considerable error.

Schweitzer and Matten (1982) described the branching pattern in their specimens of *Rellimia* as predominately "trifurcate." This somewhat confusing terminology is used to refer to the bifurcation, or "forking," of lateral branches immediately (about 1 cm) above their point of insertion. This mode of branching conflicts with the strictly pseudomonopodial branching pattern previously described for the taxon by Leclercq and Bonamo (1971) and Bonamo (1977) and gives the reconstruction of Schweitzer and Matten (1982: text figure 34) a highly branched or bushy aspect.

Rellimia has been considered to be closely related to *Textraxylopteris* (Bonamo and Banks 1967; Leclercq and Bonamo 1971; Bonamo 1975, 1977; Beck 1976; Gensel and Andrews 1984) because of similarities in the morphology of fertile organs and spores (both taxa have spores assignable to the dispersed spore genus *Rhabdosporites langii* Richardson [1960]). However, the two taxa can be distinguished on the basis of the three-ribbed primary vascular system and helical branching in *Rellimia* (figures 1.8B, C) which contrast with the four-ribbed stele and decussate branching pattern in *Textraxylopteris* (figures 1.1A, B, D). They can also be distinguished on the basis of the small, once-dichotomous, fertile organ of *Rellimia* (figures 1.6B, C) in contrast to the twice-dichotomous fertile organs of *Tetraxylopteris* (figure 1.2). Because of these differences, we consider the taxa distinct at the generic level. Furthermore, given our rudimentary understanding of character state transitions and polarities within the Aneurophytales, we see no reason at this time to consider them more closely related to each other than either is to other members of the order.

Rellimia is distinguished from *Aneurophyton* on the basis of its larger, more complex fertile organs and distinctive spore morphology. Anatomical comparisons between the two are difficult, however, because little is known about stelar architecture in either taxon. This lack of information makes comparisons between *Rellimia* and other taxa characterized by three-ribbed protosteles difficult as well. Similarities exist between *Rellimia* and *Triloboxylon ashlandicum* (Bonamo 1977), but meaningful comparisons await further data.

PHLOEM IN ANEUROPHYTES

Tissue interpreted as secondary phloem has been described in a number of aneurophytes (e.g., *Triloboxylon arnoldii* [Stein and Beck 1983]; *T. ashlandicum* [Scheckler and Banks 1971a]; *Tetraxylopteris schmidtii* [Beck 1957]; *Prokeokalon petryi* [Scheckler and Banks 1971b]. Although the relative position of the tissue in the axis and its composition support this interpretation, no conducting elements were demonstrated in any of the studies cited above.

Recently, Wight and Beck (1984) described well-preserved secondary phloem from a Middle Devonian (probably Upper Eifelian) aneurophytalean progymnosperm. The tissue is located immediately outside the secondary xylem and is comprised of cells in radial files. It is essentially identical in composition to that described in other aneurophytes, consisting of elongate thick-walled fibers, isodiametric to slightly elongate parenchyma cells, sclerieds, and elongate thin-walled cells identifiable as sieve cells by the presence of sieve pores in the walls. Wight (1985) has also demonstrated the presence of sieve elements in primary phloem of other aneurophytes.

This discovery is important for several reasons. First, it represents the oldest convincing evidence of specialized conducting elements in the phloem. Second, it lends support to arguments of phylogenetic relationship between aneurophytes and seed plants, because of the demonstrated presence in both groups of a bifacial vascular cambium. Third, it provides an opportunity to make comparisons between the phloem of progymnosperms, other pteridophytes, and seed plants. Such comparisons might be useful in a comprehensive analysis of phylogenetic relationships among vascular plants. For example, Wight and Beck (1984) emphasize the complex nature of secondary phloem in aneurophytes and note that it contrasts sharply with the anatomically simpler phloem of both archaeopteridalean progymnosperms and most Paleozoic pteridosperms. They suggest that this information might be used to support hypotheses of phylogenetic relationship between pteridosperms and archaeopterids, a position that conflicts with the suggested relationship between pteridosperms and aneurophytes (e.g., Beck 1981; Rothwell 1982).

ARCHAEOPTERIDALES

Archaeopteris was originally described by Dawson (1871b) and *Callixylon* by Zalessky (1911). Beck (1960a, 1960b) demonstrated that *Ar-*

chaeopteris is characterized by *Callixylon* anatomy. Whether all *Callixylon* secondary wood represents *Archaeopteris* is not certain, since Scheckler and Banks (1971a) described *Callixylon*-type pitting in a specimen of *Triloboxylon*, and Beck (1967) demonstrated its presence in *Eddya*. However, Stein and Beck (1983) questioned the interpretation of *Callixylon*-type pitting in *Triloboxylon*, and Beck (1967) concluded that *Eddya* might be a young plant of *Archaeopteris*. Thus, radially aligned, grouped pitting might be a unique feature of *Archaeopteris*.

Since their establishment, *Archaeopteris* and *Callixylon* have been the subject of many studies, both taxonomic and morphologic. Among the most significant studies of *Archaeopteris* are those of Nathorst (1902), Arnold (1936, 1939), Kräusel and Weyland (1941), Beck (1960b, 1971, 1981), Carluccio, Hueber, and Banks (1966), and Phillips, Andrews, and Gensel (1972). Important studies of *Callixylon* include those of Arnold (1930, 1931), Beck (1970, 1979), Beck, Coy, and Schmid (1982), and Lemoigne, Iurina, and Singerevskaya (1983).

Archaeopteris/Callixylon had a circumpolar distribution during the Upper Devonian to Lower Mississippian, as demonstrated by collections largely from northern hemisphere localities in the United States, Canada, northwestern and central Europe, the Soviet Union, and China (see Beck 1981 for literature citations). Recently it has also been reported from Australia (E. Truswell, personal communication), and additional evidence of its occurrence in China has been provided (Cai 1987; Cai, Wen, and Chen 1987; Li Cheng-sen, personal communication). Other genera that have been assigned to the order are *Eddya* (Beck 1967), *Svalbardia* (Høeg (1942), *Actinopodium* (Høeg 1942), *Siderella* (Read 1936), and *Actinoxylon* (Matten 1968). All genera assigned to the order conform—to the extent that they are known—to the ordinal description provided below. However, the characteristics presented are based largely on *Archaeopteris*, the only genus understood in any detail.

Characteristics of the Order   The order is characterized by pseudomonopodial branching (figures 1.12A, B) by three-dimensional or planated lateral branch systems (figures 1.9B; 1.10A; 1.12B) and ultimate appendages (leaves) that are laminate with dichotomous venation (figures 1.9B, C; 1.10A; 1.17A), or deeply dissected so as to appear dichotomous (figures 1.10C; 1.17C; 1.18D), and borne in a helical (figures 1.10B; 1.12B, C; 1.17) or decussate pattern (figure 1.10C). Reproduction was heterosporous, with numerous sporangia borne adaxially on sporophylls in (lax) strobili (figures 1.9C, D; 1.10C; 1.15D;

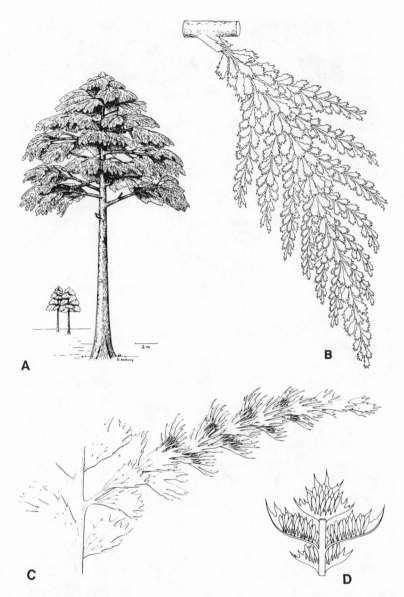

Figure 1.9. A Reconstruction of *Archaeopteris*. (From Beck 1962a, by permission.) **B** Reconstruction of a planate lateral branch system of *Archaeopteris*. (From Beck 1971, by permission.) **C** Reconstruction of a strobilus of *Archaeopteris macilenta*. (From Beck 1981, by permission.) **D** Detail of part of a strobilus of *Archaeopteris macilenta* showing decussate arrangement of sporophylls and adaxial sporangia. (From Beck 1981, by permission.)

Figure 1.10. A *Archaeopteris* lateral branch system showing planate disposition of ultimate branches, and leaves that appear to be in pairs on the main axis. × 0.8. (From Beck 1971, by permission.) **B** Lateral branch system of *Archaeopteris macilenta* illustrating helical pattern of leaf base scars on main axis. × 2.2. (From Beck 1971, by permission.) **C** Ultimate branch of *Archaeopteris macilenta* showing decussate arrangement of leaves and sporophylls. Note leaf base scars alternating with paired, opposite appendages. × 3. (From Beck 1971, by permission.)

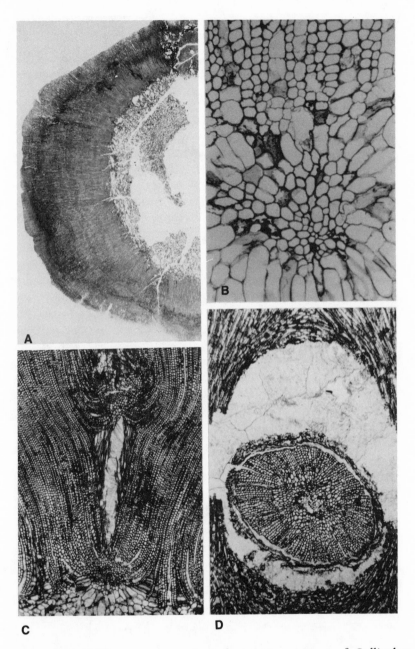

Figure 1.11. A Transverse section of a stem segment of *Callixylon brownii*. × 2.3. (From Beck 1970, by permission.) **B** Mesarch primary xylem strand at periphery of the pith in a stem of *Callixylon newberryi*. × 70. (From Beck 1970, by permission.) **C** Transverse section of stem of *Callixylon* showing radial divergence of branch trace. × 21. (From Beck 1979, by permission.) **D** Branch base of *Callixylon zalesskyi* embedded in secondary wood. × 25. (From Beck 1979, by permission.)

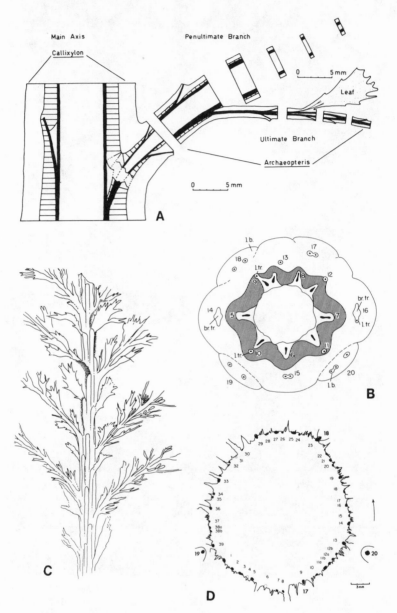

Figure 1.12. A Diagram showing relationship between a "main" axis *(Callixylon)* of *Archaeopteris* and a lateral branch system. Black regions = primary xylem; lined regions = secondary xylem. (From Scheckler 1978, by permission.) **B** Transverse section of penultimate axis of a lateral branch system of *Archaeopteris macilenta*. Black re-

1.18E). Primary xylem, mesarch in order of maturation in the stem (figure 1.11B), comprises eusteles in large axes (figure 1.12D), or protosteles or medullated protosteles in smaller axes (figures 1.11D; 1.12B; 1.17B; C) Trace divergence is radial (figures 1.12B, D), and traces to ultimate branches and leaves developed in the same ontogenetic helix in the main axes of lateral branch systems of *Archaeopteris* (figure 1.12B). Secondary xylem is compact, consisting solely of tracheids and rays (figures 1.13; 1.14A–C). Tracheids are very long, have blunt ends, and those derived from a single cambial initial are storied (figures 1.15A, B). Except in "late" wood, the tracheids are characterized by radially aligned, uniseriate to multiseriate groups of circular bordered pits in the radial walls (figures 1.13C, D; 1.15A–C; 1.16A; 1.18C). Bordered pit pairs, smaller than those in the radial walls, and ungrouped (figure 1.14B) characterize the tangential walls of tracheids with narrow radial dimensions that occur near the periphery of growth layers. Growth layers may occur in both stems and roots but are more prominent in roots. Rays are variable in width and height (figures 1.14A–C), and contain ray tracheids in some species (figures 1.13A, B, E, F). Cross-field pits are typical half-bordered pit pairs (figures 1.16D, E) with the apertures of the bordered pits varying from narrowly to broadly elliptical (figures 1.13D; 1.16D, E).

NEW CONTRIBUTIONS TO ARCHAEOPTERIDALEAN TAXONOMY AND MORPHOLOGY SINCE 1975

Taxonomy of *Callixylon* Lemoigne, Iurina, and Snigerevskaya (1983) provided a comprehensive revision of *Callixylon*. They described in detail the type species, *C. trifilievi;* provided, for the first time, specific and generic diagnoses; and designated a lectotype (Specimen No. 1415/40 in the Zalessky Collection of the F. N. Chernishev Museum [CNIGR-Museum], Leningrad). They also designated a series of isotypes.

---

gions = protoxylem; shaded regions = secondary xylem; br. tr. = branch trace; l. b. = leaf base; l. tr. = leaf trace. Numbers indicate sequence of leaf and branch traces along the ontogenetic helix. (From Beck 1971, by permission.) **C** Camera lucida drawing of part of a lateral branch system of *Archaeopteris macilenta*. (From Beck 1971, by permission.) **D** Camera lucida drawing of primary vascular system of *Callixylon brownii* as seen in transverse section. (From Beck 1979, by permission.)

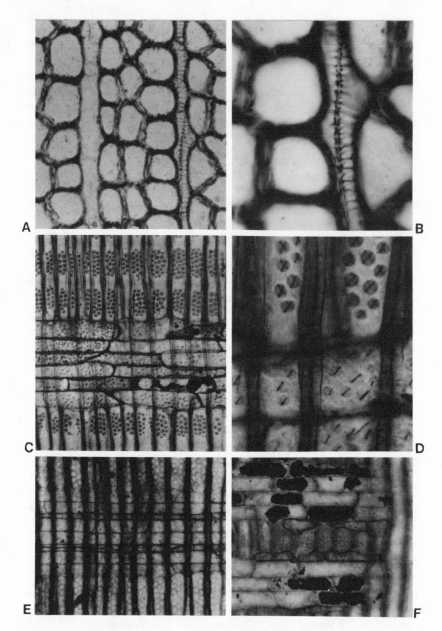

Figure 1.13. A Secondary xylem of *Callixylon erianum* in transverse section. × 210. (From Beck 1970, by permission.) **B** Detail of figure 1.13A showing ray tracheid. × 475. (From Beck 1970, by permission.) **C** Radial section of secondary xylem of *Callixylon newberryi*. × 115.

They considered *C. timanicum, C. velinense, C. zalesskyi,* and *C. whiteanum* to be synonyms of *C. trifilievi,* and *C. marshii, C. mentethense,* and *C. schmidtii* to be synonyms of *C. newberryi.* They recognized *C. henkei, C. brownii, C. arnoldii, C. bristolense, C. petryi, C. newberryi,* and *C. trifilievi* as valid species.

In their emended diagnosis of *C. newberryi,* Lemoigne et al. (1983) included the presence of ray tracheids, as did Arnold (1931) in his original diagnosis. Beck (1981) has emphasized, however, that there may be no ray tracheids in this species. He is unable, at any rate, to find any. If this is correct and ray tracheids are characteristic of *C. mentethense* (Arnold 1930), the two taxa cannot be synonymous.

We regard the possible synonomy of *C. zalesskyi, C. whiteanum,* and *C. trifilievi* with equal skepticism. Arnold was not certain about the presence of ray tracheids in *C. whiteanum* and the evidence presented for ray tracheids in *C. trifilievi* (Lemoigne et al. 1983) is unconvincing, whereas *C. zalesskyi* is characterized by numerous ray tracheids.

Another conclusion to which we take exception is the statement, by Lemoigne et al. (1983), that the characteristics of *C. arnoldii* are those of a "petite" axis, probably of *C. newberryi.* This is almost certainly incorrect, since *C. arnoldii* was collected from the Sanderson Formation (Tournaisian), and *C. newberryi,* one of the most abundant species in the New Albany Shale, has not been collected from above the Devonian in this formation or elsewhere. It should be noted further that these species are strikingly divergent in many characters. Also, it is impossible to ascertain the size of the axis from which the type material of *C. arnoldii* came, since it consists solely of fragments of secondary wood (Beck, 1962b).

New Observations on the Anatomy of *Callixylon* Secondary Wood   Beck, Coy, and Schmid (1982) investigated three species, one preserved as fusain, the second a phosphatic permineralization, and a third a siliceous permineralization. These materials, studied at rela-

---

(From Beck 1970, by permission.) **D** Radial section of secondary xylem of *Callixylon newberryi* showing crossed, slit-like apertures of bordered pit pairs, and cross-field pits (i.e., half-bordered pit pairs formed between ray parenchyma cells and contiguous tracheids). × 350. (From Beck 1970, by permission.) **E** Radial section of secondary xylem of *Callixylon erianum* showing rows of interspersed ray tracheids. × 118. (From Beck 1970, by permission.) **F** Marginal ray tracheids in secondary xylem of *Callixylon brownii.* × 118. (From Beck 1970, by permission.)

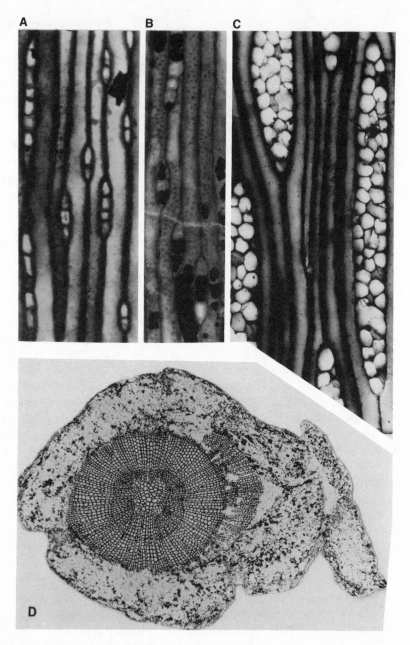

Figure 1.14. A Tangential section of secondary xylem of *Callixylon brownii* containing low, uniseriate rays. × 118. (From Beck 1970, by permission.) **B** Bordered pits in tangential walls of secondary xylem tracheids of *Callixylon brownii*. × 118. (From Beck 1970, by permission.) **C** Tangential view of tall, multiseriate rays in secondary xylem of *Callixylon newberryi*. × 118. (From Beck 1970, by permission.) **D** Transverse section of root of *Callixylon petryi*. × 18.

Figure 1.15. **A** Radial view of secondary xylem of *Callixylon arnoldii* showing radially aligned groups of uniseriate, bordered pits. UMMP No. 44714, negative No. 10048. × 75. **B** Detail of figure 1.15A, showing faceted, blunt ends of storied tracheids. × 250. (From Beck, Coy, and Schmid 1982, by permission.) **C** Bordered pits of *Callixylon arnoldii* with circular apertures. × 2000. (From Beck, Coy, and Schmid 1982, by permission.) **D** Parts of two fertile lateral branch systems of *Archaeopteris*. × l. UMMP Collection No. 375L.

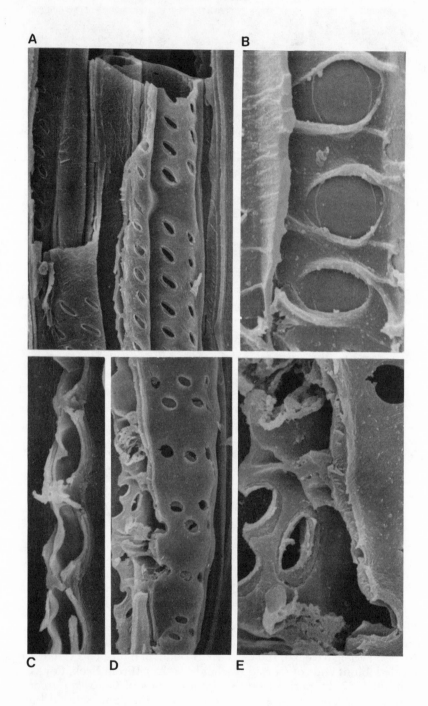

tively high magnifications, provided new information on wall structure and the morphology of bordered pits, especially the pit membranes.

In the fusinized secondary wood, pit apertures, although predominantly narrowly elliptical (figure 1.16A) are occasionally broadly ovate, exposing large areas of the pit membrane (figure 1.16B). Pit membranes in species with elliptical apertures (e.g., *C. newberryi* [see Schmid 1967], *C. erianum* and the species preserved as fusain; figure 1.16C) are homogeneous and of uniform thickness, lacking a torus. However, those of the fusinized wood may have a conspicuous, oval to circular ridge or rim, delimiting a central region of the membrane (figure 1.15B). This rim is interpreted as the possible remains of an encrustation of waste metabolites on the membrane.

C. arnoldii (Beck, 1962b) is unique among known species of *Callixylon* in possessing both circular bordered pits with essentially circular apertures (figure 1.15C), and the largest known pits in the genus ($\bar{x} = 16.6$ μm). Among possible descendants of *Archaeopteris*, conifers with elliptical apertures tend to have homogeneous pit membranes (i.e., they lack a torus), whereas those with circular apertures possess a torus. There is also a correlation between the presence of a torus in seed plants and large pit size. Since *C. arnoldii* is characterized by both circular apertures and pits of the largest size within the genus, Beck et al. (1982) carefully studied mineral layers they interpreted to

Figure 1.16. A Part of a tracheid in secondary xylem of *Callixylon*. Note groups of bordered pits with oblique, elliptical apertures; also note crassula-like region between pit groups. × 1000. (From Beck, Coy, and Schmid 1982, by permission.) **B** Broad pit apertures in *Callixylon* secondary xylem exposing pit membranes with peripheral, circular ridges interpreted as accumulations of waste metabolites. × 3500. (From Beck, Coy, and Schmid 1982, by permission.) **C** Sectional view of bordered pit pairs in secondary xylem of *Callixylon*. Note apparently uniformly thickened pit membranes. × 3000. (From Beck, Coy, and Schmid 1982, by permission.) **D** Oblique, sectional view of a ray and a contiguous tracheid wall in secondary xylem of *Callixylon*. Note cross-field pits in groups corresponding to position of ray parenchyma cells. × 1000. (From Beck, Coy, and Schmid 1982, by permission.) **E** Detail from figure 1.16D. This is a half-bordered pit pair (cross-field pit) seen from "inside" the ray parenchyma cell. Note simple pit in ray cell wall and aperture of bordered pit in contiguous tracheid wall. × 3000. (From Beck, Coy, and Schmid 1982, by permission.)

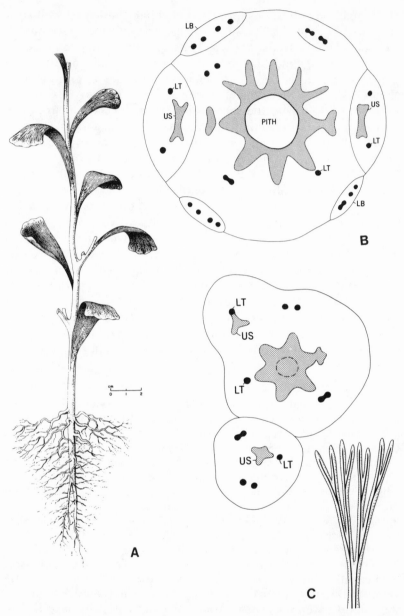

Figure 1.17. A Reconstruction of *Eddya sullivanensis*. (From Beck 1967, by permission.) **B** Diagram of a transverse section of *Siderella* showing helical arrangement of leaf traces (LT) and leaf bases (LB), and subopposite and distichous branch traces (US). Shaded regions

represent pit membranes and found that many were characterized by a central, circular region, on average about 7.1 μm in diameter, as compared with apertures of 4.4 μm in greatest mean dimension. The ratio of the diameter of the central circular regions of the pit membranes to the diameter of the apertures is 1.613, a value very similar to the ratio of torus to aperture diameters of 1.574 in four species of extant conifers measured by Beck et al. (1982). They concluded on the basis of this circumstantial evidence that the pit membranes of *C. arnoldii* might have contained a torus.

They also suggested an alternative to the interpretation of the regions between groups of pits as crassulae (Arnold 1929; Beck 1970). They believe the appearance as crassulae is caused in some cases, possibly in general, by a separation of wall layers between groups of pits, and is enhanced by the raised nature of the pit groups (figure 1.16A).

The presence of secondary wall layers that seem to conform in thickness (Schmid 1967) and micro- or macrofibrillar orientation (Beck et al. 1982) to the S1, S2, land S3 layers, characteristic of seed plants, supports the hypothesized phylogenetic relationship between *Archaeopteris* and early seed plants.

The Primary Vascular System of *Callixylon* Attempts to analyze the primary vascular system of *Callixylon* have been fraught with difficulty, primarily because of the large number of bundles in the system (figure 1.12D) and because enclosure of the primary vascular system by secondary xylem has made very difficult the recognition of small primary vascular bundles. Beck's (1979) analysis demonstrated that in major axes of the plant the system is composed of a variable number of mesarch primary vascular bundles of two contrasting sizes, with small axes having systems of fewer bundles than large axes.

He observed, as had Arnold (1930) and others, that traces diverge radially from some of the larger bundles in the system (figures 1.11C;

= primary xylem. Based on an undescribed specimen in the collections of the University of Michigan Museum of Paleontology. C Diagram of transverse section of an axis of *Actinoxylon* and its lateral branches; also an ultimate appendage. Leaf traces (LT) and branch traces (US) are arranged helically on the main axis. Leaf traces are thought to have been produced in a decussate arrangement on the ultimate branches. Shaded xylem column of penultimate axis includes both primary and secondary xylem. Based on the work of Matten (1968).

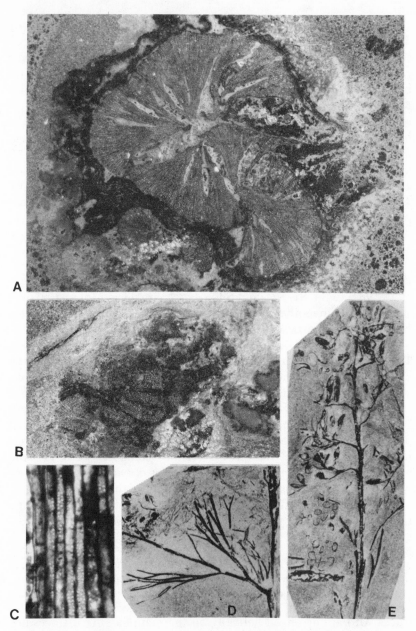

Figure 1.18. A Transverse section of root of *Eddya sullivanensis* near its base. × 16. (From Beck 1967, by permission.) **B** Distal section of root of *Eddya sullivanensis*. × 16. (From Beck 1967, by permission.) **C** Secondary xylem of *Eddya sullivanensis* showing radial banding of

1.12D). He noted further that all traces that could be mapped and studied for any distance along their length increase in size and become fluted or ribbed and medullated distally, thus acquiring the anatomical characteristics of the primary vascular system of the main axes of lateral branch systems of *Archaeopteris* (figures 1.12A, B; see also Scheckler 1978). No traces divide tangentially in the cortex as do leaf traces of *Archaeopteris* (Beck 1971; Scheckler 1978). Thus it was concluded that all traces observed are branch traces.

The absence of any evidence of leaf traces in the specimens studied suggested that, unlike the generally plagiotropic, lateral branch systems bearing leaves, the major axes of *Archaeopteris* bore only lateral branch systems. From a developmental standpoint this would mean that, in the mature plant, apical meristems of "lead" shoots in this pseudomonopodial branching system produced predominantly lateral branch systems (and, presumably, a smaller number of new "lead" shoots) and no leaves at all. In contrast, the apical meristems of the lateral branch systems produced both ultimate branches and simple leaves. Beck (1971) demonstrated that these appendages and their traces occur in the same ontogenetic helix (figure 1.12B) and thus are very probably homologous structures. He suggested further (1979) that the entire lateral branch system, commonly plagiotropic, might be homologous with a leaf. Consistent with this view is the presence of bifurcate, aphlebia-like laminate appendages at the base of the lateral branch systems (figure 1.9B) that were shed with the lateral branch systems (Nathorst 1902, 1904), or that sometimes remained attached to the axes that bore the lateral branch systems (Beck 1960b); the apparent presence of only lateral branch systems without intervening, simple leaves on major axes of the shoot system (Beck, 1979); the determinate nature of the lateral branch systems (Scheckler 1978); and the probability that they were functionally photosynthetic organs that were shed as units (Beck 1971). Indeed, it is tempting to conclude on these bases that the lateral branch systems were, essentially, compound leaves. As stated by Beck (1979), "one could reasonably conclude that evolution [of the lateral branch system] in the direction of a compound frond might have proceeded farther than has been sup-

---

grouped pits. × 115. (From Beck 1967, by permission.) **D** Ultimate branch of *Svalbardia polymorpha* bearing dichotomously divided ultimate appendages. × 1. (From Høeg 1942, by permission.) **E** Fertile ultimate branch of *Svalbardia polymorpha*. × 3. (From Høeg 1942, by permission.)

posed heretofore." One must not forget, however, that the radial symmetry of the internal anatomy of the main axes of these systems, the helical phyllotaxy of the simple, laminate appendages along the main axis and their decussate arrangement on the ultimate axes as well as the occasional variation from branch plagiotropy demonstrated by Scheckler (1978), are unleaf-like and provide a strong basis for considering them lateral branch systems—not compound leaves.

Arrangement of branch traces conforms to a typical Fibonacci pattern, and it is quite possible that the primary vascular system comprises a eustele differing from that of a seed plant such as *Lyginopteris* only in the larger number of sympodia in the system and in the direct radial divergence of traces from bundles in the stele. This latter feature was considered by Beck (1979) to reflect the origin of this stele from a protostelic ancestral type.

If the primary vascular system of *Callixylon* is a eustele, then it follows that the bundles from which leaf traces diverge might be axial bundles as suggested by Rothwell (1976) and Scheckler (1978). Because of the inability to map all bundles accurately, including the very small bundles observed by Beck (1979) in the system of *C. brownii*, it is not possible to eliminate the possibility that some of the very small bundles are axial bundles and that the large bundles from which the traces appear to diverge radially are a part of the trace complex, representing what Beck (1970) called accessory bundles in *Calamopitys foerstei*. If so, this might represent a stelar morphology intermediate between a protostele and a typical eustele. Such a condition would not be surprising in a putative seed plant ancestor.

Anatomy and Morphology of Lateral Branch Systems of *Archaeopteris*   The initial conception of the lateral branch systems of *Archaeopteris* as fronds resulted from an incorrect interpretation of their morphology by many workers and a lack of information on their anatomy prior to 1966. This interpretation was reflected in Beck's (1962a) reconstruction of a lateral branch system as a planate frond, the "rachis" of which was bilaterally symmetrical, and on which were borne a pair of "inter-rachial pinnules" between each pair of "pinnae." "Pinnules," like the pinnae on which they were borne as well as the inter-rachial pinnules, were thought to occur in a single plane.

In 1966, Carluccio, Hueber, and Banks presented evidence that demonstrated a radially symmetrical anatomy of the supposed rachis, and divergence of traces in a helical pattern, quite unlike what one would have expected on the basis of interpretations of external morphology of *Archaeopteris* fronds preserved as compressions (figure

1.10A). Beck (1971) corroborated the anatomical interpretations of Carluccio et al. (1966) and demonstrated that the external morphology as reflected in compressions of *A. macilenta* (figures 1.10B; 1.12C) does, contrary to earlier superficial interpretations, correspond to the internal anatomy (figure 1.12B). He showed that the main axis bore helically arranged, simple leaves with decurrent leaf bases (figures 1.9B; 1.10B; 1.12C)—not pairs of inter-rachial pinnules. The leaves in several specimens were more numerous (and presumably smaller) on one side of the axis than on the other. The variation in leaf size is suggested by the difference in width of the decurrent leaf base scars on the two sides of the axis. On what Beck interpreted as the adaxial side of one well-preserved specimen there are four orthostichies, whereas there are only three on the presumed abaxial side. These two groups of orthostichies, and the adaxial and abaxial sides of the axis, are delimited by two orthostichies of lateral (ultimate) branches. Although the ultimate branches of the lateral branch system are in a single plane, thus distichous, they occur in the same ontogenetic helix as the leaves (figure 1.12B). They commonly occur in opposite pairs (figures 1.10A; 1.12B), but may be displaced to such a degree as to appear alternate (figures 1.10B, 1.12C; Beck 1971). It is clear, therefore, that branching within the lateral branch systems was pseudomopodial, a pattern of branching that probably characterized the entire plant (Scheckler 1978). A map of leaf positions on both sides of the axis (Beck 1971) demonstrated a series of discontinuous parastichies, thus suggesting the early abortion during development of leaf primordia on one side of the axis. Beck hypothesized that the abaxial side bore the smaller number of apparently larger leaves, since the presence of fewer large leaves on the lower surface would probably allow an orientation whereby all leaves on that side would receive adequate light for photosynthesis. With greater density of leaves on the abaxial surface some would presumably have been deprived of adequate light, thus accounting for the postulated abortion of leaf primordia and the resulting discontinuous parastichies.

The leaves on ultimate axes had been interpreted earlier, like those on the main axes of lateral branch systems, to occur in one plane. However, both Beck (1971) and Phillips, Andrews, and Gensel (1972) demonstrated a decussate phyllotaxy on the ultimate axes of lateral branch systems of *A. macilenta* (figure 1.10C), and Phillips et al. (1972) suggested that in *A. halliana* leaves on ultimate branches "may be in four rows rather than in a spiral." It is not clear whether a decussate pattern characterizes the genus as a whole.

On the basis of comprehensive information of both anatomy and the arrangement of leaves and branches, Beck (1971) presented a detailed reconstruction of an entire lateral branch system (figure 1.9B).

Scheckler (1978) provided additional detail on the anatomy of lateral branch systems and concluded that they were characterized by determinate growth, thus supporting Beck's (1971) conclusion that they were probably shed as units. Scheckler's interpretation was based on a developmental pattern typical of determinate lateral shoots (figure 1.12A). He observed that vascular systems of both the main axis and ultimate axes of the lateral branch systems were protostelic at points of origin from axes of the next higher order, that they increased in size and became medullated distally and then began to diminish in size, becoming again protostelic in the terminal regions of ultimate axes. Throughout this system he observed a correlation between the size of the primary xylem column, or cylinder, the number of ribs, and the number of protoxylem strands.

Numerous compression specimens described and illustrated by paleobotanists since 1871 and the anatomical studies of Carluccio et al. (1966) and Beck (1971) provide evidence of only two ranks or orthostichies of ultimate branches in lateral branch systems of *Archaeopteris* (figures 1.10A, B; 1.12B, C). Scheckler (1978), however, interpreted sections of several pyritized axes to indicate the presence of traces to ultimate branches in more than two ranks—six in one specimen! He suggested that the proximal region of large main axes bore ultimate branches in several orthostichies, whereas the more commonly observed subopposite ultimate branches in two ranks is characteristic of more distal regions. If this were in fact characteristic of the genus, one would have expected some evidence of this morphology in the basal regions of the large lateral branch systems described by Nathorst (1902) and Johnson (1911), but apparently none was observed by these or other workers, and the writers have not observed any evidence to support this position in their study of specimens in their collection or of illustrations in the literature.

Morphology of Fertile Structures of *Archaeopteris*   Some lateral branch systems of *Archaeopteris* bear fertile ultimate branches (figure 1.15D). These are often the most basal ultimate branches of the lateral branch systems. They are bounded proximally by helically arranged leaves (see, e.g., Nathorst 1902) and occasionally also by vegetative ultimate branches (Johnson 1911), and distally by solely vegetative ultimate branches. Some evidence indicates that fertile and vegetative ultimate branches may be intermixed (Beck 1981). The frequency of this condition is unknown. Both Beck (1971) and Phillips et al.

(1972) describe some morphological intergradation between fertile and solely vegetative ultimate branches.

The morphology of the fertile leaves (sporophylls) and the arrangement of fertile and vegetative leaves on the ultimate branches, as well as the morphology and distribution of sporangia, are important details that have been clarified by recent work (Beck 1971, 1981; Phillips et al. 1972).

The fertile ultimate branches, designated strobili by Beck (1981), bear sporophylls, both proximal and distal to which commonly are vegetative leaves (figure 1.9C). It is not certain that such an association of vegetative leaves with the sporophylls is a generic character, since the terminal parts of fertile ultimate branches are often not preserved, and basal vegetative leaves have not been observed on all specimens (see, e.g., Phillips et al. 1972).

A. macilenta is characterized by a decussate arrangement of the sporophylls that bear, adaxially, numerous sporangia (figures 1.9C, D). The sporophylls are laminate with much-divided apical regions. The very short-stalked, fusiform sporangia are borne primarily on the broadly laminate regions (figure 1.9D), but have been observed rarely in the more dissected regions of the sporophylls (Beck 1981; Phillips et al. 1972). In contrast to *A. macilenta*, the sporophylls of *A. halliana* are thrice-dichotomous ultimate appendages. Sporangia are borne adaxially on the segments of the basal and middle dichotomies (Phillips et al. 1972). Although the exact arrangement of sporangia is unknown, Phillips et al. observed more than a single row of sporangia in basal regions of the sporophylls but illustrated only one row along dichotomous segments in more distal regions. In both of these species, and others that have been illustrated in the literature, the basal vegetative leaves and (when present) sporophylls tend to be larger than those in more distal regions, the most apical vegetative leaves becoming much reduced in size (figure 1.9C; Beck 1981).

Heterospory has been demonstrated in several species from at least four geographic localities in northeastern United States and eastern Canada (Phillips et al. 1972). Whereas early studies suggested a morphological divergence between megasporangia and microsporangia, the former thought to be relatively short and broad, the latter longer and more slender (Arnold 1939), both Pettitt (1965) and Phillips et al. (1972) demonstrated an intergradation in sporangial size. Phillips et al. (1972) emphasized, therefore, that "size distinctions of sporangia are inadequate for the presumption of heterospory in *Archaeopteris*."

The distribution of microsporangia and megasporangia within the strobili is unknown. It has been shown that in at least one species *(A.*

cf. *jacksoni)* both megasporangia and microsporangia may occur side by side on the same sporophyll (Pettitt 1965). Sporangia are fusiform and characterized by longitudinal dehiscence. They range in size in the genus from 1.0–4.0 mm long to 0.23–0.83 mm wide, and there is some size variation among species (Phillips et al. 1972).

Both megaspores and microspores from different species are similar in size and exine sculpturing and, thus, cannot be used to distinguish species (Pettitt 1965; Phillips et al. 1972). According to these workers megaspores generally fall within the concept of the dispersed spore genus *Biharisporites*, and microspores within the concept of *Cyclogranisporites*. According to Phillips et al. (1972) the reported range in diameter of *Archaeopteris* megaspores is 110–460 μm and of microspores is 33–70 μm. On the basis of estimates in the literature and our own observations of spore masses, it seems likely that megasporangia typically contained 16 to 32 spores, whereas microsporangia probably contained about 10 to 20 times as many.

Gross Morphology of *Archaeopteris*   Only one reconstruction of the whole plant (figure 1.9A) has been published (Beck 1962a) since it was determined that *Archaeopteris* and *Callixylon* represent the same plant (Beck 1960a, 1960b). However, as early as 1953, a reconstruction of *Callixylon*, for which Professor C. A. Arnold was the consultant, was published in the *Life* magazine series "The World We Live In" (see Barnett 1953). These reconstructions are basically similar, the latter differing significantly from the former only in the presence of small, conifer-like leaves rather than frond-like lateral branch systems. In both of these, no doubt, the hypothesis of an excurrent habit was influenced by the discovery of large, straight *Callixylon* logs (see Beck 1964) and to some extent by the remarkable similarity between the secondary xylem and that of conifers, and the assumption that plants with such similar wood might also have been of similar habit. Evidence of only relatively small, embedded lateral branches, in large, permineralized axes of *Callixylon*, described by Arnold (1931), gave credence to this idea.

The discovery of pseudomopodial branching in *Archaeopteris* lateral branch systems (Carluccio et al. 1966; Beck 1971; Scheckler 1978) and, apparently, also in more proximal parts of the plant *(Callixylon)* (Beck 1979) suggests that the crown of the plant was probably more diffusely branched than previously thought. The crown of *Archaeopteris* might, therefore, have had a more deliquescent aspect than indicated by the reconstruction by Beck (1962a).

It has been suggested, further, that an arborescent habit might not have characterized all species of *Archaeopteris* (Beck 1981), but at present there is no direct evidence to support this view.

OTHER ARCHAEOPTERIDALEAN TAXA

It seems unlikely that *Archaeopteris* is the only legitimate taxon in the order, yet all other taxa assigned to it have been suggested, at one time or another by one or more paleobotanists, to represent *Archaeopteris*. This may be related to the fragmentary nature of specimens assigned to other taxa, or to the fact that closely related taxa share many characters, or, of course, that one or more may, indeed, represent *Archaeopteris*.

Svalbardia After *Archaeopteris*, *Svalbardia* Høeg (1942), of Middle Devonian (Givetian) age from Spitzbergen, is the best-known genus assigned to the order. Vegetative ultimate appendages (leaves) (figure 1.18D) as well as fertile structures (figure 1.18E) closely resemble those of *Archaeopteris*. It is retained as distinct from *Archaeopteris* because of some remaining uncertainty about the arrangement of ultimate branches and lack of incontrovertible evidence of its anatomy (see section below on *Actinopodium*). Although branching in *Svalbardia* is considered by some to have been helical, Beck (1971) argued that, on the basis of Høeg's description of the plant, it was probably characterized by planate, not helical, ultimate branches and should, therefore, be considered a synonym of *Archaeopteris*. Scheckler's (1978) conclusion that ultimate branches of *Archaeopteris* may sometimes occur in more than two orthostichies, if correct, might seem to make moot the matter of the pattern of ultimate branches, and he, like Beck, suggested that *Svalbardia* seems to fall within the morphological variation exhibited by *Archaeopteris*. However, another interpretation might be that Scheckler's material represents *Svalbardia* rather than *Archaeopteris*, and that the ultimate branches of *Svalbardia* are, in fact, helically arranged. If so, one is still faced with the difficult question of whether branching pattern, as the only known distinctive feature, is a legitimate basis for generic segregation when all other known characters are those of *Archaeopteris*.

Supporting the conclusion that *Svalbardia* is probably an *Archaeopteris* is its possible stratigraphic association with *Actinopodium*, anatomically preserved axes described by Høeg (1942) (but see details below), which are apparently identical in stelar structure to the penultimate branches of *Archaeopteris*.

Actinopodium *Actinopodium nathorstii* Høeg (1942), of Middle Devonian (Givetian) age from Spitzbergen, closely resembles the main axis of an *Archaeopteris* lateral branch system. Although poorly preserved, the permineralized specimens are characterized by a medullated actinostele of apparently eight ribs (Høeg indicated "about 7"). In one specimen the ribs were interconnected, but in another they were apparently separate, wedge-shaped vascular strands. Development of primary xylem was mesarch. In both specimens illustrated by Høeg, but especially in the holotype, there is evidence of secondary xylem. Small mesarch xylem strands in the cortex resemble leaf traces of *Archaeopteris*. In fact, as emphasized by Carluccio et al. (1966), Beck (1971), and Gensel and Andrews (1984), it is possible that *Actinopodium* represents *Archaeopteris* (or *Svalbardia*, if one wishes to consider it distinct from *Archaeopteris*). *Actinopodium* and *Svalbardia* were both collected in Spitzbergen from strata of Middle Devonian age. Although the type specimens came from different localities and from strata considered by Høeg (1942) to be of slightly different ages, he observed that *Svalbardia* "also occurs along the river above Fiskeklofta," the locality from which the type material of *Actinopodium* was collected.

Actinoxylon Another genus assigned to the Archaeopteridales is *Actinoxylon* Matten (1968), based on permineralized axes of Givetian age from New York. *Actinoxylon* (figure 1.17C) exhibits several characters very similar to those of *Archaeopteris*. The penultimate branches are characterized by primary xylem in the form of a multiribbed actinostele from which diverge three- or four-ribbed ultimate branch traces and helically arranged leaf traces. In the ultimate branches, leaf traces, circular in transverse section, alternate at right angles in opposite to subopposite pairs. Ultimate appendages (leaves, according to Matten) are dichotomously divided and filiform. It differs from *Archaeopteris* in features that are considered to represent primitive character states: absence of a pith in the stele in penultimate axes (but this is also a feature of basal and apical regions of penultimate branches of *Archaeopteris macilenta* [see Scheckler 1978]), helical arrangement of ultimate branches, and dichotomy of leaf segments in at least two planes.

It should be noted that the helical branching pattern and the filiform leaves are possibly consistent with the morphology of *Svalbardia*

(Høeg 1942; see also above). One also might argue that in the absence of information on fertile structures, *Actinoxylon* could as logically be assigned to the Aneurophytales as to the Archaeopteridales. Leaf arrangement and morphology as well as the arrangement and morphology of ultimate branch traces resemble those of one or more aneurophytes. Apparently, the only reason *Actinoxylon* was assigned to the Archaeopteridales is the six-ribbed actinostelic primary vascular column, which Matten (1968) believed was more archaeopteridalean than aneurophytalean. Perhaps such a taxonomic distinction is of little importance, since, in either case, *Actinoxylon* can be considered a possible ancestor of *Archaeopteris*. It might be highly significant, however, in resolving the problem of whether the progymnosperms were monophyletic or polyphyletic—i.e., whether the Archaeopteridales evolved from an aneurophyte or whether these orders had a common ancestor.

Siderella   *Siderella* Read (1936) (figure 1.17B) is based on a permineralized axis of Lower Mississippian (Tournaisian) age from Kentucky. It is essentially identical in anatomy to that of main (penultimate) axes of *Archaeopteris* lateral branch systems (compare figures 1.12B and 1.17B), and was considered by Carluccio et al. (1966) and Beck (1971) probably to represent this genus. The type specimen is of interest because, although larger than any known permineralized axis of *Archaeopteris*, it lacks secondary xylem. Traces to ultimate branches, like those of *Archaeopteris macilenta* (Beck 1971), are four-ribbed. Interestingly, however, the first two leaf traces originate from adjacent ribs (figure 1.17B), whereas in *A. macilenta* they originate from opposite ribs (Carluccio et al. 1966; Beck 1971; Scheckler 1978). This suggests that leaves on the ultimate branches might have been two-ranked, in one plane (i.e., distichous) in contrast to the decussate arrangement of leaves in several species of *Archaeopteris* (see Beck 1971; Phillips et al. 1972). Unfortunately, available evidence is inadequate to resolve this problem. If the leaves of *Siderella* were two-ranked, this difference in phyllotaxy might reflect only a species difference. It would, however, suggest an evolutionary advance in the planation of lateral branch systems, and support the hypothesis of the evolution of compound leaves from such systems (see, e.g., Carluccio et al. 1966; Matten 1968; Beck 1971, 1979; and the following section on Evolutionary Trends).

Eddya   *Eddya sullivanensis* Beck (1967) was based on specimens of Upper Devonian (Frasnian) age from New York. These represent a small plant perhaps 0.5 m tall, consisting of a sparsely branched stem bearing helically arranged, flabelliform, dichotomously veined leaves attached to a root system (figure 1.17A). This plant is of considerable interest, since its leaves closely resemble those of *Archaeopteris*, especially *A. obtusa*, and its eustelic primary vascular system is very similar to that of *Callixylon*, although apparently consisting of only five sympodia. Pits in the outer, last-formed tracheids of the secondary xylem are grouped and radially banded (figure 1.18C). Thus, this plant exhibits some important characteristics of *Archaeopteris*, and Beck suggested that it might represent a young sporophyte of *Archaeopteris*. This suggestion was tempered by the absence of fertile structures, by the lack of planate, frond-like lateral branch systems; and especially by the unusual anatomy of the root system (figures 1.18A, B). As he emphasized, the absence of fertile structures and lateral branch systems would not be unexpected in a young plant that could be interpreted as in a "juvenile" condition, but the highly dissected, polystelic, vascular system of the root could not be so easily accommodated within the concept of *Archaeopteris*. The plant was, therefore, given a new generic name. Its taxonomic status remains equivocal.

PROTOPITYALES

In 1850 Göppert assigned the name *Protopitys buchiana* to permineralized stems of Lower Carboniferous age with conifer-like secondary wood. Other workers had earlier included them in *Pinites* or *Dadoxylon*. These specimens were of unusual interest because of the nature of their secondary xylem and because some reached a diameter of 45 cm. It was not until considerably later, however, that their central primary tissues were described (Solms-Laubach 1893). Solms-Laubach's study with subsequent descriptions (Seward 1917; Scott 1923) and investigations of new specimens (Walton 1957, 1969; Smith 1962) provided extensive information on the anatomy of the primary vascular system including trace departure, the secondary xylem, fructifications, and roots. Unfortunately, nothing is known of the plant's vegetative, photosynthetic appendages. Because of its combination of pteridophytic and gymnospermous characteristics, Beck (1960b) included Protopityales (Walton 1957) in the Progymnospermopsida, an assignment later supported by Smith (1962) and Walton (1969).

The two species *P. buchiana* and *P. scotica* (Walton 1957) are similar in known vegetative characteristics. As preserved, the parenchymatous pith is elliptical in transverse view (figures 1.19, 1.21D). Smith

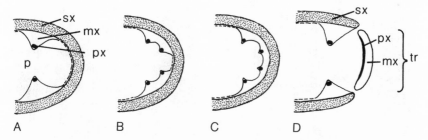

Figure 1.19. Diagrams showing the nature of trace formation in *Protopitys scotica*. mx = metaxylem, p = pith, px = protoxylem, sx = secondary xylem, tr = trace. (Based on descriptions and illustrations in Walton 1957.)

(1962) believed that, in life, the pith of the stem he studied as well as the stem itself was approximately circular in section, the elliptical shape of the pith and surrounding tissues having resulted from compression. Four conspicuous wedge-shaped bundles of primary xylem extend into the pith (figure 1.19A), and pairs of these bundles flank positions, opposite each other, from which traces to lateral appendages diverge alternately in two ranks (i.e., distichously; figures 1.19D, 1.21D). Each bundle contains an endarch to slightly mesarch protoxylem strand (figures 1.19, 1.21D). Between the bundles, a thin layer of metaxylem (thicker in *P. buchiana* than in *P. scotica*, in which it may be only one cell thick—see Walton 1957) was thought by Walton (1957) and earlier workers to have bounded the pith. Smith (1962), however, was uncertain of the presence of a continuous layer of metaxylem in the specimens of *P. scotica* he studied.

Proximal to trace formation the protoxylem strand in each of a pair of adjacent bundles divides, and the resulting strands become located in two wedge-shaped extensions of metaxylem (figure 1.19B). At a higher level these bilobed bundles of primary xylem merge into an arc (figures 1.19C; 1.21C), and the central region containing two protoxylem strands separates as an appendage trace (figures 1.19D; 1.21D). The two protoxylem strands of the trace apparently fuse, forming a thin layer of protoxylem along the adaxial surface of the trace (figure 1.19D; Walton 1957; Scott 1923).

According to Smith (1962), the protoxylem consists of helically sculptured tracheids, the metaxylem of scalariform to slightly reticulate elements. The secondary xylem tracheids of *P. buchiana* and the tracheids of the inner secondary xylem of *P. scotica* bear scalariform bordered pits (figure 1.20A, B) whereas those in the bulk of the wood

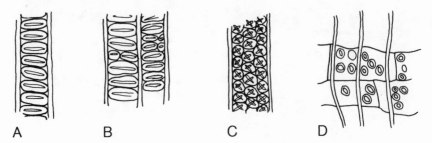

A   B   C   D

Figure 1.20. Drawings of tracheid **(A-C)** and cross-field **(D)** pitting from secondary xylem of *Protopitys scotica*. **A** and **B** represent tracheids with scalariform bordered pitting from the innermost part of the secondary xylem. **C** illustrates the multiseriate circular bordered pitting characteristic of most of the secondary xylem. × 400. (From Smith 1962, by permission.)

of *P. scotica* bear on their radial walls multiseriate circular bordered pits with crossed apertures (figures 1.20C; 1.21F). The rays of both species are uniseriate (rarely biseriate in part in *P. buchiana*—see Smith 1962) and predominantly low. In *P. buchiana* rays are commonly only one cell high, but rarely as high as 15 or 16 cells (Scott 1923; Smith 1962). Smith (1962) describes rays of *P. scotica* as mostly one cell high, with a few two or three cells high. Cross-field pits between ray cells and contiguous tracheids are of the cupressoid type (figure 1.20D; Smith 1962).

Roots associated with a stem of *P. scotica* were described by Walton (1969) as characterized by an oval, diarch xylem column and a conspicuous exodermis with thick radial walls (figure 1.21E). According to Walton (1969), the lack of evidence of any secondary xylem and the fact that "the plane in which the protoxylems of the lateral diarch rootlet lie cuts the axis of the parent root transversely" are characteristic features of pteridophytes.

The fertile organs (Walton 1967; Smith 1962) are branching systems that dichotomize twice and bear sporangia terminally on more or less pinnately arranged ultimate axes (figure 1.21A). The fertile organs described by Walton were recurved toward the axis bearing them. Smith (1961) interpreted these to be "apparently immature since the sporophylls had not unfolded and the sporangia had not dehisced." He illustrated a fertile organ of *P. scotica* spread out in one plane. All sporangia of this specimen had apparently dehisced prior to preservation. Sporangia of *P. scotica* are elongate, ending in a sharply attenuated tip. Dehiscence was by a longitudinal slit. Walton (1957) observed that all sporangia were approximately 3 mm long, but observed that some spores were conspicuously larger than others. He believed that there were three classes of spores, but Smith (1962) found only "two broad, indistinct size groups," a finding that, as he noted, supports Walton's (1957) suggestion that "we have here a preliminary stage in the evolution of [heterospory]." As Smith (1962) noted, however, aside from size there is no difference in the morphology of spores in the two groups and no way to determine whether some functioned as microspores, others as megaspores. The spores of *P. scotica* range in size from 75–355 μm with a mean of 125 μm. They consist of a central body enclosed in a perispore membrane (or pseudosaccus) (figure 1.21B), sometimes attached proximally. There is a prominent trilete scar, the rays of which are about one half the length of the radius of the central body. Central body and perispore are smooth or show a very fine, indistinct sculpturing.

Discussion   Early students of the Protopityales were impressed by its combination of pteridophytic and coniferous characteristics that made it difficult to assign to any known major taxa. Solms-Laubach (1893) grouped it with the pteridosperms; Seward (1917) compared it with conifers and cordaites, ferns, and lycopsids, but came to no conclusion as to its most likely affinities. Scott (1923) included it in the Pteridospermales because "it is in some ways Fern-like, but is not a Fern." Walton (1957) emphasized, however, that *P. scotica* has no obvious fern-like characteristics in its stem. He, nevertheless, suggested that, because of the variation in spore size, "it is probably a Pteridophyte and not a Pteridosperm. . . ." Beck (1960b) assigned Walton's order Protopityales to the Progymnospermopsida. One character considered fern-like by earlier investigators was the scalariform to oval bordered pitting of the secondary xylem. Such pitting characterizes the innermost secondary xylem of *P. scotica* and possibly all of that of *P. buchiana*, but Smith (1962) has demonstrated that, except in the inner secondary xylem of *P. scotica*, the pitting is characteristically circular bordered with crossed slit-like apertures, and often multiseriate. It is of interest to note that a similar transition from oval bordered pits in the inner wood to multiseriate circular bordered pits also characterizes the progymnosperm *Archaeopteris*. Smith (1962) emphasized that characteristics of the secondary wood of *P. scotica* strongly support Beck's assignment of Protopityales to the Progymnospermopsida. Walton (1969) noted the similarity of *Protopitys* sporangia to those of *Archaeopteris*. The sporangia are most similar, however, to those of the Aneurophytales. In fact, the entire fertile organ of

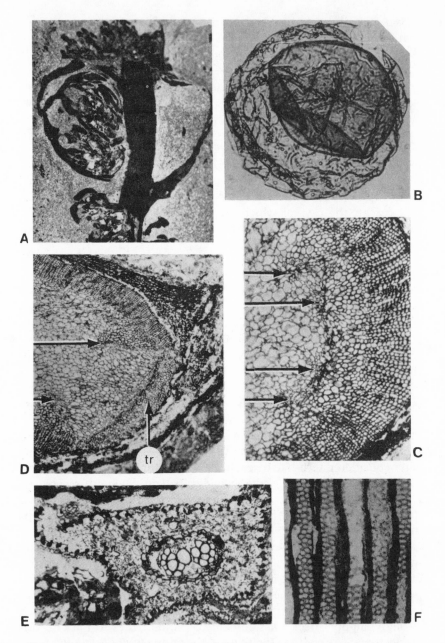

Figure 1.21. A Fertile branch of *Protopitys scotica.* × 1.63. (From Walton 1957, by permission.) **B** Spore of *Protopitys scotica.* × 370. (From Smith 1962, by permission.) **C** Transverse section of stem of

P. scotica is essentially identical in organization to that of *Tetraxylopteris*, and differs from that of *Rellimia* only in dichotomizing twice (Bonamo and Banks 1967; Leclercq and Bonamo 1971). Indeed, like all aneurophytes, the fertile organ is based on a branching plan of basal dichotomy that becomes pinnate distally. Even the spores (figure 1.21B), suggested by Smith (1962) to resemble *Remysporites*, are remarkably similar to *Rhabdosporites langii* (see Allen 1980), known to have been produced by both *Tetraxylopteris* and *Rellimia*. It should be noted, however, that *Rhabdosporites langii* has not been described above the Frasnian (McGregor 1979). Of the characteristics known at present, only the eustelic primary vascular system of *Protopitys* distinguishes it from the Aneurophytales. The distichous arrangement of fertile organs is another difference, but the distinction between distichy and tetrastichy (which characterizes *Tetraxylopteris*) is not very great. Unfortunately, we know nothing of the photosynthetic organs of *Protopitys*. Furthermore, we do not know with certainty the arrangement of vegetative appendages, since only fertile specimens with preserved lateral appendages have been discovered. One can speculate, therefore, that the vegetative appendages might have occurred in four ranks. Whether or not this was the case, *Protopitys* is very similar to members of the Aneurophytales, and might have evolved from some aneurophyte. The great significance of this group lies in the fact that it, like the aneurophytes and *Archaeopteris*, is an excellent candidate for ancestor of seed plants.

EVOLUTIONARY TRENDS WITHIN THE PROGYMNOSPERMOPSIDA

Since the recognition of the Progymnospermopsida (Beck 1960b), the group has been considered to be the ancestral plexus from which seed plants evolved. Because of this proposed relationship, morphological variation within progymnosperms has often been discussed in light of

---

Protopitys scotica showing pattern of primary xylem just proximal to trace divergence. Arrows indicate protoxylem strands. Compare with figure 1.19C. × 45. (From Walton 1957, by permission.) **D** Transverse section of stem of *Protopitys scotica* showing trace (tr) at level of divergence. Compare with figure 1.19D. × 23.6. (From Walton 1957, by permission.) **E** Root of *Protopitys scotica*. × 68. (From Walton 1969, by permission.) **F** Multiseriate circular bordered pits on radial walls of tracheids in *Protopitys scotica*. × 150. (From Smith 1962, by permission.)

the morphologies characteristic of seed plants. The result has been the proposal of evolutionary trends, within both Aneurophytales and Archaeopteridales, which often attempt to document a gradual shift in morphology toward that present in particular putative descendant groups. Generally, either the most complex morphology or the morphology most similar to that present in the presumed descendant has been interpreted as the most advanced state within the progymnosperm group under consideration. We discuss below the states and polarities within a number of characters that have been interpreted as evolutionary trends in light of evidence accumulated over the past decade.

Aneurophytales   Suggested evolutionary trends within the group include (1) increasing complexity of fertile appendages; (2) increasing complexity of primary xylem; (3) planation of lateral branching systems; and (4) planation of ultimate appendages.

Fertile Appendages.   Fertile appendages in aneurophytes have been interpreted to represent a series of increasing complexity beginning with the relatively simple *Aneurophyton*, progressing through *Rellimia*, and ending with the large, complex fertile organs of *Tetraxylopteris* (Bonamo and Banks 1967; Leclercq and Bonamo 1971; Scheckler and Banks 1971b; Bonamo 1975, 1977; Beck 1976). *Triloboxylon* (Scheckler 1975), with two basal dichotomies (like *Tetraxylopteris*) but relatively few sporangia (fewer than *Rellimia*), is difficult to fit within this linear series. An alternative interpretation is that *Triloboxylon* (Scheckler 1975) and *Tetraxylopteris* are part of a separate lineage within the Aneurophytales recognizable at this time on the basis of twice dichotomous fertile organs. This interpretation requires that similarities in spore morphology and overall size of fertile organs in *Tetraxylopteris* and *Rellimia*, characters that have been used to infer a close relationship between the two taxa (Bonamo and Banks 1967; Bonamo 1975; Beck 1976), represent retention of primitive characters or the results of parallelism.

Simple fertile organs (e.g., those of *Aneurophyton*) have been interpreted as the primitive state in this character (see references cited above). This interpretation agrees with the conventional wisdom that evolution proceeds from simple to complex. However, three of the four taxa mentioned above appear in the fossil record about mid-Eifelian, so stratigraphic evidence provides little support for this assessment of polarity. Comparison with members of the Trimerophytina (the most likely outgroup) complicates the situation further. Most species of *Psilophyton* have fertile lateral branches that dichotomize five to seven times and bear 32 to 128 sporangia (Andrews, Kasper, and Mencher 1968; Hueber 1968; Kasper, Andrews, and Forbes 1974; Banks, Leclercq, and Hueber 1975; Gensel 1979; Doran 1980). Although the condition in *Pertica* is somewhat less clear, fertile branches apparently bear similar numbers of sporangia and in some cases have comparable numbers of dichotomous divisions (Kasper and Andrews 1972; Granoff, Gensel, and Andrews 1976; Doran, Gensel, and Andrews 1978). In numbers of sporangia, trimerophyte morphology is closer to the relatively simple fertile organs of *Aneurophyton* and *Triloboxylon*. However, in terms of the number of branch orders that make up the fertile complex, the condition in trimerophytes is closer to the relatively complex fertile organ of *Tetraxylopteris*. Because of the equivocal nature of the evidence, the assessment of polarity in this character is extremely difficult. We tentatively suggest, however, that the complex condition present in *Tetraxylopteris* is primitive for the order. Our decision is based primarily on allometric considerations. It is generally assumed, and there is good indirect evidence to suggest, that aneurophytes were larger plants than trimerophytes. We would expect an evolutionary increase in size in aneurophytes to affect all parts of the organism, including fertile branching systems, and to result in plants with larger fertile organs bearing a greater number of sporangia than characterized the trimerophytes. Branching within the fertile complex would probably be unaffected by this increase in size, or at least not be reduced in complexity. This would result in a fertile complex similar in most respects to that of *Tetraxylopteris*. Interpretation of the relatively simple morphology present in *Aneurophyton* and *Triloboxylon* as primitive would require a relative decrease in the size, and an absolute decrease in the complexity, of branching within the fertile complex to be correlated with the general increase in plant size associated with the evolution of aneurophytes. We consider this alternative somewhat less likely.

Primary Xylem.   An increase in the complexity of stelar patterns (i.e., the form of primary xylem columns), correlated with changes in the symmetry of branching systems, has been suggested as an evolutionary trend in the order. This trend is reflected in the change from three-ribbed vascular systems and helical "phyllotaxy" (e.g., *Triloboxylon ashlandicum*) to four-ribbed systems and decussate "phyllotaxy" *(Tetraxylopteris)*, and ultimately to systems that change in symmetry from one axis order to the next *(Proteokalon)* (Scheckler and Banks 1971b; Bonamo 1975; Beck 1976). Interpretation of these changes as a simple linear trend requires first a change in primary xylem morphology throughout the plant (with the exception of ultimate appendages) followed by a reversal to the three-ribbed state in ultimate axes (to *Proteokalon*-type morphology). This reversal has been explained as evolutionary change toward planate lateral branching systems, and hypothesized as an intermediate stage in the evolution of pteridospermous fronds (Scheckler and Banks 1971b; see below). However, it is possible that the morphology in *Proteokalon* represents an intermediate condition between typical three-ribbed systems and the uniformly four-ribbed systems of *Tetraxylopteris*. In this case, primary xylem symmetry remained the same in ultimate axes but changed to four-ribbed in all others, resulting in *Proteokalon*-type morphology. *Tetraxylopteris*-type morphology evolved as a result of a subsequent change to four-ribbed symmetry in the ultimate axis order. This transition sequence still allows for the interpretation of *Proteokalon* as an intermediate in the evolution of fronds.

We agree that three-ribbed vascular systems are probably basic in Aneurophytales. This is supported by the occasional presence of stelar morphologies that approach the three-ribbed state in *Psilophyton dawsonii* (a member of the Trimerophytina, the likely outgroup for aneurophytes) (Banks, Leclercq, and Hueber 1975; Wight 1987). However, four-ribbed morphology *(Tetraxylopteris)* appears at about the same time as the three-ribbed state. Furthermore, at present, the Trimerophytina also includes members with four-ranked lateral appendages arranged either in "pseudowhorls" *(Pertica quadrifaria)* or in subopposite pairs and approaching a decussate pattern *(Pertica varia)*, an indication that these taxa may have four-ribbed steles (Wight 1987; see also Gensel 1984). This leaves open the possibility that the four-ribbed state is basic in the order. Information on stelar morphology in these complex trimerophytes may help resolve this problem.

In all aneurophytes, whether three- or four-ribbed, the system of protoxylem strands appears to be identical with that of *P. dawsonii*, consisting of a longitudinally continuous central strand from which the protoxylem strands of all traces to lateral appendages are ultimately derived (Beck 1957; Wight 1987). The architecture of the system in a newly discovered trimerophyte-like plant (Gensel 1984) may be fundamentally different, in that the protoxylem of traces is derived directly from protoxylem strands that persist at the ends of the three stelar ribs. If the species of *Pertica* described above are shown ultimately to have architecture similar to Gensel's plant, it would suggest that they are not related to aneurophytes and make the alternative that four-ribbed steles are primitive within Aneurophytales less likely.

If the three-ribbed state is primitive within Aneurophytales, we agree that *Proteokalon* represents the most highly derived state within the order and that *Tetraxylopteris*-type morphology represents an intermediate condition. We regard these transitions in stelar morphology as the result of changes in branching pattern (see Wight 1987) and consider it likely that a change in the developmental program resulting in decussate rather than helical branching (and thus a four-ribbed stele) would affect the entire plant rather than only a part of it. Further modification of the ultimate axis order (to *Proteokalon*-type morphology) can be envisioned as "uncoupled" from branching in the rest of the plant because of selection under a regime favoring the possible advantages of more plagiotropic photosynthetic units over three-dimensional ones. There is, however, no direct evidence to support such a scenario.

Lateral Branch Systems. The evolutionary modification of lateral branching systems of aneurophytes into the planate lateral branch systems of *Archaeopteris* and into the fronds of seed ferns (see above) are other transformations that have been considered (Beck 1957, 1976; Bonamo and Banks 1967; Scheckler and Banks 1971b; Bonamo 1975; see also section on trends in Archaeopteridales, below). No aneurophyte has truly planate lateral branch systems. However, the ultimate branch systems of *Proteokalon* are bilaterally symmetrical (Scheckler and Banks 1971b) and have been considered to represent both a derived state within Aneurophytales and an intermediate condition between primitive three-dimensional branching and the plagiotropic branches and fronds characteristic of descendant groups. *Proteokalon* does have a first occurrence later than any other aneurophyte, suggesting that its unique branch symmetry may be derived within the order. Outgroup comparison with trimerophytes (characterized by three-dimensional branching throughout) supports this interpretation as well. However, there is apparently nothing unique about the development of these branches to suggest that they are further evolved in the direction of leaf-like appendages than those of other aneurophytes (Scheckler 1976). All are characterized by indeterminate growth early in ontogeny and ultimately are determinate organs (Scheckler 1976) regardless of primary xylem morphology or symmetry. Although it remains a possibility that the lateral branches of *Proteokalon* do represent a morphological intermediate, we conclude that there is no strong evidence to support such an interpretation.

Ultimate Appendages.   Most aneurophytes have three-dimensional ultimate appendages (Scheckler and Banks 1971a; Matten 1973; Serlin and Banks 1978; Schweitzer and Matten 1982; Wight 1985). *Proteokalon,* however, has ultimate appendages that are planate and have flattened ultimate segments (Scheckler and Banks 1971b). This has been interpreted as a derived condition within Aneurophytales and an indication that the ultimate appendages of *Proteokalon* are more leaf-like than those of other aneurophytes. Stratigraphic evidence and outgroup comparison are consistent with the interpretation of this morphology as derived within the order (see above). However, as with the lateral branching systems discussed above, developmental evidence suggests that all aneurophyte ultimate appendages are determinate, appendicular organs (Scheckler 1976). Furthermore, although the ultimate appendages of *Proteokalon* may be closer morphologically to true leaves than those of other aneurophytes, there is no evidence to suggest that *Proteokalon* was in fact the direct ancestor of a plant with leaves. As a result, the conclusion that its ultimate appendages are more leaf-like than those of other aneurophytes seems empty of meaning and based primarily on predictions about the evolution of leaves made by Zimmerman (1952; see also Stewart 1983) in his telome theory.

Archaeopteridales   The recognition of legitimate evolutionary trends within the Archaeopteridales may be even more difficult than in the Aneurophytales. *Archaeopteris* is the only archaeopteridalean genus that is understood in any detail. *Svalbardia* is based solely on compressions providing information on external morphology alone that is not well understood. Although *Actinopodium* may represent the anatomy of *Svalbardia* (Carluccio et al. 1966; Gensel and Andrews 1984), this is merely a speculation based on uncertain sedimentary associations (see earlier sections on *Svalbardia* and *Actinopodium*). Furthermore, *Actinopodium,* like *Siderella,* is known only from a few transverse sections. *Actinoxylon* is also based largely on transverse sections. *Eddya,* interpreted as a possible juvenile plant of *Archaeopteris* (Beck 1967), was assigned a different generic name because of uncertainty of its identity with *Archaeopteris.*

Thus the attempt to recognize trends in a group still so poorly understood is probably futile. This problem is compounded by the fact that *Archaeopteris* largely overlaps the vertical ranges of all other taxa in the group, although *Actinoxylon, Actinopodium,* and *Svalbardia* may appear lower (Givetian) in the geological column than *Archaeopteris* (late Givetian to early Frasnian).

Several trends that have been proposed are (1) planation of lateral branch systems; (2) planation and lamination of leaves; (3) evolution of compound leaves from lateral branch systems; and (4) evolution of the eustele through medullation and dissection of the protostele. In the evolution of planate lateral branch systems, it has been suggested that the primitive condition is represented in an aneurophyte or in *Actinoxylon* in which ultimate branches are helically arranged (Matten 1968) and the derived conditions represented in *Archaeopteris* and *Siderella*, characterized by largely or solely two-ranked (distichous) ultimate branches. Leaves in the Givetian taxon *Actinoxylon* are non-laminate and dichotomously divided, and interpreted by Matten (1968) to dichotomize "in more than one plane." In geologically younger species of *Archaeopteris*, leaves are either nonlaminate but planate, or laminate. It has also been proposed (Carluccio et al. 1966; Matten 1968; Beck 1971) that in *Siderella* from the Tournaisian, leaves may have been produced in only two ranks on the ultimate branches, unlike the decussate arrangement of leaves on ultimate branches of several Frasnian species of *Archaeopteris*. On the basis of this relatively meager evidence, these workers have hypothesized the possible evolution of compound leaves from the lateral branch systems of *Archaeopteris* (see the section on *Siderella*, and caveats in the section on the primary vascular system of *Callixylon*).

The evolution of the eustele from a protostele is an attractive idea that has been hypothesized as having occurred in several different major groups, and the Archaeopteridales must be included among these (Beck 1979). Whether *Archaeopteris* evolved from a trimerophyte, an aneurophyte, or a primitive archaeopteridalean such as *Actinoxylon*, there are protostelic forms in taxa that preceded *Archaeopteris* in the geological column that could serve as the primitive type. Interestingly, the stele of large axes of *Archaeopteris (Callixylon)* produced traces along radial planes, and thus seems to represent an intermediate morphology between the protostelic condition and the eustelic morphology of most seed plants characterized by an initial tangential divergence of the leaf trace (Beck 1979).

The traditional explanation of the evolution of the eustele by medullation and dissection of the protostele (Bower 1908) is based on the concept of the adaptive value of increase in surface area to volume with overall increase in size of the plant axis. Wight (1987), however, has hypothesized that early changes in the configuration of the protostele—from haplostele to actinostele—were probably passive, nonadaptive, and related to a change in branching patterns—from dichotomy to pseudomonopodial to monopodial branching.

PHYLOGENETIC SIGNIFICANCE OF THE PROGYMNOSPERMS

Progymnosperms have been widely accepted as the ancestral group from which gymnosperms evolved. Perhaps the most significant single character indicative of relationship is the bifacial vascular cambium that produced gymnospermous secondary tissues (Scheckler and Banks 1971a; Wight and Beck 1984; Stein and Beck, in press). Many other features that support this proposed relationship have been described in detail elsewhere (e.g., Beck 1957, 1970, 1971, 1976, 1981; Meeuse 1963; Bonamo and Banks 1967; Scheckler and Banks 1971a, 1971b; Bonamo 1975).

Despite the acceptance of a general ancestor-descendent relationship between progymnosperms and seed plants, determination of the specific ancestral taxon or taxa from which the major groups of gymnosperms evolved has remained the subject of some controversy. In a series of papers, Beck (1957, 1966, 1970, 1971, 1981) has developed the idea that pteridosperms and their descendants evolved from some aneurophyte-like progymnosperm, whereas the cordaites, conifers, and related taxa had their origin in the Archaeopteridales. This hypothesis suggests that seed plants are a polyphyletic group—i.e., that the seed habit evolved twice. In contrast, Rothwell (1981, 1982) has suggested that seed plants are monophyletic. He believes that aneurophytes represent an ancestral group from which pteridosperms evolved (a similarity with Beck's hypothesis); that "coniferopsids" are derived from within the seed fern complex, presumably from a taxon similar to *Callistophyton;* and that *Archaeopteris,* therefore, is unrelated to any group of seed plants. Meyen (1984, 1986) has presented a third alternative, in which seed plants are polyphyletic, as in the Beck hypothesis, but he considers both initial groups of seed plants to be derived from archaeopteridaleans.

Aneurophytales are attractive candidates for ancestor of the earliest seed plants. Many of the early pteridosperms—e.g., *Stenomyelon, Tristichia, Tetrastichia, Microspermopteris, Heterangium, Galtiera,* etc.— are characterized by protosteles and secondary xylem similar to those of aneurophytaleans. In addition, similarities between aneurophytalean fertile organs and the pollen organs (Stidd and Hall 1970; Rothwell 1981) and cupulate systems of early lyginopterids (Doyle and Donoghue 1986a) have been cited as evidence of relationship. Although Meyen (1984) argues for an archaeopteridalean ancestry of seed ferns, we regard this as a somewhat less likely alternative and tentatively interpret similarities between these two groups (e.g., relatively simply secondary phloem in both and similarities between the leaf-like lateral branch systems of *Archaeopteris* and fronds of pteridosperms) to have resulted from parallel evolution.

Understanding the relationships of coniferopsids, however, has proven to be a more difficult problem. As evidence in support of an archaeopteridalean ancestry for coniferopsids, Beck (1966, 1970, 1971, 1981) has emphasized similarities in habit, lateral branch systems, and secondary xylem between *Archaeopteris* and early conifers. In contrast, Rothwell (1981, 1982) has primarily emphasized similarities in reproductive biology between coniferopsids and the pteridosperm *Callistophyton* in support of his hypothesis. With respect to the Beck hypothesis, Rothwell notes that similarities between *Archaeopteris* and coniferopsids are solely vegetative, and that known features of reproductive biology are "extremely different." He also cites the stratigraphic hiatus between the most recent occurrences of *Archaeopteris* (Tournaisian) and the earliest occurrence of a cordaite (Namurian) or a conifer (Westphalian), and the difference in divergence of traces from cauline stelar bundles between *Archaeopteris* (radial) and early seed plants (tangential) as evidence against their relationship (Rothwell 1976, 1982). Meyen (1984) agrees with Beck in suggesting *Archaeopteris* as the ancestor of coniferopsids and the polyphyletic status of gymnosperms.

Several recent workers attempting to resolve the question of seed plant mono- or polyphyly and the position of aneurophytes and/or archaeopterids as ancestors have assembled data within the context of an analysis based on cladistic principles (e.g., Crane 1985a, 1985b; Doyle and Donoghue 1986a, 1986b, 1987; Stein 1986a, 1986b, 1987). These analyses are noteworthy in several respects, but primarily for the explicit framework within which characters are analyzed and taxa assigned to states. We shall consider the three hypotheses presented above (Beck, Rothwell, and Meyen) primarily within the context of the analysis of Doyle and Donoghue (1986a) because it is the most recent and, in many respects, the most explicit and complete, and because it provides a useful framework for discussion.

Doyle and Donoghue performed a strict parsimony analysis that produced a total of thirty-six equally parsimonious cladograms, or trees (their term). Most of the variation among the thirty-six trees was the result of what they considered relatively minor rearrangements of a few taxa (Doyle and Donoghue 1986a: 352–360) and pointed to the need for more morphological data on Paleozoic seed ferns. Of the thirty-six, Doyle and Donoghue chose one tree (1986a: figure 4) as their preferred phylogenetic hypothesis because, for various reasons, they consider it more likely than the others. One interesting feature of their results is that their most parsimonious cladograms support the status of gymnosperms as a monophyletic group and show coniferopsids nested somewhere within the seed fern complex, thus providing support for Rothwell's hypothesis. According to their cladograms and the character transitions shown, seed plants are descendants of a heterosporous plant that was otherwise no different from an aneurophytalean progymnosperm.

Although we believe that these results are significant, we feel, as do Doyle and Donoghue (1986a), that an archaeopteridalean origin for coniferopsids remains a viable alternative. As part of their analysis, they investigated the effects of forcing particular groups together in order to evaluate competing hypotheses of relationship among progymnosperms and seed plants. Although tree topologies consistent with Meyen's hypotheses for the origin of seed plants were considerably less parsimonious (150 steps), a cladogram consistent with Beck's hypothesis was only very slightly less parsimonious (124 steps) than their most parsimonious trees (123 steps). Given the amount of homoplasy apparently present in the data (124 character state transitions in a cladogram based on a data matrix containing 62 characters implies that, on average, each character undergoes two transitions: either two parallel evolutionary events or a single evolutionary event followed by a reversal), we consider the difference of one character state transition between the two hypotheses insignificant and an insufficient basis upon which to choose the more viable alternative. This view is reinforced by several other considerations discussed below. Some of these were also discussed by Doyle and Donoghue (1986a).

Lack of Information on Critical Taxa   Doyle and Donoghue (1986a) and Beck (1981) noted that information on diversity among Lower Carboniferous taxa is critical to the resolution of gymnosperm origins. The current lack of information has two manifestations, which are discussed below.

Missing Information on Taxa Included in the Analysis.   The data matrix used by Doyle and Donoghue (1986a, Table II) includes several characters (numbers 13, 24, 55, 57–62) that are noteworthy for the lack of available information on fossil taxa. They remove these characters (along with several others that they feel are rarely preserved in Mesozoic fossils) and obtain a tree that is essentially identical to one of the most parsimonious trees. On that basis they suggest that the inferred relationship between angiosperms and Gnetales is not an artifact of incomplete information on fossils. This particular inference may be true, but removal of characters does not necessarily duplicate results that will be obtained when information on those characters is discovered. In addition, removal of the characters cited above apparently produced a single tree (not 36) that differed from their preferred tree in the arrangement of several taxa near the base. This indicates that the basal nodes may be more sensitive to the deleted characters than distal portions. Information on these characters, therefore, may be critical to resolving the question of the origin of seed plants.

Taxa Excluded from the Analysis. Doyle and Donoghue (1986a) excluded a number of taxa from their analysis because of a lack of available information and/or their lack of potential relevance to angiosperm relationships. This is an operational constraint present in any phylogenetic analysis. Not included in their analysis, however, were members of the Noeggerathiopsida, a group of Carboniferous (and perhaps Permian) heterosporous pteridophytes that Beck (1981) considers a possible continuation of the archaeopteridalean line. If this is true, there would be no stratigraphic hiatus between the youngest occurrence of archaeopteridaleans and the oldest occurrence of coniferopsids. Similarly, poorly understood taxa of the lower Carboniferous, such as the *Archaeopteris*-like vegetative remains *Rhacopteris* and *Anisopteris* and isolated, platyspermic seeds such as *Lyrasperma*, were omitted from the analysis. If these or other similar taxa were found to have a combination of primitive progymnosperm-like features and coniferopsid advances, they would provide support for the Beck hypothesis (Doyle and Donoghue, 1986a).

To emphasize the effect that missing data and/or taxa can have on phylogenetic reconstruction, in terms of relationships between groups as well as character transitions on a tree, we cite recent studies by Stein (1987), Doyle and Donoghue (1987), and Galtier, Kluge, and Roe (personal communication, cited in Doyle and Donoghue 1987).

X-Coding of Characters In an attempt to avoid an a priori bias in favor of any one hypothesis of evolutionary relationship, Doyle and Donoghue (1986a) devised a system of coding for multistate characters (X-coding) that allows for alternative interpretations of relationships of the several derived states to the ancestral state and to each other. Their coding scheme takes advantage of the ability of software for phylogenetic inference to deal with missing data, and scores the presence or absence of particular intermediate states as unknown in some taxa if there are major alternative morphological theories on homologies that result from competing hypotheses. This is a note-worthy attempt to test the relative merits of alternative hypotheses such as those of Beck and Rothwell. However, this coding scheme has the undesirable consequence of effectively eliminating the influence of heavily X-coded characters on tree topology. Tree topology is deter-mined primarily by non-X-coded characters, and transitions in the X-coded ones are placed on the tree in positions that give the most parsimonious result for a given position of a taxon. Consequently, the analysis might be viewed as a test of alternative hypotheses based on the non-X-coded characters. We do not believe, therefore, that it is accurate to say that the test is based on the entire set of characters. Furthermore, we disagree with the idea that overall parsimony is the best or ultimate arbiter in cases of alternative interpretations of ho-mology (see Stein 1987). Rather, we feel that conflict such as this is an invitation for more detailed investigation of the character in ques-tion to determine whether or not evidence can be brought to bear on which character hypothesis is more likely correct. In the end, it is the characters themselves that are the critical elements and the basis of any phylogenetic analysis. X-coding is a reasonable approach when no other alternative is available, but results generated in this manner should be viewed with caution.

As an indication of the importance of X-coded characters in gener-ating the results obtained by Doyle and Donoghue (1986a), we note that of the 62 characters used in the analysis, 19 either have X-codings for at least 5 taxa or are part of multistate characters that do. Another 15 characters have equal amounts of missing data for the taxa in the analysis. Furthermore, if one looks at the basal portions of the tree where the origin of gymnosperms is resolved, one can observe that, of the 18 characters used to define the basal six nodes of the tree, 9 characters are heavily X-coded or are part of multistate characters that are. As we stated above with respect to missing data, new infor-mation on transformations in these characters might have a profound effect on branching at the base of the tree.

Effect of New Information on Character Coding   Recently, several workers have presented data that may alter the character coding of some taxa. Rothwell (1986) has argued on several bases that seed symmetry (i.e., radial versus platyspermic—Doyle and Donoghue characters 41 and 42), one of the primary characters used to discrim-inate between cycadophytic and coniferophytic gymnosperms, is poorly defined in most analyses and therefore contains little useful phylogenetic information. More critical analyses of seed morphology may produce interpretations that more accurately reflect homologies.

Similarly, recent work on the primary vascular systems of early seed plants and progymnosperms (Beck and Stein 1985, 1987; Wight 1985, 1986, 1987; Stein and Beck in press) has shown that protoxylem architecture can be a useful morphological feature to consider in interpreting homologies in stelar morphology. This work, along with recognition of the diversity of morphologies present among early Carboniferous "protostelic" seed plants, demonstrates that the simple transition from protostele to eustele, proposed in most phylogenetic analyses treating the origin of seed plants, may be inaccurate.

In their preliminary analysis of relationships among vascular plants, Doyle and Donoghue (1986) scored Cladoxylales, primitive ferns (a questionable assemblage including *Stauropteris*, zygopterids, and iridopterids [Stein 1981, 1982a]), sphenopsids (potentially a polyphyletic group [Stein, Wight, and Beck 1984]), a Lower Devonian plant recently described by Gensel (1984), and protostelic seed plants as having homologous stelar morphologies based on the presence of actinosteles with several mesarch protoxylem points in all. This was viewed as an advance over the condition present in trimerophytes. Stein (1981, 1982a) and Stein, Wight, and Beck (1983) have shown that stelar morphology in iridopterids is considerably different from that of early zygopterids and have suggested that the use of "peripheral loops" as a homology uniting these groups is highly questionable. Indeed, some workers consider iridopterids to be early sphenopsids (e.g., Stewart 1983), although Stein (1981, 1982a) considered their relationship problematic. Aneurophytalean progymnosperms have been shown to have stelar morphology that is fundamentally similar to trimerophytes (Wight 1985, 1986, 1987). Comparison of this morphology with that present in iridopterids and zygopterids, as well as Gensel's (1984) new plant, reveals striking differences. Gensel's plant seems to have a unique protoxylem architecture, while cladoxylopsids have some similarities to iridopterids. Recoding stelar morphologies to account for this diversity would produce a much different pattern of outgroup relationships. In fact, if raw similarity of stelar morphologies is used, iridopterids might be classified more reasonably with the actinostelic lycopsids than as trimerophyte descendants.

An additional character of the primary vascular system cited by Rothwell (1976; but not used by Doyle and Donoghue; see their Appendix III) as suggesting that *Archaeopteris* was unrelated to seed plants is its radial pattern of trace divergence that contrasts with the tangential trace divergence of eustelic seed plants. The condition in

Archaeopteris can be interpreted with equal ease as an intermediate condition between a protostelic state and the typical seed plant eustele (see Beck 1979).

Recent morphological studies on the microsporangiate cones of some early conifers—e.g., *Darneya* and *Sertostrobus* (Grauvogel-Stamm 1978), *Sashinia* (Meyen 1981, 1984), and certain walchian specimens (Mapes 1983)—have shown that the microsporophylls bear adaxial sporangia and are remarkably similar to sporophylls in strobili of *Archaeopteris* (Beck 1981). Thus the abaxial position of microsporangia characteristic of extant conifers was not a feature of all Paleozoic species, and the adaxial position may represent the basic condition within conifers (Mapes, personal communication), a difference from some pteridosperms.

Further Consideration of the Beck and Rothwell Hypotheses
Rothwell (1982) has criticized the Beck hypothesis for its reliance on vegetative characters alone. In analyses using subsets of their data, Doyle and Donoghue (1986a) showed that use of vegetative characters alone supports coniferopsids nested within a group including *Archaeopteris*. However, use of "macro-reproductive" characters alone showed a division of seed plants into two major groups that parallel the classical separation of coniferophytic from cycadophytic gymnosperms. Use of "micro-reproductive" characters alone produced a tree closest to the Rothwell hypothesis, indicating that it too is primarily dependent upon a small subset of available characters. Rothwell (1982, 1986) has argued that these characters are the most accurate reflection of "true" phylogenetic relationships. Doyle and Donoghue (1986a) also argue in favor of this viewpoint (at least within their study group) on the basis of higher consistency indices for trees based upon reproductive characters. Higher consistency indices are also found for the subsets of reproductive characters on the preferred, 123-step tree of Doyle and Donoghue. However, they go on to state that "all sets of characters show some congruent patterns and thus provide significant information on relationships, and a priori elimination of characters could not be defended" (Doyle and Donoghue 1986a: 379).

As a further caution against a priori weighting of particular subsets of characters, we make the following observations. Consistency indices are based upon the amount of homoplasy present in characters given a particular tree topology. Taxa with missing data, whether actually unknown or X-coded, are assigned to states of these characters to maximize the parsimony (i.e., minimize homoplasy) of the tree. Therefore, characters that are X-coded or unknown for a number of taxa, because they have this "built-in" flexibility, may contribute to a misleading impression of character agreement within the data set. Examination of the data matrix used by Doyle and Donoghue (1986a: Table II) shows that the distribution of characters that are X-coded for at least five taxa or are part of multistate characters that are, or have an equal amount of missing data, is not uniform in the three subsets of characters. Within the vegetative subset, approximately 46 percent of the characters fall into this category; within macro-reproductive, 72 percent; and within micro-reproductive, 55 percent. On this basis, we suggest that the higher consistency indices for micro- and macro-reproductive characters are, at least in part, an artifact of the character-coding scheme and missing information. Consequently, we do not regard it as an adequate basis upon which to evaluate the relative merits of particular subsets of characters.

Rothwell (1982) cited the stratigraphic hiatus between *Archaeopteris* and coniferopsids as a problem with the Beck hypothesis (see above). We agree that this is a complicating factor. On the other hand, Doyle and Donoghue (1986a) note that many of their most parsimonious trees also imply stratigraphic ranges for taxa that are inconsistent with known distributions.

Rothwell's (1982) hypothesis requires the evolution of the simple leaves and strobiloid fructifications characteristic of cordaites and conifers from within a group characterized by large compound fronds and lacking any similar reproductive structures. He has suggested heterochrony as a possibly mechanism for at least the former transformation. Although this certainly is a feasible hypothesis, at present there is no direct evidence to support it.

The discussions above, although in most cases presented within the context of the analysis of Doyle and Donoghue (1986a), are not meant to be taken as criticism of their study. Rather, as stated earlier, their study provided a useful framework within which to evaluate competing hypotheses of seed plant origins. These discussions do, however, point to the need for more basic morphological data in order to resolve the question of gymnosperm origins and argue strongly for the continued consideration of alternative hypotheses of relationship.

ACKNOWLEDGMENTS

We are especially grateful to Dr. William Stein of the University of Michigan for stimulating discussion about many aspects of this paper, and for providing several photographs. We acknowledge with thanks the loan of slides by Dr. Karl Niklas of Cornell University, and express our gratitude to Dr. Stephen Scheckler of Virginia Polytechnic Institute and State University for permission to use several of his unpublished photographs. We are also grateful to Dr. Patricia Bonamo of the State University of New York, Binghamton, who provided several photographs. Shayne Davidson of the University of Michigan drew figures 1.17 and 1.20. This paper was supported in part by National Science Foundation grants BSR 81 13542 to C. B. Beck, BSR 83 06893 to W. E. Stein, Jr., and C. B. Beck, and BSR 8600660 to G. W. Rothwell and D. C. Wight.

LITERATURE CITED

Allen, K. C. 1980. A review of *in situ* late Silurian and Devonian spores. *Rev. Palaeobot. Palynol.* 29:253–270.

Andrews, H. N. 1940. On the stelar anatomy of pteridosperms, with particular reference to the secondary wood. *Ann. Missouri Bot. Gard.* 27:51–119.

Andrews, H. N., A. E. Kasper, and E. Mencher. 1968. *Psilophyton forbesii,* a new Devonian plant from northern Maine. *Bull. Torrey Bot. Club* 95:1–11.

Arber, E. A. 1921. *Devonian Floras: A Study of the Origin of the Cormophyta.* Cambridge: Cambridge University Press.

Arnold, C. A. 1929. On the radial pitting in *Callixylon. Am. Jour. Bot.* 16:391–393.

—— 1930. The genus *Callixylon* from the Upper Devonian of central and western New York. *Pap. Mich. Acad. Sci.* 11:1–50.

—— 1931. On *Callixylon newberryi* (Dawson) Elkins et Wieland. *Contr. Mus. Paleont. Univ. Michigan* 3:207–232.

—— 1935. Some new forms and new occurrences of fossil plants from the Middle and Upper Devonian of New York State. *Bull. Buffalo Soc. Nat. Sci.* 17:1–12.

—— 1936. Observations on fossil plants from the Devonian of eastern North America. 1. Plant remains from Scaumenac Bay, Quebec. *Contr. Mus. Paleont. Univ. Michigan* 5:37–48.

—— 1939. Observations on fossil plants from the Devonian of eastern North American. 4. Plant remains from the Catskill Delta deposits of northern Pennsylvania and southern New York. *Contr. Mus. Paleont. Univ. Michigan* 5:271–313.

—— 1940. Structure and relationships of some Middle Devonian plants from western New York. *Am. Jour. Bot.* 27:57–63.

—— 1947. *An Introduction to Paleobotany.* New York: McGraw-Hill.

Banks, H. P. 1966. Devonian flora of New York State. *Empire State Geogram* 4:10–24.

Banks, H. P., S. Leclercq, and F. M. Hueber. 1975. Anatomy and morphology of *Psilophyton dawsonii*, sp. n. from the late Lower Devonian of Quebec (Gaspé) and Ontario, Canada. *Paleontogr. Am.* 8:77–127.

Barnard, P. D. W. and A. G. Long. 1975. *Triradioxylon*—a new genus of lower Carboniferous petrified stems and petioles together with a review of the classification of early Pterophytina. *Trans. Roy. Soc. Edinburgh.* 69:231–249.

Barnett, L. 1953. The world we live in. Part 5. The pageant of life. *Life* magazine 35 (September 7):54–71.

Beck, C. B. 1957. *Tetraxylopteris schmidtii* gen. et sp. nov., a probable pteridosperm precursor from the Devonian of New York. *Am. Jour. Bot.* 44:350–367.

—— 1960a. Connection between *Archaeopteris* and *Callixylon*. *Science* 131:1524–1525.

—— 1960b. The identity of *Archaeopteris* and *Callixylon*. *Brittonia* 12:351–368.

—— 1960c. Studies of New Albany Shale plants. 1. *Stenokoleos simplex* comb. nov. *Am. Jour. Bot.* 47:115–124.

—— 1962a. Reconstructions of *Archaeopteris*, and further consideration of its phylogenetic position. *Am. Jour. Bot.* 49:373–382.

—— 1962b. Plants of the New Albany Shale. 2. *Callixylon arnoldii* sp. nov. *Brittonia* 14:322–327.

—— 1964. The woody, fern-like trees of the Devonian. *Mem. Torrey Bot. Club* 21:26–37.

—— 1966. On the origin of gymnosperms. *Taxon* 15:337–339.

—— 1967. *Eddya sullivanensis*, gen. et sp. nov., a plant of gymnospermic morphology from the Upper Devonian of New York. *Palaeontographica* 121B:1–22.

—— 1970. The appearance of gymnospermous structure. *Biol. Rev.* 45:379–400.

—— 1971. On the anatomy and morphology of lateral branch systems of *Archaeopteris*. *Am. Jour. Bot.* 58:758–784.

—— 1976. Current status of the Progymnospermopsida. *Rev. Palaeobot. Palynol.* 21:5–23.

—— 1979. The primary vascular system of *Callixylon*. *Rev. Palaeobot. Palynol.* 28:103–115.

—— 1981. *Archaeopteris* and its role in vascular plant evolution. In K. J. Niklas, ed., *Paleobotany, Paleoecology, and Evolution*, vol. 1. New York: Praeger.

Beck, C. B., K. Coy, and R. Schmid. 1982. Observations on the fine structure of *Callixylon* wood. *Am. Jour. Bot.* 69:54–76.

Beck, C. B. and W. E. Stein. 1985. Eustelic protosteles in early seed plants? *Am. Jour. Bot.* 72:889–890.

—— 1987. *Galtiera bostonensis*, gen. et sp. nov., a protostelic calamopityacean from the New Albany Shale of Kentucky. *Can. Jour. Bot.* 65:348–361.

Bonamo, P. M. 1975. The Progymnospermopsida: Building a concept. *Taxon* 24:569–579.

—— 1977. *Rellimia thomsonii* (Progymnospermopsida) from the Middle Devonian of New York State. *Am. Jour. Bot.* 64:1272–1285.

—— 1983. *Rellimia thomsonii* (Dawson) Leclercq and Bonamo (1973): The only correct name for the aneurophytalean progymnosperm. *Taxon* 32:449–454.

Bonamo, P. M. and H. P. Banks. 1967. *Tetraxylopteris schmidtii:* Its fertile parts and its relationships within the Aneurophytales. *Am. Jour. Bot.* 54:755–768.

Cai, Chong-yang. 1987. On the occurrence of *Callixylon* Zalessky and petrified axis of *Leptophloeum rhombicum* Dawson in Xinjiang, China. *Kexue Tongbao* (Science Bulletin) 32:718–719.

Cai, Chong-yang, Wen Yao-guang and Chen Pei-quan. 1987. *Archaeopteris* florula from Upper Devonian of Xinhui County, central Guangdong and its stratigraphical significance. (In Chinese, with an English summary.) *Acta Palaeont. Sinica* 26:55–64.

Carluccio, L. C., F. M. Hueber, and H. P. Banks. 1966. *Archaeopteris macilenta,* anatomy and morphology of its frond. *Am. Jour. Bot.* 53:719–730.

Carruthers, W. 1873. On some lycopodiaceous plants from the Old Red Sandstone of the north of Scotland. *Jour. Bot.,* n. s. 2:321–327.

Chirkova-Zalesskaja, E. F. 1957. *Delenie Terrigennogo Devona Uralo-Povolz ja na Asnovanii Iskopaemych Raslenij.* Moscow: Akad. Nauk. UdSSr., Institut nefti.

Crane, P. R. 1985a. Phylogenetic analysis of seed plants and the origin of angiosperms. *Ann. Missouri Bot. Gard.* 72:716–793.

—— 1985b. Phylogenetic relationships in seed plants. *Cladistics* 1:329–348.

Dannenhoffer, J. M. and P. M. Bonamo. 1984. Secondary anatomy of *Rellimia thomsonii,* an aneurophytalean progymnosperm from the Middle Devonian of New York. *Am. Jour. Bot.* 71(5), part 2:114.

Dawson, J. W. 1871a. On new tree ferns and other fossils from the Devonian. *Quart. Jour. Geol. Soc. London* 27:269–275.

—— 1871b. The fossil plants of the Devonian and Upper Silurian formations of Canada. *Geol. Surv. Can.,* pp. 1–92, Montreal.

—— 1878. Notes on some Scottish Devonian plants. *Can. Naturalist and Quart. Jour. Sci.* 8:379–389.

—— 1882. The fossil plants of the Erian (Devonian) and Upper Silurian Formation of Canada, part 2, *Geol. Surv. Can.,* pp. 95–142, Montreal.

—— 1888. *The Geological History of Plants.* London: Kegan Paul, Trench.

Doran, J. B. 1980. A new species of *Psilophyton* from the Lower Devonian of northern New Brunswick, Canada. *Can. Jour. Bot.* 58:2241–2262.

Doran, J. B., P. G. Gensel, and H. N. Andrews. 1978. New occurrences of trimerophytes from the Devonian of eastern Canada. *Can. Jour. Bot.* 56:3052–3068.

Doyle, J. A. and M. J. Donoghue. 1986a. Seed plant phylogeny and the origin of angiosperms: An experimental cladistic approach. *Bot. Rev.* 52:321–431.

—— 1986b. Relationships of angiosperms and Gnetales: A numerical cladistic analysis. In R. A. Spicer and B. A. Thomas, eds., *Systematic and Taxonomic Approaches in Palaeobotany.* Syst. Assoc. Special Vol. 31:177–198. Oxford: Oxford University Press.

—— 1987. The importance of fossils in elucidating seed plant phylogeny and macroevolution. *Rev. Palaeobot. Palynol.* 50:63–95.

Gensel, P. G. 1979. Two *Psilophyton* species from the Lower Devonian of eastern Canada with a discussion of morphological variation within the genus. *Palaeontographica* 168B:81–99.

—— 1984. A new Lower Devonian plant and the early evolution of leaves. *Nature* 309:785–787.

Gensel, P. G. and H. N. Andrews. 1984. *Plant Life in the Devonian.* New York: Praeger.

Gillespie, W. H., G. W. Rothwell, and S. E. Scheckler. 1981. The earliest seeds. *Nature* 293:462–464.

Goldring, W. 1924. The Upper Devonian forest of seed ferns in eastern New York. *New York State Mus. Bull.* 251:49–72.

Göppert, H. R. 1850. *Monographie der fossilen Coniferen.* Naturwerkundige Verhand. Holland. maatschap. Leiden: Wettenschappen Haarlem.

Granoff, J. A., P. G. Gensel, and H. N. Andrews. 1976. A new species of *Pertica* from the Devonian of eastern Canada. *Palaeontographica* 155B:119–128.

Grauvogel-Stamm, L. 1978. *La flore du Grès à Voltzia (Bundsandstein supérieur) des Vosges du Nord (France): Morphologie, anatomie, interprétations phylogénetique et paléogéographique.* Mémoire 50. Strasbourg: Institut de Géologie. Université Louis Pasteur.

Halle, T. G. 1936. On *Drepanophycus, Protolepidodendron* and *Protopteridium*, with notes on the Paleozoic flora of Yunnan. *Palaeont. Sinica* 1:1–38.

Høeg, O. A. 1942. The Downtonian and Devonian flora of Spitzbergen. *Norges Svalbard-Og Ishavs-undersøkelser.* 83:1–228.

Hoskins, J. R. and A. T. Cross. 1951. The structure and classification of four plants from the New Albany Shale. *Am. Midl. Nat.* 46:684–716.

Hueber, F. M. 1968. *Psilophyton:* The genus and the concept. In D. H. Oswald, ed., *International Symposium on the Devonian System,* vol. 1. Calgary: Alberta Society of Petroleum Geologists.

Iurina, A. L. 1969. *Devonian Flora of Central Kazakhstan.* A. A. Bogdanova, ed. Moscow State University.

Johnson, T. 1911. Is *Archaeopteris* a pteridosperm? *Sci. Proc. Roy. Dublin Soc.,* n. s. 13:114–136.

Kasper, A. E. and H. N. Andrews. 1972. *Pertica,* a new genus of Devonian plants from northern Maine. *Am. Jour. Bot.* 59:897–911.

Kasper, A. E., H. N. Andrews, and W. H. Forbes. 1974. New fertile species of *Psilophyton* from the Devonian of Maine. *Am. Jour. Bot.* 61:339–359.

Kidston, R. 1903. The fossil plants of the Carboniferous rocks of Canonbie, Dumfriesshire and of parts of Cumberland and Northumberland. *Trans. Roy. Soc. Edinburgh.* 40:741–833.

Kräusel, R. and H. Weyland. 1923. Beiträge zur Kenntnis der Devonflora 1. *Senckenbergiana* 5:154–184.

—— 1926. Beiträge zur Kenntnis der Devonflora 2. *Abh. Senckenb. Naturforsch. Ges.* 40:114–155.

——— 1929. Beiträge zur Kenntnis der Devonflora 3. *Abh. Senckenb. Naturforsch. Ges.* 41:315–360.

——— 1932. Pflanzenreste aus dem Devon 2. *Senckenbergiana* 14:185–190.

——— 1933. Die Flora des Böhmischen Mitteldevons. *Palaeontographica* 78B: 1–46.

——— 1935. Pflanzenreste aus dem Devon. 9. Ein Stamm von *Eospermatopteris*-Bau aus dem Mitteldevon des Kerberges, Elberfeld. *Senckenbergiana* 17: 9–20.

——— 1938. Neue Pflanzenfunde im Mittledevon von Elberfeld. *Palaeontographica* 83B:172–195.

——— 1941. Pflanzenreste aus dem Devon von Nord-Amerika. 2. Die Oberdevonischen Floren von Elkins, West Virginia, und Perry, Maine, mit Berücksichtigung einiger Stück von der Chaleur-Bai, Canada. *Palaeontographica* 86B:3–78.

Krejči, J. 1880. Notiz über die Reste von Landpflanzen in der böhmishcen Silurformation. *Sitz. Ber. K. Böhm. Ges. Wiss. Prague*, pp. 201–204.

——— 1881. Über ein neues Vorkommen von Landpflanzen und Fucoiden in der böhmishcen Silurformation. *Sitz. Ber. K. Böhm. Ges. Wiss. Prague*, pp. 68–69.

Lang, W. H. 1925. Contributions to the study of the Old Red Sandstone flora of Scotland. 1. On plant remains from the fish-beds of Cromarty. *Trans. Roy. Soc. Edinburgh* 54:253–272.

——— 1926. Contributions to the study of the Old Red Sandstone flora of Scotland. 3. On *Hostimella (Ptilophyton) thomsonii*, and its inclusion in a new genus *Milleria*. *Trans. Roy. Soc. Edinburgh* 54:785–790.

Leclercq, S. 1940. Contribution a l'étude de la flore du Dévonien de la Belgique. *Mém. Acad. Roy. Belgique Cl. Sci.* 12:1–65.

——— 1970. Classe des Cladoxylopsida Pichi-Sermolli, 1959. In E. Boureau, ed., *Traité de Paléobotanique*, Vol. 4, fasc. I, pp. 119–177, *Filicophyta*. Paris: Masson.

Leclercq, S. and P. M. Bonamo. 1971. A study of the fructification of *Milleria (Protopteridium) thomsonii* Lang from the Middle Devonian of Belgium. *Palaeontographica* 136B:83–114.

——— 1973. *Rellimia thomsonii*, a new name for *Milleria (Protopteridium) thomsonii* Lang 1926 Emend. Leclercq and Bonamo 1971. *Taxon* 22:435–437.

Lemoigne, Y., A. Iurina, and N. Snigerevskaya. 1983. Révision du genre *Callixylon* Zalessky 1911 *(Archaeopteris)* du Dévonien. *Palaeontographica* 186B:81–120.

Long, A. G. 1963. Some specimens of *Lyginorachis papilio* Kidston associated with stems of *Pitys. Trans. Roy. Soc. Edinburgh* 65:211–224.

McGregor, D. C. 1979. Spores in Devonian stratigraphical correlation. Palaeontological Association, London. *Special Papers in Palaeontology* 23:163–184.

Mapes, G. K. 1983. Permineralized *Lebachia* pollen cones. *Am. Jour. Bot.* 70(5), part 2:74.

Matten, L. C. 1968. *Actinoxylon banksii* gen. et sp. nov.: A progymnosperm from the Middle Devonian of New York. *Am. Jour. Bot.* 55:773–782.

—— 1973. The Cairo flora (Givetian) from eastern New York. 1. *Reimannia,* terete axes, and *Cairoa lamanekii* gen. et sp. n. *Am. Jour. Bot.* 60:619–630.

—— 1974. The Givetian flora from Cairo, New York: *Rhacophyton, Triloboxylon,* and *Cladoxylon. Bot. Jour. Linn. Soc.* 68:303–318.

—— 1975. Additions to the Givetian Cairo flora from eastern New York. *Bull. Torrey Bot. Club* 102:45–52.

Matten, L. C. and H. P. Banks. 1966. *Triloboxylon ashlandicum* gen. and sp. n., from the Upper Devonian of New York. *Am. Jour. Bot.* 53:1020–1028.

—— 1967. Relationship between the Devonian progymnosperm genera *Sphenoxylon* and *Tetraxylopteris. Bull. Torrey Bot. Club* 94:321–333.

—— 1969. *Stenokoleos bifidus* sp. n. in the Upper Devonian of New York State. *Am. Jour. Bot.* 56:880–891.

Matten, L. C. and H.-J. Schweitzer. 1982. On the correct name for *Protopteridium (Rellimia) thomsonii* (fossil). *Taxon* 31:322–326.

Meeuse, A. D. J. 1963. From ovule to ovary: A contribution to the phylogeny of the megasporangium. *Acta Biotheoretica* 16:127–182.

Meyen, S. V. 1981. Some true and alleged Permotriassic conifers of Siberia and Russian Platform and their alliance. *Paleobotanist* 28–29:161–176.

—— 1984. Basic features of gymnosperm systematics and phylogeny as evidenced by the fossil record. *Bot. Rev.* 50:1–111.

—— 1986. Gymnosperm systematics and phylogeny: A reply to commentaries by C. B. Beck, C. N. Miller, and G. W. Rothwell. *Bot. Rev.* 52:300–320.

Mustafa, H. 1975. Beiträge zur Devonflora 1. *Argumenta Palaeobot.* 4:101–133.

—— 1978. Beiträge zur Devonflora 2. *Argumenta Palaeobot.* 5:91–132.

Nathorst, A. G. 1902. Zur Oberdevonischen Flora der Bären-Insel. *K. Svenska Vetenkaps-Akad. Handl.* 36:1–60.

Obrhel, J. 1959–1961. Die Flora der Srbsko-Schichten (Givet) des mittelböhmischen Devons. *Sborník Ústridního Ústavu Geologickeho* 26:7–46.

—— 1966. *Protopteridium hostinense* Krejči und Bemerkungen zu den übrigen Arten der Gattung *Protopteridium. Casopsis pro Mineralogii a Geologii* 4:441–443.

—— 1968a. Rekonstruktion einiger Devonpflanzen Mittelböhmens. *Paläont. Abh. Geol. Ges. DDR* IIB:709–715.

—— 1968b. Die silur und Devonflora des Barrandiums. *Paläont. Abh. Geol. Ges. DDR* IIB:663–701.

Pettitt, J. M. 1965. Two heterosporous plants from the Upper Devonian of North America. *Bull. Br. Mus. Nat. Hist. (Geol.)* 10:83–92.

Phillips, T. L., H. N. Andrews, and P. G. Gensel. 1972. Two heterosporous species of *Archaeopteris* from the Upper Devonian of West Virginia. *Palaeontographica* 139B:47–71.

Piedboeuf, J. L. 1887. Über die jungsten Fossilienfunde in der Umgebung von Düsseldorf. *Mitt. Naturwiss. Ver Düsseldorf* 1:9–57.

Potonié, H., and C. Bernard. 1904. *Flore dévonienne de l'étage H-h1 de Barrande.* Leipzig.

Read, C. B. 1936. The flora of the New Albany Shale. Part 1. *Diichnia kentuckiensis*, a new representative of the Calamopityeae. *U.S. Geol. Surv. Prof. Paper* 185-H:149–161.

—— 1937. The flora of the New Albany Shale. Part 2. The Calamopityeae and their relationships. *U.S. Geol. Surv. Prof. Paper* 186-E:81–104.

Read, C. B. and G. Campbell. 1939. Preliminary account of the New Albany Shale flora. *Am. Midl. Nat.* 21:435–453.

Richardson, J. B. 1960. Spores from the Middle Old Red Sandstone of Cromarty, Scotland. *Palaeontology* 3:45–63.

Rothwell, G. W. 1976. Primary vasculature and gymnosperm systematics. *Rev. Palaeobot. Palynol.* 22:193–206.

—— 1981. The Callistophytales (Pteridospermopsida): Reproductively sophisticated Paleozoic gymnosperms. *Rev. Palaeobot. Palynol.* 32:103–121.

—— 1982. New interpretations of the earliest conifers. *Rev. Palaeobot. Palynol.* 37:7–28.

—— 1986. Classifying the earliest gymnosperms. In B. A. Thomas and R. A. Spicer, eds., *Systematic and Taxonomic Approaches in Palaeobotany*, pp. 137–161. Oxford: Oxford University Press.

Scheckler, S. E. 1974. Systematic characters of Devonian ferns. *Ann. Missouri Bot. Gard.* 61:462–473.

—— 1975. A fertile axis of *Triloboxylon ashlandicum*, a progymnosperm from the Upper Devonian of New York. *Am. Jour. Bot.* 62:923–934.

—— 1976. Ontogeny of progymnosperms. 1. Shoots of Upper Devonian Aneurophytales. *Can. Jour. Bot.* 54:202–219.

—— 1978. Ontogeny of progymnosperms. 2. Shoots of Upper Devonian Archaeopteridales. *Can. Jour. Bot.* 56:3136–3170.

—— 1982. Anatomy of the fertile parts of *Tetraxylopteris schmidtii* (Progymnospermopsida). *Bot. Soc. Am. Misc. Publ.* 162:64. (Abstract.)

Scheckler, S. E. and H. P. Banks. 1971a. Anatomy and relationships of some Devonian progymnosperms from New York. *Am. Jour. Bot.* 58:737–751.

—— 1971b. *Proteokalon*, a new genus of progymnosperms from the Devonian of New York State and its bearing on phylogenetic trends in the group. *Am. Jour. Bot.* 58:874–884.

Schmid, R. 1967. Electron microscopy of wood of *Callixylon* and *Cordaites*. *Am. Jour. Bot.* 54:720–729.

Schweitzer, H.-J. 1974. Zur mitteldevonischen Flora von Lindlar (Rheinland). *Bonner paläobot. Mitt.* 1:1–9.

Schweitzer, H.-J. and L. C. Matten. 1982. *Aneurophyton germanicum* and *Protopteridium thomsonii* from the Middle Devonian of Germany. *Palaeontographica* 184B:65–106.

Scott, D. H. 1923. *Studies in Fossil Botany*, vol. 2. 3rd ed. London: Black.

Serlin, B. S. and H. P. Banks. 1978. Morphology and anatomy of *Aneurophyton*, a progymnosperm from the late Devonian of New York. *Paleontogr. Am.* 8:343–359.

Seward, A. C. 1917. *Fossil Plants,* vol. 3. Cambridge: Cambridge University Press.

Smith, D. L. 1962. Three fructifications from the Scottish Lower Carboniferous. *Palaeontology* 5:225–237.

Solms-Laubach, H. 1893. Über die in den Kalksteinen des Culm von Glätzisch-Falkenberg in Schlesien enthaltenen strukturbietenden Pflanzenreste, 2. *Bot. Zeitung* 51:197–210.

Stein, W. E. 1981. Reinvestigation of *Arachnoxylon kopfii* from the Middle Devonian of New York State, USA. *Palaeontographica* 147B:90–117.

—— 1982a. *Iridopteris eriensis* from the Middle Devonian of North America, with systematics of apparently related taxa. *Bot. Gaz.* 143:401–416.

—— 1982b. The Devonian plant *Reimannia,* with a discussion of the class Progymnospermopsida. *Palaeontology* 25:605–622.

—— 1986a. Conflict and agreement in major views of gymnosperm relationships. *Am. Jour. Bot.* 73:706–707.

—— 1986b. Two important characters in the origin of seed plants. *Am. Jour. Bot.* 73:707.

—— 1987. Phylogenetic analysis and fossil plants. *Rev. Palaeobot. Palynol.* 50:31–61.

Stein, W. E. and C. B. Beck. 1983. *Triloboxylon arnoldii* from the Middle Devonian of western New York. *Contr. Mus. Paleont. Univ. Michigan* 26:257–288.

—— 1987. The early pteridosperm *Bostonia* as a potential ancestor of Medullosales. *Am. Jour. Bot.* 74:689.

—— (In press). Paraphyletic groups in phylogenetic analysis: Progymnospermopsida and Prephanerogames in alternative views of seed plant relationships. *Bull. Soc. Bot. de France.*

Stein, W. E., D. C. Wight, and C. B. Beck. 1983. *Arachnoxylon* from the Middle Devonian of southwestern Virginia. *Can. Jour. Bot.* 61:1283–1299.

—— 1984. Possible alternatives for the origin of Sphenopsida. *Syst. Bot.* 9:102–118.

Stewart, W. N. 1983. *Paleobotany and the Evolution of Plants.* Cambridge: Cambridge University Press.

Stidd, B. M. and J. W. Hall. 1970. *Callandrium callistophytoides,* gen. et sp. nov., the probable pollen-bearing organ of the seed fern *Callistophyton. Am. Jour. Bot.* 57:394–403.

Stockmans, F. 1948. Végétaux Dévonien Supérieur de la Belgique. *Mus. Roy. Hist. Natur. Belgique Mém.* 110:1–85.

—— 1968. Végétaux Mésodévoniens récoltes aux confins du Massif du Brabant (Belgique). *Mus. Roy. Hist. Natur. Belgique Mém.* 159:1–49.

Streel, M. 1964. Une association des spores du Givétien inférieur de la Vésdre, à Goé (Belgique). *Ann. Soc. Géol. Belgique* 87:1–30.

Stur, D. 1881. Die Silur-Flora der Etage H-h1 in Böhmen. *Sitz. Ber. K. Akad. Wiss. Wien* 84:330–391.

Takhtajan, A. L. and S. G. Zhilin. 1976. Rehabilitation of the genus *Ptilophyton* J. W. Dawson, 1878. *Taxon* 25:577–579.

Termier, H. and G. Termier. 1950. Le Flore Eifelienne de Dechra Ait Abdallah (Maroc-Central). *Soc. Géol. France, Bull.*, 5th ser. 20:197–224.

Thomas, D. E. 1935. A new species of *Calamopitys* from the American Devonian. *Bot. Gaz.* 97:334–345.

Walton, J. 1957. On *Protopitys* (Göppert): with a description of a fertile specimen *Protopitys scotica* sp. nov. from the Calciferous Sandstone Series of Dunbartonshire. *Trans. Roy. Soc. Edinburgh* 63:333–340.

—— 1969. On the structure of a silicified stem of *Protopitys* and roots associated with it from the Carboniferous limestone, Lower Carboniferous (Mississippian) of Yorkshire, England. *Am. Jour. Bot.* 56:808–813.

Wight, D. C. 1985. *Aneurophytalean Progymnosperms from the Middle Devonian Millboro Shale of Southwestern Virginia.* Ph.D. dissertation, University of Michigan, Ann Arbor. Ann Arbor, Mich.: University Microfilms.

—— 1986. Primary vascular architecture of aneurophytalean progymnosperms and its bearing on evolutionary relationships of the group. *Am. Jour. Bot.* 73:714.

—— 1987. Non-adaptive change in early land plant evolution. *Paleobiology* 13:208–214.

Wight, D. C. and C. B. Beck. 1982. A distinctive aneurophytalean progymnosperm from the Middle Devonian of southwestern Virginia. *Bot. Soc. Am. Misc. Publ.* 162:67.

—— 1984. Sieve cells in phloem of a Middle Devonian progymnosperm. *Science* 225:1469–1471.

Wilcox, M. S. 1967. *An Upper Devonian Flora from Central New York State.* Ph.D. dissertation, Cornell University, Ithaca, N.Y. Ann Arbor, Mich.: University Microfilms.

Zalessky, M. D. 1911. Étude sur l'anatomie du Dadoxylon tchihatcheffi Goeppert. *Trudy geol. Kom.* 68:18–29.

Zimmermann, W. 1952. Main results of the "Telome Theory." *Paleobotanist* 1:456–470.

2

Biology of Ancestral Gymnosperms

GAR W. ROTHWELL
STEPHEN E. SCHECKLER

▦

The most ancient of gymnosperm fossils occur in sediments that were deposited about 350 million years ago, during the Famennian stage of the Upper Devonian (figure 2.1; Pettitt and Beck 1968; Gillespie et al. 1981; Fairon-Demaret 1986). Their appearance marks the occurrence of fundamental changes in the composition and distribution of the land flora. Among the most significant of these changes were the development of a new level of complexity in reproductive biology, a rapid increase in species diversity, and the origin of new modes (i.e., "Architectural Models" of Halle and Oldeman 1970) of vegetative growth among vascular plants. By the end of the Paleozoic, these changes had established the dominance of a rapidly radiating assemblage of gymnosperms and had dramatically expanded the ecological amplitude of land plants.

In this study we present fossil evidence for the most ancient gym-

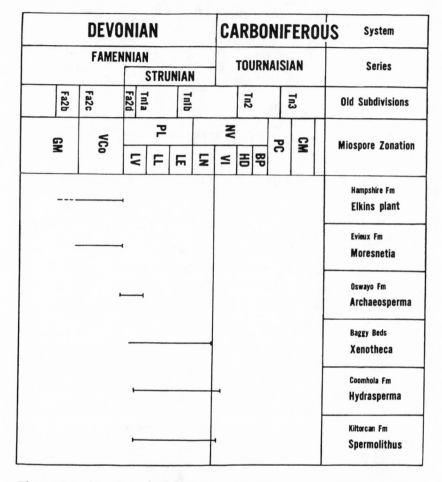

Figure 2.1. Stratigraphic chart showing the occurrence of Upper Devonian gymnosperms and their correlations with the European spore zonation. (Modified from Fairon-Demaret 1986).

nosperms and attempt to characterize them as whole plants, as biological organisms, and as members of plant communities. Included are characterizations of reproductive and vegetative structures, interpretations of growth and reproductive biology, and evaluations of sedimentological and associational data. Much of the discussion is centered on seeds and seed-bearing structures, with particular emphasis on the function of the megasporangium in pollination and post-pollination biology; on the structure and function of the seed integument; and on specializations for effecting pollination. Also stressed are the means of propagation by sexual reproduction, the nature of the propagule, and the biology of the propagule. All of these are interpreted in terms of the ecological parameters within which primitive gymnospermous biology arose and the earliest seed plants diversified.

REDEFINING THE PROBLEM

By employing evidence from the fossil record to interpret the sequence in which the events took place, it is apparent that the biological changes that most clearly separate gymnospermous reproduction from pteridophytic reproduction occurred earlier than morphological evolution of the integument (Rothwell 1987a). These include (1) the evolution of pollination (i.e., the pollen chamber); (2) a change from dehiscent to indehiscent megasporangia (Stewart 1983); and (3) the origin of propagule abscission (Rothwell 1986). Therefore, the evolution of gymnospermous biology predated the evolution of the seed and thus of seed plants (Rothwell 1987a). Within this context, the question of gymnosperm origins is effectively redefined as the evolution of a reproductive syndrome with accompanying structural and developmental changes, and the evolution of the seed (viz., the integument) is recognized as having occurred among plants that already had achieved the level of gymnospermous reproduction.

From this perspective, we are encouraged to address a much broader spectrum of biological questions in our assessment of gymnosperm origins. Among these are questions about the ecological setting, pollination biology, fertilization biology, propagule dispersal, and embryogeny and seedling establishment, and about the ontogeny and evolution of vegetative tissues, organs and shoot systems of the most primitive gymnosperms.

FOSSILS OF PRIMITIVE GYMNOSPERMS

As is characteristic of tracheophyte fossils in general, remains of the most ancient gymnosperms are preserved primarily as disarticulated organs and fragments of organs. Only seeds (or ovules) and seed-bearing structures are diagnostic of gymnosperms. Other remains are interpreted as gymnospermous because they occur in association with the seeds and compare favorably with organs of known gymnospermous affinities from more recent strata (Rothwell 1980, 1985). Direct evidence for Devonian gymnosperms consists of dispersed seeds, isolated ovulate cupules, and cupules attached to forking systems of terete axes. Indirect evidence for gymnospermy includes clustered sporangia or synangia bearing trilete grains, isolated segments of protostelic stems and roots, and frond fragments consisting of pinnae and pinnules. These fossils may be used for inferring the generalized whole-plant features of the most ancient gymnosperms, and for interpreting the reproductive functions for several tissues and organs. When considered in concert with careful studies of sedimentological, associational, and taphonomic evidence, these data dramatically expand the parameters for an evolutionary analysis and provide a broader basis for understanding the most primitive gymnosperms.

Specimens from most Devonian localities are preserved primarily as coalified compressions or impressions, but a few also have areas that are anatomically preserved by pyrite, marcasite, or iron hydroxides. The latter specimens allow one to correlate external morphology with internal anatomy, and provide crucial evidence for interpreting ontogeny and reproductive biology. Fossils from other localities are preserved primarily as cellular permineralizations, of which some specimens are exposed at the surface of the rock where external morphology also can be determined.

Fossils from basal Carboniferous strata tend to be preserved either as compression/impressions or cellular permineralizations within a rock matrix. It is, therefore, somewhat more difficult to correlate internal anatomy with external morphology for these species. The Carboniferous fossils are basically similar to those from the Devonian strata, but they display greater structural and taxonomic diversity.

Diagnostic Characters   Perhaps the most prominent feature that distinguishes gymnosperms from free-sporing pteridophytes is sexual reproduction involving a megasporangium that is enclosed in a protective integument, and that contains only a single functional megaspore (figure 2.2a). This "integumented megasporangium" traditionally has been defined as a seed if the egg has been fertilized and it contains an embryo (Andrews 1963; Stewart 1983), and has been identified as an ovule if unfertilized. However, the megasporangium in many of most ancient species (Rothwell 1985) is not completely surrounded by an integument, and none of these has been found with an enclosed embryo (Rothwell 1982a; Stewart 1983). Therefore, it can be argued that the most ancient species only approach the structure and function of a modern seed (Stewart 1983; Rothwell 1985). Nevertheless, such fossils are widely recognized as the megasporangiate remains of the earliest gymnosperms, and are termed seeds by most

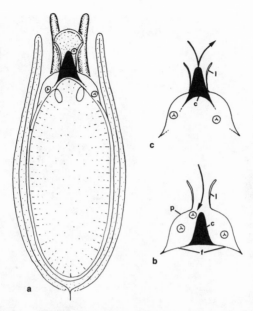

Figure 2.2. Diagrammatic reconstructions showing diagnostic features of ovules with hydrasperman reproduction. **a** Longitudinal view of mature generalized ovule. Note the vascularized integumentary lobes surrounding the lagenostome, the central column that is sealing the pollen chamber at the base of the lagenostome, and the cellular megagametophyte that is no longer separated from the microgametophytes by the pollen chamber floor. **b** and **c** Longitudinal views of generalized hydrasperman pollen chamber showing (at **b**) pollination stage, where prepollen can enter pollen chamber, and (at **c**) post-pollination stage, where growth of the megagametophyte has pushed the central column up into the base of the lagenostome to seal the pollen chamber and rupture the pollen chamber floor. Arrows indicate path of objects falling on the nucellar apex at the pollination and post-pollination stages of development. (**b** and **c** redrawn and modified from Rothwell 1987b. c = central column, f = pollen chamber floor, 1 = lagenostome, p = pollen chamber wall.)

authors. In this paper we frequently refer to attached specimens as ovules (i.e., those within cupules) and to isolated specimens as seeds. It should be stressed, however, that this usage is arbitrary with respect to whether the specimens are in a pre- or post-fertilization stage of ontogeny, and in this context the terms may be used interchangeably.

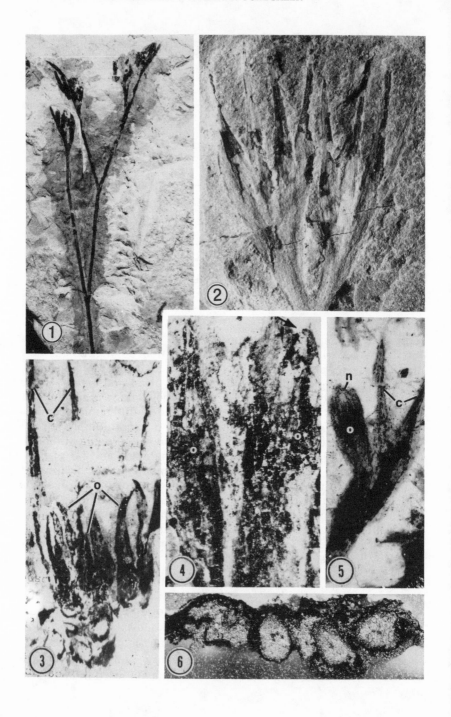

Fossils and Facies For almost ten years the ovulate cupules of *Archaeosperma arnoldii* Pettitt and Beck (1968: figures 10–12) stood as the only *bona fide* evidence for a Devonian gymnosperm. Since 1977, however, there has been a greatly expanded interest in earlier gymnosperms, so that now we recognize several new genera and have reinterpreted some previously described, but poorly understood, fossils as seeds or seed-bearing cupules. These are derived from Famennian and Tournaisian Devonian deposits of North America, western Europe and Asiatic Europe (figure 2.1). Included are numerous specimens from near Elkins, West Virginia (Gillespie et al. 1981) that we shall refer to as the "Elkins" fossils; *Moresnetia zalesskyi*, from several deposits in Belgium; *Xenotheca devonica*, from southern England; and cupulate specimens described as *Hydrasperma tenuis*, from southern Ireland. In addition, several other fossils may represent seeds (e.g., *Spermolithus devonicus*), but their taxonomic or organographic identity remains equivocal.

The most ancient ovules are the Elkins specimens and *Moresnetia*, which are both middle Famennian (lower Fa2c; figure 2.1). They are followed by *Archaeosperma* (Fa2d), then *Xenotheca* (Fa2d–lower Tn1b); the youngest are Irish *Hydrasperma* and *Spermolithus*, which are both Tournaisian (Tn1a–lower Tn1b).

Figure 2.3. 1-6 Ovulate cupules ("Elkins gymnosperm") from the Upper Hampshire Formation near Elkins, Randolph County, West Virginia (c = cupule lobes, n = nucellus, o = ovule, in all subsequent figures.) **1** Portion of a cupuliferous branch system showing the terminal clusters of cupules in a shallow, corymb-like arrangement. Ohio University Paleobotanical Herbarium (OUPH) 74-2, part. × 1. **2** Exterior view of a cupule half showing the symmetrical forks of the quarters. OUPH 1-92 counterpart. × 5. **3** Interior view of a cupule prepared by interrupted and precision cleaved transfer that now shows cupule lobes, and the apical portions of three ovules, the free integument lobes of which arch over each nucellus. Virginia Polytechnic Institute and State University Paleobotanical Collections (VPISUPC) 61.2B(+). × 9. **4** Two ovules from interior view of a cupule half, to show the ridged ovule integument with free apical lobes that curve over the nucellar apex (arrow at right) or that are pressed together (left). VPISUPC 80.1C. × 9. **5** Interior view of a cupule showing a partial cupule quarter bearing an ovule from which the free integument tips have been removed to show the nucellar apex. VPISUPC 80.1A. × 7. **6** Transverse section of a permineralized cupule quarter to show its terete segments. OUPH 1-88, 9 Bot.

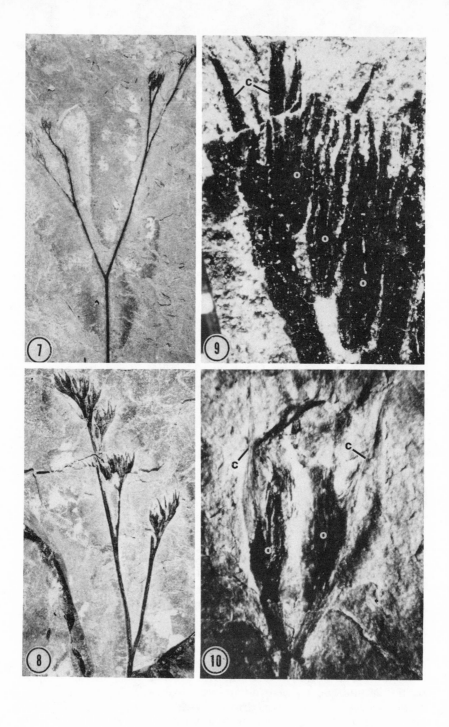

The Elkins fossils consist of more than two hundred ovulate cupules (figure 2.3), cupuliferous branches, and thousands of dispersed seeds preserved in the Upper Hampshire Formation near Elkins, Randolph County, West Virginia (Gillespie et al. 1981). This horizon is pre-Strunian Famennian (lower Fa2c = VUi = lower VCo spore zone; Gillespie et al. 1981; Fairon-Demaret 1986) and falls within a delta margin complex of shoaled delta lobes and lower delta plains interbedded and buried by near shore marine storm beds (Scheckler 1986a, 1986b).

Moresnetia zalesskyi (Stockmans 1948) was originally described as a leafy branch system, the leaves of which forked and were pointed so as to falsely resemble cupules. Many authors have commented on the similarity of Stockman's plant to *Archaeosperma* or other cupulate organs (Pettitt and Beck 1968; Pettitt 1970; Gillespie et al. 1981; Gensel and Andrews 1984), but proof of its identity was obtained only when ovules were found inside the cupules (figures 2.3:7–9; Fairon-Demaret and Scheckler 1986; Fairon-Demaret and Scheckler, in press). Several hundred specimens of *Moresnetia* have been collected, and these consist of ovulate cupules and large cupuliferous branch systems. The fossils have been collected at several localities in the Evieux Formation of south central and eastern Belgium and are also pre-Strunian Famennian (lower Fa2c = lower VCo spore zone; Fairon-Demaret 1986; Fairon-Demaret and Scheckler, in press). These beds are mostly marine and were deposited along the landward side of a near shore barrier, often as lagoon-fills (Thorez and Dreesen 1986).

Archaeosperma arnoldii Pettitt and Beck (1968) is known from several isolated ovulate cupules from the Oswayo Formation, a nearshore marine sandstone of northern Pennsylvania, near Port Allegany (Arnold 1935). Based on adjacent stratigraphical sections in New York, the Oswayo Formation is not older than late Famennian or basal

Figure 2.3. continued. **7-9** *Moresnetia zalesskyi* from the Evieux Formation of the Bocq Valley, Province of Namur, Belgium. **7** Portion of an asymmetrically forked cupuliferous branch system bearing terminal raceme-like clusters of cupules. IRSNB b 12.935. × 0.6. **8** Closer view of the terminal raceme-like clusters of cupules. ULLPP 12.952. × 1.5. **9** Interior view of fractured cupule to show portions of 3 ovules each with 8 to 10 long, free integument lobes. Note the oblique vertical arrangement of ovules in this cupule and their size differences. IRSNB b 12.995. × 15. **10** Holotype of *Archaeosperma arnoldii* Pettitt and Beck from the Oswayo Formation near Port Allegany, Pennsylvania (Pettitt and Beck 1968; also from Arnold 1935). Impression/compression of cupule with two ovules. UMMP 16069. × 7.

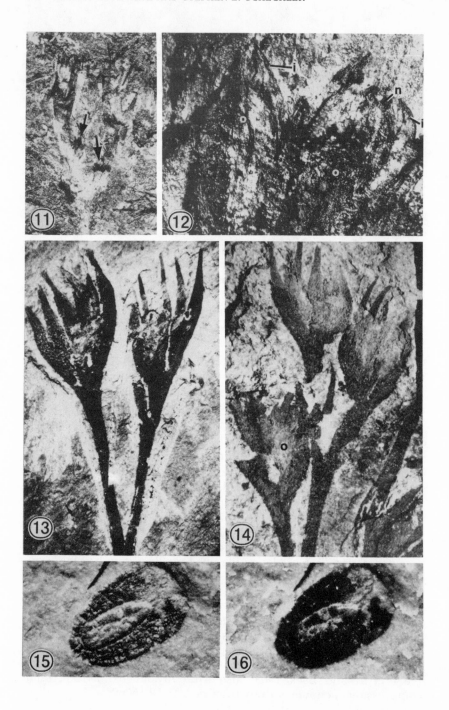

Strunian (Fa2d = PLi = lower LV spore zone; Gillespie et al. 1981; Fairon-Demaret 1986).

Xenotheca devonica (Arber and Goode 1915) was originally interpreted as a late Devonian cupule, but no evidence for seeds was found. Rogers (1926) described additional specimens (figures 2.3:13, 14) from the same quarry (Goldring 1970, 1971), and several of these can now be demonstrated to bear portions of ovules (e.g., figure 2.3:14; Scheckler 1985; Fairon-Demaret and Scheckler, in press). These fossils come from a thin lens of Hoe facies, interpreted by Goldring (1971) as a freshwater fluvial bed within a marginal marine sequence, from a quarry in the Baggy Beds (Marwood Formation) of southwestern England, near Barnstaple. Palynological studies (Fairon-Demaret 1986) suggest that the flora is Strunian (late Devonian = Fa2d to lower Tn1b = PLi to NVi = lower LV to LL, LE, and LN spore zones).

Hydrasperma tenuis (figures 2.3:17–21) of Matten, Lacey, and Lucas (1980) and Matten, Fine, Tanner, and Lacey (1984) consists of numerous ovulate cupules. It should be stressed that the type specimen of *H. tenuis* is an isolated ovule from another, more recent (viz., late Tournaisian or Visean of Scotland; Long 1961; Scott, Galtier, and Clayton 1984) locality, and that the Irish specimens are generically distinct from other ovuliferous cupules assigned to *Hydrasperma* (Long

Figure 2.3. continued. **11** Paratype transfer preparation of *Archaeosperma arnoldii* Pettitt and Beck, from the Oswayo Formation, near Port Allegany, Pennsylvania, showing the forked cupules and cupule segments, and stalks (at arrows) from which ovules have been shed. These cupules were illustrated as plates 1 and 2 of Pettitt and Beck (1968). UMMP 57289. × 3.5. **12** Paratype transfer preparation of *Archaeosperma arnoldii* Pettitt and Beck, showing two ovules from inside a cupule. Note the short, free integument lobe tips (i) that arch above the fused, ridged integument of these two ovules. UMMP 57289. × 15. **13** Portions of two asymmetrically divided cupules of *Xenotheca devonica*. Specimen from the Baggy beds of Croyde Hoe Farm, near Barnstaple, North Devon, England; from the Rogers collection. BM(NH) v. 31136. × 3.5. **14** *Xenotheca devonica* from Rogers (1926, plate X, figure 2). Asymmetric cupules in a slightly overtopped, cupuliferous branch system. BM(NH) v. 21480. × 2. **15** *Spermolithus devonicus* of Chaloner, Hill, and Lacey (1977) from the Kiltorcan Beds, County Kilkenny, Ireland. Low angle illumination of dry *Spermolithus* "seed" to show impression surface of integument and nucellus. BM(NH) v. 59650, "seed" no. 1, figures 1b, 2a. × 11. **16** Same specimen as in **15** but with polarized light to show the apparent opening of the "integumentary micropyle." × 11.

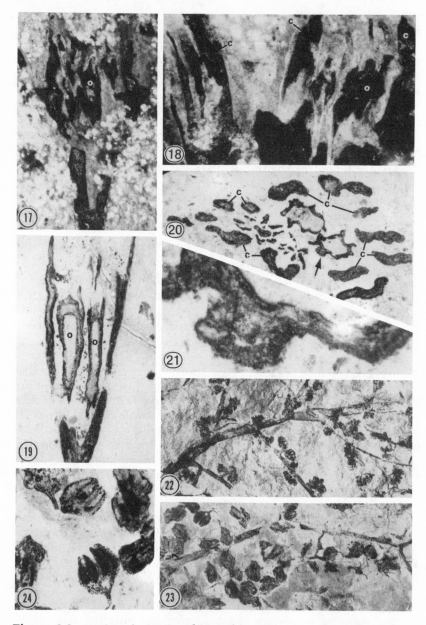

Figure 2.3. continued. **17** On the surface of a wet rock slab, cupule pair of *Hydrasperma tenuis* of Matten, Lacey, and Lucas (1980), from the Coomhola Formation of Ballyheigue, near Kerry Head, County Kerry, Ireland; from specimens given to Scheckler by Matten. × 6.6.

1977; Rothwell and Wight 1987). Nevertheless, for the purposes of this study, we will refer to the Irish specimens as *H. tenuis* (*sensu* Matten, Lacey, and Lucas 1980). The ovulate cupules come from the Coomhola Formation of Kerry Head at Ballyheigue, County Kerry, Ireland (Matten, Lacey, May, and Lucas, 1980). This formation represents Devonian Tournaisian or late Strunian time (Tn1a–lower Tn1b = PLs to NVi = upper LV to LL, LE, and LN spore zones; Fairon-Demaret 1986) and is a sandier correlative of the Kiltorcan Formation, apparently deposited nearer the shore.

Another Devonian fossil that has received considerable attention as a possible seed is *Spermolithus devonicus* (Johnson 1917). This species consists of isolated seed-like bodies (figures 2.3:15, 16) from the Devonian Tournaisian or late Strunian (Tn1a–lower Tn1b = PLs to NVi = upper LV to LL, LE, and LN spore zones; Fairon-Demaret 1986) Kiltorcan Formation. At Kiltorcan, County Kilkenny, these beds are interpreted as a lacustrine fill sequence within an alluvial fan system (Holland 1981).

Ovules and Seeds Pre-Carboniferous ovules are all quite small and fall into a relatively narrow size range, mostly 3–7 mm long and 1–2 mm wide. Specimens consist of a nucellus (megasporangium) that is surrounded by integumentary lobes or telomes that are more or less fused to the nucellus and to one another (figures 2.2a; 2.3:3; 2.3:10; 2.3:19, 20). The nucellus is characterized by an elaborate apical region

---

18 Another cupule pair on the wet surface of a rock slab. These specimens show the cupule lobes and ovules. Note the free, curved integument lobe tips and fused, ridged integument surface of the right ovule (o) and the ridged integument of the other ovule (extreme left). × 12. **19** Longitudinal peel section showing portions of two ovules inside a cupule. × 13. **20** Transverse peel section of a cupule with portions of four ovules visible. The upper right two ovules are cut at level of integument fusion, while the lower left two are cut at level of free integument lobes. Arrow indicates segment of integument in **21**. × 13. **21** Transverse peel section of integument of ovule in **20**, to show fusion of its nucellus to the integument at this level and the xylem strand and cortical hypodermis of an integument ridge. × 100. **22** *"Aneurophyton" olnense* branch system with fertile organs; specimen from the Evieux Formation of Olne, Belgium. Plate 3, figure 1 of Stockmans (1948, his no. 30410). × 0.8. **23** Synangia of gymnospermous type from Upper Hampshire Formation near Elkins, West Virginia, with associated branched axis. × 2. **24** Closer view of synangia. × 12.

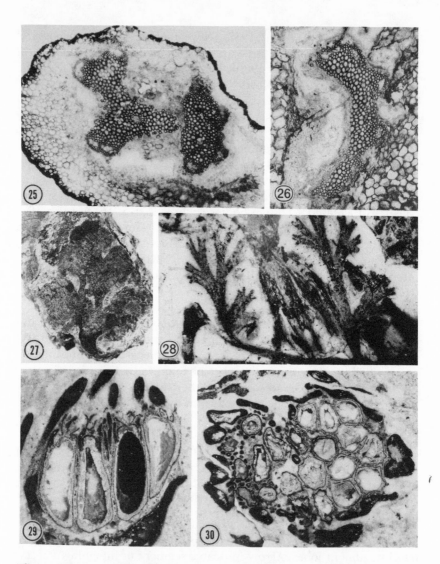

Figure 2.3. continued. **25-28** Gymnospermous axes and foliage associated with the Elkins plant from Upper Hampshire Formation near Elkins, West Virginia. **25** Stem with three-ribbed protostele from which a leaf trace (right) has diverged. OUPH 196-1(1a). × 19. **26** Frond axis with a *Lyginorachis*-like xylem strand. OUPH 49-1 (B) #3. × 28. **27** Lobed pinnule from pinna rachis with permineralized trace; similar to some of Stockmans' (1948, plate V) figures of *Sphenopteris boozensis*. OUPH 192-22. × 3. **28** Isolated cupule of the Elkins plant and associated foliage similar to some *Sphenopteris flaccida* of Stock-

that has been termed a hydrasperman pollen chamber (figures 2.2b, c; Rothwell 1985). This consists of a hemispherical chamber (viz., the pollen chamber proper) with a uniseriate wall and membranous floor. At the apex of the pollen chamber, the wall extends distally to form a specialized hollow region that has been termed a lagenostome if it is relatively short, and a salpinx if it is tubular (Rothwell 1985). The pollen chamber floor is thickened at the center to form a structure called the central column. Proximally, the nucellus contains the membrane of a large, functional megaspore, and may also display at its apex three abortive megaspores of the tetrahedral tetrad (Pettitt and Beck 1968; Fairon-Demaret and Scheckler 1986; Fairon-Demaret and Scheckler, in press). In the most mature ovules the megaspore membrane may contain a cellular megagametophyte (Matten, Fine, Tanner, and Lacey, 1984). In the species that show internal anatomy, a terete vascular strand enters the base of the ovule and divides to produce several strands, each of which enters an integumentary lobe and extends to near its apex (figure 2.2a).

Moresnetia (figure 2.3:9) and the Elkins ovules (figure 2.3:3) show the least fusion of integumentary lobes, and *Archaeosperma* (figure 2.3:12) shows the most. Lobes of the Irish *Hydrasperma* specimens (figures 2.3:18–20) are intermediate. Where known, all of the species have integumentary lobes that are fused to the nucellus distal to where the lobes separate from one another. The integument of *Moresnetia* consists of 8 to 10 thin cylindrical lobes that are fused to one another only at the chalaza and often flare away from the ovule apex (e.g., figures 2.3:9; 2.4b; 2.5e–g); its nucellus is entirely free above the chalaza (Fairon-Demaret and Scheckler 1986; Fairon-Demaret and Scheckler, in press). The integument of the Elkins ovules (Gillespie et al. 1981; figures 2.4a; 2.5a–d) consists of 4 to 5 thicker lobes that are fused to each other in the basal one-third of the ovule, where they

---

mans (1948, plate V, figures 7–7a). VPISUPC 80.1C and 80.1B. × 4. **29** Longitudinal section of a permineralized *Hydrasperma longii* cupule with numerous ovules; specimen from late Tournaisian (Tn 3) Cementstone Group sediments at Oxroad Bay, Scotland. Note the crowded arangement of the ovules, the lagenostomes of which project among their integument lobes into the space defined by the surrounding cupule lobes. OUPH 632 E$_3$ Top #3 Bot. × 7. **30** Transverse slice through a permineralized cupule of *Hydrasperma longii* with numerous ovules. Note how the tightly packed ovules are closely surrounded by the cupule lobes at this level. OUPH 632 Epv$_3$ Top #6 Bot. × 7

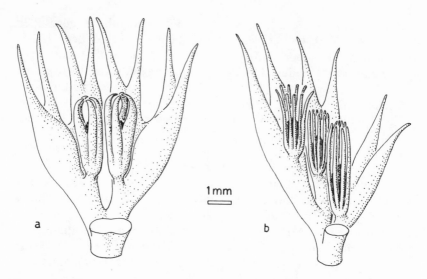

Figure 2.4. Reconstructions of portions of ovulate cupules at the inferred time of pollination. **a** New plant of Gillespie et al. (1981), shown with one half of the cupule removed. **b** *Moresnetia zalesskyi*, shown with one cupule quarter removed from the right half.

form 4 to 5 ridges. Distally, the free lobes curve inward over the ovule apex (figures 2.3:3, 4) and are fused only to the basal half of the nucellus. *Moresnetia* and the Elkins ovules resemble reconstructions of *Genomosperma kidstonii* and *G. latens* (Long 1960; Andrews 1963; Taylor and Millay, 1979), respectively. The lagenostome of *Moresnetia* is large and conical, while that of the Elkins specimens is smaller and less tapered (figure 2.5). The integument of the Irish *Hydrasperma* is fused for half the total ovule length into 8 to 10 ridges (figure 2.3:20). Above that level the 8 to 10 integument parts are free from each other, cylindrical (figure 2.3:20), and often curve inward over the ovule apex (figures 2.11:18, 19). The integument is fused to the nucellus for about four-fifths of its length, so that only the protruding pollen chamber and lagenostome are free (Matten, Lacey, and Lucas 1980). The lagenostome is long and tubular and flares apically.

We know much less about *Archaeosperma* and *Xenotheca*. In the former the integument is five-six parted and covered by prominent trichomes at the base, and is fused extensively so that only the apical one-fifth extends as free lobes (figure 2.3:12, at left) that curve inward (figure 2.3:12, at right). Features of the nucellar apex in *Archaeosperma* are unknown, but the middle and base were apparently fused

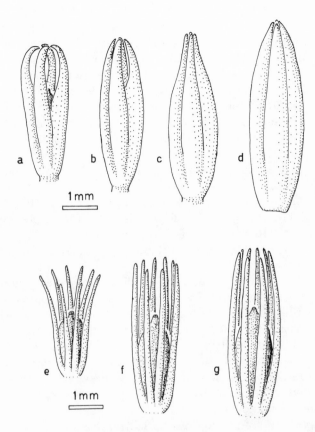

Figure 2.5. Reconstructions of ovules to show their sizes and integument morphologies at various stages inferred as pre- and post-pollination. **a–d** Elkins plant: **a** Pre-pollination to pollination configuration. **b, c** Post-pollination configurations. **d** Dispersed seed. **e–g** *Moresnetia zalesskyi:* **e** Pre-pollination to pollination configuration. **f, g** Post-pollination configurations.

to the integument (Pettitt and Beck 1968). Ovules are incompletely preserved or exposed in Rogers' (1926) *X. devonica* specimens (e.g., figure 2.3:14) and cannot be observed at all in Arber and Goode's (1915) syntypes. The integument seems to consist of 4 to 5 stout lobes (figure 2.3:14 at "O") that appear fused at their base and free for the apical one-third to one-half. Details of the nucellus are unknown.

All of these ovules are relatively round or polygonal in cross sections. Arranged stratigraphically, they demonstrate that numerous, free, terete integument lobes that flare from or that surround a dome-shaped nucellar apex from which protrudes a conical lagenostome are archaic features of the earliest gymnosperms, all of which lack micropyles. Similar conclusions were reached earlier (Long 1960, 1961; Andrews 1963) but were deduced from comparisons of a suite of much younger (Latest Tournaisian to Visean) ovules and dispersed seeds that showed a full range of integument morphologies. These early hypotheses seem now to be fully corroborated by the stratigraphic sequence in which the hypothesized ancestral morphologies are the oldest and the hypothesized derived forms are younger. However, the serious search for ancestral gymnospermous plants has only just begun. And, just as the recent past has exposed an unexpected level of diversity among Famennian gymnosperms (Scheckler 1985), continued study of these and older beds promises to demonstrate even more. When the transition from Frasnian progymnospermy to Famennian gymnospermy is clearer, we will be better able to evaluate these hypotheses more critically.

Spermolithus devonicus appears to differ from these cupuliferous ovules in several respects. The species was first interpreted as a flattened seed-like body with a central ovoid region and a peripheral wing (Johnson 1917). More recently, it was redescribed (figures 2.3:15, 16) from additional collections (Chaloner, Hill, and Lacey 1977). The new specimens also were interpreted as dispersed platyspermic seeds, but the presence of a megaspore membrane was not demonstrated and the anatomical structure of the putative nucellus and integumentary tissues is unknown.

Spermolithus "seeds" (figures 2.3:15, 16) are difficult to interpret. The apparent flattening could be a feature of the living specimens, or most or all of it could be diagenetic. Chaloner et al. (1977) felt that it was not diagenetic, since the fossils have a uniform appearance which implies that their shape led to a preferred orientation on bedding planes. If this is so, and the inner body (figure 2.3:15) is correctly interpreted as nucellus, then these "seeds" would seem to have two flaps of integument (figure 2.3:16) that partially surround, but are not fused to, the nucellus. Up to now, the chlorite-replaced compression/impressions have not been prepared further, such as by sectioning, to demonstrate the accuracy of this interpretation.

Ovules of *Archaeosperma*, the Elkins specimens, *Moresnetia*, and Irish *Hydrasperma* are all proven to be correctly identified as gymnosperms by the demonstration of a single functional megaspore (e.g., figures 2.3:19, 20) inside a nucellus that is at least partly covered by, even if not always fused to, an integument (Pettitt and Beck 1968; Gillespie et al. 1981; Fairon-Demaret and Scheckler 1986; Fairon-

Figure 2.6. *Moresnetia zalesskyi:* reconstruction of raceme-like clusters of cupules found at the terminations of the cupuliferous branch system.

Demaret and Scheckler in press; Matten, Lacey, and Lucas 1980; Matten, Fine, Tanner, and Lacey 1984). *Spermolithus* (figures 2.3:15, 16) however, is interpreted as a seed (Chaloner et al. 1977) because it resembles some compressed platyspermic seeds from the Carboniferous. Although this is a logical and perhaps appropriate inference, we lack evidence, comparable to that listed for the other taxa, that these are correctly identified as seeds or that they are truly platyspermic.

At the present time there is no evidence that "flat" versus "round"

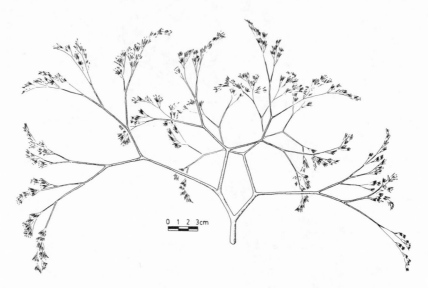

Figure 2.7. *Moresnetia zalesskyi:* reconstruction of a whole cupuliferous branch system. (Redrawn from Fairon-Demaret and Scheckler, in press.)

shape of the seed has any phylogenetic significance among primitive gymnosperms (Rothwell 1985), nor do we know how or to what *Spermolithus* was attached in life (Rothwell 1985). We similarly lack complete proof that *Xenotheca devonica* (figures 2.3:13, 14) is an ovulate cupule. This attribution, however, seems reasonable, since objects similar to the proven ovules of other taxa are attached in the same fashion to the inner surfaces of cupule quadrants (figure 2.3:14 at "o"; Rogers' [1926] specimens BM[NH] V. 21480 and BM[NH] V. 31134). These objects also show apparent integument and nucellus like those of the other taxa.

Ovulate Cupules   With the exception of *Spermolithus*, where the mode of attachment is unknown, pre-Carboniferous ovules are borne within forking systems of axes that form dichotomously organized cupules. We define a cupule as the unit that encircles or encloses a single space into which one or more ovules protrude (e.g., figure 2.3:20). Because successive forks typically are at right angles to one another, forming axes that extend off into four directions, this has been referred to as cruciate forking. When attached, the cupules are terminal (figures 2.3:1; 2.3:7, 8; 2.3:13, 14; 2.6; 2.7). They occur either singly or in pairs on the shorter axes of overtopped, unequally forked branching systems. The branching systems show cruciate forking throughout, and the angles formed by the basal forks are relatively wide (i.e., 60°–80°; figure 2.5). The three-dimensional morphology of these systems produces a characteristic arrangement of groups of cupules that varies in a consistent fashion from species to species (e.g., figure 2.3:7, 8).

Cupules of the Elkins ovules, *Moresnetia,* and *Xenotheca* (figures 2.3:2; 2.3:9; 2.3:13, 14) dichotomize cruciately to form four main parts (quarters). Each quarter may bear an ovule on the inner surface proximal to the level where one or two more forks in the same plane form long tapered cupule tips (figures 2.3:5; 2.3:10; 2.3:11, 14; 2.3:18; 2.4 Gillespie et al. 1981; Fairon-Demaret and Scheckler in press).

Archaeosperma was originally interpreted (Pettitt and Beck 1968) as having pairs of two-seeded cupules, the interiors of which faced each other (their text figure 1). However, subsequent studies of other Devonian cupules, preserved by permineralization as well as by compression, led others to interpret *Archaeosperma* as isolated cupules that each enclosed four seeds (Matten, Lacey, and Lucas 1980; Matten and Lacey 1981; Gillespie et al. 1981; Fairon-Demaret and Scheckler in press). Irish *Hydrasperma* (figures 2.3:17–21) is similar to these other cupules, but divides into 4 to 6 main parts initially before bearing 4 to 6 ovules. The individual cupule parts then fork like the cupule quarters of the others (Matten, Lacey, and Lucas 1980).

Where cupules are permineralized, the cupule segments are cylindrical and often expand above the level of ovule insertion before tapering into terete pointed tips (Matten, Lacey, and Lucas 1980; Gillespie et al. 1981; Matten, Fine, Tanner, and Lacey 1984; Fairon-Demaret and Scheckler in press). Even the expanded portions, just above ovule insertions and below the terminal forks, are terete, despite their "laminar" appearance in compressions (figures 2.3:2; 2.3:11, 13; Pettitt and Beck 1968). Partially permineralized compressions of the Elkins cupules and of *Moresnetia* show that the nonpermineralized portions of cupule segments became flattened and coalified by diagenetic compaction of the enclosing sediment. However, the permineralized areas of these same cupules show conclusively that the original structures were cylindrical (figure 2.3:6). We interpret *Archaeosperma* as also having had cylindrical cupule segments.

Permineralized cupules of the Elkins specimens, *Moresnetia,* and Irish *Hydrasperma* (figures 2.3:6; 2.3:20) all have cortex with a pronounced hypodermal layer of fibers. Impressions of the first two taxa and of *Archaeosperma* (figures 2.3:2; 2.3:9, 10; Arnold 1935; Pettitt and Beck 1968) show surface patterns of fine longitudinal striations that probably correspond to the hypodermal fibers. Similar striations also occur on impressions of cupuliferous branches and integument lobes that anatomically had a fibrous hypodermis (figure 2.3:21). We suspect that this hypodermis was important for maintaining the shapes and arrangements of integuments, cupules, and cupuliferous branch systems so that their configuration was quite uniform within each taxon.

Cupules of these early gymnosperms are remarkably uniform in size. All are about 10 mm long (measured from the basal fork to the cupular tips) and flare to about 7–10 mm wide at their distal opening. They are less uniform, though, in the patterns of their divisions. Those of *Archaeosperma* and the Elkins plant (figure 2.3:2; 2.3:10; 2.3:11) show near-equal, or only slightly overtopped, basal forks. This produces cupule halves that extend to nearly the same level (figure 2.3:11). The cupule halves divide symmetrically into equal quarters (figures 2.3:2; 2.3:10), all of which bear their ovules at nearly the same level (figure 2.4a). Cupule quarters of these two plants divide twice more, in one plane, above the insertion of their ovules, so that a complete cupule has 16 tapered tips. The Irish *Hydrasperma* cupules divide less regularly, into 16 to 24 tips and bear 4 to 6 ovules, but are overall rather symmetrical (figure 2.3:17, 18). Cupule tips of these three appear to be arranged in a ring (figure 2.4a) that is oriented perpendicular to the basal axis of the cupule.

Cupules of *Xenotheca* are less symmetrical (figure 2.3:13, 14). One half is larger than the other, and one quarter of each half is longer. Cupule quarters are also less regularly divided, so that a complete cupule has 8 to 16 tips. Since the longer cupule parts are on one side, the cupule tips appear to be arranged in an ellipse that is obliquely tilted with respect to a cupule axis.

Cupules of *Moresnetia* bear 1 to 4 ovules and are even more asymmetrically divided (figure 2.3:8, 9), so that the cupule tips encircle a markedly elliptical opening. This opening is obliquely vertical to the cupular axis (figure 2.6). Because cupule quarters vary significantly in length, ovules are borne in an oblique vertical series (figures 2.3:9; 2.4b) within a strongly overtopped cupule, and only some of the cupule quarters bear ovules. The ovuliferous quarters divide less (once) than those that are sterile (which divide twice), so that complete cupules can have 8 to 16 tips depending on the number of ovules they bear (those with 16 tips are sterile; Fairon-Demaret and Scheckler, in press).

In addition to these five genera of ovulate cupules, Soviet authors for years have identified *Moresnetia* (*M. zalesskyi*, *M. krystofovichii*, and *M. sibirica)* from their late Devonian strata (e.g., Lepekhina, Petrosyan, and Radchenko 1962). Another example of these attributions is the "*M. zalesskyi*" illustrated by Pettitt (1970, his plate 6, figure 1) from the British Museum (Natural History) collections. This specimen and several other hand samples were collected from Upper Devonian strata (Touran Series of Lennoi Log) near Krasnoyarsk, Siberia, (Vologdine 1937). None of the "*Moresnetia*" from this locality (nor any other Soviet locality) has been shown to bear ovules, although one unillustrated specimen (BM[NH] V. 44784) seems to contain a fragmentary ovule, so that the identity of at least some of these specimens as ovulate cupules seems correct (Fairon-Demaret and Scheckler, in press). Far more important, that specimen, as well as Pettitt's (1970), have smaller and differently arranged cupules than the Belgian *Moresnetia*. We can tentatively identify the age of this collection, based on its floral associates, as probably late Famennian to basal Strunian (Fa2c–Fa2d; Fairon-Demaret and Scheckler, in press). However, these Russian specimens are not the same as Stockmans' *Moresnetia*, nor are they the same as any of the other genera currently known, and thus may represent additional genera of Late Devonian gymnosperms.

Other identifications are even more equivocal. For example, Bassett and Edwards (1982) referred a microsporangiate cupule from Upper Old Red Sandstone beds (? late Strunian) from near Cardiff, Wales, to *Xenotheca* sp. This cupule, however, lacks the characters of Arber and Goode's plant and should be placed into another genus. Similarly, Stockmans' (1948) *Xenotheca bertrandii* was found (Fairon-Demaret and Scheckler, in press) to be either indeterminable or possible cupule parts or abortive cupules of the *Moresnetia* type. We recommend that *X. bertrandii* be abandoned.

Cupuliferous Branching Systems   Variations in cupule morphology among the various genera, subtle though they may seem, are consistent among the thousands of specimens available for study. We find that cupule symmetry is correlated with the way individual cupules are borne within the whole cupuliferous branch system, particularly with the arrangement of cupule clusters at the tips of the branchlets. This is shown especially well by *Moresnetia*. In this genus there are pairs of cupules or clusters of three cupules borne at the ends of unequally forked, short axes that arise from racemose terminal branchlets (figures 2.3:7, 8; 2.6). The branchlets are part of a larger, unequally and cruciately forked, essentially hemispherical cupuliferous branching system (figure 2.7).

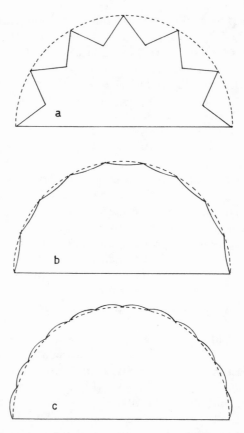

Figure 2.8. Hypothetical longitudinal views of the hemispherical cupuliferous branch systems (dotted) of some early gymnosperms to show the topographies of the "surfaces" (solid lines) on which cupules were borne. **a** *Moresnetia zalesskyi*. **b** Elkins plant. **c** *Xenotheca devonica*. **b** and **c** are extrapolated from smaller pieces, while **a** is known for several whole specimens.

The asymmetrical cupules are oriented within each "raceme," so that their open ends are directed toward a funnel-like space (figures 2.3:7, 8; 2.6; 2.7; 2.8a) formed by immediately proximal branches of the cupuliferous branching system. The cupule-bearing "racemes" are thus disposed so as to form what can be thought of as funnel-shaped "indentations" (figure 2.8a) into the surface of the "hemisphere" formed by the cupuliferous branching system.

The cupuliferous branching system of the Elkins plant is less completely known. Distal segments fork unequally like those of *Moresnetia* but are overtopped to a lesser degree (figure 2.3:1). The shorter distal portions also terminate in one, two, or more cupules. At the terminus of the system, we interpret groups of cupules to be borne in loose, corymbose tufts. The circular opening of each cupule faces upward toward the distal "surface" of the tuft and of the cupuliferous branching system as a whole. Even though individual cupules vary considerably in the width or taper of their segments, they are all quite symmetrical and bear their ovules at the same levels (figures 2.3:2–4; 2.4a).

Only the smaller terminal parts of the cupuliferous branches of *Xenotheca* are known. Specimens are less symmetrical than the Elkins fossils, but significantly more symmetrical than *Moresnetia*. *Xenotheca* cupules are borne in pairs that terminate the shorter halves of the unequally forked cupuliferous branchlets (figures 2.3:13, 14). We interpret them to form a "domed," corymb-like tuft (figures 2.3:14; 2.8c). Within this tuft, the individual cupules are oriented by their asymmetry to face the outer (distal) "surface" of the cupuliferous branching system. So far, only solitary cupules of *Archaeosperma* are known, and only solitary or paired cupules of the Irish *Hydrasperma* (figure 2.3:17, 18) have been discovered. When in pairs, the *Hydrasperma* cupules terminate halves of an equal fork, and the distal opening of each points upward.

Taxonomy of Early Gymnosperms   The observed close correlation between cupule shapes and their arrangements on the terminal branchlets of the cupulate system suggests that, although the differences between cupule morphologies of these taxa may seem subtle, they are consistent within each taxon. Coupled with the observed differences among their ovules, which do not intergrade, we interpret these five taxa of gymnosperms *(Archaeosperma,* the Elkins plant, *Moresnetia, Xenotheca,* and the Irish *Hydrasperma)* to be distinct genera. The possibility that the taxonomic distinctions may reflect subtle differences in aerodynamic requirements for abiotic pollination or some degree of habitat partitioning is explored in a later section of this paper.

Should *Spermolithus* prove to be a seed, then it would represent a sixth Late Devonian gymnosperm. We have already commented on the possibility that the Russian *"Moresnetia"* also represents one or more late Devonian gymnosperm taxa. If so, these are quite different from the Belgian *Moresnetia* (Fairon-Demaret and Scheckler, in press) and would represent additional genera.

Primitive Gymnospermous Microsporangia   Several types of aggregated sporangia or synangia with partly fused sporangia also occur at the Devonian localities that yield ovulate cupules. By their similarities to pteridospermous synangia from Carboniferous strata, we suspect that some or all of these Devonian fossils also may be gymnospermous pollen organs. The first of these occurs with *Moresnetia* cupules in the Evieux flora from Belgium and has been described as *Aneurophyton olnense* (Stockmans 1948). Specimens consist of small, fertile organs that fork and bear pinnate pairs of elongate sporangia (figure 2.3:22) similar to the fertile organs of Aneurophytales (Scheckler 1975, 1987). The fertile organs are helically arranged on pinnate laterals of a larger branch system. The laterals sometimes bear distal sterile forked "ultimate appendages," similar to those of Aneurophytales, that resemble the pinnules of *Sphenopteris olnensis* from the same locality (Stockmans 1948).

Although similar to *Aneurophyton* in some features, these fertile axes, in our opinion, should be placed in a separate genus. Elsewhere on the Old Red Continent (Laurussia) Aneurophytales underwent an abrupt decline in the Frasnian and appear to have been extinct before the onset of the Famennian (Scheckler 1986c). The *Aneurophyton olnense* plant could be a late remnant of aneurophyte progymnosperms, or it could be the pollen organ of an early gymnosperm. Should the latter prove correct, then the similar organization of *A. olnense* to the fertile parts of Aneurophytales would constitute additional evidence for an ancestor/descendant link between these progymnosperms and early gymnosperms (Rothwell 1982b; Rothwell and Erwin 1987).

More convincing gymnospermous pollen organs come from the same beds as the Elkins cupules and were briefly mentioned by Gillespie et al. (1981) as branch systems with synangia like *Aneurophyton olnense.* However, further study of the Hampshire synangia as well as of Stockmans' plant shows that they are only superficially similar. The Elkins synangia (figures 2.3:23, 24) consist of up to three pairs of sporangia that are fused by their bases to an enlarged pad of tissue that terminates thin, unequally forked branchlets (figure 2.3:24). The branchlets appear to be borne on a larger cruciately forked branch (figure 2.3:23). Sporangial apices are free (figure 2.3:24) and extend into a thin curved tip that points toward the center of the cluster. All the synangia appear to be at the same state of maturation. Further study will clarify many details of their arrangement, construction, dehiscence, and miospore content, but we can already say that they are similar to *Telangiopsis* (Eggert and Taylor 1971), from the late Mississippian (Chesterian = Visean to Namurian A) of Arkansas; to the reconstructions of *Burnitheca* and *Phacelotheca* (Meyer-Berthaud and Galtier 1986a, 1986b), from the late Tournaisian and late Visean of Scotland, respectively; and to other Carboniferous pollen organs that are usually attributed to lyginopterid pteridosperms (Millay and Taylor 1979; Taylor and Millay 1981).

Most recently, permineralized synangia of probable gymnospermous affinities were briefly described from the Coomhola Tournaisian Devonian deposits (Tn1a–Tn1b) at Kerry Head, Ireland (Fine and Matten 1987). The specimens consist of 4 to 8 sporangia that are fused together (or to an inflated pad of tissue) at their bases, and that terminate in awned tips. Dehiscence is via a longitudinal slit along the surface of the wall that faces the center of the synangium. The synangia are aggregated into clusters of 2 to 4 on terete branchlets that terminate a cruciately forked branching system (Matten, personal communication).

All other possible gymnospermous pollen organs known from Late Devonian strata are mere scraps of sporangial clusters, such as Arber and Goode's (1915) *"Telangium"* sp. (their plate IV, figures 8, 9) and Stockmans' (1948) cf. *"Calathiops"* sp. (his plate XI, figures 20, 21).

Vegetative Remains of Possible Gymnospermous Affinity As we stressed earlier, stems, leaves, or roots have not been found in attachment to any of the Devonian ovulate "fructifications." As a result, the attribution of such organs is based on their association with ovules and on a structural similarity to gymnospermous organs from more recent strata.

The Elkins Assemblage. The most ancient evidence of this type occurs with the Elkins cupules in the Hampshire Formation of West Virginia (Rothwell and Erwin 1987). Specimens consist of anatomically preserved stem segments and frond fragments, with some pinnules also preserved by compression. The slender stems are 2–3 mm wide, and have a mesarch three-ribbed protostele that is surrounded by a prominent zone of cortex (figure 2.3:25). A narrow zone of secondary xylem is present on some stems. There is a central protoxylem strand that divides repeatedly to produce protoxylem in the ribs of the stele and in the leaf traces (figure 2.3:25). *Lyginorachis*-type petiole traces of a C shape diverge from the ribs of the stele in a helical (apparent one-third) phyllotaxis, and are separated from each other by internodes of 2–3 cm. Segments from proximal levels of the frond fork at a broad angle that is reminiscent of the basal portions of the ovulate branching systems. Internally, they have a C-shaped vascular strand (figure

2.3:26) like those at the periphery of the stem cortex (Rothwell and Erwin 1987). Branching at more distal levels is unequal and alternate, and may form a three-dimensional frond. Broadly lobed pinnules of the *Sphenopteris* type (figure 2.3:27) are alternately arranged along the pinna rachis (Rothwell and Erwin 1987).

Because only one type of gymnospermous ovulate cupule, pollen organ, stem, and basal frond segment has been recovered from the Elkins assemblage, one might be tempted to hypothesize that they represent parts of a single type of plant. That proposal is supported by a common type of branching (viz., wide-angle, often unequal forking, usually cruciate) and similar anatomical and histological features among all of the organs. Because of their antiquity, the individual features of the hypothesized "plant" can be used to suggest the general sporophyte structure of ancestral gymnosperms. Such a "plant" was relatively small (i.e., stems only a few millimeters in diameter), and probably was propagated primarily or exclusively by sexual reproduction. We envisage that such "plants" formed relatively low patches or thickets of ground cover. Our evidence for its ecology and community structure will be presented in a later section.

Irish Fossils. A far more diverse assemblage of putatively gymnospermous vegetative remains is preserved among the *Hydrasperma tenuis* cupules in the late Devonian flora from Ballyheigue, Ireland. These include shoots described as *Laceya, Tristichia, Tetrastichia, Buteoxylon,* and *Kerryoxylon* (Matten, Tanner, and Lacey 1984). Like the Elkins stems, all these shoots have protosteles from which *Lyginorachis*-type petioles diverge. They also have a prominent cortex that may include sclerenchyma fibers near the periphery, and varying amounts of secondary xylem. Stems described as *Laceya hibernica* (May and Matten 1983) and *Tristichia* sp. have three-ribbed steles similar to the stems from Elkins (figure 2.3:25), while specimens of *Tetrastichia* have four ribs like *T. bupatides* from Lower Carboniferous strata. Other stems such as *Kerryoxylon hexalobatum* have more complex, multiribbed steles to which large segments of fronds remain attached (Matten, Fine, Tanner, and Lacey 1984; Matten, Tanner, and Lacey 1984). The mixed protostele of *Buteoxylon* is broader and less distinctly ribbed than the others.

In addition to the petioles of all these, specimens similar to *Lyginorachis papilio* are found (Matten, Lacey, May, Lucas 1980). Isolated woody roots described as *Rhizoxylon ambiguum* also occur in the assemblage (Matten, Fine, Tanner, and Lacey 1984). These specimens are similar to some species of the form-genus *Amyelon* from more recent strata (Barnard 1962).

These putatively gymnospermous remains from the uppermost Devonian of Ireland display considerably more diversity than the Famennian Devonian remains from Elkins, and the stems range to somewhat larger diameters (viz., over 1 cm). However, the plants appear to be of the same basic type. They all have protosteles that produce fronds with a single trace, as is characteristic of lyginopteridalean seed ferns. They also have secondary vascular tissues and varying numbers of fibrous bundles in the cortex, but little or no periderm is produced. Like the remains from Elkins, they show no apparent specializations for vegetative propagation, and were probably disseminated primarily by seeds.

If all of these uppermost Devonian vegetative remains from Ireland are from gymnosperms, then their occurrence with only one species of ovulate cupules (viz., *Hydrasperma tenuis*) is paradoxical. Nevertheless, the vegetative remains could all be gymnosperms. We reach this conclusion because similar, if not identical, shoots occur in late Tournaisian to Visean strata of Scotland and France (Scott, Galtier, and Clayton 1984), where ovulate cupule and seed diversity is far greater. The lack of additional ovules at Ballyheigue may be due to taphonomic processes.

Plants of the New Albany Shale. A rather different circumstance prevails with the plants from the New Albany Shale of Indiana, Kentucky, and adjacent states. No seeds or other reproductive structures have yet been found in these beds. However, several stems and petioles that are important to our perceptions of gymnospermous evolution do occur (i.e., *Calamopitys, Diichnia, Stenomyelon, Bostonia, Galtiera, Plicorachis*, and others). These fossils are interpreted as gymnosperms because of their similarity to other, better documented gymnosperms from Lower Carboniferous strata.

In general, the stems are larger than those from Elkins and Ballyheigue (i.e., several centimeters) and grade from protostelic to eustelic. They typically have a broad cortex with prominent fibrous bundles (i.e., "sparganum" or "dictyoxylon" cortex), and more secondary xylem than the Devonian fossils described earlier. Petioles also are larger and have several traces, as is characteristic of seed ferns assignable to the Calamopityaceae.

A troubling problem with the New Albany Shale is that its age(s) is (are) unknown. The "formation" is apparently a transgressive series of black phosphatic shales that were deposited under anoxic conditions within a "starved" basin (Ettensohn and Barron 1981; Ettensohn 1985). The "formation" is interpreted as late Devonian by American geologists. If so, the plants would be remarkable, since similar plants elsewhere have been dated as late Tournaisian to Visean. The age of the New Albany Shale cannot be precisely correlated with either the Appalachian or European strata because spores have not yet been recovered in sufficient quantities nor from close enough stratigraphic intervals (Fairon-Demaret 1986).

These black shales represent unknown levels within a condensed section that can at best be bracketed as possibly as old as middle Famennian to as young as late Tournaisian. The common presence of *Callixylon* suggests that some beds are no younger than Latest Strunian (lower Tn1b), while the occurrence of *Calamopitys* and *Stenomyelon* suggests that other beds are much younger, not older than late Tournaisian (Tn3), since these genera are this age or younger elsewhere in the world (Fairon-Demaret 1986; Scott, Galtier, and Clayton 1984).

Compressed Fern-like Foliage.   In addition to the vegetative remains discussed above, compressed fern-like foliage (e.g., figure 2.3:28) occurs at several horizons where ovulate cupules have been discovered (Hampshire, Evieux, Baggy, and Touran beds). For example, Stockmans (1948) recognized *Sphenopteris maillieuxi, S. modavensis, S. olnensis, S. flaccida, S. boozensis, S. mourloni, Diplothmema pseudokeilhaui,* and *Sphenocyclopteridium belgicum* in association with *Moresnetia* in the Evieux Formation of Belgium. And *Sphenopteridium rigidum* occurs in the latter two beds with *Xenotheca* and the Russian *"Moresnetia."* The appearance of these types of foliage and their increase in diversity and abundance parallel, stratigraphically, those of gymnosperm cupules and seeds. By late Tournaisian this type of foliage is generally assumed to be pteridospermous because of its gymnospermous anatomy and consistent association with dispersed seeds, but this assumption is much less certain for the middle to late Famennian foliage.

An additional complication in interpreting this foliage as gymnospermous and reconstructing whole plants is that some horizons contain many taxa of *Sphenopteris*-like foliage or gymnospermous axes, but have only one kind of ovulate cupule (Hampshire Formation, Evieux Formation, Coomhola Formation), while at other horizons the same foliage (e.g., *Sphenopteridium rigidum*) occurs with two different types of ovulate cupules (Baggy beds, Touran series). Possible cupule/

foliage correspondence becomes even more uncertain when individual localities are compared. For example, *Moresnetia* in the Evieux Formation (Stockmans 1948) occurs with only *Sphenopteris flaccida* at two localities (Assesse, Villers-le-Temple), but occurs with other foliage types at two others (Hamois, Moresnet) and without any foliage at three more (Hun-Annevoie, Evrehailles, Strud-Haltinne). Similar foliage, however, occurs without *Moresnetia* at eight localities (Modave, Esneux, Aywaille, Neche-Bolland, Olne, Booze-Trembleur, Val-Dieu-Charneux, Dison). In fact, the greatest diversity of foliage in all the Evieux localities occurs at these last four localities.

Although some of these foliage types were definitely produced by gymnosperms at higher stratigraphic levels, other types might not have been. Therefore, these Devonian occurrences must be carefully scrutinized for supporting evidence before gymnospermous affinities can be interpreted with confidence. For example, the *Sphenopteris*-type fronds from the Hampshire Formation at Elkins (figure 2.3:27) that show *Lyginorachis*-type anatomy (Rothwell and Erwin 1987) suggest that they are gymnospermous. Another *Sphenopteris* from this locality (figure 2.3:28) has ovules, cupules, and cupuliferous axes with similar surface patterns, and may also be gymnospermous (Rothwell, Scheckler, and Gillespie, in preparation).

REPRODUCTIVE BIOLOGY
OF ANCESTRAL GYMNOSPERMS

Because seeds may be used to interpret the reproductive biology of the plants that produced them and are relatively easy fossils to recognize, they have been the principal focus of many previous studies that discuss the origin of seed plants (e.g., Andrews 1963; Camp and Hubbard 1963; Smith 1964; Niklas 1981a). This approach has provided a focus for interpreting the evolution of the integument (Andrews 1963; Niklas 1981a) and of other changes in the heterosporous life cycle that culminated in the evolution of the seed (Stewart 1983).

As has been realized for many years, the earliest gymnosperms had indehiscent megasporangia (Stewart 1983) with an apex that is specialized for pollen reception (Andrews 1963; Camp and Hubbard 1963; Smith 1964; Taylor and Millay 1979). The megasporangia also were more or less surrounded by integumentary lobes, and contained only one functional megaspore. In order to function effectively in sexual reproduction, such megasporangia had to be (1) pollinated to bring together the micro- and megagametophytes, (2) sealed after pollination to allow for gametophyte development, fertilization, and embryogeny, and (3) abscised to liberate the propagule (Rothwell 1986). The structure/function relationships for each of these facets are discussed below, using evidence from late Devonian strata.

Hydrasperman Structure and Reproduction   The suite of plants discussed here demonstrates an archaic form of gymnospermous reproduction termed "hydrasperman biology" (Rothwell 1986, 1987a, 1987b). Hydrasperman biology differs from the reproduction of later gymnosperms in that the ovules lack a true integumentary micropyle. Consequently, the nucellus itself has an elaborate mechanism for pollen capture and post-pollination sealing of the megasporangium that facilitated gametophyte development, fertilization, and embryogeny. Although this mechanism was first deduced from permineralized Carboniferous ovules, late Devonian gymnosperms also illustrate the concept and allow us to extend it to include the mode of pollination biology as well.

The key to recognizing hydrasperman reproduction is to understand the ontogeny and functional morphology of the pollen chamber and accessory structures (e.g., integumentary lobes, cupule, cupulate branching system). As described earlier, the hemispherical pollen chamber is surmounted by a hollow lagenostome, and has a membranous floor with a prominent central column (figure 2.2). Aided by integumentary lobes, cupule lobes, and the entire morphology of the cupulate branching system, wind-borne prepollen was directed into the lagenostome and the pollen chamber proper (figure 2.2b). Following pollination, growth of the megagametophyte pushed the central column into the base of the lagenostome, tightly sealing the pollen chamber (figure 2.2c). At the same time, the membranous part of the floor ruptured, to bring the megagametophyte into contact with the microgametophytes. Thus, the functions of pollination and post-pollination sealing of the ovule, which are usually performed by the integument of modern ovules, were conducted by the nucellus in ovules of the most primitive gymnosperms. At about the time of fertilization, the seeds were shed. This occurred by abscission of the seed and/or cupule in some taxa, and may have been accomplished by disaggregation of the senescent cupuliferous branching systems in others.

Small, presumably young ovules of the Elkins plant have a rounded nucellar dome (figure 2.5a) from which protrudes a small and tubular lagenostome, which may have a slight apical flare. Larger, presumably older ovules have an enlarged nucellar apex that appears quite conical. Lagenostomes of *Moresnetia* are larger and more conical than those of the Elkins plant. By dissecting ovules of differing sizes and integument morphologies, Fairon-Demaret was able to show that the form of the nucellar apex varied with apparent ovule maturity (Fairon-Demaret and Scheckler, in press). Smaller, presumably younger ovules have smaller, rounded nucellar apices from which extend lagenostomes with a slight apical flare (figure 2.5e). Larger, presumably older ovules have larger nucellar apices that appear conical (figure 2.5f–g). Whether the expansion and change of shape of the nucellar apex involved only the pollen chamber, from which the lagenostome was lost, or included the lagenostome too, is unknown. Regardless of the histological details, by comparison with permineralized *Hydrasperma* and Carboniferous ovules, these changes of the lagenostome and nucellar apex in the Elkins plant and *Moresnetia* appear to reflect the post-pollination sealing mechanism (figure 2.2).

Irish *Hydrasperma* shows similar changes in the form of the nucellar apex (compare figures 2, 7 of Matten, Lacey, and Lucas 1980 with figures 6, 7 of Matten, Fine, Tanner, and Lacey 1984). In this species, the expansion of the nucellar apex may be due to the growth and cellularization of the megagametophyte underneath. These stages of lagenostome form and nucellar apex shape correlate with ovule size and integument form in the Elkins plant and *Moresnetia* and support our supposition, discussed later, that some ovule morphologies represent pre-pollination states, while others are post-pollination.

Megagametophyte. Matten, Fine, Tanner, and Lacey (1984) have described remarkably preserved megagametophytes in their Irish *Hydrasperma*. From the remnant cell pattern in these mature gametophytes and the apparent free nuclear condition of presumed younger gametophytes described earlier (Matten, Lacey, and Lucas 1980), one can infer that cellularization resulted from the subdivision of centripetally developing alveoli, just as in all other gymnosperms except *Gnetum* and *Welwitschia*. Growth was less, or was restricted, at the chalaza, so that mature gametophytes were obconically elongate and tapered from a point at the base to a rounded apex that projected into the pollen chamber. Cellularization was complete by the time several (up to three) jacketed archegonia were produced at the apex. Apparent archegonial canals opened into a shallow, ring-like depression that surrounded a low "tent pole" at the gametophyte apex (figure 2.2a; Matten, Fine, Tanner, and Lacey 1984). This recalls the archegonial chambers of *Ginkgo* and modern cycads (Foster and Gifford 1974) that develop just before fertilization by swimming sperm.

The *Hydrasperma* ovules in which these megagametophytes occurred are attached inside cupules and are the same sizes and shapes as most others without gametophytes, which are presumably less mature. If pollination occurred near the time of megasporocyte meiosis, as it does in most living gymnosperms, then the ovules and detached seeds that comprise our principal evidence for early gymnosperms might be mostly preserved at stages far beyond this time. Whether integument form changed markedly between the time of pollination and the time of dispersal (discussed later), as seems likely for the Elkins plant and *Moresnetia*, is unknown for other taxa. Nor is there any assurance that pollination occurred near the time of megasporo-cyte meiosis in these hydrasperman gymnosperms, since they lack modern analogs.

That fully mature gametophytes were found in attached ovules not otherwise different from many others, however, suggests that game-tophyte growth was rapid and commenced soon after the expansion of the functional megaspore. From the optical thickness of the megas-pore wall of early gymnosperms (e.g., *Archaeosperma*, Pettitt and Beck 1968) compared to younger seed plants (Singh 1978), we conclude that transfer of nutrients may have been impeded and gametophyte growth thus restricted. Further growth, such as that of an embryo, might have been limited by the food reserves already available, since little more is likely to have passed through this thick megaspore wall. Therefore, we suspect that seeds were dispersed before significant embryo growth had occurred, because little could be gained by delay-ing abscission.

Pollination Biology and Ovule Ontogeny   Recent studies of func-tional morphology in pollination have focused attention on possible selective pressures for the evolution of the integument (Niklas 1981a). That the form of the integument of early ovules affected pollen cap-ture had been proposed earlier (Taylor and Millay 1979; Taylor 1981, 1982). However, not until experiments with the impact of pseudopol-len on models of Early Carboniferous ovules and cupules was it sug-gested that a complete integumentary jacket (with a micropyle) pro-vided a selective advantage related to increasing efficiency of pollen capture (Niklas 1981a, 1981b, 1983a). Prior to these experiments, fusion of integumentary lobes was attributed to selection for increas-ing protection of the nucellus (e.g., Andrews 1963).

We embrace the hypothesis that the integument of early gymnos-perms may have evolved in response to its effect on pollination biol-ogy (Niklas 1981a). We likewise agree, for reasons stated below, that the integument was also probably selected for its role in protection (e.g., Andrews 1963). In order to understand the subtleties of integument form and function, however, one must have fuller knowledge of the pre- and post-pollination form of integuments (Rothwell and Taylor 1982), the three-dimensional form of cupules, the disposition of ovules within cupules, and the arrangements of cupules within the cupuliferous branch system.

Of perhaps equal importance, one must consider contributions of these and other organs to the overall reproductive biology of the plant (Rothwell and Taylor 1982). The dramatic increase of evidence for late Devonian and early Carboniferous gymnosperms that has accumulated in the last few years now allows us to interpret reproductive biology more satisfactorily for several late Devonian gymnosperms than for the early Carboniferous plants used by Andrews (1963), Niklas (1981a, 1981b, 1983b), and others (e.g., Taylor and Millay 1979).

Integument. Because of the large number of ovulate cupules prepared for the Elkins plant and for *Moresnetia*, we have some information about the possible changes in ovule form during pre- to post-pollination ontogeny. In the Elkins plant most cupules contain ovules of similar size and integument shape (figures 2.3:3–5), but these can vary among different cupules. Some have smaller ovules in which the integument lobes are highly curved and arch over the nucellar apex (figures 2.3:3; 2.4a) so that the lagenostome is fully exposed. Slightly larger ovules have integument lobes that are less curved (figure 2.5b) so that the lagenostome is partly covered. Often in those same cupules are other ovules in which the free, apical portions of the integument lobes are tightly pressed together (figures 2.3:4; 2.5c) and completely enclosed the lagenostome. Dispersed, isolated seeds are the largest, often about 6–7 mm long, and have the apical portion of the integument tightly pressed together also, but the whole ovule apex is wider (figure 2.7d). We interpret these integument shapes as indicative of the pre-pollination or pollination configurations (figure 2.5a), early post-pollination configurations (figure 2.5b, c), and late post-pollination to fertilization configurations (figure 2.5d). The uniformity of ovule form in any particular cupule suggests that ovule maturation was simultaneous (for that cupule) but may have been acropetal for the whole cupuliferous truss. Dispersed seeds occur by the thousands in layers interbedded with hypoautochthonous mats of cupuliferous axes at Elkins. This suggests that abscission and dispersal of seeds was also a seasonally dictated and simultaneous event in this plant.

Moresnetia shows similar changes in ovule form that we also interpret as suggesting pre- to post-pollination ontogeny. Where cupules have several ovules (figure 2.3:9), these always occur in an oblique vertical series parallel to the axis of the asymmetric cupule (figure 2.4b). The distal ovule of a series is the smallest and has diverged integument lobes (figures 2.4b, 2.5e), so that the lagenostome is maximally exposed. The proximal ovule is the largest, and its integument lobes converge toward the nucellar apex (figures 2.4b, 2.5g), so that the lagenostome is maximally covered. The middle ovule is intermediate (figures 2.4b; 2.5f). We interpret figure 2.5e as the pre-pollination to pollination states, and figure 2.5f, g as showing post-pollination to fertilization configurations. The evident gradient of apparent maturity within a single cupule (seen many times by Fairon-Demaret and Scheckler) suggests that ovules matured and were pollinated sequentially. We presume that they were also dispersed sequentially rather than simultaneously, although there is no direct evidence for this. We do note, however, that mats of dispersed *Moresnetia* seeds have not yet been found in the Evieux, even from some hypoautochthonous cupule beds at Bocq Valley.

From the evidence presented above, and by comparison with the ovules of living gymnosperms (Rothwell 1971), it seems likely that integument form does change with apparent ovule maturation in early gymnosperms. For the Elkins plant and *Moresnetia*, at least, the likely integument shape at pollination seems to be quite different from that observed in more mature ovules or dispersed seeds. This suggests that previous experiments on the pollination biology of early gymnosperms, in which models were constructed from descriptions of dispersed seeds, should be viewed with caution (Rothwell and Taylor 1982; Niklas 1981a, 1981b, 1983a, 1983b, 1985).

Cupule. Niklas (1985) discussed the apparent analogy between cupule form and ovule integument with regard to their effects on pollen flow. He was unable to place these into a larger context of the whole plant because such information was largely unknown. We are now coming much closer to this goal.

Moresnetia (Fairon-Demaret and Scheckler, in press) and the Elkins plant are preserved mainly by compression, but include many specimens with partial permineralization of cupules, ovules, and cupuliferous branches. Study of hundreds of specimens of each taxon shows that the cylindrical, swollen, and then tapered cupule segments enclose a cavity into which the ovules protrude (figure 2.4). Where ovules are attached at the same level in symmetrical cupules, such as those of the Elkins plant, the swollen basal parts of cupules surround the chalazal ends of the ovules, while the thinner and tapered cupule tips diverge to define a space into which the free, curved integument tips of the ovules project (figure 2.4a). Irish *Hydrasperma* appears to be similar (figures 2.3:19, 20; Matten, Lacey, and Lucas 1980), as does also the younger *H. longii* (figures 2.3: 29, 30).

The strongly overtopped cupule segments of *Moresnetia* are also cylindrical but are not so swollen at their bases (figure 2.4b). Ovules are borne in an oblique vertical series, rather than at one level, and project outward toward the elliptical opening of the asymmetric cupule (figure 2.4b). Thus the shapes of the cupule parts operate with their branching pattern to produce a precisely defined space in which pollination occurred (e.g., figures 2.3:29, 30; 2.4).

Cupuliferous Branch Systems. Although known for only a few taxa, these cupule shapes and ovule dispositions seem to correlate with the overall arrangements of cupules on the terminal branchlets of larger, three-dimensional branch systems (figures 2.3:12; 2.3:7–9; 2.3:13, 14; 2.4; 2.6; 2.7). We thus envision a pollen receptive surface of great complexity, with at least three and possibly four levels of pollen interception. The first tier would be the overall arrangement of cupules on the terminal branchlets of the cupuliferous system. This was a shallow, bowl-like surface for the Elkins plant (figure 2.3:1), funnel-like for *Moresnetia* (figures 2.3:7, 8; 2.6), and a small dome for *Xenotheca* (figures 2.3:13, 14). The shapes of individual cupules defines a second tier (figure 2.4), while the morphology of individual ovule apices defines a third (figure 2.5). Where larger portions of the cupuliferous branch system are known, as in *Moresnetia* (figure 2.3:7, 8), the whole hemispherical branching system with its funnel-like "indentations" (differently shaped in the other taxa; figure 2.7) may have formed yet another complex surface for a fourth tier (figures 2.7; 2.8). How these large hemispherical systems were borne on the whole plant might have defined yet further levels of pollen interaction. Similar consideration might also prove insightful for the study of the cupuliferous systems of early Carboniferous gymnosperms. Those of *Stamnostoma huttonense*, for example, seem similar to portions of *Moresnetia* (Andrews, 1963). Cupuliferous branches of *Lagenospermum imparirameum* (Gensel and Skog 1977), on the other hand, are quite different in that they bear pinnate laterals, the divisions of which terminate in cupule clusters.

Protection and Integument Evolution With the growing popularity of the hypothesis that the integument evolved as a result of selection for increased efficiency in wind pollination, the older hypothesis that the integument provides protection for the propagule has received less attention. The new information discussed earlier about the ontogeny of late Devonian ovules and the ovule/cupule interactions of permineralized early Carboniferous species now indicates that both played important roles in the origin of the seed.

The structural diversity that we have interpreted to be ontogenetic variation in both *Moresnetia* and the Elkins plant shows important parallels. Although the structural details are different in the two genera, the less mature ovules of both have integumentary lobes that we infer are spread apart to expose the lagenostome for pollination and that more tightly enclose the megasporangium in more mature ovules or seeds (figure 2.5). If the integumentary lobes had only a pollination function, this ontogenetic change would not be expected.

Likewise, protection appears to be the function of the integumentary lobes of the late Tournaisian *Hydrasperma longii*. Recently discovered specimens of this species demonstrate that the ovules are tightly packed together within the lobes of the cupule (figures 2.3:29, 30). In specimens of this type, potentially pollinating wind currents could have operated only in the apical portion of the cupule—i.e., between the lobes of the cupule and among the integumentary lobes of the ovules (figure 2.3:29). On the other hand, the stubby lobes of the integuments appear to form a dense armor of bristles that together could have effectively held herbivores away from the tissues of the nucellus and megagametophyte (figures 2.3:29, 30). Sclerenchyma fibers that characterize the outer cortex of the integumentary lobes of all these most ancient ovules (e.g., figure 2.3:21) would have provided the mechanical support to make them effective for this function. It seems clear from these fossils that both protection and pollination were important selective forces in the evolution of the integument.

Abscission/Dispersal   In the transition from pteridophytic reproduction to gymnospermous reproduction, there was a change in the nature of the propagule from a naked megaspore to a megasporangium (and associated structures) with its enclosed megaspore, megagametophyte, and (presumably) embryo. In pteridophytes, dehiscence of the megasporangium liberates the propagule, but another mechanism is necessary to separate a seed from the parental sporophyte. In gymnosperms this mechanism is abscission. Therefore, abscission is just as essential to gymnospermous reproduction as is pollination. Despite this, we have remarkably little evidence for seed abscission among Devonian gymnosperms, except for the numerous dispersed Elkins seeds and *Spermolithus* (figure 2.3:15, 16), if the latter is a seed.

Many cupules of Devonian gymnosperms are exposed by fractures that pass over the outside of the cupule segments, so that no ovules are seen (e.g., figure 2.3:2). Nevertheless, very few cupules are empty. Some have a reduced number of ovules by the failure of some cupule portions to produce them (e.g., *Moresnetia*), but fractures that pass through the interior of Devonian cupules usually reveal the presence of attached ovules (e.g., figure 2.3:3). Only Elkins specimens and some *Archaeosperma* cupules (e.g., Pettitt and Beck 1968, plates 1–2; our figure 2.30:11, at arrows) show ovule pedicels that end abruptly, suggesting that they had abscised seeds. This interpretation is also consistent with the fact that dispersed seeds were common among Devonian gymnospermous remains only in the Elkins assemblage.

By contrast, cupules and fragments of cupuliferous branching systems are relatively common fossils in the Elkins assemblage. Most of the cupules still contain ovules, some of which appear as mature as those also found dispersed. Also, Devonian cupuliferous branching systems have not been found attached to more proximal structures (presumably fronds). Together, these data suggest that fertile systems of the earliest gymnosperms may have withered at senescence. If so, then gymnospermous propagule dissemination may have been at first facilitated by fragmentation of the fertile systems, with ovule abscission evolving later. In this regard, the isolated seeds of the Elkins assemblage may represent a derived character that facilitated more effective dissemination of propagules than in *Moresnetia* and *Xenotheca*.

ECOLOGY OF EARLY GYMNOSPERMS

In order to understand more fully the biology of early gymnosperms, we must consider them within the broader context of their whole ecology. By this we mean the habitat partitioning of the ancient landscapes and the resultant interactions among the neighboring plant communities. Great strides have been made in paleoecological reconstructions of Paleozoic terrestrial communities in the last 10 years (e.g., Phillips et al. 1974; Scott 1977), and some of these results bear directly on our studies of early gymnosperms. We summarize below the general methodology of paleoecological studies and the specific results that apply to this discussion.

Most fossil assemblages are dramatically different from the plant communities they represent. Species composition and spatial distribution are severely distorted by pre- and post-burial decay, by differential survival and winnowing during transport, and by diagenesis of sediment to rock. These taphonomic processes are well studied in modern clastic and volcanoclastic sediments of fluvial, lacustrine, and marginal marine depositional systems (Spicer 1980; Spicer and Greer 1986). Identification of similar ancient depositional systems requires intense scrutiny of many individual horizontal and vertical rock exposures within a regional context. When these efforts include analyses of the beds for floral and faunal content, for indirect traces of biogenic activity, and for the probable taphonomic histories of the fossils, then we can often infer the original composition of plant communities, their relative spatial distribution over a landscape, and the partitioning of habitats.

Limitations of such paleoecological reconstructions, however, need to be stressed. Only the sequence and kinds of geological structures, the floral and faunal content, and the degree of completeness of those fossils are facts. From these we can reconstruct the depositional histories of the individual sediment units, of the local assemblages of units, and, finally, of the region itself. We can then interpret the taphonomic histories of the various taxa of plants and animals and from these infer the original structure of ancient communities. Correlation of particular communities with individual sediment units may then yield information on habitat partitioning. The success of these reconstructions can be judged in part by our ability to use them to predict the occurrences of fossils in similar rocks. Autochthonous rocks, the contents of which have been transported little, or not at all, offer the highest probability of successful reconstruction. Allochthonous rocks, the contents of which have been transported, have a lower probability of yielding successful reconstructions of source vegetation, the probability varying largely with the distances and conditions of transport. The study of autochthonous plant assemblages was pioneered by Phillips (e.g., Phillips et al. 1974; Phillips and DiMichele 1981) and that of allochthonous assemblages by Scott (e.g., 1977, 1979). Their techniques have been subsequently adopted and modified by others (e.g., Spicer and Greer 1986; Spicer 1987).

Late Devonian Gymnosperms   The richly fossiliferous deltaic complex of Upper Hampshire Formation sediments exposed near Elkins, West Virginia (Gillespie et al. 1981) was studied by Scheckler (1986a) and compared with stratigraphically adjacent marine and upland sections at other localities. Scheckler concluded that apparent autochthonous to hypoautochthonous concentrations of fossils occurred at Elkins only on the distributary levee tops of the leading edges of small delta lobes that were prograding into a shallow, barrier-protected bay. Here the gymnosperm cupules, cupuliferous axes, pieces of *Sphenopteris,* and dispersed seeds occur as thick, nearly pure mats. Immediately above these beds the gymnosperm was ecologically succeeded and replaced by autochthonous *Rhacophyton* thickets, as the interdistributary areas were filled by marshes. Eventual abandonment of the delta lobe and subsequent subsidence led to its drowning and burial by marine storm beds.

The allochthonous marine storm beds that buried the abandoned delta lobes contain a far more diverse flora, of which the gymnosperm cupules are just a small part. They and the other plants (*Archaeopteris* and *Rhacophyton* were dominant, followed by axes of a *Protolepidodendropsis*-like lycopod) occur as disarticulated fragments. This cycle was repeated at least once as the marine bay-fill shoaled, distributaries switched, and new delta lobes were built and were also eventually abandoned. Upland sites had only *Rhacophyton* and *Archaeopteris* except at one backswamp oxbow (Rawley Springs, Virginia), where the lycopod occurred nearby, so that its remains were common in a crevasse splay that intruded into the *Rhacophyton* swamp. Analysis of this swamp-fill uncovered only a single seed, of the same kind seen at Elkins, in the crevasse splay.

From these analyses we conclude that the Elkins gymnosperm was a pioneer colonizer of newly emerged, primary successional habitats near shorelines. These habitats were characterized by immature soils with poor organic content, high illuminance, and little or no interspecific competition. Although present in a least one upland site, this gymnosperm was far more abundant in shoreline habitats.

Fairon-Demaret and Scheckler (in press) have reached similar tentative conclusions for *Moresnetia.* This gymnosperm occurs exclusively in allochthonous, back barrier and lagoon-fill deposits. The relative completeness of *Moresnetia* in these deposits, compared to the other taxa, suggests its proximity to these sites—i.e., along shore edges. Upland beds are very rare in the Evieux formation (Lejeune 1986) and have not yet yielded *Moresnetia.*

All known *Xenotheca* comes from a thin lens of Hoe facies (Baggy Beds) enclosed within beds of Timber facies in one quarry at Croyde Hoe Farm, near Barnstaple (Rogers 1926). The Baggy Beds are considered to be freshwater fluvial deposits laid down under conditions of low flow intensity. The Timber facies are interpreted as delta-like, channel-fill sediments of distributaries draining into a transgressive, shallow water sea (Goldring 1971). The *Xenotheca* remains are all fragmentary and show strong evidence of size sorting. Aside from proximity to a shoreline, no other ecological inferences can be made.

At Kiltorcan, the late Devonian beds from which *Spermolithus* comes are interpreted as lacustrine-fills of a lake on an alluvial plain, cut by braided streams, on which forests of *Archaeopteris* and *Cyclostigma* grew nearby (Colthurst 1978; Holland 1981). No finer ecological statements can be drawn here. The equivalent-aged Tunheim Series coal measures (Fairon-Demaret 1986) of Bear Island, in the Norwegian Arctic, have a similar, but more diverse, flora (Scheckler 1986c). Gymnosperm cupules or seeds are unknown, but *Sphenopteridium keilhaui* might be gymnospermous foliage, although Scheckler found that other pieces, also called *S. keilhaui* by Nathorst (1902), were attached to his *Cephalopteris mirabilis*, which is very similar, if not identical, to *Rhacophyton*. Consistent facies relations suggests that *S. keilhaui*, regardless of its affinity, grew in drier habitats with *Archaeopteris fimbriata*.

Early Carboniferous Gymnosperms   Results similar to those just discussed are not yet known for the very early Carboniferous. But by late Tournaisian (Tn3), pteridospermous gymnosperms had become dominant on upland levees and the drier parts of floodplains adjacent to *Lepidodendropsis*-filled backswamps of the prograding deltaic facies of the Upper Price Formation of southwestern Virginia (Scheckler 1986b, 1986c). Similar habitats of this age were apparently much more ephemeral on the other (southestern) side of the Acadian Mountain Belt that divided the Old Red Continent (Laurussia; Bambach, Scotese, and Ziegler 1980). Coeval sediments of Scotland at Oxroad Bay reveal a volcanic terrain in which Late Tournaisian gymnosperms are interpreted as colonizers of volcanogenic ash deposits (Rothwell and Scott 1985).

Study of slightly younger Visean beds near Pettycur, Scotland (Rex and Scott 1987) suggests that a diverse gymnosperm flora grew along stream margin levees and drier floodplains in a volcanogenically disturbed upland. Frequent fires swept the area, especially where pteridosperms grew, keeping it at early states of succession. Preservation of the gymnosperms as fossils occurred mainly when their burnt or shed parts were swept into nearby lakes and ponds, where they were incorporated into the lacustrine sediments as compressions, permineralizations, or fusinized fragments.

Ecological Conclusions   Although the data are still few, a highly consistent image of early gymnosperm ecology and habitat preference emerges. In the Late Devonian, when upland floodplains were dominated by *Archaeopteris* and *Rhacophyton*, the earliest gymnosperms occupied the temporary environments, characterized by primary succession, on newly emerged portions of coastlines. Only after the demise of *Archaeopteris* and *Rhacophyton* by early Carboniferous (upper Tn1b) did pteridosperms become common on upland floodplains (Scheckler 1986c). By Tn3 time, levees and dry floodplains were dominated by various pteridospermous gymnosperms, while wet floodplains were dominated by just two species of tree lycpods *(Lepidodendropsis).*

Common to the vegetation in all of these environments was ephemeral exploitation of newly emerged or frequently disturbed local or patchy habitats. The adaptive breakthroughs of early gymnosperms that allowed them to thrive under those conditions can be inferred from their vegetative organization and structure and from their reproductive biology. These are as follows:

1. Rapid growth (small stems with little or no secondary growth).
2. Early onset of reproduction (few growth layers on most stems).
3. Xeric adapted shoots with large photosynthetic surfaces.
4. Extensive and highly branched root systems.
5. Cupule and integument morphology designed for abiotic capture of wind-borne pollen by a modified megasporangial apex (lagenostome).
6. Independence from soil moisture for direct gametophyte growth.
7. Production of large numbers of small, readily dispersed, diaspore units (seeds).

SUMMARY AND SIGNIFICANCE OF THE EARLIEST GYMNOSPERMS

1. The sequence of ovule and cupule morphologies now known to have occurred in the late Devonian confirms the power of comparative morphology as a tool for deducing ancestral states of organization.

2. The unity of organization of all known early gymnosperms implies monophylesis of gymnospermy (Rothwell 1980, 1982b). Hydrasperman reproduction (Rothwell 1986) characterizes these plants. Megagametophyte cellularization of the most ancient gymnosperms is also by partitioning of centripetally growing alveoli, just as it is in all other gymnosperms except for *Gnetum* and *Welwitschia*. This pattern of cellularization is unique to gymnosperms and differs from the acropetal (if the tetrad suture is held to be proximal) cellularization that occurs in the megagametophytes of all other heterosporous plants.

3. The diversity of middle to late Famennian (Fa2c–Fa2d) gymnosperms implies an earlier time for the origins of this group (Scheckler

1985). Based on the increase of gymnosperm diversity seen in the Tournaisian and on our interpretation that the adaptive breakthrough for gymnospermy was the set of biological parameters that allowed them to exploit primary successional habitats near streams and shores that were free of competition, we project backward to a suspected origin for gymnospermy in the Frasnian, consistent with the first appearance of *Cystosporites devonicus*-type megaspores in nearshore marine sediments.

4. Reproductive biology of early gymnosperms varied slightly among the taxa but was fundamentally driven by the requirements for successful wind pollination, dissemination of numerous diaspore propagules, and rapid growth needed to survive in ephemeral habitats. Lagenostomes, unfused apical integument lobes, cupules, and cupuliferous branch systems were all coordinated for interception of pollen. The uniformity of cupule sizes (abut 10 mm long and 7–10 mm wide) and ovule sizes (about 3–6 mm long and 1–2 mm wide) suggests that these early gymnosperms were all adapted to similar habitats along deltaic/shore margins. Only when gymnosperms became dominant on upland levees and dry floodplains, by the late Tournaisian, is there an increase in diversity in cupule size (e.g., *Calathospermum scoticum*) and number of ovules per cupule (e.g., *Hydrosperma longii*, figures 2.3:29, 30; *C. scoticum* and *Lagenospermum imparirameum*, Gensel and Skog 1977; Taylor and Millay 1979). We interpret these changes to be responses to the diversity of habitats available to gymnosperms as they radiated in the uplands.

5. The adaptive radiation of gymnosperms in the late Devonian and early Carboniferous may also correlate with opportunities provided by the demise of other groups. The Frasnian decline and apparent extinction of aneurophytalean progymnosperms (Scheckler 1986c), previously common in deltaic sediments, may have left vacant the habitats critical to the success of early gymnosperms. The often noticed anatomical similarities between Aneurophytales and early gymnosperms might have resulted from similar ecological selection. Similarly, the extinction of the *Archaeopteris/ Rhacophyton* community at the end of the Devonian Tournaisian (lower Tn1b) may have left vacant the upland habitats critical to the radiations of gymnosperms by the late Tournaisian. We suspect that early gymnosperms succeeded less by their ability to compete against other plant groups (they were replaced during succession by mid-Famennian *Rhacophyton;* Scheckler 1986a) than by their ability to thrive in habitats underutilized by other plant groups.

ACKNOWLEDGMENTS

Several individuals graciously allowed us to study fossils entrusted to their care. We thank Dr. P. Sartenaer and Dr. F. Martin, Musée Royal d'Histoire Naturelle de Belgique; Dr. M. Fairon-Demaret and Professor M. Streel, Department of Paleobotany and Paleopalynology, Université de Liège; Dr. D. Edwards, Department of Plant Science, University College, Cardiff; Dr. C. Hill and C. Shute, Department of Palaeontology, British Museum (Natural History); Dr. C. Beck, Museum of Paleontology, University of Michigan; and Dr. A. Scott, Royal Holloway and Bedford New College, University of London. Dr. L. Matten, Department of Botany, Southern Illinois University, generously supplied sections of the Irish *Hydrasperma tenuis* and *Laceya*. Drs. Gene Mapes and D. C. Wight, Ohio University, made helpful suggestions in revision of the manuscript, and Dr. Wight provided preparations of *Hydrasperma longii*. This work was supported in part by research grants from the National Science Foundation (BSR 83-15254 to S. E. Scheckler and BSR 86-00660 to G. W. Rothwell and D. Wight) and the National Geographic Society (NGS 2409-81 to S. E. Scheckler). Additional support was provided by the Ohio University Baker Fund (to G. W. Rothwell), and by the College of Arts and Sciences and Core Research Program, Virginia Polytechnic Institute and State University, and the Virginia Center for Coal and Energy Research (to S. E. Scheckler).

LITERATURE CITED

Andrews, H. N. 1963. Early seed plants. *Science* 142 (3594):925–931.

Arber, E. A. N. and R. H. Goode. 1915. On some fossil plants from the Devonian rocks of North Devon. *Proceedings of the Cambridge Philosophical Society* 18(3):89–104, pl. 4–5.

Arnold, C. A. 1935. On seedlike structures associated with *Archaeopteris*, from the Upper Devonian of Northern Pennsylvania. *Contr. Mus. Paleont. Univ. Michigan* 4:283–286.

Bambach, R. K., C. R. Scotese, and A. M. Ziegler. 1980. Before Pangea: The geographies of the Paleozoic world. *American Scientist* 68:26–38.

Barnard, P. D. W. 1962. Revision of the genus *Amyelon*. *Palaeontology* 5:213–224

Bassett, M. G. and D. Edwards. 1982. Fossil plants from Wales. *National Museum of Wales, Cardiff. Geological Series* 2:1–42.

Camp, W. H. and M. M. Hubbard. 1963. On the origins of the ovule and cupule in lyginopterid pteridosperms. *American Journal of Botany* 50:235–243.

Chaloner, W. G., A. J. Hill and W. S. Lacey. 1977. First Devonian platyspermic seed and its implications in gymnosperm evolution. *Nature* 265(5591:233–235.

Colthurst, J. R. J. 1978. Old Red Sandstone rocks surrounding the Slievenamon Inlier, Counties Tipperary and Kilkenny. *Journal of Earth Sciences of the Royal Dublin Society* 1:77–103.

Eggert, D. A. and T. N. Taylor. 1971. *Telangiopsis* gen. nov., an Upper Mississippian pollen organ from Arkansas. *Botanical Gazette* 132:30–37.

Ettensohn, F. R. 1985. The Catskill Delta complex and the Acadian Orogeny: A model. *Geological Society of America, Special Paper* 201:39–49.

Ettensohn, F. R., and L. S. Barron. 1981. Depositional model for the Devonian–Mississippian black shales of North America: A paleoclimatic–paleogeographic approach. In T. G. Roberts, ed., *Geological Society of America, Cincinnati '81 Field Trip Guidebooks*. 2. *Economic Geology, Structure*, pp. 344–361. *American Geological Institute*.

Fairon-Demaret, M. 1986. Some Uppermost Devonian megafloras: A stratigraphical review. *Annales de la Societé géologique de Belgigue* 109:43–48.

Fairon-Demaret, M. and S. E. Scheckler. 1986. A propos de Moresnetia. *L'evolution des gymnosperms, approche biologique et paléobiologique*. Colloque organise par la Fondation Louis Emberger–Charles Sauvage. A l'Université des Sciences et Techniques du Languedoc. Montpellier, France. Resumés des communications, p. 19.

—— (In press). Typification and redescription of *Moresnetia zalesskyi* Stockmans 1948, an early seed plant from the late Devonian (Famennian) of Belgium. *Bulletin de l'Institut Royal des Sciences Naturelle de Belgigue*.

Fine, T. I. and L. C. Matten. 1987. Sporangial remains from the uppermost Devonian of Ireland. *American Journal of Botany* 74(5):683. (Abstract.)

Foster, A. S. and E. M. Gifford. 1974. *Comparative Morphology of Vascular Plants*, 2nd ed. San Francisco: W. H. Freeman.

Gensel, P. G. and H. N. Andrews. 1984. *Plant Life in the Devonian*. New York: Praeger.

Gensel, P. G. and J. E. Skog. 1977. Two Early Mississippian seeds from the Price Formation of southwestern Virginia. *Brittonia* 29(3):332–351.

Gillespie, W. H., G. W. Rothwell, and S. E. Scheckler. 1981. The earliest seeds. *Nature* 293(5832):462–464.

Goldring, R. 1970. The stratigraphy about the Devonian-Carboniferous boundary in the Barnstaple area of North Devon, England. *Sixième congrès international de stratigraphie et de géologie du Carbonifère. Compte Rendu* 2:807–816. Sheffield, England, 1967.

—— 1971. Shallow-water sedimentation as illustrated in the Upper Devonian Baggy Beds. *Memoirs of the Geological Society of London* 5:1–80, 12 plates.

Halle, F. and R. A. A. Oldeman. 1970. *Essai sur l'architecture de la dynamique de croissance des l'arbres tropicaux*. Paris: Masson.

Holland, C. H. 1981. Devonian. In C. H. Holland, ed., *A Geology of Ireland*, pp. 121–146. Edinburgh: Scottish Academic Press.

Johnson, T. 1917. *Spermolithus devonicus* gen. et sp. nov., and other pteridosperms from the Upper Devonian beds of Kiltorcan, C. Kilkenny. *Royal Society of Dublin, Science Proceedings*, n. s. 15:245–254.

Lejeune, V. 1986. Sedimentation aluviale et paléosols associés de la Formation d'Évieux (Synclinorium de Dinant). *Mémoire pour l'obtention du grade de Licencié en Sciences Géologiques et Minéralogiques.* Institut de Minéralogie, Laboratoire Géologie des Argiles. Université de Liège, Belgium.

Lepekhina, V. G., N. M. Petrosyan, and G. P. Radchenko. 1962. *Materi k fitostratigrafii devonskikh otlozh. Altae-Sayanskoi gornoi oblasti* (The most important Devonian plants of the Altai-Sayan mountain region). 70:61–189, plates 1–24. Leningrad. (in Russian).

Long, A. G. 1960. On the structure of *Calymmatotheca kidstonii* Calder (emended) and *Genomosperma latens* gen. et sp. nov. from the Calciferous Sandstone Series of Berwickshire. *Transactions of the Royal Society of Endinburgh* 64:29–44.

—— 1961. Some pteridosperm seeds from the Calciferous Sandstone Series of Berwickshire. *Transactions of the Royal Society of Edinburgh* 64:401–419.

—— 1977. Some Lower Carboniferous pteridosperm cupules bearing ovules and microsporangia. *Transactions of the Royal Society of Edinburgh* 70: 1–11.

Matten, L. C., T. I. Fine, W. R. Tanner, and W. S. Lacey. 1984. The megagametophye of *Hydrasperma tenuis* Long from the Upper Devonian of Ireland. *American Journal of Botany* 71(10):1461–1464.

Matten, L. C. and W. S. Lacey. 1981. Cupule organization in early seed plants, In R. C. Romans, ed., *Geobotany II.*, pp. 221–234. New York: Plenum.

Matten, L. C., W. S. Lacey, and R. C. Lucas. 1980. Studies on the cupulate seed genus *Hydrasperma* Long from Berwickshire and East Lothian in Scotland and County Kerry in Ireland. *Botanical Journal of the Linnean Society* 81:249–273.

Matten, L. C., W. S. Lacey, B. I. May, and R. C. Lucas. 1980. A megafossil flora from the uppermost Devonian near Ballyheigue, Co. Kerry, Ireland. *Review of Palaeobotany and Palynology* 29:241–251.

Matten, L. C., W. R. Tanner, and W. S. Lacey. 1984. Additions to the silicified Upper Devonian/Lower Carboniferous flora from Ballyheigue, Ireland. *Review of Palaeobotany and Palynology* 43:303–320.

May, B. I. and L. C. Matten. 1983. A probable pteridosperm from the Uppermost Devonian near Ballyheigue, Co. Kerry, Ireland. *Botanical Journal of the Linnean Society* 86:103–123.

Meyer-Berthaud, B. and J. Galtier. 1986a. Une nouvelle fructification de Carbonifère Inférieur d'Écosse: *Burnitheca*, Filicinée ou Ptéridospermale? *Compte Rendu de l'Académie des Sciences de Paris* 303, Série 2 (13):1263–1268.

—— 1986b. Studies on a Lower Carboniferous flora from Kingswood near Pettycur, Scotland. 2. *Phacelotheca*, a new synangiate fructification of pteridospermous affinities. *Review of Palaeobotany and Palynology* 48:181–198.

Millay, M. A. and T. N. Taylor. 1979. Paleozoic seed fern pollen organs. *Botanical Review* 45:301–375.

Nathorst, A. G. 1902. Polarländer. Zur Oberdevonischen Flora der Bären-Insel. *K. Svenska Vetenskaps-Akademiens Handlingar* 36:1–60.

Niklas, K. J. 1981a. Simulated wind pollination and airflow around ovules of some early seed plants. *Science* 211(4479):275–277.

—— 1981b. Airflow patterns around some early seed plant ovules and cupules: Implications concerning efficiency in wind pollination. *American Journal of Botany* 68(5):635–650.

—— 1983a. The influence of Paleozoic ovule and cupule morphologies on wind pollination. *Evolution* 37:968–986.

—— 1983b. Early seed plant wind pollination studies: A reply. *Taxon* 32(1):99–100.

—— 1985. The aerodynamics of wind pollination. *Botanical Review* 51(3):328–386.

Pettitt, J. M. 1970. Heterospory and the origin of the seed habit. *Biological Reviews of the Cambridge Philosophical Society* 45(3):401–415, plates 1–6.

Pettitt, J. M. and C. B. Beck. 1968. *Archaeosperma arnoldii*—a cupulate seed from the Upper Devonian of North America. *Contr. Mus. Paleont. Univ. Michigan,* 22:139–154.

Phillips, T. L. and W. A. DiMichele. 1981. Paleoecology of Middle Pennsylvanian age coal swamps in southern Illinois/Herrin Coal Member at Sahara Mine No. 6. In K. J. Niklas, ed., *Paleobotany, Paleoecology, and Evolution,* vol. 1, pp. 231–284. New York: Praeger.

Phillips, T. L., R. A. Peppers, M. J. Avcin, and P. F. Laughnan. 1974. Fossil plants and coal: Patterns of changes in Pennsylvanian coal swamps of the Illinois basin. *Science* 184:1367–1369.

Rex, G. M., and A. C. Scott. 1987. The sedimentology, paleoecology and preservation of the Lower Carboniferous plant deposits at Pettycur, Fife, Scotland. *Geology Magazine* 124:43–66.

Rogers, I. 1926. On the discovery of fossil fishes and plants in the Devonian rocks of North Devon. *Transactions of the Devonshire Association for the Advancement of Science, Literature, and Art* 58:223–234.

Rothwell, G. W. 1971. Ontogeny of the Paleozoic ovule *Callospermarion pusillum. American Journal of Botany* 58:706–715.

—— 1980. Permineralized parts of putative primitive gymnosperms. *Botany 80. Botanical Society of America Miscellaneous Series Publication* 158:98.

—— 1982a. *Cordaianthus duquesnensis* sp. nov., anatomically preserved ovulate cones from the Upper Pennsylvanian of Ohio. *American Journal of Botany* 69:239–247.

—— 1982b. New interpretations of the earliest conifers. *Review of Palaeobotany and Palynology* 37:7–28.

—— 1985. The role of comparative morphology and anatomy in interpreting the systematics of fossil gymnosperms. *Botanical Review* 51:319–327.

—— 1986. Classifying the earliest gymnosperms. In R. A. Spicer and B. A. Thomas, eds., *Systematic and Taxonomic Approaches in Palaeobotany.* Systematics Association Special Vol. 31: Oxford: Oxford University Press. 137–161.

—— 1987a. Biology of early gymnosperm reproduction: An overview. *Abstracts of the 14th International Botanical Congress*, Berlin, p. 283.

—— 1987b. Origin of gymnosperms. In *McGraw-Hill Yearbook of Science and Technology, 1987*, pp. 341–343. New York: McGraw-Hill.

Rothwell, G. W. and D. M. Erwin. 1987. Origin of seed plants: An aneurophyte/seed fern link elaborated. *American Journal of Botany* 74(6):970–973.

Rothwell, G. W. and A. C. Scott. 1985. Ecology of the Lower Carboniferous plant remains from Oxroad Bay, East Lothian, Scotland. *American Journal of Botany* 2(6):899. (Abstracts.)

Rothwell, G. W. and T. N. Taylor. 1982. Early seed plant wind pollination studies: A commentary. *Taxon* 31(2): 308–309.

Rothwell, G. W. and D. C. Wight. 1987. Taxonomic diversity among Devonian and Lower Carboniferous fructifications with *Hydrasperma tenuis*-type ovules. *American Journal of Botany* 74(5):688. (Abstracts.)

Scheckler, S. E. 1975. A fertile axis of *Triloboxylon ashlandicum*, a progymnosperm from the Upper Devonian of New York. *American Journal of Botany* 62:923–934.

—— 1985. Seed plant diversity in the late Devonian (Famennian). *American Journal of Botany* 72(6):900. (Abstracts.)

—— 1986a. Geology, floristics, and paleoecology of Late Devonian coal swamps from Appalachian Laurentia (U.S.A.). *Annales de la Societé géologique de Belgique* 109:209–222.

—— 1986b. Old Red Continent facies in the late Devonian and early Carboniferous of Appalachian North America. *Annales de la Societé géologique de Belgique* 109:223–236.

—— 1986c. Floras of the Devonian-Mississippian transition. In T. W. Broadhead, ed., *Land Plants: Notes for a Short Course*, organized by R. A. Gastaldo. University of Tennessee, Knoxville, Department of Geological Sciences. *Studies in Geology* 15:81–96.

—— 1987. Reproduction of progymnosperms and their relations with early seed plants. *Abstracts of the 14th International Botanical Congress*, Berlin, p. 283.

Scott, A. C. 1977. A review of the ecology of Upper Carboniferous plant assemblages, with new data from Strathclyde. *Palaeontology* 20:447–473.

—— 1979. The ecology of some coal measure floras from northern Britain. *Proceedings of the Geological Association* 90:97–116.

Scott, A. C., J. Galtier, and G. Clayton. 1984. Distribution of anatomically preserved floras in the Lower Carboniferous in western Europe. *Transactions of the Royal Society of Edinburgh: Earth Sciences* 75:311–340.

Singh, H. 1978. Embryology of gymnosperms. *Handbuch der Pflanzenanatomie*. Berlin and Stuttgart: Gebrüder Borntraeger.

Smith, D. L. 1964. The evolution of the ovule. *Biological Review* 39:137–159.

Spicer, R. A. 1980. The importance of depositional sorting to the biostratigraphy of plant megafossils. In D. L. Dilcher and T. N. Taylor, eds., *Biostratigraphy of Fossil Plants*, pp. 171–183. Stroudsburg, Pa.: Dowden, Hutchinson, and Ross.

—— 1987. Quantitative sampling of plant megafossil assemblages. In W. A. DiMichele and S. L. Wing, eds., *Methods and Applications of Plant Paleoecology: Notes for a Short Course*, pp. 41–71. Paleobotanical Section, Botanical Society of America.

Spicer, R. A. and A. G. Greer. 1986. Plant taphonomy in fluvial and lacustrine systems. In T. W. Broadhead, ed., *Land Plants: Notes for a Short Course*, organized by R. A. Gastaldo. University of Tennessee, Knoxville, Department of Geological Sciences. *Studies in Geology* 15: 10–26.

Stewart, W. N. 1983. *Paleobotany and the Evolution of Plants.* Cambridge: Cambridge University Press.

Stockmans, F. 1948. Végétaux Dévonien Supérieur de la Belgique. *Mémoires du Musée Royal d'Histoire Naturelle de Belgique* 110:1–85, plates 1–14.

Taylor, T. N. 1981. *Paleobotany: An Introduction to Fossil Plant Biology.* New York: McGraw-Hill.

—— 1982. Reproductive biology in early seed plants. *BioScience* 32(1):23–28.

Taylor, T. N. and M. A. Millay. 1979. Pollination biology and reproduction in early seed plants. *Review of Palaeobotany and Palynology* 27:329–355.

—— 1981. Morphologic variability of Pennsylvanian lyginopterid seed ferns. *Review of Palaeobotany and Palynology* 32:27–62.

Thorez, J. and R. Dreesen. 1986. A model of a regressive depositional system around the Old Red Continent as exemplified by a field trip in the Upper Famennian "Psammites du Condroz" in Belgium. *Annales de la Societé géologique de Belgique* 109:285–323.

Vologdine, A. 1937. La vallée de Lennoi Log dans la crête du Touran. In M. Tetiaev, ed., *Excursion siberienne: Le pays de Krasnoiarsk*, pp. 69–73. Congrès géologique international, XVII session. Leningrad and Moscow: ONTI NKTP URSS.

3

Morphology and Phylogenetic Relationships of Early Pteridosperms

JEAN GALTIER

▦

Much has been already written about the historical development and evolutionary significance of the concept of pteridosperms, and this group has been widely used as a taxonomic unit at different ranks —e.g., division Pteridospermophyta (Taylor 1981); class Pteridospermopsida (Emberger 1968); order Pteridospermales (Sporne 1974; Stewart 1983).

Today it is believed that gymnospermous reproduction originated during the late Devonian as a result of an evolutionary transition from aneurophytalean progymnosperms to plants that are identified as early pteridosperms (Beck 1976; Rothwell 1982).

Two recent phylogenetic analyses of seed plants (Meyen 1984; Crane 1985) strongly emphasized the heterogeneity of the pteridosperms.

Within the classification proposed by Meyen, the pteridosperms disappear and are replaced by several orders, of which two occupy a key
phylogenetic position: the Calamopityales, which are considered
primitive to the Ginkgoopsida clade, and the Lagenostomales, primitive to both the Cycadopsida and the Pinopsida clades. Crane (1985),
using cladistic analysis, came to the conclusion that "for phylogenetic
purposes the current concept of the pteridosperms is valueless and
has been a major source of confusion in attempts to analyse the
phylogenetic relationship of seed plants."

A reader persuaded by Meyen's or Crane's conclusions will certainly
question the interest of the present paper, and I agree that this is
perhaps a "mission impossible" in the present state of our knowledge.
The aim of this paper, however, is to review recent research and to
present some new information on early Carboniferous plants that are
currently assigned to the families Calamopityaceae, Lyginopteridaceae, and Buteoxylonaceae and that must be recognized as the first
gymnosperms. The interpretation of early seeds is described in a
separate paper in this book by G. Rothwell and S. Scheckler. Here I
am mainly concerned with the characterization of the vegetative
anatomy and morphology of the first gymnosperms, with some new
information on male reproductive structures.

DISTRIBUTION OF EARLY PTERIDOSPERM TAXA

In this review I will discuss plant fossils of late Devonian to Lower
Namurian age—i.e., a time span of approximately 30 million years
that corresponds approximately to the Mississippian.

As a result of many recent studies—particularly by Long, Barnard,
Beck, Stein, Matten et al., Jennings, Mapes, Rothwell, Meyer-Berthaud,
and the present author—our understanding of early gymnosperms
has been greatly enhanced. At the same time a precise chronology of
the Lower Carboniferous deposits in Europe in which they occur has
recently been established (Scott et al., 1984). All this research has
enabled the construction of a table (figure 3.1) showing the stratigraphic distribution of early pteridosperms. Figure 3.1 includes only
taxa (genera) based on permineralized or anatomically preserved stems
and ovules. In addition we have included selected representatives of
the progymnosperms and well-known stem genera of the Pennsylvanian families Lyginopteridaceae, Medullosaceae and Callistophytaceae. This nonexhaustive chart demonstrates that, by Mississippian
times, at least 20 genera of permineralized stems representing putative early pteridosperms had evolved above the progymnosperm level.

This number of taxa largely exceeds those known at any younger period, particularly during Pennsylvanian times, when the pteridosperms were a successful and dominant element of some floras (Phillips 1981). Knoll (1986) also illustrated the dramatic increase of pteridosperm taxa in fossil assemblages of mesic tropical and subtropical floras preserved as compressions. The concomitant augmentation of plant taxa in both states of preservation may be interpreted as the radiation of the seed plants. This conclusion is supported by evidence from the distribution of early seeds, as shown in the upper part of figure 3.1, which correlates exactly with the increase in the number of putative early pteridosperm stems preserved as permineralizations.

PROVISIONAL SYSTEMATIC TREATMENT OF MISSISSIPPIAN PTERIDOSPERMS

The taxa attributed to Mississippian pteridosperms correspond to fragments of permineralized stems and ovules recorded in figure 3.1. Other pteridosperm taxa are based on either permineralizations or compressions of frond fragments and pollen organs. All these taxa may be tentatively classified as follows:

1. Calamopityaceae. Stems, restricted to manoxylic genera (figure 3.1): *Calamopitys, Stenomyelon, Diichnia, Bostonia, Galtiera;* petioles and frond fragments: *Kalymma* (pro parte); *Lyginorachis, Calamopteris, Megalorhachis, Chapelia;* pollen organs: none; ovules: none (but the platyspermic ovule *Lyrasperma* is frequently found in association with *Stenomyelon* [Long 1960], and it has been suggested that it belongs to this plant, a view supported by Meyen [1984] and discussed by several authors [Beck 1985; Miller 1985]).

2. Lyginopteridaceae and putative Lyginopteridaceae. Stems currently attributed to this family: *Lyginopteris, Heterangium, Rhetinangium, Tetrastichia, Tristichia;* stems of putative Lyginopteridaceae (see reference): *Pitus* (Long 1963, 1979), *Lyginopitys* (Galtier 1970), *Laceya* (May and Matten 1983); petioles and frond fragments: *Lyginorachis* (pro parte); pollen organs: *Telangium,* as illustrated by Jennings (1976) and Long (1979); pollen organs of putative Lyginopteridaceae: *Phacelotheca* Meyer-Berthaud and Galtier (1986), *Melissiotheca* Meyer-Berthaud (1986); radiospermic ovules currently assigned to the Lagenostomales (Seward 1917)—i.e., related to the Lyginopteridaceae: *Lagenostoma, Physostoma, Sphaerostoma, Geminitheca, Calathospermum, Salpingostoma, Anasperma, Dolichosperma, Eurystoma, Genomosperma, Stamnostoma, Tantallosperma, Hydrasperma, Rhynchosperma.* Several platyspermic ovules are recorded from the

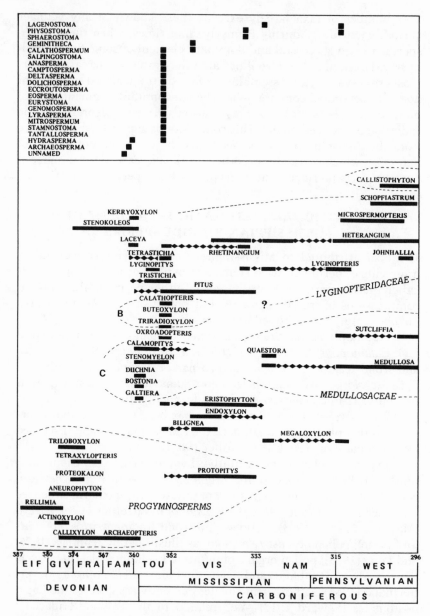

Figure 3.1. Stratigraphical range of some early gymnosperm genera represented by permineralized stems (below) and ovules (above), and comparison with a range of selected Devonian progymnosperms and Pennsylvanian pteridosperms of the families Lyginopteridaceae, Med-

Mississippian strata and are generally classified with the Lagenosto-males: *Lyrasperma, Eosperma, Deltasperma, Eccroutosperma, Camptosperma, Mitrospermum* (however, see Long 1960 and Meyen 1984 for *Lyrasperma*).

3. Buteoxylonaceae. Stems (figure 3.1B): *Buteoxylon, Triradioxylon,* (?) *Calathopteris*—however, the affinities of this genus with the Buteoxylonaceae are unclear; petioles or frond fragments: *Lyginorachis* (pro parte); pollen organs and ovules: none.

4. Medullosaceae. Stems: *Quaestora, Medullosa;* pollen organs and ovules: none.

5. *Incertae sedis* and putative pteridosperms. Stems: *Kerryoxylon, Stenokoleos;* decorticated stems: *Eristophyton, Bilignea, Endoxylon,* all previously considered as pycnoxylic Calamopityaceae (Scott 1923; Read 1937; Lacey 1953); *Megaloxylon* (Mapes 1985); petioles, rachises: *Periastron.*

6. Fern-like frond compressions. Genera attributed to the pteridosperms correspond to fronds often forked near the base and occasionally fertile: *Sphenopteridium, Rhodea, Diplothmema, Adiantites, Aneimites, Triphyllopteris, Sphenocyclopteridium, Alcicornopteris, Spathulopteris, Sphenopteris* (pro parte), *Diplopteridium.*

Below are a series of selected comments on the stratigraphic occurrence and systematic position of some of the taxa listed above.

1. The families Calamopityaceae and Buteoxylonaceae are restricted to early Mississippian taxa and are based only on vegetative remains. The assigning of the Buteoxylonaceae to the pteridosperms instead of to the progymnosperms has been questioned by Barnard and Long (1975). Furthermore, Beck (1985) considers that "there seems to be little reason for segregating these groups (Calamopityaceae and Lyginopteridaceae) and the Buteoxylonaceae into separate families."

2. The problem of delimiting the family Lyginopteridaceae is cru-

---

ullosaceae, and Callistophytaceae. **B** and **C** Genera attributed to the families Buteoxylonaceae and Calamopityaceae. The distribution of progymnosperms and of Pennsylvanian pteridosperms is based respectively on data from Beck (1976) and from Phillips (1981). Unpublished data included in this figure are kindly acknowledged—e.g., the extension of *Lyginopteris* down into the Upper Visean (U. Bertram, personal communication), the description of the new calamopityan genus *Galtiera* (Beck and Stein 1987), and the discovery of *Lagenostoma* in Upper Visean–Lower Namurian of Morocco (Galtier, Phillips and Chalot-Prat 1986).

cial in any attempt to understand early phylogenetic relationships of pteridosperms.

3. The Medullosaceae have only two late Mississippian representatives: *Quaestora* (Mapes and Rothwell 1980) and *Medullosa* sp. (Taylor and Eggert 1967).

4. The monogeneric family Callistophytaceae is not known in Mississippian deposits.

5. The large number of Upper Tournasian ovules described (figure 3.1) from Berwickshire is mainly the result of the outstanding research of A. G. Long. However, it is surprising that the slightly older deposits of the New Albany Shale (United States), Montagne Noire (France), and Saalfeld (East Germany), where vegetative remains of the Calamopityaceae are very common, have never yielded seeds.

CHARACTERIZATION OF EARLY PTERIDOSPERMS

This group of plants combines primitive features shared with the ancestral Aneurophytales and a number of advanced features. Both types of features are listed below:

1. Primitive or generalized features shared with the progymnosperms. Protostelic organization; tracheidal pitting generally of the multiseriate, elliptical bordered type; bifacial cambium producing secondary xylem and phloem; outer cortex with longitudinally orientated bands of sclerenchyma; clusters of terminally borne, elongate, exannulate male sporangia; trilete prepollen with proximal dehiscence.

2. Advanced features. Protostele with mixed pith to eustele; loss of central protoxylem strands; tangential leaf trace divergence; manoxylic secondary xylem; pitting restricted to radial walls in wood tracheids; axillary branching; megaphyllous leaves generally in the form of a bipartite frond but sometimes with a median fertile rachis; dorsiventral petiole anatomy; pinnule morphology varying from nonlaminate to broad, undissected; synangiate organization of male organs; single functional megaspore within an ovule; incipient tegument and micropyle.

These characters will be discussed below, together with quantitative data on some features—such as overall size, stem diameter, and primary vascular system diameter—that must also be taken into account.

It is significant that some of the apomorphic characters listed by Crane (1985) in the Pennsylvanian Medullosaceae and Callistophytaceae (linear tetrad of spores, saccate pollen, double vascular supply to ovules, siphonogamy) have not yet been recognized in Mississippian pteridosperms.

SIZE OF EARLY PTERIDOSPERMS

Although it is not primarily utilizable in terms of phylogeny, the size of the plant is of biological significance and may be useful for systematic purposes. The height of a plant broadly relates to its stem diameter; the latter measurement is readily available in fossil plants (Chaloner and Sheerin 1979) and will be used in this discussion.

A recent investigation of *Calamopitys* of Tournaisian age from the Montagne Noire and Saalfeld has provided additional information confirming the ontogenetic correlation between stem diameter, primary xylem diameter, maximum diameter of tracheid metaxylem, and the width of petiolar bases. All the nondecorticated specimens had a well-developed primary cortex (expressed in the ratio of total diameter:wood diameter equal to or greater than 3), no evidence of a periderm and a relatively small diameter of the primary vascular system (ratio of total diameter:primary vascular system diameter approximately 10). This indicates that partly decorticated *Calamopitys* stems with a large primary vascular system (approximately 15 mm in diameter) actually had a total diameter up to 10 cm (figure 3.2: 4); they would exceed in stem diameter all the Mississippian and even Pennsylvanian pteridosperms (except the medullosans), as shown on figure 3.2.

Figure 3.3 clearly illustrates the general trend in stem diameter amongst Mississippian and Pennsylvanian pteridosperms and shows the comparison with some progymnosperm taxa. It is significant that among the 44 plotted taxa, only 5, representing 4 genera, *(Callixylon, Pitus, Protopitys, Medullosa)* exceed 20 cm in diameter and that 3 of these taxa are Upper Devonian–Lower Carboniferous progymnosperms *(Callixylon, Protopitys)* or putative pteridosperms *(Pitus)*. Among the pteridosperms only the Medullosaceae show a phylogenetic trend toward arborescence during Pennsylvanian times, which continued into the Permian. It is interesting to note that all the other pteridosperms show very little (Calamopityaceae) or no increase in size with respect to the supposed ancestral aneurophytalean progymnosperms. The large diameter (10 cm) extrapolated above for *Calamopitys* has not been plotted on figure 3.3. Clearly an evolutionary relationship between the Calamopityaceae and the Lyginopteridaceae and/or the Callistophytaceae would require a reduction in size. *Pitus* is the only known putative arborescent pteridosperm; it is comparable in both

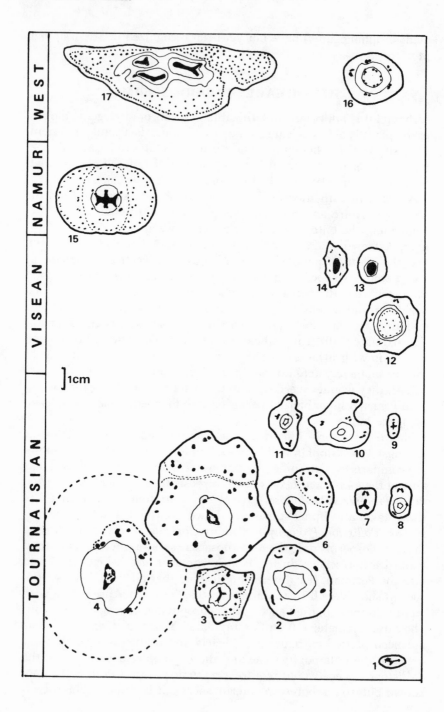

maximum stem diameter and primary vascular system diameter to the progymnosperm *Callixylon*.

THE PRIMARY VASCULAR SYSTEM

The organization of the primary vascular system in relation to leaf trace divergence and anatomy represents a very distinctive vegetative feature of the pteridosperms. As a result, it has been widely used for systematic purposes. Studies on a series of Mississippian species of *Calamopitys* demonstrated that the gymnosperm eustele evolved through longitudinal dissection of a protostele (Namboodiri and Beck 1968; Beck 1970), a view supported by other authors (Galtier 1973; Rothwell 1976).

Stelar Organization In interpreting stelar organization in early pteridosperms in terms of primitive versus advanced characters, one must take into consideration the ontogenetic and interspecific variability.

An attempt to summarize the main types of stelar organization, with respective sizes, in some aneurophytalean progymnosperms and early pteridosperms is shown in figure 3.4. Three categories are recognized.

1. Lobed protostele or actinostele (in black). They are generally solid protosteles or have a few parenchyma cells intermixed with the metaxylem tracheids. The protoxylem is always mesarch in the "arms" of the protostele. This type is common in the Aneurophytales, in the problematic *Stenokoleos*, in some Calamopityaceae such as *Stenomyelon*, in some putative early lyginopterids (e.g., *Laceya*, *Tristichia*, and *Tetrastrichia*), and in *Triradioxylon*, attributed to the Buteoxylonaceae. The primitiveness of this stelar organization beginning with the aneurophytaleans fits with Beck's model of stelar evolution. The lobed

Figure 3.2. Stratigraphic distribution of some representative members of the early pteridosperm families Calamopityaceae, Buteoxylonaceae, Lyginopteridaceae, Medullosaceae. All are transverse sections of stems drawn at the same magnification. **1** *Laceya*. **2** *Diichnia*. **3** *Galtiera*. **4** *Calamopitys americana*. **5** *C.* cf. *embergeri*. **6** *Stenomyelon bifasciculare*. **7** *Tristichia longii*. **8** *Lyginopitys*. **9** *Tetrastichia*. **10** *Calathopteris*. **11** *Buteoxylon*. **12** *Pitus dayi*. **13** *Rhetinangium*. **14** *Heterangium grievii*. **15** *Quaestora*. **16** *Lyginopteris*. **17** *Medullosa*. Stelar xylem and leaf traces in black. (Data from references listed in figure 3.3.)

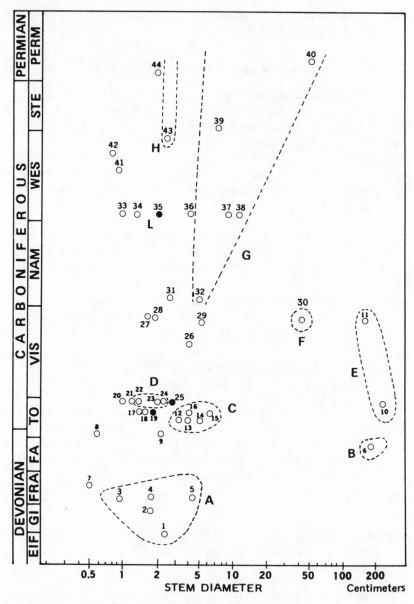

Figure 3.3. Maximum observed stem diameter plotted on a log scale for some late Devonian progymnosperms, Mississippian gymnosperms, and selected Pennsylvanian pteridosperms. The geological stages, in abbreviated form, are shown on the vertical axis. **A** Aneurophytales. **B** Archaeopteridales. **C** Calamopityaceae. **D** Buteoxylona-

anatomy reflects cladotaxy in the aneurophytaleans and phyllotaxy in early pteridosperms: this is normally 2/5 even in three-lobed protosteles. It is significant that the maximum size recorded for lobed protosteles was in the Frasnian *Tetraxylopteris* (from data in Scheckler 1976) and that there is no record of this stelar type in Visean or younger specimens.

2. Parenchymatized protostele (stipled on figure 3.4, using the terminology of Beck et al. 1982). This is in fact a lobed to cylindrical protostele as typified by *Heterangium*, showing a regular mixture of parenchyma cells separating clusters of metaxylem tracheids or showing tracheids with regularly interspersed parenchyma. The protoxylem is mesarch to exarch. Generally there is a lack of evidence showing continuity to form sympodial strands.

This type has not been recognized among aneurophytaleans, and therefore it is considered that a parenchymatized protostele is a derived feature that evolved in some Calamopityaceae *(Stenomyelon*

---

ceae. **E** *Pitus*. **F** Protopitys. **G** Medullosaceae. **H** Callistophytaceae. **L** Lyginopteridaceae. Filled black dots represent branching stems. Data from references as follows: **1** *Rellimia* (Leclercq and Bonamo 1971). **2** *Aneurophyton* (Serlin and Banks 1978). **3** *Proteokalon* (Scheckler and Banks 1971). **4** *Triloboxylon* (Matten and Banks 1966). **5** *Tetraxylopteris* (Beck 1957). **6** *Callixylon* (Chaloner and Sheerin 1979). **7** *Stenokoleos* (Matten and Banks 1969). **8** *Kerryoxylon* (Matten et al. 1984). **9** *Laceya* (May and Matten 1983). **10** *Pitus primaeva*. **11** *Pitus withamii* (Long 1979). **12** *Galtiera* (Beck and Stein 1986). **13** *Diichnia* (Read 1936). **14** *Calamopitys americana* (Read 1937). **15** *Calamopitys annularis* (new data). **16** *Stenomyelon bifasciculare* (Meyer-Berthaud), **17** *Lyginopitys* (Galtier 1970). **18** *Tristichia longii* (Galtier 1977). **19** *Calamopitys* branching (Galtier 1974). **20** *Oxroadopteris* (Long 1984). **21** *Tetrastichia* (Gordon 1938). **22** *Triradioxylon* (Barnard and Long 1975). **23** *Calathopteris* (Long 1976). **24** *Buteoxylon* (Barnard and Long 1973). **25** *Pitus primaeva* branching (Long 1979). **26** *Eristophyton* (Scott 1902). **27** *Rhetinangium* (Gordon 1912). **28** *Heterangium* (Williamson and Scott 1895). **29** *Endoxylon* (Scott 1924). **30** *Protopitys* (Solms-Laubach 1893). **31** *Megaloxylon* (Mapes 1985). **32** *Quaestora* (Mapes and Rothwell 1980). **33** *Microspermopteris* (Taylor and Stockey 1976). **34** *Heterangium* (Williamson and Scott 1895). **35** *Lyginopteris* branching. **36** *Lyginopteris* (Blanc-Louvel 1966). **37** *Sutcliffia* (Scott 1923). **38** *Medullosa anglica* (Scott 1923). **39** *Sutcliffia* (Phillips and Andrews 1963). **40** *Medullosa* (Delevoryas 1955). **41** *Schopfiastrum* (Rothwell and Taylor 1972). **42** *Johnhallia* (Stidd and Phillips 1982). **43** *Callistophyton* (Rothwell 1975). **44** *Heterangium* (Renault 1896).

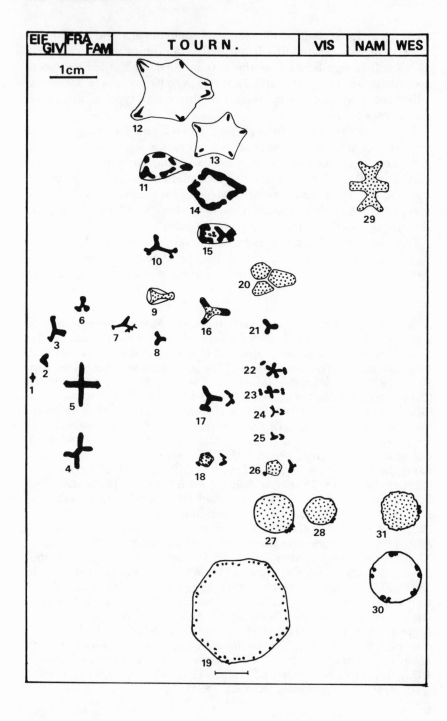

muratum, S. tuedianum, Galtiera), Buteoxylonaceae *(Buteoxylon)*, Lyginopteridaceae *(Rhetinangium, Heterangium)*, and Medullosaceae *(Quaestora)*. It is interesting to note that this was a successful stelar organization within the pteridosperms until Permian times in both the Lyginopteridaceae *(Heterangium)* and the Medullosaceae *(Medullosa)*.

3. Protostele with mixed pith and eustele. This stelar type is characterized by a ring of discrete or longitudinally interconnected sympodial strands at the periphery of a parenchymatous pith or of a mixed pith with scattered tracheids; the primary xylem comprising the sympodia is mesarch. This definition corresponds in part to the protostele with mixed pith or superparenchymatized protostele proposed by Beck et al. (1982). In fact the gradation to eustele observed in the French material of *Calamopitys* (Galtier, 1973, 1975) makes it difficult to separate strictly protostele and eustele. On the other hand, five sympodial strands corresponding to a 2/5 phyllotaxy are generally recognizable, and I consider that the term eustele really applies to some species of *Calamopitys* and *Diichnia* (figure 3.4: 11–15). The putative early lyginopterid, *Lyginopitys* (figure 3.4: 18), has a mixed pith and medullary strands as in *Pitus*, and may be referred to as having a protostele with a mixed pith or a eustele.

Figure 3.4. Diagrammatic record of the primary vascular system (xylem) in representative aneurophytalean progymnosperms and early pteridosperms; all drawn at the same scale except item **19**. Three main types are figured: solid actinostele (black), parenchymatized protostele (stippled), and eustele (black sympodia). Data and references as follows: **1** *Aneurophyton,* **2** *Rellimia,* **3** *Triloboxylon,* **4** *Proteokalon,* and **5** *Tetraxylopteris* (all data from Beck 1976 and Scheckler 1976). **6** *Stenokoleos bifidus* (Matten and Banks 1969). **7** *Laceya* (May and Matten 1983). **8** *Stenomyelon* sp. (Beck 1970). **9** *Stenomyelon muratum* (Read 1937). **10** *Galtiera* (Beck and Stein 1986). **11** *Calamopitys* sp. (Beck 1970). **12** *C. foerstei* (Read 1937). **13** *Diichnia* (Read 1936). **14** *Calamopitys saturni.* **15** *Calamopitys* sp. (new data). **16** *Stenomyelon bifasciculare* (Meyer-Berthaud 1984). **17** *Tristichia longii* (Galtier 1977). **18** *Lyginopitys* (Galtier 1970). **19** *Pitus antiqua* (Scott 1902). **20** *Stenomyelon tuedianum* and **21** *S. primaevum* (Long 1964). **22, 23** *Tetrastichia* (Gordon 1938). **24** *Tristichia ovensi* (Long 1961). **25** *Triradioxylon* (Barnard and Long 1975). **26** *Buteoxylon* (Barnard and Long 1973). **27** *Rhetinangium* (Gordon 1912). **28** *Heterangium grievii* (Williamson and Scott 1895). **29** *Quaestora* (Mapes and Rothwell 1980). **30** *Lyginopteris oldhamia* (Scott 1923) **31** *Heterangium* (Blanc Louvel 1969).

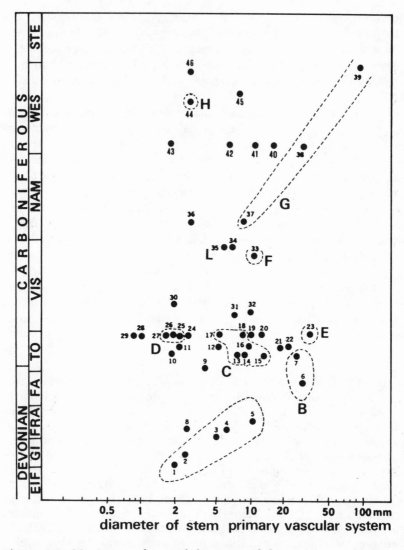

Figure 3.5. Maximum observed diameter of the primary xylem plotted on a log scale for Mississippian gymnosperms, selected progymnosperms, and Pennsylvanian pteridosperms. Same taxa and references as for figure 3.3. **1** *Rellimia.* **2** *Aneurophyton.* **3** *Triloboxylon.* **4** *Proteokalon.* **5** *Tetraxylopteris.* **6** *Callixylon newberryi.* **7** *Callixylon brownii.* **8** *Stenokoleos bifidus.* **9** *Laceya.* **10** *Stenokoleos simplex.* **11** *Tristichia longii.* **12** *Lyginopitys.* **13** *Diichnia.* **14** *Galtiera.* **15** *Calamopitys americana.* **16** *Calamopitys annularis.* **17** *Stenomyelon primaevum.* **18** *S. heterangioides.* **19** *Rhetinangium.* **20** *Eristophyton beinertianum.* **21** *Apo-*

Interestingly, the two stelar forms discussed above, eustele (3) and parenchymatized protostele (2), seem to have evolved simultaneously and independently in at least two pteridosperm families (Calamopityaceae and Lyginopteridaceae). Figure 3.4 though clearly suggests, however, that two types of broad eusteles had already evolved during Tournaisian times, long before *Lyginopteris:* one in the Calamopityaceae *(Calamopitys foerstei, Diichnia)* with only five sympodia (figure 3.4: 12 and 3.4: 13); the other in *Pitus* (figure 3.4: 19), with up to 40 or more xylem strands.

Maximum size of the primary vascular system   A graph of the stelar diameter of 44 taxa plotted on a log scale is shown in figure 3.5. In order to facilitate comparison with figure 3.3, the same taxa of progymnosperms, Mississippian, and Pennsylvanian pteridosperms have been plotted. Interestingly the distribution of taxa broadly correlates with that observed for stem diameter. The same trends are recognizable.

1. There is no increase in size or even reduction in stelar diameter within the Mississippian Buteoxylonaceae, the lyginopterids, and the Pennsylvanian *Callistophyton* with respect to the Devonian aneurophytaleans.

2. There is a small increase in stelar size within the Calamopityaceae.

3. There is a well-marked increase in stelar size during Pennsylvanian times in the Medullosaceae that had a considerable volume of primary vascular tissue.

4. As for overall dimensions, *Pitus* stands apart from other Mississippian gymnosperms with a eustele exceeding 20 mm in diameter; the maximum was recorded from a Tournaisian species, *P. antiqua,* with a diameter of 34 mm (Scott 1902).

Ratio of Primary Stelar Diameter to Stem Diameter   There is a significant difference between the Calamopityaceae and the Lyginop-

---

roxylon. **22** unnamed plant (see figure 3.10F). **23** *Pitus antiqua.* **24** *Tetrastichia.* **25** *Calathopteris.* **26** *Buteoxylon.* **27** *Triradioxylon.* **28** *Tristichia ovensi.* **29** *Oxroadopteris.* **30** *Eristophyton fasciculare.* **31** *Bilignea resinosa.* **32** *Eristophyton waltonii.* **33** *Protopitys buchiana.* **34** *Rhetinangium.* **35** *Heterangium grievii.* **36** *Megaloxylon wheelerae.* **37** *Quaestora.* **38** *Sutcliffia.* **39** *Medullosa.* **40** *Megaloxylon scottii.* **41** *Lyginopteris oldhamia.* **42** *Heterangium shorense.* **43** *Microspermopteris.* **44** *Callistophyton.* **45** *Schopfiastrum.* **46** *Johnhallia.*

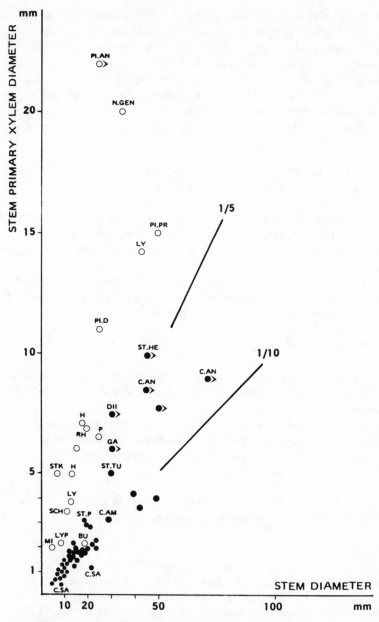

Figure 3.6. Primary stelar diameter plotted against stem diameter for some members of the Calamopityaceae (black dots) and comparison with other pteridosperms. Black dots without lettering = *Calamopitys embergeri* assemblage. **C.SA** *Calamopitys saturni*. **C.AM** *C.*

teridaceae in regard to this parameter. In a large collection of *Cala-mopitys* specimens, ranging from 5 mm to 5 cm in diameter, with their cortex perfectly preserved, the ratio averages 1:10 and always exceeds 1:5, as shown in figure 3.6. In *Pitus* and the Lyginopterida-ceae, of Mississippian and Pennsylvanian age, the ratio varies from 1:2 to 1:4.

The biological significance of this difference in ratio is most important. It indicates that Lyginopteridaceae and *Pitus* may produce stems with a large primary stelar diameter but with a relatively narrow cortex. In contrast, the Calamopityaceae had a small vascular procambium but produced a massive cortex; the latter is related to the large size of the petiole and the multifascicular nature of the leaf trace in the Calamopityaceae. The Lyginopteridaceae and *Pitus* had relatively small petioles and a generally less divided petiolar strand. Data on *Buteoxylon* indicate a ratio rather similar to that of the Calamopityaceae.

LEAF TRACE DIVERGENCE

Namboodiri and Beck (1968) and Beck (1970) emphasized that leaf trace divergence is a very important feature in gymnosperm evolution. These authors suggested that the tangential leaf trace divergence characteristic of modern conifers evolved from the primitive radial trace divergence common in the progymnosperms. This transition, concomitant with the gradation from protostele to eustele, was documented by Beck (1970) and supported by Galtier (1973) in a series including Lower Mississippian Calamopityaceae and the Pennsylvanian *Lyginopteris*.

Contemporaneous forms related to the lyginopterids—for example, *Lyginopitys* (Galtier 1970)—had a protostele with mixed pith and well-marked sympodial strands. This may represent a further parallel evolutionary trend toward a eustele, but with attainment of tangential divergence earlier than in the Calamopityaceae. If this is the case,

---

americana. **C.AN** *C. annularis.* **ST.P** *Stenomyelon primaevum.* **ST.TU** *S. tuedianum.* **ST.HE** *S. heterangioides.* **GA** *Galtiera;* **DII** *Diichnia.* **BU** *Buteoxylon.* **MI** *Microspermopteris.* **LYP** *Lyginopitys.* **SCH** *Schopfias-trum.* **LY** *Lyginopteris.* **H** *Heterangium.* **STK** *Stenokoleos.* **RH** *Rhetin-angium.* **PI.D** *Pitus dayi.* **PI.PR** *P. primaevum.* **PI.AN** *P. antiqua.* **N.GEN** Unnamed plant. Arrow on the right of the dot designates decorticated specimens with a higher value of stem diameter.

then the apomorphic character of tangential divergence would have evolved independently in the lyginopterids and the calamopityans, a view slightly different from the model proposed by Beck (1970).

Unfortunately we have no clear information concerning leaf trace divergence in *Pitus*. Reexamination of Gordon's slides of *Pitus dayi* suggests, however, that divergence was tangential in this species.

LEAF TRACE AND PETIOLE ANATOMY

In the systematics of pteridosperms based on permineralized stems, the leaf trace and petiole anatomy are considered to be important characters. Barnard and Long (1975) even distinguished the families Lyginopteridaceae, Calamopityaceae, and Medullosaceae solely on the basis of the types of petiole and rachis anatomy of *Lyginorachis*, *Kalymma*, and *Myeloxylon*, respectively. They also erected a fourth family, Buteoxylonaceae, for stems with a T-shaped petiole strand, that could not be assigned to the previously described types.

Figures 3.7 and 3.8 represent, at almost the same magnification, the stele and leaf trace of more than 20 taxa of pteridosperms belonging to these four families. For comparison, two aneurophytalean progymnosperms *(Proteokalon* and *Triloboxylon)*, the problematic *Stenokoleos*, and the Pennsylvanian *Callistophyton* have been added.

Leaf Trace Origin   Some significant differences in leaf trace origin exist. In the Lyginopteridaceae, Buteoxylonaceae, Callistophytaceae, and most Calamopityaceae, the leaf trace originates as a single bundle. In contrast, within three calamopityans—*Diichnia* (figure 3.7:7), *Galtiera* (figure 3.7:6), and *Stenomyelon bifasciculare* (figure 3.7:5)— leaf trace origin is double. In *Quaestora* (figure 3.7:8) and other Medullosaceae the leaf trace origin is even more complex. The double trace in *Diichnia* apparently originated from two distinct sympodia. Its evolutionary significance has been interpreted in several ways by a comparison with *Ginkgo* and some modern conifers in which a double leaf trace is also present (Read 1936; Namboodiri and Beck 1968; Meyen 1984). In *Stenomyelon bifasciculare* the two bundles resulted from the precocious division of a leaf trace that originated from one sympodium (Meyer-Berthaud 1984), and it appears that a similar situation exists in a species of *Calamopitys* from the Montagne Noire (figure 3.7:1). Beck and Stein (1986) considered that stelar organization of *Galtiera* was derived from a eustele of the *Diichnia* type, the two leaf trace bundles being connected to two distinct sympodia.

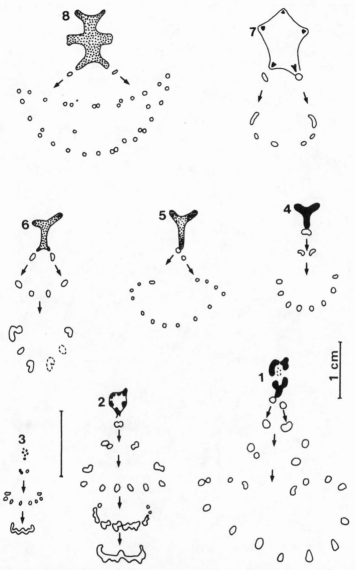

Figure 3.7. Schematic diagram of stele and leaf trace with subsequent modification to petiole strand. **1** *Calamopitys* sp. **2, 3** *Calamopitys embergeri.* **4** *Stenomyelon primaevum.* **5** *S. bifasciculare.* **6** *Galtiera.* **7** *Diichnia.* **8** *Quaestora.*

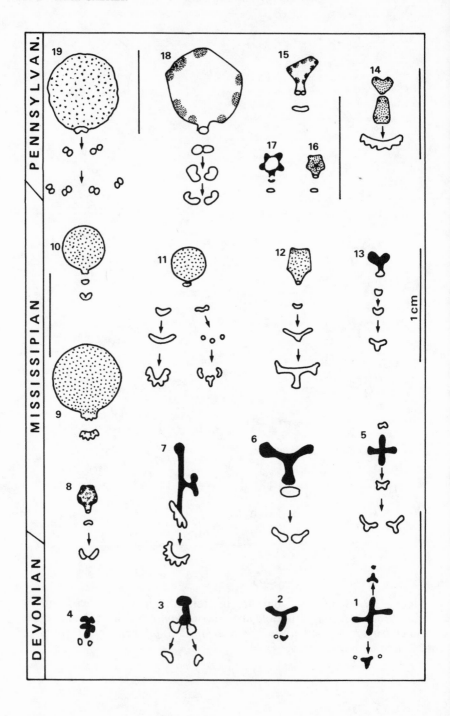

In *Stenokoleos* (figure 3.8:3 and 4) the interpretation is different, as the two bundles correspond to a pair of organs.

Leaf Trace Configuration and Variability   Generally the leaf trace begins as a single xylem strand with mesarch development and is oval to circular in cross section.

In the actinostelic taxa *Tetrastichia, Tristichia, Laceya, Triradioxylon* (figure 3.8:5, 6, 7, and 13) the leaf trace is very similar at first to the branch trace in aneurophytaleans—for example, *Proteokalon* (figure 3.8:1) and *Triloboxylon* (figure 3.8:2). This similarity has been used as an argument to support the homology of the pteridosperm frond and the lateral branch system of the aneurophytaleans. In the Buteoxylon-aceae and the Callistophytaceae the leaf trace is undivided (figure 3.8:12, 13, and 15). In the Calamopityaceae it divides into at least four or eight bundles (*Kalymma* type) (figure 3.7:1–7). In the Lyginopteri-daceae the trace remains undivided but shows several protoxylem strands, as in *Tetrastichia, Laceya, Rhetinangium, Heterangium, Schop-fiastrum and Microspermopteris,* (figure 3.8:5, 7, 9, 10, 14, and 16); or the leaf trace divides into two bundles that subsequently fuse in a V- or W-shaped bundle—i.e., the true *Lyginorachis* type (*Tristichia, Lygi-nopitys, Lyginopteris;* figure 3.8:6, 8, 18); or the leaf trace divides repeatedly into eight bundles similar to the *Kalymma* type (some *Heterangium* species; figure 3.8:19) (Blanc-Louvel 1969; Shadle and Stidd 1975). Finally, in *Calathopteris* there are two distinct types of petiole anatomy (figure 3.8:11): one undivided U-shaped trace resem-bling some Lyginopteridaceae; the other that divides into three bun-dles as in *Pitus* (figure 3.9G–I).

Additional modifications affect the foliar trace higher within the petiole. Such modifications have been demonstrated in attached petioles of *Calamopitys embergeri* (Galtier 1974), in which the original leaf trace divides into 10 bundles that fuse higher into a single C-

Figure 3.8. Schematic diagram of stele and leaf trace (except **1** and **2**), with subsequent modification to petiole strand (arrows). **1** *Proteo-kalon.* **2** *Triloboxylon.* **3** *Stenokoleos bifidus.* **4** *Stenokoleos simplex.* **5** *Tetrastichia.* **6** *Tristichia.* **7** *Laceya.* **8** *Lyginopitys.* **9** *Rhetinangium.* **10** *Heterangium grievii.* **11** *Calathopteris.* **12** *Buteoxylon.* **13** *Triradioxylon.* **14** *Schopfiastrum.* **15** *Callistophyton.* **16** *Microspermopteris.* **17** *Johnhal-lia.* **18** *Lyginopteris.* **19** *Heterangium shorense.*

shaped strand (figure 3.7:2 and 3) superficially resembling *Lyginorachis*.

The important variability illustrated above demonstrates that petiolar anatomy and, still more important, the anatomy of detached rachises must be used with caution as a taxonomic character, especially in short fragments where this variation cannot be recognized.

Volume of Vascular System in Petiole Base In the Calamopityaceae (figure 3.7:1–7) the total volume of the multifascicular petiole trace largely exceeds the volume of the cauline primary vascular system at the same level. In contrast, in the Lyginopteridaceae, with the exception of *Schopfiastrum*, the volume of the petiolar trace never exceeds the stem stele volume (figure 3.8:5–10, 18, and 19). The situation is intermediate in the Buteoxylonaceae. In each case, however, there is a distinct increase in volume from the original small leaf trace to the petiolar vascular system. This can be compared with the epidogenetic phase of growth demonstrated by Scheckler (1976, 1978) in the proximal part of the lateral branch systems in progymnosperms.

THE PROBLEM OF PITUS

Reexamination of the extremely well-preserved original sections of *Pitus dayi* (Gordon 1935) has confirmed that near the apex the vascular system was composed of at least fifteen sympodia and a number of central protoxylem strands (figure 3.9C); this may be interpreted as a protostele with mixed pith or a eustele identical to that observed in more proximal levels (figure 3.9A).

Two interpretations of the origin of the *Pitus* eustele are proposed that have a very different significance in terms of taxonomic relationships.

1. The *Pitus* eustele may have evolved from the aneurophytalean type of lobed protostele through an intermediate form such as *Lyginopitys* in the Middle Tournaisian (Galtier 1970). The latter has a protostele with a mixed pith, at least eight sympodial strands, and tangential divergence of the leaf traces, but it also retained central medullary strands. The transition from *Lyginopitys* to the Upper Tournaisian *Pitus* would require only a concomitant increase in size and number of sympodia. Such a model is, in fact, parallel to Beck's (1970) proposal and would indicate that *Pitus* belongs with *Lyginopitys* in a group of early "lyginopterids" that evolved directly from the Aneurophytales and independent of the Calamopityaceae.

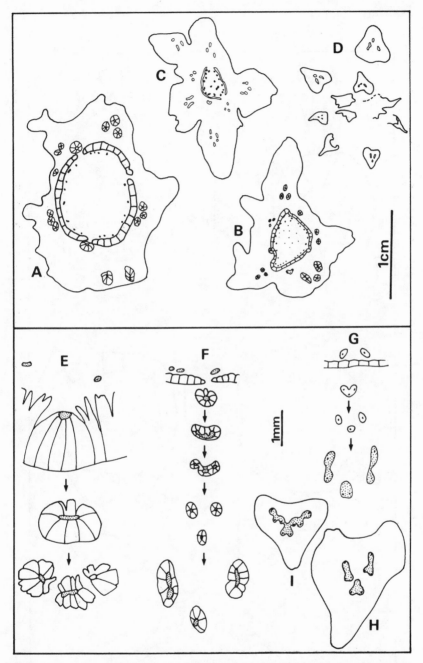

Figure 3.9. *Pitus dayi*, drawings of transverse sections from the original slides of Gordon. **A–D** Successive sections through the apex of a twig with petioles attached. **E–G** Successive stages of three leaf traces at different levels in the twig. **H, I** Petioles showing two configurations of vascular strand.

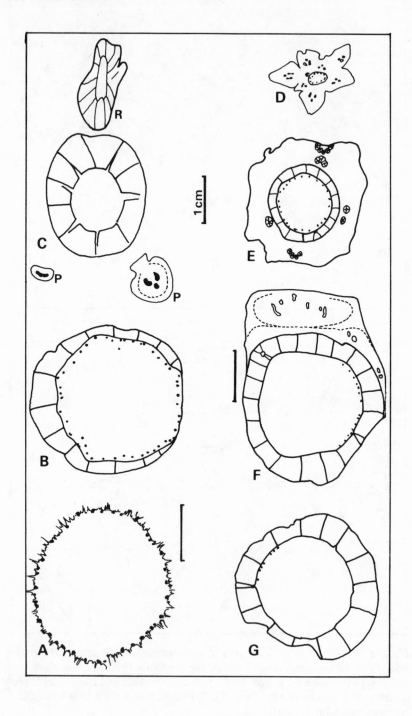

2. A second hypothesis is that *Pitus* evolved from the Archaeopteridales. Both overall size (figure 3.3) and primary vascular system diameter (figure 3.5) show *Pitus* to be closer to *Callixylon-Archaeopteris* than to any other early gymnosperms. This may be purely convergence, but the similarity is confirmed by the stelar anatomy. *Callixylon brownii* (Beck 1979) has a eustele composed of a large number of sympodia surrounding a wide parenchymatous pith (figure 3.10A) similar to that of *Pitus antiqua* (figure 3.10B) and *Pitus dayi* (figure 3.10D, E). In addition, *Callixylon* and most species of *Pitus* have pycnoxylic wood. Indeed, petioles found associated (figure 3.10C) or attached to *Pitus* stems (figures 3.9H, I and 3.10D) are suggestive of pteridosperm-like fronds (Long 1979). According to this hypothesis, *Pitus* would represent an arborescent pteridosperm with a broad eustele and pycnoxylic to manoxylic wood. The existence of such a group of plants, distinct in origin from other pteridosperms, is supported by the discovery of a still unnamed taxon from the Tournaisian of the Montagne Noire. A transverse section of this plant is illustrated in figure 3.10F. Like *Pitus* it shows a typical eustele with a large number of sympodia (over 40) and a broad parenchymatous pith without medullary strands but with pycnoxylic wood. However, leaf bases contain six to eight bundles with a *Kalymma*-type anatomy. On the basis of its petiole anatomy, this plant would be classified with the Calamopityaceae, but it differs from the calamopityans in all other features. Similarly, *Pitus* is attributed to the lyginopterids only because of its *Lyginorachis*-like petiole anatomy, but it differs in other characters. Another contemporaneous eustelic plant with a large number of sympodia and pycnoxylic wood was described by Unger (1856) from the Tournaisian of Saalfeld under the name of *Aporoxylon*. This plant (figure 3.10G) is currently under investigation. Unfortunately all the specimens are decorticated and its affinities may be difficult to elucidate.

Figure 3.10. Transverse section of stems with a broad eustele comprising a large number of sympodia (black dots) surrounding a wide parenchymatous pith. **A** *Callixylon brownii* (adapted from Beck 1979). **B** *Pitus antiqua* (Scott 1902). **C** *Pitus primaeva* (from Long 1979); stem with branch (R) and associated petioles (P). **D, E** *Pitus dayi*, sections at different levels in the axis. **F** New unnamed stem from the Montagne Noire. **G** *Aporoxylon* from Saalfeld. All about the same magnification. Secondary xylem lined. Compare with figures 3.7, 3.8.

PITUS-LYGINORACHIS-TRISTICHIA RELATIONSHIPS

The *Pitus dayi* specimens described by Gordon (1935) are the only known specimens of *Pitus* within which the entire cortex is preserved and leaf bases attached. Four serial sections near to and at the apex of one specimen are illustrated in figure 3.9A–D. Examination of leaf traces at different levels in the branch (figure 3.9E–G) confirms that they originated as a single bundle that quickly trifurcated. In the proximal part of the petiole, three bundles were arranged in an arc, with the median bundle clearly in an abaxial position (figure 3.9H); occasionally they were fused into a single V-shaped bundle with eight protoxylem strands and an abaxial ridge (figure 3.9I). Long (1979) described partly decorticated stems of *Pitus primaeva* showing identical three fascicular leaf traces; the same author (1963, 1979) also described under the name *Lyginorachis papilio* detached rachises with three to six proximal bundles that fused higher forming a single "papilionoid" strand (figure 3.11A–C) resembling *Lyginorachis papilio* (figure 3.11D) described by Crookall (1931). Long considered that these rachises belonged to *Pitus*. However, the abaxial ridge in the *Pitus dayi* petiole strand (figure 3.11F) is absent in the *Lyginorachis papilio*-type specimen (figure 3.11D) and in Long's specimens (figure 3.11A–C), where there is a groove suggestive of the future dichotomy of the petiole. Indeed, Long (1963) reconstructed a dichotomous petiole of *Lyginorachis papilio* with a "papilionoid" strand. He also assigned, to the same species, another rachis with three bundles (figure 3.11E); it is clear, however, that the arrangement of the bundles here is different from the petioles of *Pitus dayi* (figure 3.11F). More important, the three bundles do not fuse higher, as in the other specimens (figure 3.11A–C), but each represents the vascular strand of a distinct organ resulting from the trifurcation of the rachis. The median axis has a three-lobed xylem strand that Long identified as *Tristichia* (figure 3.11E). Unfortunately this organ is not preserved distally. It is on the basis of this that Long identified *Pitus* fronds as *L. papilio* trifurcated fronds bearing a *Tristichia* median axis.

In conclusion, the available information confirms that the petioles of *Pitus* had three vascular strands developed proximally that could fuse distally into a single V-shaped bundle with an abaxial ridge. There is no known species of *Lyginorachis* with the same anatomy. *Megalorhachis elliptica* (Unger 1856), from the Tournaisian of Saalfeld, certainly has the most similar, although distinct, vascular organization (figure 3.11G). The trifurcate rachis bearing a median *Tristichia*-

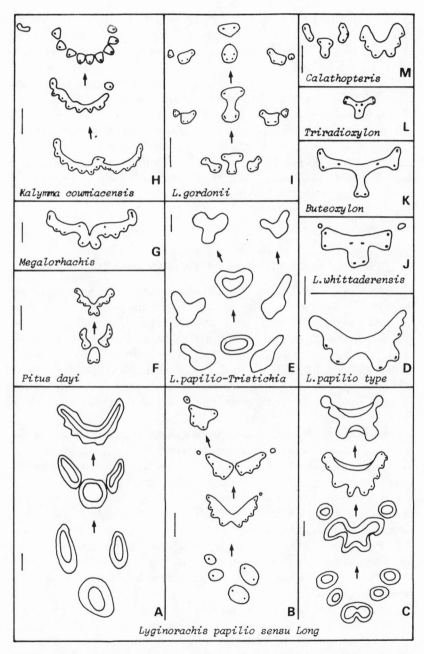

Figure 3.11. Comparison of the vascular anatomy of some pteridosperm petioles in connection and of detached rachises of the *"Lyginorachis"* type. All about the same magnification; scale represents 1 mm.

like axis (figure 3.11E) described by Long (1963) is difficult to reconcile either with the petioles of *Pitus* or *L. papilio* rachis in which the three bundles differ in arrangement and fuse higher instead of separating into different organs.

Nevertheless, I believe that Long's specimen is related to *L. gordonii* (figure 3.11I), from the Visean of Scotland (Galtier and Scott 1986). This is another trifurcate rachis with three proximal bundles, of which the median is trilobed. The median organ of *L. gordonii* was perhaps cauline in nature (as it should be in *Tristichia*) and its first division was in a plane perpendicular to frond branching. Such an organization fits with the model of a tridimensional frond suggested for compressions *Diplopteridium teilianum*, as described by Walton (1931) and *D. holdenii* (Rowe 1986).

THE PROBLEM OF THE DELIMITATION OF LYGINORACHIS

This organ genus was erected for detached rachises having a papilionoid vascular anatomy similar to that seen in the petioles of *Lyginopteris oldhamia* and therefore indicative of lyginopterid affinities. The concept of this genus has been greatly expanded, and, even though the type species *L. papilio* has a V-shaped strand, it includes species with a U-shaped bundle *(L. taitiana)* or with a T-shaped bundle (*L. whittaderensis;* figure 3.11J), which may correspond to petioles of the Buteoxylonaceae (Barnard and Long 1975) or to species with three vascular strands *(L. trinervis, L. gordonii).*

Currently *Lyginorachis* has no taxonomic significant above the generic level. Several categories may be recognized, but only in the first are there any indications of close affinities with the Lyginopteridaceae *sensu stricto*, as represented by *Lyginopteris*.

1. Petioles with a V-shaped bundle resulting from the enlargement or fusion of two bundles, as in *Lyginopteris, Lyginopitys, Tristichia, Tetrastichia,* and *Heterangium*.

2. Petioles with four or more proximal strands that fuse higher up, forming a single papilionoid strand: *Heterangium shorense, L. papilio sensu lato,* and *L. boehmii.* These represent bifurcated fronds also related to lyginopterids.

3. Petioles with a U-shaped bundle resulting from the enlargement of an undivided leaf trace: *Laceya* and *Calathopteris*.

4. Petioles with three bundles that may or may not fuse within the petiole: *Calathopteris, L. gordonii, Pitus?,* and *"L. papilio"* p.p. These clearly represent trifurcate fronds with a median (fertile?) rachis.

Their assignment to the Lyginopteridaceae has yet to be demonstrated.

5. Petioles with a T-shaped strand: *L. whittaderensis*. These may belong to unbranched rachises related not to the lyginopterids but to the Buteoxylonaceae.

Evidence from *Calathopteris* suggests that both bifurcate and trifurcate fronds may be borne on the same plant. Such a condition has recently been confirmed from compression material described by Rowe (1986), who demonstrated that both bifurcate and trifurcate fronds of *Diplopteridium* can be found attached to the same stem.

In practice, the calamopityan petioles and *Kalymma*-type rachises are distinguished by overall dimensions that exceed those of the petioles and rachises of the Lyginopteridaceae and Buteoxylonaceae. The difference is even more marked if the volume of the vascular strand is considered (figures 3.7, 3.8).

FROND MORPHOLOGY

A bifurcate petiole or primary rachis is characteristic of pteridosperm fronds in general. It has been demonstrated in *Kalymma* (Matten and Trimble 1978; Galtier 1974, 1981), *Lyginorachis* (Long 1963; Galtier 1981), and for a long time in many compressions. Some of the latter are partially permineralized, yielding information on anatomy, as in *Rhodea* and *Adiantites* (Jennings 1976, 1985). However, all the fronds were not bifurcate; some were trifurcate with a median fertile rachis. Anatomically preserved specimens of this type may correspond to *Lyginorachis trinervis*, *L. gordonii*, some *L. papilio*, and *Calathopteris*, as discussed above. Anatomy and trifurcation of *Lyginorachis gordonii* (Galtier and Scott 1986) suggest that it belonged to a frond with the same morphology as *Diplopteridium teilianum*. On the other hand, the T-shaped petiole trace of the Buteoxylonaceae may be indicative of a nonbifurcate frond (a characteristic of this family, according to Barnard and Long 1973).

Chapelia campbellii (Beck and Bailey 1967) is another example of a tridimensional frond attributed to the Calamopityaceae, of which the main rachis divides into four organs—two lateral and two median ones that might represent fertile rachises.

FOLIAGE

Several fronds with well-developed pinnules have been attributed to the Calamopityaceae—for example, *Triphyllopteris* (see Taylor 1981) or *Adiantites antiquus,* of which a partly permineralized specimen (Jennings 1985) had a vascular strand resembling the petiole of *Calamopitys embergeri* (Galtier 1974). *Kalymma coumiacensis* also had broad pinnules, with a thick lamina borne on a third order rachis (Galtier 1981). In contrast, a large frond compression with dissected pinnules of the *Sphenopteridium* type has been attributed to *Stenomyelon* by Long (1964).

Fronds with diverse foliage have been referred to early Lyginopteridaceae. Jennings (1976) described an upper Mississippian frond of *Rhodea* devoid of lamina that, he suggested, belongs to *Heterangium.* Previously, fronds of *Diplothmema* type and of *Sphenopteris obtusiloba* had been attributed to different species of *Heterangium* (Scott 1923; Shadle and Stidd 1975). In contrast, the Tournaisian *Lyginorachis boehmii* had broad opposite pinnules resembling *Rhacopteris* (Galtier 1981).

Foliage of members of the Buteoxylonaceae is unknown. Current information indicates that there evolved among the early pteridosperms highly dissected fronds devoid of lamina, fronds with dissected pinnules, and fronds with broad undissected pinnules; the foliage of the earliest (late Devonian) pteridosperms has to be documented.

CORTICAL FEATURES

An outer cortex with alternating strands of sclerenchyma and parenchyma (sparganum structure) is a primitive feature of common occurrence in pteridosperms and has no taxonomic significance.

Sclerotic nests regularly distributed in the inner cortex of stems and petioles are also very common; I consider that their absence may be a specific character (e.g., in *Calamopitys*).

Secretory cells either isolated or arranged in vertical files and resin ducts are relatively common, but secretory cavities such as those characteristic of the Pennsylvanian *Callistophyton* are not well documented.

As discussed above, the volume of cortex of primary growth was greater in calamopityans. There is no evidence of well-developed periderm in any early pteridosperm.

SECONDARY XYLEM

Manoxylic secondary xylem is generally considered characteristic of the pteridosperms, whereas their supposed aneurophytalean ancestors had pycnoxylic wood. This indicates that manoxylic wood is a derived feature. In fact, no satisfactory definition of these terms exists, and descriptions reveal wide differences of interpretation between one author and another.

Barnard and Long (1975) emphasized the resemblance between the wood of *Triradioxylon* and the aneurophytaleans, especially in terms of the distribution and size of rays but also of the presence of pitting on all tracheid walls. Pitting is generally restricted to radial walls of secondary tracheids in most pteridosperms. However, the protostelic *Laceya, Tristichia, Tetrastichia, Triradioxylon,* and *Stenomyelon bifasciculare* also show pitting on all tracheid walls, as in aneurophytaleans. This may be interpreted as additional evidence supporting the direct evolution of these early pteridosperms from aneurophytaleans.

Are there pycnoxylic pteridosperms? This question arises in the case of a number of genera with dense wood and narrow rays but of various height—e.g., *Eristophyton* or *Pitus,* where ray width varies depending on species. Zalessky created the genus *Eristophyton* for two species of *Calamopitys* with pycnoxylic wood. Later, Lacey (1953) interpreted *Eristophyton, Bilignea,* and *Endoxylon* to have a closer relationship with the Cordaitales. Unfortunately all these genera and *Megaloxylon* are known only from decorticated stems. Until their leaves and reproductive organs are known, the affinity of these genera must remain in doubt. However, the large size of their leaf traces and the phyllotaxy suggest that their leaves were not small and crowded and that these plants might be related to the pteridosperms rather than to the younger cordaitaleans.

The case of *Pitus* is different, as its secondary xylem has been described as pycnoxylic or manoxylic depending on the species. This would indicate that the genus is not a natural one or that the manoxylic versus pycnoxylic character has no taxonomic significance. Several characteristics of *Pitus* that distinguish this genus from other early gymnosperms have already been discussed. At present there is no clear argument against considering it as an arborescent pteridosperm.

PHLOEM

This delicate tissue has been recorded in some early pteridosperms: *Laceya, Tristichia longii, Lyginopitys, Pitus dayi, Rhetinangium, Heterangium grievii, Calamopitys annularis,* and *C. embergeri.* The only detailed description of this tissue is from *Calamopitys embergeri,* in which it consists of alternating rows of sieve cells and parenchyma cells

(Galtier and Hébant 1973). In fact, the older aneurophytaleans possessed a more complex secondary phloem with four types of constituent cells (Scheckler and Banks 1971; Wight and Beck 1984). The presence or absence of fibers and tanniniferous cells may be a character to be considered carefully in future studies as of possible taxonomic significance among early pteridosperms.

BRANCHING AND HABIT

Branching has been documented in a very small number of pteridosperms: the Pennsylvanian *Lyginopteris*, *Heterangium* (one record in Williamson and Scott 1895), *Microspermopteris*, *Callistophyton*, *Johnhallia* and the Mississippian *Pitus* (Long 1979), and *Calamopitys* (Galtier and Holmes 1981). In all these taxa, except *Pitus*, it has been shown that branching was axillary but with some variation in the spatial relationship of the branch trace to the axillant leaf trace. *Calamopitys* and, to a lesser degree, *Lyginopteris* seem to represent the primitive condition in the evolution of the axillary branch. Branching in the arborescent *Pitus* was demonstrated as being rhythmic and similar to that of some modern conifers (Long 1979); however, it is not clear if branching was axillary or pseudomonopodial, as in the progymnosperms.

Taxa with slender stems have the largest internodes (2–20 cm, or 3 to 10 times their diameter)—e.g., *Laceya*, *Tristichia*, *Triradioxylon*, *Calamopitys*, *Stenomyelon*, and the Pennsylvanian *Schopfiastrum*. Probably these plants were not strong enough to support their large fronds, and it is supposed that they had a scrambling or lianescent habit. This supposition is also supported by the wide angle of attachment and swollen base of the petioles, as in present lianas. With the exception of *Pitus*, all the examples of branching stems recorded correspond to plants having this habit; this certainly represents an adaptive strategy, as opposed to the monocaulous habit typical of most pteridosperms.

POLLEN ORGANS

Our knowledge of pollen organs attributed to early pteridosperms is poor in comparison with the substantial information on female structures. The Lower Carboniferous fructifications *Zimmermannitheca*, *Schuetzia*, *Alcicornopteris*, *Staphylotheca*, *Simplotheca*, *Paracalathiops*, *Protopitys*, and *Geminitheca* are all composed of compressed clusters of terminal sporangia and are considered by Millay and Taylor (1979) as the male organs of early members of seed ferns. *Protopitys* is, however, generally referred to progymnosperms (Beck 1976). In fact, evidence from permineralized fructifications indicates that the synangiate stage was already established in the early Mississippian.

Telangium (permineralization) and *Telangiopsis* (compression) from the Lower Carboniferous are represented by different types of simple erect synangia of six to eight sporangia more or less comparable to the Pennsylvanian species *Telangium scottii* that Benson (1904) described as a lyginopterid fructification. In the *Telangium* attributed to *Pitus* by Long (1979), eight sporangia are arranged in a single, oval whorl (figure 3.12C) and, basally, embedded in a parenchymatous pad vascularized by dichotomizing strands that end at the base of each sporangium. The *Telangium* that Jennings (1976) described, associated with a *Heterangium*, had eight sporangia organized in a circle around a basal, central column of parenchyma with a unique central vascular strand (figure 3.12D). In contrast, *Telangium scottii* (figure 3.12E) was clearly bilateral, with two rows of four sporangia, all of which were fused to each other. *Telangiopsis bifidum* (Kidston 1923) represented another type of synangium, with a large number of sporangia (twenty-five) attached on the margin of a disc.

Two new genera of synangiate pollen organs have recently been described from the Visean of Scotland. They extend the range of characters defining the earliest synangiate pteridosperm fructifications. *Melissiotheca* (figure 3.12B) is a broad, simple synangium composed of more than one hundred sporangia partly fused and incorporated basally in a massive parenchymatous cushion. Vascular strands divide repeatedly and terminate at each sporangium base (Meyer-Berthaud 1986). *Phacelotheca* (Meyer-Berthaud and Galtier 1986) is interpreted as aggregated synangia of two to four sporangia, terminally borne on foreshortened tridimensional dichotomizing axes (figure 3.12A). The vascularization is similar to that of *Melissiotheca*.

Figure 3.12 illustrates the synangiate pollen organs of the Lower Carboniferous and those of the early Pennsylvanian considered to be "probable lyginopterid fructifications" on the basis of their sporangia, with distally free tips, and on their common spherical and trilete prepollen. In this group *Vallitheca* (figure 3.12F), *Potoniea* (figure 3.12G), and *Schopfiangium* (figure 3.12H) are larger and more complex than *Telangiopsis bifidum*, *Phacelotheca*, or *Melissiotheca*. Work in progress by Meyer-Berthaud on Lower Carboniferous pollen organs suggests that two main trends may be recognized among these Mississippian and Pennsylvanian pteridosperm pollen organs. The first, reflected in *Phacelotheca* (figure 3.12A), *Telangium scottii* (figure 3.12E), and *Valli-*

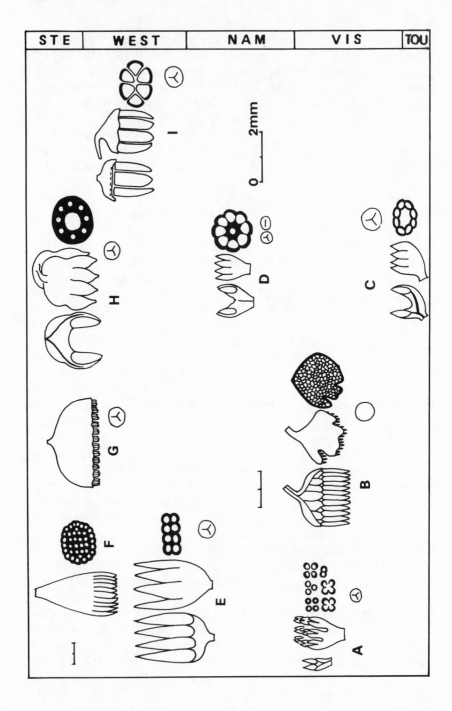

theca (figure 3.12F), with a basic bilateral symmetry, may have resulted from aggregation of paired sporangia terminally borne on dichotomizing axes. *Potoniea* is a broad compound synangium that may be derived from aggregated synangia. A second trend would include synangia with a small number of sporangia (up to eight) arranged in a uniseriate ring, mostly radial in symmetry, as in *Telangium sp.* Long (figure 3.12C) and *Telangium sp.* Jennings (figure 3.12D). *Schopfiangium* (figure 3.12H) may also be interpreted as an aggregation of synangia of this type. *Melissiotheca* may belong to a third category, in which fusion has affected large fertile portions of aneurophytalean plants such as *Tetraxylopteris*. The Pennsylvanian *Feraxotheca* differs from all the previous fructifications by its free sporangia attached to a proximal, common pad (figure 3.12I).

The representation of sporangia as erect or pendent, in figure 3.12, is speculative and results from comparison with compressions of similar size and morphology. Our knowledge of frond architecture is based only on compressions. Several categories of fertile fronds can be recognized: fronds with their distal part entirely fertile and lacking pinnules—e.g., *Triphyllopteris* and *Telangium arkansanum;* fronds with a median fertile rachis borne on the dichotomy *(Diplopteridium);* and fronds with fertile pinnae intermixed with sterile pinnae *(Telangiopsis).*

All of the fructifications discussed above have spherical and trilete prepollen except *Telangium sp.* Jennings, the prepollen of which is sometimes monolete, and *Melissiotheca*, in which the trilete mark has not been recognized with confidence. The prepollen grains are small, ranging from 25 μm *(Telangium* sp. Jennings) to 65 μm *(Feraxotheca)* in diameter. This corresponds to the "lyginopterid morphology" as defined by Millay and Taylor (1979) and contrasts with the large size

Figure 3.12. Diagram of the main types of permineralized pollen organs of early pteridosperms compared with Pennsylvanian pollen organs producing trilete prepollen. In each case, a transverse section, a longitudinal section, and prepollen are schematized. Note the different magnifications. **A** *Phacelotheca* (from Meyer-Berthaud and Galtier 1986). **B** *Melissiotheca* (Meyer-Berthaud 1986). **C** *Telangium* sp. (from Long 1979). **D** *Telangium* sp. (from Jennings 1976). **E** *Telangium scottii* (from Benson 1904). **F** *Vallitheca* (from Mapes and Schabilion 1981). **G** *Potoniea* (Stidd 1978). **H** *Schopfiangium* (Stidd et al. 1985). **I** *Feraxotheca* (Millay and Taylor 1979). (Figure prepared by B. Meyer-Berthaud).

of pseudosaccate spores found in compressions like *Paracalathiops* or *Simplotheca* (up to 250 μm), which are comparable in size and morphology to the spores of the supposed aneurophytalean ancestors.

Ultrastructure of the prepollen wall is an important systematic character. In the Lower Carboniferous fructifications where it has been studied, *Melissiotheca* and *Phacelotheca*, the sporoderm is composed of a thin (lamellated?) nexine and a thick heterogeneous sexine; it is comparable to the wall of the spores that have been attributed to *Schopfiangium*. It has no feature in common with the sporoderm of *Potoniea*, which consists of a thick nexine and a thin sexine lined by conspicuous orbicules. It is also different from the prepollen of *Crossotheca* with a homogeneous sexine and a filicinean aperture that Millay, Eggerts, and Dennis (1978) doubtfully assigned to the lyginopterids.

CONCLUSION

While there is an increasing amount of data on both vegetative and reproductive structures attributed to early pteridosperms, we lack evidence of connection between the two, and at the present time not a single example of a species has been reconstructed as a whole plant. The present review emphasizes the difficulty in interpreting the evolutionary significance of some characters and, as a consequence, in defining relationships among the current pteridosperm families Calamopityaceae, Buteoxylonaceae, and Lyginopteridaceae. However, the large number of advanced features shared by these plants supports their interpretation as a natural group (e.g., Lyginopteridales of Barnard and Long 1975; Taylor 1981). Features such as eustele, tangential leaf trace divergence, broad pinnules, and axillary branching apparently evolved independently and synchronously in calamopityans and lyginopterids. This may be interpreted as evidence of very close affinity if not of identity between these two families, but I suggest that this, in fact, represents convergence. This conclusion is supported by the differences in leaf trace and petiole anatomy and stelar organization. The latter must be used as taxonomic characters at the familial and generic level, but their interpretation is problematic because of ontogenetic variability and convergence. The ratio of primary vascular system diameter to stem diameter clearly separates the Calamopityaceae from the Lyginopteridaceae. Similarly, the cumulative volume of the petiolar vascular system exceeds the stelar volume in calamopityans but never in lyginopterids. These differences parallel the diversity recognized in contemporaneous fertile structures. We have demonstrated an early diversification of synangiate pollen organs, but we have no basis on which to attribute these organs to one family or another. The same uncertainty exists with regard to female organs, where the attribution of radiospermic ovules to the lyginopterids is generally accepted while the attribution of platyspermic ovules to the calamopityans has been controversial.

I consider that the diversification that existed among early pteridosperms, in both vegetative and fertile structures, supports the hypothesis that they were composed of at least two and probably more groups of familial rank. A revision of the systematics of these plants depends on the discovery of new evidence demonstrating direct or indirect connection of fertile and vegetative structures.

In conclusion, at the present time, I prefer to continue to use the current families Calamopityaceae, Buteoxylonaceae, and Lyginopteridaceae. *Pitus* certainly belongs to a fourth group, either directly derived from the Archaeopteridales or with closer affinities to the lyginopterids, as has been suggested by Long.

ACKNOWLEDGMENTS

I am particularly indebted to B. Meyer-Berthaud and G. Rex for helpful criticism and improvement of the manuscript and for technical assistance. Figure 3.12 was prepared by B. Meyer-Berthaud. I thank J. Courbet for assistance with the illustrations.

LITERATURE CITED

Barnard, P. D. W. and A. G. Long. 1973. On the structure of a petrified stem and some associated seeds from the Lower Carboniferous rocks of East Lothian, Scotland. *Trans. Roy. Soc. Edinburgh* 69:91–108.

—— 1975. *Triradioxylon*, a new genus of Lower Carboniferous petrified stems and petioles together with a review of the classification of early Pterophytina. *Trans. Roy. Soc. Edinburgh* 69:231–250.

Benson, M. 1904. *Telangium scotti*, a new species of *Telangium* (*Calymmatotheca*) showing structure. *Ann. Bot.* 18:161–177.

Beck, C. B. 1957. *Tetraxylopteris schmidtii* gen. et sp. nov., a probable pteridosperm precursor from the Devonian of New York. *Am. J. Bot.* 44:350–367.

—— 1970. The appearance of gymnospermous structure. *Biol. Rev.* 45:379–400.

—— 1976. Current status of the Progymnospermopsida. *Rev. Palaeobot. Palynol.* 21:5–23.

—— 1979. The primary vascular system of *Callixylon. Rev. Palaeobot. Palynol.* 28:103–115.

—— 1981. *Archaeopteris* and its role in vascular plant evolution. In K. J. Niklas, ed., *Paleobotany, Paleoecology, and Evolution,* vol. 1, pp. 193–229. New York: Praeger.

—— 1985. Gymnosperm phylogeny—a commentary on the views of S. V. Meyen. *Bot. Rev.* 51:273–294.

Beck, C. B. and R. Bailey. 1967. Plants of the New Albany Shale. 3. *Chapelia campbellii gen. n. Am. J. Bot.* 54:998–1007.

Beck, C. B., R. Schmid, and G. R. Rothwell. 1982. Stelar morphology and the primary vascular system of seed plants. *Bot. Rev.* 48:691–815.

Beck, C. B. and W. E. Stein. 1987. *Galtiera bostonense,* gen. et sp. nov., a protostelic calamopityacean from the New Albany Shale of Kentucky. *Can. J. Bot.* 64: 348–361.

Blanc-Louvel, C. 1966. Etude anatomique comparée des tiges et des pétioles d'une Ptéridospermée du Carbonifère du genre *Lyginopteris Potonié. Mèm. Mus. Natl. Hist. Nat. Paris,* n.s., C 18.

—— 1969. Sur la phyllotaxie d'*Heterangium shorense* Scott. *Ann. Paléont. (Invert.)* 55:143–154.

Chaloner, W. G. and A. Sheerin. 1979. Devonian macrofloras. In M. B. House, C. T. Scruton, and M. G. Basset, eds., *The Devonian system.* Special paper on Paleontology 23:145–161. London: Palaeontological Association.

Crane, P. 1985. Phylogenetic analysis of seed plants and the origin of angiosperms. *Ann. Missouri Bot. Gard.* 72:716–793.

Crookall, R. 1931. The genus *Lyginorachis* Kidston. *Roy. Soc. Edinburgh Proc.* 51:27–34.

Delevoryas, T. 1955. The Medullosae—structure and relationships. *Palaeontographica* 97 B:114–167.

Emberger, L. 1968. *Les Plantes Fossiles dans leurs Rapports avec les Végétaux Vivants.* Paris: Masson.

Galtier, J. 1970. Recherches sur les végétaux à structure conservée du Carbonifère inférieur Français. *Paléobiol. Continentale* 1:1–221.

—— 1973. Remarques sur l'organisation et la signification phylogénétique de la stèle des Calamopityacées. *C. R. Acad. Sci. Paris* 276:2147–2150.

—— 1974. Sur l'organisation de la fronde de *Calamopitys,* Ptéridospermales probables du Carbonifère inférieur. *C. R. Acad. Sci. Paris* 279:975–978.

—— 1975. Variabilité anatomique et ramification des tiges de *Calamopitys. C. R. Acad. Sci. Paris* 280:1967–1970.

—— 1977. *Tristichia longii,* nouvelle Ptéridospermale probable du Carbonifère de la Montagne Noire. *C. R. Acad. Sci. Paris* 284:2215–2218.

—— 1981 Structures foliaires de fougères et Ptéridospermales du Carbonifère inférieur et leur signification évolutive. *Palaeontographica* 180B:1–38.

Galtier, J. and C. Hébant. 1973. Sur le phloème et le cambium d'une Calamopityacée, Ptéridospermale probable du Carbonifère inférieur Français. *C. R. Acad. Sci. Paris* 276:2257–2259.

Galtier, J. and J. C. Holmes. 1982. New observations on the branching of Carboniferous ferns and pteridosperms. *Ann. Bot.* 49:737–746.

Galtier, J., T. L. Phillips, and F. Chalot-Prat. 1986. Euramerican coal-swamp plants in Mid-Carboniferous of Morocco. *Rev. Palaeobot. Palynol.* 49:93–98.

Galtier, J. and A. C. Scott. 1986. *Lyginorachis gordonii*, nouvelle Ptéridospermale probable du Carbonifère inférieur d'Ecosse. *C. R. Acad. Sci. Paris* 302:251–256.

Gordon, W. T. 1912. On *Rhetinangium arberi*, a new genus of Cycadofilices from the Calciferous Sandstone Series. *Trans. Roy. Soc. Edinburgh* 48:813–825.

—— 1935. The genus *Pitys* Witham emend. *Trans. Roy. Soc. Edinburgh* 58:279–311.

—— 1938. On *Tetrastichia bupatides:* a Carboniferous pteridosperm from East Lothian. *Trans. Roy. Soc. Edinburgh* 59:351–370.

Jennings, J. R. 1976. The morphology and relationships of *Rhodea, Telangium, Telangiopsis* and *Heterangium*. *Am. J. Bot.* 63:1119–1133.

—— 1985. Fossil plants from the Mauch Chunk Formation of Pennsylvania: Morphology of *Adiantites antiquus*. *J. Paleontol* 59:1146–1157.

Kidston, R. 1923–1925. Fossil plants of the Carboniferous rocks of Great Britain. *Mem. Geol. Surv. Palaeontology* 2:1–670.

Knoll, A. M. 1986. Patterns of change in plant communities through geological time. In J. Diamond and J. J. Case, eds., *Community Ecology*, 126–141. New York: Harper & Row.

Lacey, W. S. 1953. Scottish Lower Carboniferous plants: *Eristophyton waltoni* sp. nov. and *Endoxylon zonatum* (Kidst.) Scott from Dumbartonshire. *Ann. Bot.* 17:579–596.

Leclercq, S. and P. M. Bonamo. 1971. A study of the fructification of *Milleria (Protopteridium) thomsonii* Lang from the Middle Devonian of Belgium. *Palaeontographica* 136B:83–114.

Long, A. G. 1960. On the structure of *"Samaropsis scotica"* Calder (emended) and *"Eurystoma angulare"* gen. et sp. nov., petrified seeds from the Calciferous Sandstone Series of Berwickshire. *Trans. Roy. Soc. Edinburgh* 64:261–280.

—— 1961. *Tristichia ovensi* gen. et sp. nov.: A protostelic Lower Carboniferous pteridosperm from Berkwickshire and East Lothian, with an account of some associated seeds and cupules. *Trans. Roy. Soc. Edinburgh* 64:477–489.

—— 1963. Some specimens of *"Lyginorachis papilio"* Kidston associated with stems of *"Pitys."* *Trans. Roy. Soc. Edinburgh* 65:211–224.

—— 1964. Some specimens of *Stenomyelon* and *Kalymma* from the Calciferous Sandstone Series of Berwickshire. *Trans. Roy. Soc. Edinburgh* 65:435–447.

—— 1969. *Eurystoma trigona* sp. nov., a pteridosperm ovule borne on a frond of *Alcicornopteris* Kidston. *Trans. Roy. Soc. Edinburgh* 68:171–182.

—— 1976. *Calathopteris heterophylla* gen. et sp. nov., a Lower Carboniferous pteridosperm bearing two kinds of petioles. *Trans. Roy. Soc. Edinburgh* 69:327–336.

—— 1979. Observations on the Lower Carboniferous genus *Pitus* Witham. *Trans. Roy. Soc. Edinburgh* 70:111–127.

Mapes, G. 1985. *Megaloxylon* in midcontinent North America. *Bot. Gaz.* 146:157–167.

Mapes, G. and G. W. Rothwell. 1980. *Quaestora amplecta* gen. et sp. n., a structurally simple medullosan stem from the Upper Mississippian of Arkansas. *Am. J. Bot.* 67:636–647.

Mapes, G. and J. T. Schabilion. 1981. *Vallitheca valentia* gen. et sp. nov., permineralized synangia from the Middle Pennsylvanian of Oklahoma. *Amer. J. Bot.* 68:1231–1239.

Matten, L. C. and H. P. Banks. 1966. *Triloboxylon ashlandicum* gen. and sp. n. from the Upper Devonian of New York. *Am. J. Bot.* 53:1020–1028.

—— 1969. *Stenokoleos bifidus* sp. n. in the Upper Devonian of New York State. *Am. J. Bot.* 56:880–891.

Matten, L. C., W. S. Lacey, B. I. May, and R. C. Lucas. 1980. A megafossil flora from the uppermost Devonian near Ballyheigue, Co. Kerry, Ireland. *Rev. Palaeobot. Palynol.* 29:241–251.

Matten, L. C., W. R. Tanner, and W. S. Lacey. 1984. Additions to the silicified Upper Devonian/Lower Carboniferous flora from Ballyheigue, Ireland. *Rev. Palaeobot. Palynol.* 43:303–320.

Matten, L. C. and L. J. Trimble. 1978. Studies on *Kalymma*. *Palaeontographica* 167B:161–174.

May, B. I. and L. C. Matten. 1983. A probable pteridosperm from the uppermost Devonian near Ballyheigue, Co. Kerry, Ireland. *Bot. J. Linn. Soc.* 86:103–123.

Meyen, S. W. 1984. Basic features of gymnosperm systematics and phylogeny as evidenced by the fossil record. *Bot. Rev.* 50:1–111.

Meyer-Berthaud, B. 1984. *Stenomyelon* from the Upper Tournaisian of the Montagne Noire (France). *Can. J. Bot.* 62:2297–2307.

—— 1986. *Melissiotheca:* A new pteridosperm pollen organ from the Lower Carboniferous of Scotland. *Bot. J. Linn. Soc.* 93:277–290.

Meyer-Berthaud, B. and J. Galtier. 1986. Studies on a Lower Carboniferous flora from Kingswood near Pettycur, Scotland. 2. *Phacelotheca*, a new synangiate fructification of pteridospermous affinities. *Rev. Palaeobot. Palynol.* 48:181–198.

Millay, M. A. and T. N. Taylor. 1979. Paleozoic seed fern pollen organs. *Bot. Rev.* 45:301–375.

Millay, M. A., D. A. Eggert, and R. L. Dennis. 1978. Morphology and ultrastructure of four Pennsylvanian prepollen types. *Micropaleont.* 24:303–315.

Miller, C. N. 1985. A critical review of S. V. Meyen's "Basic features of gymnosperm systematics and phylogeny as evidenced by the fossil record." *Bot. Rev.* 51:295–318.

Namboodiri, K. K. and C. B. Beck. 1968. A comparative study of the primary vascular system of conifers. III. Stelar evolution in gymnosperms. *Am J. Bot.* 55:464–472.

Phillips. T. L. 1981. Stratigraphic occurrence and vegetational patterns of Pennsylvanian pteridosperms in Euramerican Coal Swamps. *Rev. Palaeobot. Palynol.* 32:5–26.

Phillips, T. L. and H. N. Andrews. 1963. An occurrence of the medullosan seed-fern *Sutcliffia* in the American Carboniferous. *Ann. Missouri Bot. Gard.* 50:29–51.

Read, C. B. 1936. The flora of the New Albany Shale. Part 1. *Diichnia kentuckiensis*, a new representative of the Calamopityeae. *U.S. Geol. Surv. Prof. Paper* 185-H:149–161.

—— 1937. The flora of the New Albany Shale. Part 2. The Calamopityeae and their relationships. *U.S. Geol. Surv. Prof. Paper* 186-E:81–104.

Renault, B. 1893–1896. Bassin houiller et permien d'Autun et d'Epinac. 4. Flore fossile; 2. Atlas (1893), plates, 28–89; text (1896). Paris: Études Gîtes Minéraux de la France.

Rothwell. G. W. 1975. The Callistophytaceae (Pteridospermopsida): 1. Vegetative structures. *Palaeontographica* 151B:171–196.

—— 1976. Primary vasculature and gymnosperm systematics. *Rev. Palaeobot. Palynol* 22:193–206.

—— 1982. New interpretations of the earliest conifers. *Rev. Palaeobot. Palynol.* 37:7–28.

—— 1985. The role of comparative morphology and anatomy in interpreting the systematics of fossil gymnosperms. *Bot. Rev.* 51:319–327.

Rothwell, G. W. and T. N. Taylor. 1972. Carboniferous pteridosperm studies: Morphology and anatomy of *Schopfiastrum decussatum. Can. J. Bot.* 50:2649–2658.

Rowe, N. P. 1986. *The fossil flora of the Drybrook sandstone (Lower Carboniferous) from the Forest of Dean, Gloucestershire.* Ph.D. dissertation, University of Bristol.

Scheckler, S. E. 1976. Ontogeny of progymnosperms. 1. Shoots of Upper Devonian Aneurophytales. *Can. J. Bot.* 54:209–219.

—— 1978. Ontogeny of progymnosperms. 2. Shoots of Upper Devonian Archaeopteridales. *Can. J. Bot.* 56:3136–3170.

Scheckler, S. E. and H. P. Banks. 1971. *Proteokalon, a new genus of progymnosperms from the Devonian of New York State and its bearing on phylogenetic trends in the group. Am. J. Bot.,* 58:874–884.

Scott, A. C., J. Galtier, and G. Clayton. 1984. Distribution of anatomically preserved floras in the Lower Carboniferous in western Europe. *Trans. Roy. Soc. Edinburgh* 75:311–340.

Scott, D. H. 1902. On the primary structure of certain paleozoic stems with the *Dadoxylon* type of wood. *Trans. Roy. Soc. Edinburgh* 40:331–365.

—— 1923. *Studies in Fossil Botany*, vol. 2. 3rd ed. London: Black.

—— 1924. Fossil plants of the *Calamopitys* type, from the Carboniferous rocks of Scotland. *Trans. Roy. Soc. Edinburgh* 53:569–596.

Serlin, B. S. and H. P. Banks. 1978. Morphology and anatomy of *Aneurophyton,* a progymnosperm from the late Devonian of New York. *Palaeontographica americana* 8:343–359.

Seward, A. C. 1917. *Fossil Plants*, vol. 3. Cambridge: Cambridge University Press.

Shadle, G. L. and B. M. Stidd. 1975. The frond of *Heterangium*. *Am. J. Bot.* 62:67–75.

Solms-Laubach, H. 1893. Über die in den Kalksteinen des Culm von Glätzisch-Falkenberg in Schlesien enthaltenen strukturbietenden Pflanzenreste. *Bot. Zeitung* 51:197–210.

Sporne, K. R. 1974. *The Morphology of Gymnosperms: The Structure and Evolution of Primitive Seed Plants*, 2nd ed. London: Hutchinson.

Stein, W. E., Jr. and C. B. Beck. 1978. *Bostonia perplexa* gen. et sp. nov., a calamopityan axis from the New Albany Shale of Kentucky. *Am. J. Bot.* 65:459–465.

Stewart, W. N. 1983. *Paleobotany and the Evolution of Plants*. Cambridge: Cambridge University Press.

Stidd, B. M. 1978. An anatomically preserved *Potoniea* with *in situ* spores from the Pennsylvanian of Illinois. *Am. J. Bot.* 65:677–683.

Stidd, B. M. and T. L. Phillips. 1982. *Johnhallia lacunosa* gen. et sp. no.: A new pteridosperm from the middle Pennsylvanian of Indiana. *J. Paleont.* 56:1093–1102.

Stidd, B. M., M. O. Rischbieter, and T. L. Phillips. 1985. A new lyginopterid pollen organ with alveolate pollen exines. *Am. J. Bot.* 72:501–508.

Taylor, T. N. 1981. *Paleobotany: An Introduction to Fossil Plant Biology*. New York: McGraw-Hill.

Taylor, T. N. and D. A. Eggert. 1967. Petrified plants from the Upper Mississippian (Chester Series) of Arkansas. *Trans. Am. Micros. Soc.* 86:412–416.

Taylor, T. N. and R. A. Stockey. 1976. Studies of paleozoic seed ferns: Anatomy and morphology of *Microspermopteris aphyllum*. *Am. J. Bot.* 63:1302–1310.

Unger, F. 1856. Schiefer and Sandsteinflora: Zweiter theil. In R. Richter and F. Unger, Beitrag zur Paläontologie des Thüringer waldes. *Denkschr. Kaiser. Akad. Wiss. Wien.* 11:139–186.

Walton, J. 1931. Contributions to the knowledge of Lower Carboniferous plants. Part III. On the fossil flora of the black limestones in Teilia Quarry, Gwaen-ysgor, near Prestatyn, Flintshire, with special reference to *Diplopteridium teilianum* Kidston sp. (gen. nov.) and some other fern-like fronds. *Philos. Trans. Roy. Soc. London* 219B:347–379.

Wight, D. C. and C. B. Beck. 1984. Sieve cells in phloem of a Middle Devonian progymnosperm. *Science* 225:1469–1471.

Williamson, W. C. and D. H. Scott. 1895. Further observations on the organization of the fossil plants of the Coal Measures. 3. *Lyginodendron* and *Heterangium*. *Philos. Trans. Roy. Soc. London*, 186:703–779.

4

Pollen and Pollen Organs of Fossil Gymnosperms: Phylogeny and Reproductive Biology

THOMAS N. TAYLOR

⣿

Almost all fossil gymnosperm groups have at one time or another played a key role in discussions of the evolution of the angiosperms. Many of these studies have focused on features relating to the closure of an organ (e.g., cupule or megasporophyll) around seeds, thus satisfying what has been historically suggested as the precursor to the angiosperm carpel (Zavada and Taylor 1986). By comparison, the pollen organs have received relatively little attention, in part because it has been difficult to homologize the pollen-producing organs among diverse groups of fossil gymnosperms. In this paper pollen organs and pollen of all major groups of gymnosperms are discussed. In some groups like the Ginkgophytes and Gnetophytes relatively little is known about these organs from the fossil record. In other groups like the fossil conifers only selected taxa will be considered because of space limitations and other contributions in this volume. For each group several of the major pollen organs are considered together with details about *in situ* pollen. Where information is available, data on pollen ultrastructure and features of the microgametophyte are included.

LYGINOPTERIDALES

The Lyginopteridales are Paleozoic plants that were first characterized as producing seeds on fern-like foliage (Oliver and Scott 1904). However, despite the historical importance of the discovery of pteridosperms, the group remains today one of the least understood in terms of whole plant biology (Taylor and Millay 1981b). The order is restricted to sediments of Carboniferous age, and no doubt many of the isolated ovules (e.g., Long 1961, 1966) known from Mississippian deposits were produced by lyginopterid seed ferns.

In general the pollen organs of the group are poorly known and few have been conclusively demonstrated to belong to particular stem taxa. In addition, few are known from structurally preserved specimens, thus making it difficult to understand the evolution of organs within the order. One of the few structurally preserved pollen organs that is believed to have been produced by a lyginopterid seed fern is *Feraxotheca* (Millay and Taylor 1977, 1978). This fructification consists of a parenchymatous pad that bears four to eight laterally attached sporangia (figure 4.1A). The organization of the entire synangium may be radial or bilateral, depending on the number of sporangia that are present. Dehiscence of each sporangium takes place through the thin inner facing wall (figure 4.2B). Morphologically *Feraxotheca* appears quite similar to the compressed pollen organ *Crossotheca* (figure 4.2A), some forms of which have been long regarded as belonging to the lyginopterid seed ferns. Specimens of *Crossotheca* are variable in the number of sporangia per synangium and also in the position and number of synangia borne on the frond (Stubblefield et al. 1982).

Another pollen organ that has been traditionally associated with the lyginopterid seed ferns is *Telangium* (figure 4.1B). This generic name is today used only for anatomically preserved specimens. *Telangium scottii* (Benson 1904) is characterized by bilateral synangia with six to eight sporangia arranged in parallel rows. Like *Feraxotheca*,

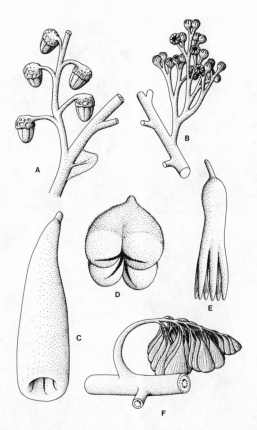

Figure 4.1. **A** *Feraxotheca.* **B** *Telangium.* **C** *Halletheca.* **D** *Rhetinotheca.* **E** *Codonotheca.* **F** *Parasporotheca.*

sporangia were fused laterally and dehiscence occurred along the inner facing wall. Morphologically similar synangia that are preserved as impression/compressions are assigned to *Telangiopsis* (Eggert and Taylor 1971). In *T. arkansanum,* an Upper Mississippian form, the synangia are borne terminally on either dichotomously or monopodially branched axes lacking planated foliar structures.

Potoniea is an interesting pollen organ that has been placed in both the lyginopterid and medullosan seed fern groups. Although the discovery of structurally preserved specimens clarifies the nature of the organ (Stidd 1978a), the taxonomic position continues to remain speculative. The fructification is a compound unit consisting of concentric rings of radial sporangia arranged in clusters of four to six (figure

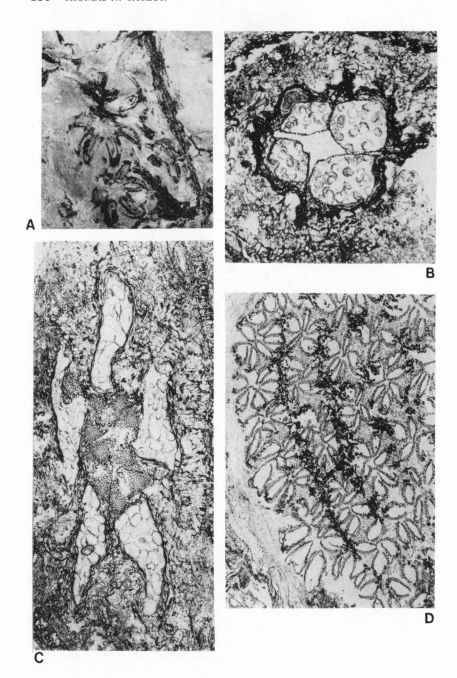

4.2D), with the free tips of the sporangia extending from the distal surface of the organ. Pollen was shed toward the center of each radial cluster, a feature that appears to be consistent for all pteridosperm pollen organs. Tubular sporangia arranged in small clusters that are in turn organized in concentric cycles also characterize the Mississippian pollen organ *Vallitheca* (Mapes and Schabilion 1981). The assignment of this microsporangiate organ to the lyginopterid seed ferns is perhaps even more tentative, since pollen was not preserved.

Two recently described structurally preserved pollen organs that may have affinity with the lyginopterid seed ferns are *Schopfiangium* (Stidd et al. 1985) and *Phacelotheca* (Meyer-Berthaud and Galtier 1986). In *S. varijugatus* each synangium consists of seven to nine sporangia, with the sporangial tips elongated. Synangia of the older (Visean) pollen organ *P. pilosa* are constructed of two to four sporangia fused at their bases and attached to a reduced frond axis. Both *Schopfiangium* and *Phacelotheca* are morphologically similar to other lyginopterid pollen organs except for features of the pollen.

Several evolutionary trends have been suggested for the pollen organs of the lyginopterid pteridosperms (Millay and Taylor 1979). These include the shift from paired, nonsynangiate terminal sporangia of some trimerophyte or progymnosperm to the radial clustering of sporangia and finally to distinct synangia arranged in a ring with a hollow central region. Microsporangia were initially borne on three-dimensional branches, with the branching systems of geologically younger types more planated and associated with lamina. Until recently the pollen of lyginopterid pteridosperms was considered to be rather homogeneous, consisting of trilete grains with a homogeneous sporoderm. The recent discovery of lyginopterid pollen possessing an alveolate infrastructure serves to demonstrate the diversity within the group or to underscore the polyphyletic origin of the order.

Pollen   Pollen of the lyginopterid seed ferns is simple and generally indistinguishable from pteridophyte spores (figure 4.3A, B). The tri-

Figure 4.2. **A** *Crossotheca kentuckiensis* showing several reflexed sporangia. Longitudinal lines on sporangia are believed to be dehiscence slits. × 6.5. **B** Transverse section of *Feraxotheca culcitaus* synangium. × 65. **C** Transverse section of *Halletheca reticulatus* showing solid central column surrounded by ring of five pollen sacs. × 19. **D** Transverse section of *Potoniea illinoiensis* pollen organ. Note that the clusters of pollen sacs are arranged in concentric rings. × 12.

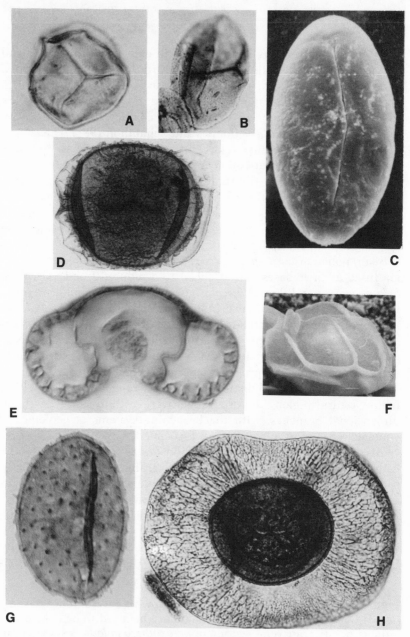

Figure 4.3. Fossil gymnosperm pollen. **A** *Crossotheca hughesiana.* ×
500. **B** *Crossotheca kentuckiensis.* × 500. **C** *Monoletes* sp. × 200. **D**
Parasporites sp. × 200. **E** *Vesicaspora* sp. × 1500. **F** *Cycadeoidea dacotensis.* × 1000. **G** *Nanoxanthiopollenites* sp. × 375. **H** *Felixipollenites
macroreticulata.* × 500.

lete grains are 40–80 μm in diameter and are ornamented by relatively small sculptural elements. The exine is two-parted, consisting of an inner homogeneous nexine and a sculptured sexine. Thin sections of the pollen of several species of *Crossotheca* indicate that the nexine extends through the sporoderm and lines the inner surface of the suture (Taylor and Smoot, in press). Depending upon the stage of sporoderm development, nexine lamellae may be present. Some stages in exine development are known for *Potoniea carpentieri* and include the initiation of nexine lamellae (figure 4.4A) from the protoplast of the grain. As pollen wall development continues, a substantial amount of sporopollenin is added to the surface of the grain in the form of orbicules produced by a secretory tapetum (Taylor 1982).

The ultrastructure of the pollen of *Schopfiangium* is quite different from all lyginopterid pollen described to date. The exine is approximately 4 μm thick and two-parted. The interesting feature of this sporoderm is the apparent alveolate organization of the sexine, a feature that is unknown in all lyginopterid pollen but consistently present in the pollen of the medullosan seed ferns. Although ultrathin sections of *Phacelotheca* pollen have not been published to date, fractured surfaces of grains suggest that the sporoderm is also alveolate.

Pollination in lyginopterid seed ferns is believed to be anemophilious, based on the highly ornate organization of the pollen-receiving mechanism (salpinx) present in many seeds thought to have been produced by lyginopterid pteridosperms. In a few grains delicate walls have been described inside the sporoderm, suggesting the presence of an endosporally developed microgametophyte (Florin 1937). These have not been critically examined, but superficially appear similar to folds in the pollen walls that were once thought to be the gametophyte stage in *Cycadeoidea* pollen (Taylor 1973). In another report, Benson (1908) described the occurrence of sperm in the pollen chamber of the lyginopterid ovule *Lagenostoma ovoides*.

MEDULLOSALES

The Medullosales include a large and diverse group of seed ferns that can be traced from the Upper Mississippian into the Permian. The group is a conspicuous component of many coal swamp floras and, apparently like most, possibly all, Paleozoic pteridosperms, had a liana-type growth habit.

The pollen organs of the Medullosales are the largest of any pteridosperm group, ranging up to several centimeters in diameter, and represented by both impression/compression fossils and perminerali-

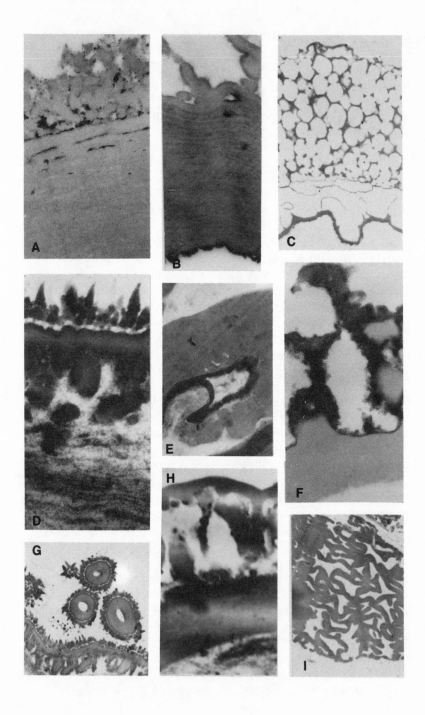

zations. In addition to the size and general organization of the organs, the distinctive ultrastructure of the pollen grain wall of this group has been used as a basis for assigning detached organs to the order.

The majority of the pollen organs of the Medullosales are constructed of elongate pollen sacs (sporangial tubes) arranged in a ring or aggregated into concentric rings that in some taxa form compound units (figure 4.5). 'Because of the large number of permineralized taxa, it has been possible to demonstrate a series of suggested homologies within the order (Dennis and Eggert 1978; Millay and Taylor 1979). Millay and Taylor (1979) have grouped the anatomically preserved medullosan pollen organs into three basic types: simple, aggregate, and compound. Each of these is considered below.

Perhaps the most simple and basic medullosan pollen organ is best illustrated by *Halletheca* (figure 4.1C), a pyriform, radially symmetrical fructification approximately 1.5 cm long that may have been borne at the tips of branches. In *H. reticulatus* (Taylor 1971; Taylor and Millay 1981a) each synangium is constructed of five to seven elongate sporangia embedded in a parenchymatous ground tissue. At the distal end the center of the organ is hollow; in the basal half is a fibrous central column (figure 4.2C). Sporangial dehiscence takes place toward the center of the organ and is accomplished through the longitudinal splitting of the fibrous central column. A single vascular strand enters the stalk and branches several times so that each sporangium is associated with a peripheral vascular bundle. A smaller synangium consisting of up to 12 sporangia arranged in a ring has been described from the Middle Pennsylvanian (Wewoka Formation) of Oklahoma under the binomial *Halletheca conica* (Mapes 1981). Although they were preserved as compressions, the basic morphology of such organs

Figure 4.4. Ultrastructure of gymnosperm fossil pollen. **A** *Potoniea carpentieri* showing a few nexine lamellae. × 30,000. **B** Nexine lamellae in sporoderm of *Monoletes (Bernaultia sclerotica)* pollen grain. × 30,000. **C** Exine of *Monoletes* grain showing alveolate sexine and thin, homogeneous nexine. × 8,000. **D** Exine of *Classopollis* pollen grain showing lamellate nexine and complex sexine with spinules. × 10,000. **E** Sporoderm of *Sahnia* pollen grain showing homogeneous–granular organization. × 15,000. **F** *Lasiostrobus polysacci* showing columellate-like organization. × 40,000. **G** Three orbicules associated with *Classopollis* pollen grain showing the same ornament. × 7,000. **H** *Willsiostrobus cordiformis* pollen wall. × 15,000. **I** Sexinous threads within the saccus of *Willsiostrobus rhomboides*. × 5000.

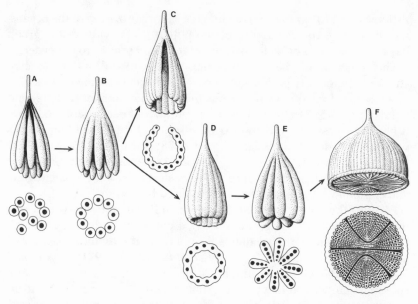

Figure 4.5. Suggested trends in the evolution of medullosan pollen organs. **A** Cluster of terminal sporangia of some progymnosperm or trimerophyte. **B** Sporangia (pollen sacs) arranged in a ring like *Codonotheca*. **C** Fusion of some pollen sacs to form bilateral organ similar to *Parasporotheca*. **D** Fused ring of pollen sacs like *Halletheca*. **E** Pollen organ showing plication resulting in an increased number of pollen sacs like *Sullitheca*. **F** Compound pollen organ like *Bernaultia*.

as *Codonotheca* (figure 4.1E), *Aulacotheca*, *Boulayatheca*, and *Schopfitheca* suggest that they may have been of the simple, solitary type like *Halletheca*.

The simple organization of *Halletheca* is also present in such forms as *Stewartiotheca* (Eggert and Rothwell 1979) a bell-shaped organ with up to 80 elongate sporangia that surround a fibrous central column. Like *Halletheca*, each sporangium is associated with a vascular bundle. Another Middle Pennsylvanian medullosan pollen organ that possesses a central core of sclerenchyma is *Sullitheca dactylifera* (Stidd et al. 1977). In this large (3.0 cm long) pollen organ the distal end is organized into a ring of solid finger-like projections and the fibrous central core is H-shaped in the base, with each of the fibrous arms bifurcated at more distal levels.

Aggregate medullosan pollen organs consist of simple synangia that are clustered in groups (Millay and Taylor 1979). The best known of this type is *Rhetinotheca* (Leisman and Peters 1970), a small radial synangiate pollen organ. Each synangium has the same basic organization as the *Halletheca* organ except for the smaller number of pollen sacs (four) and reduced size (up to 3.6 mm long). Sporangial dehiscence takes place toward the hollow center of the organ. In *R. patens* (Rothwell and Mickle 1982) each ellipsoidal pollen sac (figure 4.1D) is associated with a delicate vascular strand of sclariform tracheids. One of the interesting structural features of aggregate pollen organs like *Rhetinotheca* is the presence of numerous peg-like trichomes that arise from the surface of the organ and that are believed to interdigitate with hairs of adjacent syanagia, thus forming compact clusters of pollen organs.

Another pollen organ of the aggregate type is *Parasporotheca* (Dennis and Eggert 1978). Unlike other medullosan synangia that are radial, *P. leismanii* is constructed of elongate pollen sacs that form a fan or scoop-shaped organ (figure 4.1F). Each sporangium alternates with elongate lacunae in the ground tissue and is associated with a delicate vascular strand on the outer surface. Covering each scoop-shaped synangium was a delicate epidermis ornamented by interlocking peg-like trichomes. *Parasporotheca* was described from an incomplete cluster of tightly compacted synangia 20.0 cm long and 3.0 cm wide. The absence of a common ground tissue has been used to suggest that the individual synangia became disassociated, perhaps during drying.

Whittleseya is an impression/compression synangium that was probably of the aggregate type. Each synangium is fan-shaped and measures 5.0 cm long and 3.0 cm wide. It has long been speculated that the organ was radial in organization, consisting of a ring of elongate uniseriate sporangia surrounding a central hollow (Halle 1933). Another interpretation characterizes the synangium of *W. elegans* as a flattened unit consisting of paired sporangia much like *Parasporotheca* (Schopf 1948). Based on the known morphology of other medullosan pollen organs and features of the synangia, Millay and Taylor (1979) supported the bilateral organization of the *Whittleseya* synangium, and reconstructed the fructification as a series of closely associated flattened units. The synangia of *Whittleseya* are typically found in large numbers, but, to date, specialized trichomes in the form of interlocking pegs have not been identified on the surface of the pollen sacs.

The third major type of medullosan pollen organ is the compound synangium and is best illustrated by some specimens assigned to *Bernaultia* (Rothwell and Eggert 1986). The older generic name *Doler-*

otheca has been retained for fragments of these complex, compound units in which the exact organization cannot be determined. Some specimens of *Bernaultia* range up to 4.0 cm in diameter and consist of up to 1,000 elongate sporangia that may be arranged in pairs or rows of four (figure 4.6A). In recent years there have been several interpretations of the basic organization of the *Bernaultia* campanulum. These include Halle's (1933) suggestion that the fructification is homologous with a ring of fused sporangia, and Schopf's (1948) hypothesis that the campanulum is homologous with a single telome in which the elongate sporangia have arisen through septation. More recently, Stidd et al. (1977) have proposed that the fructification is represented by a cluster of *Codonotheca*-like pollen organs (Stidd 1978b) that have become fused. A slightly modified suggestion based on the pattern of vascular tissue interprets the unit as a series of pinnately arranged pendulous sporangia (Dufek and Stidd 1981; Stidd 1981). Still another interpretation regards the *Bernaultia* reproductive unit as homologous with one, two, or four highly plicated synangia that have fused evolutionarily (figure 4.5) through the reduction of a small planated branching system (Dennis and Eggert 1978; Millay and Taylor 1979; Rothwell and Eggert, 1986).

There are a number of pollen organs that are not structurally preserved but that are associated with the medullosan seed ferns on the basis of a distinctive type of pollen. Some of these may represent different preservation modes of anatomically preserved taxa, while others underscore the diversity of fructification types within the order. Several of the more common Pennsylvanian taxa include *Aulacotheca* (Halle 1933), *Boulayatheca* (Kurmann and Taylor 1984; Taylor and Kurmann 1985) (=*Boulaya*, Halle, 1933), *Codonotheca* (Sellards 1903, 1907), and *Schopfitheca* (Delevoryas 1964). All consist of elongate tubular sporangia containing pollen of the *Monoletes* type.

Despite the large number of medullosan pollen organs that have been described, information regarding the number and where the fructifications were borne on the plants is sparse. Eggert and Kryder (1969) have described a fertile frond member of *Aulacotheca iowensis* in which synangial units are alternately borne on nonplanated ultimate laterals. Earlier authors suggested that the fructifications of *Bernaultia* were borne in the place of pinnules (e.g., Kidston and Jongmans 1911). However, Ramanujam et al. (1974) have suggested that the *Bernaultia*-type pollen organ was borne in the position of a penultimate pinna. This suggestion is based on the number and arrangement of vascular bundles in the peduncle of the fructification and those of the frond.

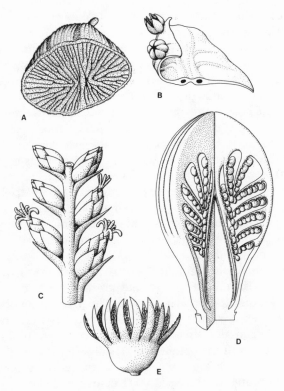

Figure 4.6. **A** *Bernaultia.* **B** *Idanothekion.* **C** *Gothania.* **D** *Cycadeoidea.* **E** *Weltrichia.*

Within the medullosan, and perhaps some lyginopterid *(Potoniea)* pollen organs, the clustering of the pollen sacs was accomplished by mechanical (specialized interlocking trichomes) and developmental means (e.g., *Parasporotheca*, figure 4.1F). The large number of pollen sacs in the compound organ *Bernaultia* (figure 4.6A) is accomplished by the plication of several uniseriate rings of elongate sporangia, thereby maintaining the pattern of inwardly directed sporangial dehiscence.

Pollen Despite the morphological diversity demonstrated by the large number of medullosan pollen organs, the pollen produced by these organs is quite uniform, consisting of two basic types. With the exception of *Parasporotheca*, all medullosan pollen organs that have been described with *in situ* pollen contain grains of the *Monoletes* type (figure 4.3C). Mature *Monoletes* pollen is prolate and contains a single, angled suture on the proximal surface. The deflection of the suture represents the position of the third laesura, indicating that *Monoletes* pollen was borne in tetrahedral tetrads (Schopf 1948). In *A. collicola* the suture is bordered by a narrow ridge (Mickle and Leary 1984). Mature *Monoletes* grains range in length from less then 80 μm in some species of *Aulacotheca* to more than 500 μm in *Bernaultia* (Taylor 1978). Because of their large size, *Monoletes* grains were among the first fossil grains to be examined with the electron microscope. The sporoderm ranges from 6 to 22 μm in thickness and is constructed of two principal zones (figure 4.4C). In some pollen grains the inner, nonsculptured layer or nexine is constructed of up to 30 lamellae (figure 4.4B), each 80–120 nm in thickness. The sexine, or outer sculptured zone, is constructed of a series of anastomosing muri that give the sporoderm a chambered, or alveolate, organization (figure 4.4C). Studies tracing the development of the pollen wall suggest that the alveolate sexine asises through the separation of the nexine lamellae (Taylor and Rothwell 1982), and that grain expansion results in the rupturing of many sporopollenin units. The presence of orbicules and stacks of bilayered membranes attests to the fact that grains of this type were associated with a secretory type of tapetum (Taylor 1976). The fine structure of the pollen wall has also been a useful diagnostic feature in demonstrating the taxonomic limits of certain taxa that differ in the manner of preservation (Kurmann and Taylor 1984).

Monoletes pollen has been found in the pollen chambers of several species of the seed *Pachytesta* (Taylor 1965). Anatomical details in the region of the micropyle and the simple organization of the pollen chamber indicate that a pollination droplet was produced. The discovery of several *Monoletes* grains at the base of a leg segment of *Arthropleura*, a large Carboniferous arthropod, and in presumed arthropod coprolites may suggest that some medullosans possessed an entomophilous pollination syndrome (Scott and Taylor 1983). With the exception of two dense structures regarded as sperm found in a *Monoletes* grain in the pollen chamber of *P. hexangulata* (Stewart 1951), nothing is known about the microgametophyte phase, including whether or not pollen tubes were produced.

The presence of *Monoletes* grains in reproductive organs of Carboniferous age has been the primary basis for the assignment of impression/compression fructifications to the Medullosales. The only exception, to date, is the occurrence of *Parasporites* pollen (figure 4.3D) in the pollen sacs of *Parasporotheca* (Dennis and Eggert 1978). Pollen of this type was initially described from Middle and Upper Pennsylvanian coals from the Eastern Interior Basin (Schopf 1938) and assigned to the *sporae dispersae*. The grains are large (approximately 280 μm) and characterized by a crescent-shaped saccus that extends from opposite ends of the elongate grain (figure 4.3D). On the proximal surface is a single suture that shows the same angular deflection of the laesura as that present in *Monoletes*. The grains are ornamented on the proximal surface by conspicuous interconnected rugulae (Millay et al. 1978). The fine structure of the sporoderm consists of an inner lamellate nexine and outer sculptured sexine.

The alveolate sexine that characterizes the sporoderm of *Monoletes* and *Parasporites* is also present in *Nanoxanthiopollenites* Clendening and Nygreen (Taylor 1980), a pollen type known only from dispersed grains. This pollen type is also bilateral, monolete, and ornamented by prominent spinae (figure 4.3G). No grains of this type have been found in pollen sacs to date.

The medullosan seed ferns are believed by many to represent the progenitors of the modern Cycadales. One of the similarities between the two groups is the alveolate organization of the pollen wall (Audran and Masure 1977). Developmental studies of the pollen wall in both groups suggest some basic differences in the manner in which the alveolate exine is formed. In cycadalean pollen, sporopollenin accumulates on radially oriented microtubles (Audran 1981), while in *Monoletes* the alveolate organization of the exine appears to result from the separation of nexine lamellae (Taylor and Rothwell 1982). In addition, there are several features in which the grain types differ that involve the appearance of the nonsculptured (nexine) component of the wall.

CALLISTOPHYTALES

The Callistophytales comprises a small family of Pennsylvanian seed plants that may extend into the Permian (Rothwell 1975, 1981). The plants are reconstructed as having a small, scrambling, shrub-like habit with pinnately compound leaves bearing sphenopterid pinnules.

The pollen organs are sessile and borne on the abaxial surface of unmodified pinnules and are organized into synangia (figure 4.6B). Although two genera of pollen organs were initially described (*Idanothekion glandulosum* Millay and Eggert 1970 and *Callandrium callistophytoides* Stidd and Hall 1970), subsequent studies have confirmed that the two genera are identical and that *Callandrium* must be regarded as a synonym (Rothwell 1980).

Synangia of *Idanothekion* are radially symmetrical and consist of five to nine partially fused, elongate sporangia that surround a vascularized central column. At more distal levels the central region is hollow and the sporangial tips are free (figure 4.6B). The externally oriented sporangial walls are thickened and include large secretory cavities. Dehiscence occurs through a longitudinal slit along the inward-facing wall of each sporangium.

Pollen Much is known about the pollen and microgametophyte stages in the Callistophytales. Pollen of the *Vesicaspora*-type (figure 4.3E) has been found *in situ* in *Idanothekion* (=*Callandrium*) pollen sacs (Hall and Stidd 1971). Mature grains are bilateral, monosaccate and range between 40 and 55 μm long and up to 49 μm wide (polar view). Hall and Stidd (1971) have described several stages in the ontogeny of the grains including the formation of isobilateral, decussate, and intermediate tetrad configurations. On the distal surface of the corpus is a well-defined furrow that may be slightly overarched by the saccus lobes. The saccus is of the eusaccate type (figure 4.3E) in which the endoreticulations of the inner saccus wall are not in contact with the corpus (Scheuring 1974; Millay and Taylor 1974). Small rods of spherical subunits arranged in a patchwork pattern ornament the surface of both the saccus and corpus (Millay and Taylor 1970).

The microgametophyte of *Callistophyton* is understood in detail (Millay and Eggert 1974). Several developmental stages of the microgametophyte have been described from pollen extracted from the pollen organs of the Callistophytales. Microgametophyte development involved the production of a series of axially oriented prothallial cells (figure 4.3E). An embryonal cell, including what the authors interpret as a nucleus, is also present in some grains. In a few grains there is evidence of a still later stage in which a tube cell and antheridial cell are present in the corpus. The presence of a pollen tube extending from the sulcus of a *Vesicaspora* pollen grain embedded in the nucellus of a seed has also been documented in this group of pteridosperms (Rothwell 1972). The tube is branched and up to 70 μm long. It is not known whether the tube was siphonogomous or haustorial in function.

CORDAITALES

This order of fossil gymnosperms can be traced from the early Pennsylvanian well into the Permian. It is now known that the plants range from small trees with stilt-like roots and large strap-shaped leaves to small shrubs (Rothwell and Warner 1984). The reproductive parts have been known for a long time, with the ovule-producing organs playing a pivotal role in discussions of the homology of the modern conifer cone scale complex.

By comparison, the pollen cones have received relatively little attention, although several structurally preserved specimens have been reported. Like the ovule-producing organs, the pollen cone *Cordaianthus concinnus* is a compound structure consisting of a primary axis bearing bracts in four ranks (Rothwell 1977). In the axils of each bract is a secondary axis bearing helically arranged scales. Attached to several of the more distal scales are elongate pollen sacs. In *C. concinnus* each fertile scale bears up to six elongate pollen sacs that are fused in a ring at the base (Delevoryas 1953). In *C. penjoni* the pollen sacs are interspersed with the sterile scales. Another cordaitean pollen cone that closely resembles *Cordaianthus* is *Gothania* (figures 4.6C, 4.7A; Daghlian and Taylor 1979). Several features of the cones—including the number and arrangement of the pollen sacs, manner in which the sporangia dehisce, features of the pollen, and stratigraphic age—are currently used to distinguish *Gothania* from *Cordaianthus* (Trivett and Rothwell 1985).

Pollen Although the monosaccate pollen of the Cordaitales is structurally similar, there are several features that distinguish pollen within the group. Pollen produced by *Cordaianthus* is of the *Florinites* type (figure 4.8A; Millay and Taylor 1974). The grains are bilaterally symmetrical, with the saccus attached to both the proximal and distal surfaces. The saccus is of the eusaccate type, with endoreticulations extending a short distance from the inner surface of the saccus wall. Germination is believed to be from the distal surface based on the ultrastructure of the pollen wall on the distal surface (Millay and Taylor 1974).

Another monosaccate pollen type extracted from cordaitean pollen sacs is *Sullisaccites* (Millay and Taylor 1974). The bilaterally symmetrical trilete pollen is monosaccate with the eusaccus attached to both surfaces of the corpus. An asymmetrical trilete suture occurs on the proximal surface of the corpus. The sporoderm is two-parted, consisting of an inner nonlamellate nexine and sculptured sexine. Grains of this type also have been reported recently from the pollen cone of the Upper Pennsylvanian cordaitean *Mesoxylon priapi* (Trivett and Rothwell 1985).

Large (115–180 μm) monosaccate grains originally described as *Felixipollenites* (figure 4.3H; Millay and Taylor 1974) were produced

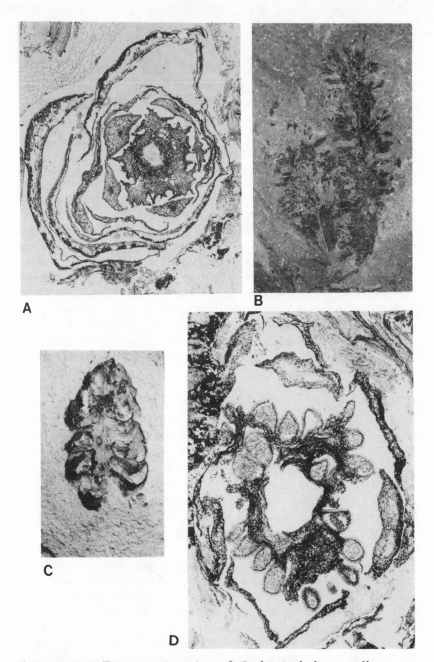

Figure 4.7. A Transverse section of *Gothania lesliana* pollen cone showing arrangement of scales. × 15. **B** Portions of two microsporophylls of *Pteruchus dubius* showing clusters of pollen sacs. × 3. **C** Pollen cone of *Trisacocladus tigrensis*. × 5. **D** Transverse section of *Lasiostrobus polysacci* cone showing arrangement of microsporophylls and sporangia. × 11.

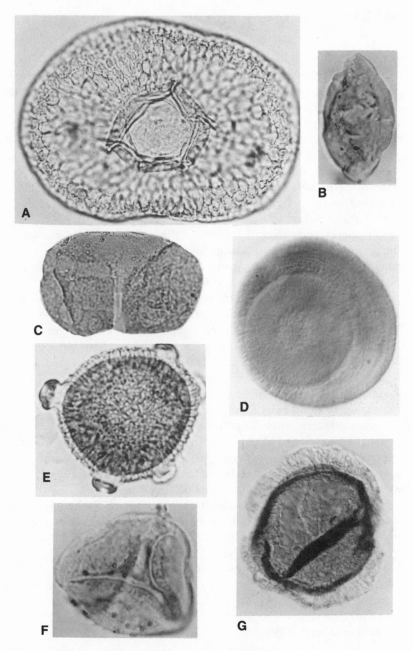

Figure 4.8. Fossil gymnosperm pollen. **A** *Florinites* sp. (*Cordaianthus* sp.). × 1000. **B** *Sahnia* sp. × 1450. **C** *Pteruchus dubius*. × 440. **D** *Classopollis* sp. *(Classostrobus comptonensis)*. × 1500. **E** *Lasiostrobus polysacci*. × 1500. **F** *Trisaccites* sp. *(Trisacocladus tigrensis)*. × 1000. **G** *Willsiostrobus denticulatus (Masculostrobus denticulatus)*. × 500.

by the cordaitean pollen cone *Gothania lesliana* (Daghlian and Taylor 1979). Both radial and bilaterally symmetrical forms were recovered from the same pollen sacs, with approximately 20 percent of the forms characterized by trilete haptotypic marks; 9 percent possessed monolete sutures on the proximal surface of the corpus. The remaining grains showed a variety of suture configurations between the two types, suggesting that in *Felixipollenites* there was considerable variation in tetrad configuration and/or grain abortion. The suture is complex, consisting of a narrow groove from which arises a median ridge. Delicate ribs that extend at right angles from the inner surface of the groove are attached to the ridge (Taylor and Daghlian 1980). A thin membrane extends over the suture. Ultrastructural studies of both immature and mature *in situ* grains indicate that germination was from the proximal surface. At maturity the saccus is attached only to the proximal surface (Taylor and Daghlian 1980).

A number of cordaitean pollen grains have been described with endosporally developed microgametophytes (e.g., Renault 1896, 1902; Florin 1936). Florin's illustrations are perhaps the most convincing, especially in light of the stages in microgametophyte development now known to have been present in the Callistophytales (Millay and Eggert 1974). Within the central body of several pollen grains extracted from sporangia of *Cordaianthus saportanus*, Florin figured an inner layer of small cells along the periphery of the central body that he suggested were antheridial in origin. In addition, the central body contains an axial row of nuclei (cells) that were interpreted as either spermatogenous cells or sperm. Millay and Taylor (1974) have suggested that the peripheral cells are folds in the corpus wall, while the nuclei are prothallial cells like those of *Vesicaspora*.

Relatively little is known about the pollination syndrome of the Cordaitales, or whether pollen tubes were produced. However, the saccate organization of the grains and large number of Paleozoic ovules containing foreign pollen of the *Florinites* type would suggest that the plants were anemophilous. Additional support for this hypothesis may be found in the recent ontogenetic study of structurally preserved cordaitean ovulate fructifications in which mature ovules were elevated above the level of the subtending scales, perhaps to facilitate pollination (Rothwell 1982).

CYCADEOIDALES

Members of this group of extinct gymnosperms formed a conspicuous component of the landscape from the Triassic well into the Cretaceous. In the Cycadeoidaceae many plants had short squat trunks and in habit closely resemble some modern members of the Cycadales. In the Williamsoniaceae the trunks are believed to have been slender and branched several times (Delevoryas and Hope 1976). Both groups had pinnate leaves and persistent leaf bases.

Cycadeoidaceae The reproductive organs of *Cycadeoidea* were tightly compacted cones borne on short lateral branches situated among the leaf bases; in *Monanthesia* each leaf bore a cone in its axile. Although opinions have varied through the years as to the structural organization of the cones (e.g., Wieland 1906; Delevoryas 1963 1968), it now appears that all were bisporangiate (Crepet 1974). Microsporophylls in the form of pinnae were arranged in a whorl and attached at the base of the ovule-bearing receptacle (figure 4.6D). The fronds were recurved so that the distal tips were adjacent to the base. Attached to the microsporophylls were thick-walled synangia that contained from 8 to 20 elongate sporangia arranged along the periphery of the unit (Delevoryas 1965).

Pollen. *Cycadeoidea* pollen extracted from *C. dacotensis*, a common species, is up to 24 μm long and 16 μm wide (figure 4.3F). On one surface is an elongate sulcus. The sporoderm is two-layered, with no appreciable thinning in the region of the sulcus (Taylor 1973). Grain ornamentation consists of a series of closely spaced punctae.

Stages in the development of the *Cycadeoidea* microgametophyte were described and illustrated by Wieland (1906). In some grains up to five cells were observed lining the inner surface of the sporoderm. Subsequent examination based on ultrathin sections of similar cell contents indicates that what Wieland interpreted as stages in the development of a cellular microgametophyte were merely folds in the wall (figure 4.3F).

Crepet's (1974) study of *Cycadeoidea* cones has not only provided a wealth of information about the development of the cones but has also clarified the pollination syndrome in these plants. Since the cones of *Cycadeoidea* are now known to have been bisporangiate (figure 4.6D) and the microsporophylls never opened, the once held view that these plants were anemophilous appears to be no longer tenable. Rather, the structural organization of the cones indicates that they were, predominately, functionally self-pollinated. Pollination may have been accomplished through the disintegration of sporangial tissues on the cone, but perhaps also through the activities of boring insects. Crepet (1974) observed that in a high percentage of cones with insect borings, these extend through all tissues, suggesting that some pollen transfer may have been accommodated by insect vectors. Based on the high probability that the cycadeoids were self-pollinated, it is suggested that the resultant highly homozygous genome was a potential cause of extinction during the latter stages of the Cretaceous, when climatic variability was high (Crepet 1974).

Williamsoniaceae Although the habit of several Mesozoic cycadophytes has been reconstructed, Delevoryas and Hope (1976) suggested that the basis upon which these reconstructions were made is uncertain. *Wielandiella angustifolia* is known from specimens that suggest that at least some Mesozoic cycadeoids were slender plants bearing tufts of leaves at the base of branches (Nathorst 1909).

The microsporangiate organs of the Williamsoniaceae are described under the generic name *Weltrichia* (Harris 1969). Several specimens from the Jurassic of Yorkshire, England, consist of a shallow cup of up to 10 cm wide from which arise a series of finger-like microsporophylls (figure 4.6E). In the depressed portion of the cup are numerous sterile scales in some species; in others, resinous sacs may be present (Harris 1969). On the inner surface of each microsporophyll are two rows of synangia (figure 4.6E). Because the microsporangiate parts of the Williamsoniaceae are known only from impressed/compressed specimens, details of the anatomy and arrangement of the pollen sacs are unknown. In *W. sol* the synangia are described as semicircular and organized into two valves. Each synangium consists of from 12 to 15 slightly elongate pollen sacs.

The frond-borne pollen organ of some Paleozoic seed fern has been postulated as the progenitor of the Mesozoic cycadeoid cone (Delevoryas, 1968a, 1968b). According to this foliar hypothesis, there has been a progressive reduction of a portion of a fertile branch until it assumes the axillary position characteristic of *Cycadeoidea*. Another view suggests that the pollen-bearing parts of *Cycadeoidea* have arisen through neotony (Stidd 1980). Here the pollen organ of the cycadeoid progenitor represents a modified fertile frond that was developmentally arrested while still circinately folded. According to this idea, other cycadeoid pollen organs represent modifications of a basic theme and not intermediate evolutionary stages.

Pollen. Pollen of *Weltrichia* is oval and monosulcate. In *W. setosa* the grains measure up to 37 μm; slightly larger grains (60 μm) have been reported in *W. spectabilis* (Harris 1969). Nothing is known about the fine structure of the pollen or features of the microgametophyte phase.

CYCADALES

In many features members of the true cycads resemble cycadeoids. The majority consist of short, unbranched trunks with persistent leaf bases (Delevoryas and Hope 1971). Leaves are pinnately compound and all living cycads are dioecious. Seeds are borne on modified leaves; pollen sacs are attached to the abaxial surface of microsporophylls arranged into small, compact cones.

Information on the reproductive parts of late Paleozoic cycads is based on compressed megasporophylls-bearing seeds (e.g., Mamay 1973; Zhu and Du 1981). Structurally preserved vegetative remains have been identified from the Triassic, based on the characteristic girdling configuration of the leaf traces (e.g., Archangelsky and Brett 1963; Gould 1971; Ash 1985). The recent discovery of a silicified cycad stem from the early–middle Triassic of Antarctica underscores the fact that the group was well established by the end of the Paleozoic (Smoot et al. 1986).

To date, relatively little information is available about the pollen-producing organs of fossil cycads. *Androstrobus* is the generic name used for microsporangiate pollen cones that are believed to have been borne by some Jurassic cycads (Harris 1964). Morphologically these cones are similar to those of modern cycads, consisting of helically arranged microsporophylls that are distally upturned. Numerous sporangia are scattered over the abaxial surface. Pollen is small (30–40 μm), oval, and monosulcate. The occurrence of glandular bodies on the megasporophylls of *Phasmatocycas* has been suggested as evidence for a potential entomorphilous syndrome in this Permian cycad (Mamay 1976). A cone that is morphologically similar to a modern cycad pollen cone was borne on the Triassic cycad *Leptocycas* (Delevoryas and Hope 1971). No pollen was recovered from this cone.

Fossil evidence contributes little toward an elucidation of the organization of the microsporangiate organs of modern cycads. Most believe that the Cycadales can be traced to the medullosan seed ferns. However, this hypothesis would require loss of the synangiate organization along with the extreme reduction of the microsporophyll that characterizes the modern cycad strobilus.

Nothing is known about fossil cycad pollen ultrastructure, nor have any grains been identified that contain evidence of stages in the development of the microgametophyte.

An interesting pollen cone that has been suggested as having affinity with the cycads is *Lasiostrobus* (Taylor 1970). Attached to the axis

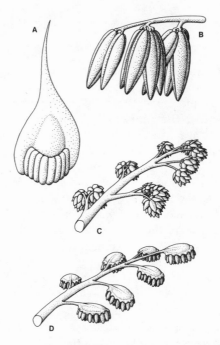

Figure 4.9. A *Lasiostrobus* microsporophyll showing abaxial pollen sacs. **B** *Caytonanthus* pollen sacs. **C** *Antevsia* microsporophyll showing clustered pollen sacs. **D** *Pteruchus* pollen organ showing abaxial attachment of pollen sacs.

of this Upper Carboniferous strobilus are helically arranged microsporophylls (figure 4.7D) bearing elongate pollen sacs on the abaxial surface (figure 4.9A). One of the features that has made the assignment of *Lasiostrobus* to the Cycadales problematical is the pollen. The pollen grains are circular and range from 20 to 30 μm in diameter. Most are alete, but a few specimens possess a trilete mark. Extending from the surface of the grain in a slightly subequatorial orientation are several (three to eight) reduced sacci (figure 4.8E). The wall of the grain is complex and consists of an inner homogeneous layer from which arise closely spaced radial columns (figure 4.4F). Extending across the ends of the columns is a delicate layer that is analogous to a tectum in many angiosperm pollen grains.

Despite the uncertain taxonomic position of this gymnospermous pollen cone, the occurrence of inclusions in many of the grains has important implications for the reproductive biology of this group

(Taylor and Millay 1977). The inclusions are characterized by a homogeneous central region surrounded by a more granular matrix and represent the nucleus and cytoplasm, respectively. Since many of the grains with cell contents lie outside of the pollen sacs, it is suggested that *Lasiostrobus* shed its grains in the microspore stage. Based on studies of microgametophyte development in extant gymnosperms, it is believed that the absence of prothallial cells is an advanced character (Sterling 1963). Thus, unless prothallial cell production was appreciably delayed in *Lasiostrobus*, this gymnospermous pollen cone was reproductively advanced in microgametophyte development (Taylor and Millay 1977).

MESOZOIC PTERIDOSPERMS

Mesozoic seed ferns are incompletely known. Nearly all of the taxa are from just a few localities, and most are preserved as impression/compressions. Thus the level of resolution that can be used in discussing the pollen organs of these plants is far from adequate. Four orders are recognized: Caytoniales, Corystospermales, Peltaspermales, and Glossopteridales.

Caytoniales The ovule-bearing cupules of the Caytoniales led to the suggestion that this group of Mesozoic seed plants had enclosed ovules and thus possessed an angiospermous reproductive system (Thomas 1925). Although the gymnospermous affinity of the Caytoniales has been repeatedly documented (e.g., Reymanowna 1973), the organization of the cupule has continued to fascinate evolutionary biologists as a possible precursor to the angiosperm carpel (e.g., Doyle 1978; Dilcher 1979). Specimens believed to represent members of the Caytoniales are known from the Triassic into the Cretaceous, but the most thoroughly documented remains come from the Middle Jurassic sediments of Cayton Bay, Yorkshire.

The pollen organ *Caytonanthus* (Harris 1941) consists of a branched, pinnate microsporophyll to which are attached synangia of three to five elongate pollen sacs (figure 4.9B). Sporangial dehiscence occurred toward the center of the pollen sac cluster much like that in the synangiate pollen organs of Paleozoic pteridosperms. The pollen is small and bisaccate, with germination occurring from the distal surface. The ultrastructure of the sporoderm has recently been described as alveolate in several species (Zavada and Crepet 1986; Pedersen and Friis 1987). Nothing is known about the microgametophyte, although the structure of the cupules and seeds suggests that the plants possessed a pollination droplet and were anemophilous.

Corystospermales   The corystosperms are a small, predominately Triassic group of gymnosperms that were initially described from the Molteno beds of Natal (Thomas 1933). In recent years additional specimens have extended both the geographic and stratigraphic range (see Crane 1985). The plants are thought to have been woody, with the frond either pinnate or bipinnate.

Pteruchus is one of the more common microsporangiate organs assigned to the Corystospermales (Townrow 1962). Microsporophylls were helical to alternately arranged (figure 4.9D), with numerous pollen sacs attached to the lower surface (figure 4.7B). There is distinct morphological similarity between *Pteruchus* and some lyginopterid microsporangiate organs like *Feraxotheca* (figure 4.1A) and *Crossotheca* (figure 4.2A).

In *P. dubius* the pollen is bisaccate (figure 4.8C) with the bladders slightly distally inclined (Taylor et al. 1984). Germination apparently took place through a thin distal sulcus. Grains range from 80 to 115 μm in the primary plane; the height of the corpus is approximately 50 μm. The sporoderm is homogeneous and the sacci is of the protosaccate type (Taylor et al. 1984; Zavada and Crepet 1985). In other accounts of *Pteruchus* pollen based on light microscopy (e.g., *P. thomasii* and *P. gopadensis* Pant and Basu 1979), internal saccus ornamentation (endoreticulations of eusaccate grains or sexine elements of protosaccate grains) is not described.

Another probable corystospermalean microsporangiate organ consisting of microsporophylls and abaxially borne pollen sacs is *Pteroma* (Harris 1964). The inclusion of this pollen organ with the corystosperms is based on its association with the foliage genus *Pachypteris*. The vesiculate pollen is in the 80 μm size range, with the sacci ornamented by minute pits. Pits in the wall of the saccus have also been reported in several species of *Pteruchus* as well in the pollen of *Caytonanthus* (Krassilov 1977a; Zavada and Crepet 1986). Nothing is known about the microgametophyte in this group.

Peltaspermales   The peltasperms are a small group of seed ferns known from a limited number of localities of Permian and Triassic age (Townrow 1960). Although several different types of megasporophylls have been described (Kerp 1982), the pollen-producing organs are of one basic type. The microsporophylls of *Antevsia* are considered to be bipinnate, with irregular secondary branches (figure 4.9C). Attached to the distal ends of these branches are several (4–12) elongate pollen sacs that dehisced longitudinally. One of the unusual features of these seed plants is the absence of saccate pollen. In both *A. zeilleri* and *A. extans* the grains are boat-shaped and characterized by a conspicuous sulcus. Nothing is known about the pollination system, the organization of the pollen wall, or features of the microgametophyte in the peltasperms.

Glossopteridales Despite its historically early appearance in paleobotanical literature and abundance in Triassic and Permian sediments of the southern hemisphere, the plants comprising the *Glossopteris* flora continue to be far from adequately known. Based on associations and the discovery of some structurally preserved specimens, Gould and Delevoryas (1977) suggested that at least one of the plants that produced foliage of the *Glossopteris*-type was a large tree. Anatomically preserved leaves from the Permian of Antarctica indicate that there are several leaf types that differ significantly in the pattern of venation and histology of the tissues (Pigg and Taylor 1985). Thus a picture appears to be emerging that the *Glossopteris* foliage type was produced by several different kinds of plants, a fact that is also confirmed by the several types of seed-bearing organs that have been described. One of these, based on silicified material from the Permian of Queensland, consists of small ovoid seeds attached to one side of a megasporophyll (Gould and Delevoryas 1977).

The pollen organs that have been attributed to the glossopterids show some basic morphological uniformity. The pollen organ *Glossotheca* consists of a small branching axis that arises from the petiole of a leaf. At the end of ultimate axes are clusters of elongate pollen sacs (Surange and Chandra 1974). On the surface of the microsporangia are longitudinally oriented striations. *Eretmonia* is another microsporangiate organ that consists of clusters of pollen sacs that are attached to a branching axis that extends from the base of a leaf (Lacey et al. 1975). The small, stalked pollen sacs are typically in two clusters, one on either side of the main axis (figure 4.10A). White (1978) described a small pollen cone attached to a shoot bearing leaves of *Glossopteris linearis*. Attached to a cone axis are scales that protected clusters of pollen sacs of the *Arberiella*-type. Elongation of the scales may have facilitated the release of the pollen.

Information about the pollen of the glossopterids comes from grains extracted from a number of isolated pollen sacs (e.g., *Arberiella*). The pollen grains are large (up to 85 μm), vesiculate, and characterized by prominent striations on the cappus (Pant and Nautiyal 1960). Germi-

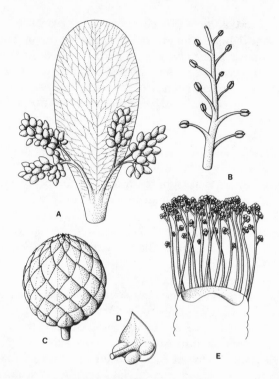

Figure 4.10. **A** *Eretmonia.* **B** *Ginkgo huttoni* pollen cone. **C** *Classostrobus comptonensis* pollen cone. **D** Microsporophyll of *C. comptonensis* showing position of three pollen sacs. **E** *Sahnia* microsporophylls.

nation was from a distal sulcus; nothing is known about the structure of the pollen wall or details of the microgametophyte.

Despite the fact that the Mesozoic seed ferns represent a diverse group, there are some morphological similarities that suggest that some members represent a reduction product of geologically earlier forms. Synangia are absent in two groups (peltasperms and glossopterids) and are present in a morphologically simple form in the Corystospermales and Caytoniales. In the corystosperms, peltasperms, and *Caytonanthus*, the fertile frond is pinnate and resembles a reduced vegetative leaf. All possess pollen sacs that dehisced longitudinally. In the glossopterids, sporangia are borne in tufts on what appears to be a small sterile leaf and may be organized in the form of a strobilus.

PENTOXYLALES

Many of the plants included in this order are known from structurally preserved specimens collected principally from Jurassic rocks of India, and knowledge of the complete plant has been recently presented (Bose et al. 1985). *Pentoxylon sahnii* is thought to have been a small shrub with primary axes up to 4 cm in diameter. Arising from these were several types of branches on which the leaves and reproductive organs were borne. What have been interpreted as growth rings are present in the endocentrically developed secondary xylem of some specimens.

The pollen organs *(Sahnia)* are attached to short shoots and consist of a ring of branching axes (microsporophylls) that arise from the margin of a receptacle (figure 4.10E). Each of the axes branches several times and contains numerous stalked pollen sacs at the distal tips. Recently, Crane (1985) has suggested a close relationship between members of the Pentoxylales and the Cycadeoidales. In his interpretation the microsporangiate "flower" is similar in organization to the microsporophylls in *Weltrichia* (figure 4.6E). In *Cycadeoidea* the pollen-producing sporophylls are also arranged in a ring, but recurved, adnate, and apparently never becoming extended. One major difference between the pollen organs of the Cycadeoidales and the Pentoxylales is the apparent absence of synangia in the Pentoxylales.

Pollen grains recovered from *Sahnia* are monosulcate and approximately 25 μm long (figure 4.8B). The sporoderm is two-parted (figure 4.4E), consisting of a thin inner zone with delicate lamellae, and a thicker outer layer that ranges from homogeneous to granular. Nothing is known about the microgametophytes or pollination in this group.

GINKGOPHYTA

The majority of our information about the geological record of ginkgophytes comes from several different types of leaves from Mesozoic sediments (Harris and Millington 1974). Many of these show the same basic morphological form and, in some instances, cuticular structure, as the leaves of *Ginkgo biloba*. Few pollen organs are known for this group. Associated with leaves of *Ginkgo huttoni* (Jurassic) are small pollen cones (figure 4.10B) consisting of microsporophylls, each bearing two pollen sacs (Van Konijnenburg-Van Cittert 1971). The pollen is described as monosulcate and up to 42 μm in length. Nothing is known about the microgametophyte or pollen ultrastructure.

CZEKANOWSKIALES

This is a small group of Jurassic and Cretaceous plants that had ginkgophyte foliage but unusual seed-bearing organs. Some authors regard the plants as gymnospermous (e.g., Harris and Miller 1974), while others regard them as proangiosperms (Krassilov 1977b). To date, there appears to be only one pollen cone that has been associated with the Czekanowskiales. *Ixostrobus* is a microsporangiate organ approximately 10.0 cm long and consisting of widely spaced microsporophylls, each bearing a synangium of four pollen sacs. Pollen is circular and contains an elongate germinal aperture.

CONIFEROPHYTA

Other authors in this volume will detail various aspects of coniferophyte evolution. Even considering only the pollen cones and pollen, the coniferophytes are far too large a group to deal with here; consequently only a few examples will be presented.

Voltziales   In general it is believed that modern conifers originated among some late Carboniferous plants included in the Voltziales, a group that coexisted with the Cordaitales. These transition conifers, as they have come to be known, are often separated into two families, the Lebachiaceae, which generally includes Upper Carboniferous and early Permian forms, and the Voltziaceae, a family that is principally Mesozoic. Much of the early work on these plants was carried out by Florin (1951), who suggested the homologies between the ovule-bearing organ of *Cordaianthus* and the ovuliferous scale of modern conifers. In comparison, the pollen cones of the early conifers have received relatively little attention. Early voltzialean pollen cones consist of helically arranged microsporophylls, each with two elongate pollen sacs borne on the abaxial surface. In contrast, most of the Mesozoic voltzialean pollen cones (e.g., *Masculostrobus* [figure 4.11A], *Voltziostrobus, Sertostrobus*) have microsporophylls with numerous pollen sacs that are borne on pedicels on the inner surface of a distal lamina (figure 4.11B; e.g., Grauvogel-Stamm 1969, 1972).

Pollen.   The pollen of many lebachicean conifers is monosaccate, with germination occurring from a trilete suture on the proximal surface.

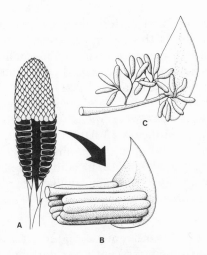

Figure 4.11. A *Masculostrobus* pollen cone. **B** Microsporophyll showing pollen sacs attached to the inner surface of the distal lamina. **C** Microsporophyll of *Voltziostrobus* sp. showing attachment of pollen sacs to delicate branches arising from pedicle.

It is not known with certainty whether these grains are proto- or eusaccate, but a thick section of a *Potonieisporites* grain in the pollen chamber of *Lebachia lockardii* suggests that the saccus may be of the latter type (Mapes and Rothwell 1984). Pollen of several voltzialean cones is typically bisaccate (figure 4.8G), the sacci filled with sexinous threads (=protosaccate; figure 4.4I) and the sporoderm alveolate–columellate (figure 4.4H). Nothing is known about the microgametophyte of the voltzialean coniferophytes. However, an anemophilous pollination syndrome associated with a pollination droplet is present in at least one species of *Lebachia* (Mapes and Rothwell 1984).

Beginning with the compound organization of the cordaitean pollen cone, the simple pollen cone of a modern conifer can be viewed as the homologue of the fertile axillary short shoot. Thus, in *Pinus*, for example, the cluster of axillary pollen cones near a stem tip is considered comparable to the cordaitean compound structure. In this transition the determinate cordaitean reproductive structure has become an indeterminate long shoot that also may bear vegetative leaves. There has also been a shift from the few terminally borne pollen sacs of the cordaiteans to the abaxial, protected pollen sacs of modern conifers. In the voltzialeans, the pollen sacs may arise from delicate branches that extend from the narrow petiole of the sporophyll (figure 4.11C).

Cheirolepidiaceae   Information about the stratigraphic and geographic distribution of this group of extinct coniferophytes is based principally on the occurrence of its unique pollen. In recent years, megafossil evidence has been presented which indicates that the family was large and diverse (Watson 1982; Alvin 1982). The pollen cones are typically small (5 mm long) and consist of microsporophylls with abaxial pollen sacs. In *Classostrobus* (figure 4.10C; Wealden) three pollen sacs are present (figure 4.10D). In other microsporangiate cones thought to belong to this group, the number of pollen sacs per microsporophyll may be as high as 12. In both the number and manner in which they are arranged, they appear like those of some voltzialean microsporangiate organs.

Pollen.   Because of several unique features *Classopollis* pollen has been extensively studied in recent years with both light and electron microscopy (e.g., Srivastava 1976; Medus 1977). The grains are spherical and characterized by a triradiate structure on the proximal surface and a cryptopore on the distal face (figure 4.8D). Encircling the grain slightly below the equator is a thickened band and distally oriented narrow, shallow trough. The sporoderm of *Classopollis* (figure 4.4D) is perhaps the most complex structurally of any pollen type, extant or fossil, and the pattern of sporoderm development (figure 4.4G) is known for one *Classopollis* grain type (Taylor and Alvin 1984). An anemophilous pollination syndrome has been suggested based on the widespread occurrence of *Classopollis* pollen. Nothing is known about the microgametophytes in this interesting group of conifers.

Pinaceae   Much of the information on the diversity in this group of conifers comes from the remains of ovulate cones that first appear in the Triassic (Delevoryas and Hope 1973). Several modern genera had evolved by the middle Cretaceous (Miller 1977). Relatively few pollen cones are known in detail, but in several features they resemble those of modern forms. *Amydrostrobus* is a small Triassic pollen cone that morphologically resembles the pollen cones in the Pinaceae, but the pollen grains are nonvesiculate (Harris 1935). *Millerostrobus* is another pollen cone from the late Triassic that may belong to the Pinaceae or Podocarpaceae (Taylor, Delevoryas, and Hope, 1987). The cones are about 1.0 cm long and have microsporophylls with two pollen sacs each. The protosaccate grains are 80 μm across the equatorial plane, and germination is believed to have been distal.

Podocarpaceae Podocarp remains are abundant in the fossil record beginning in the Triassic (Miller 1977). The pollen cones are simple, with helically arranged microsporophylls, each bearing two pollen sacs. Pollen cones of the Triassic genus *Rissikia* are 1.0 long and contain bisaccate pollen approximately 50 μm long (Townrow 1967). Germination occurs through a distal leptoma. Another podocarpaceous pollen organ of slightly younger geologic age is *Trisacocladus* (figure 4.7C) from the Cretaceous of Argentina (Archangelsky 1966). The number of pollen sacs per microsporophyll is variable, with two being the most common number. Pollen is characterized by three reduced sacci that are distally inclined (figure 4.8F). The sporoderm is complex, with the sexine constructed of sporopollenin rods, some of which are fused across their tips (Baldoni and Taylor 1982). The general ultrastructural organization of the pollen wall is similar to that of some extant podocarp pollen types, except that the fossil shows features of the saccus that appear to be intermediate between the protosaccate and eusaccate condition.

ACKNOWLEDGMENTS

I wish to thank Michael A. Millay for his incisive comments on many aspects pertaining to fossil gymnosperm pollen organ evolution, and David M. Dennis for preparing the line drawings in this paper. This study was supported in part by funds from the National Science Foundation (BSR 84 02813).

LITERATURE CITED

Alvin, K. L. 1982. Cheirolepidiaceae: Biology, structure and paleoecology. *Rev. Palaeobot. Palynol.* 37:71–98.

Archangelsky, S. 1966. New gymnosperms from the Tico Flora, Santa Cruz Province, Argentina. *Bull. British Museum (Natural History) Geol.* 13:261–295.

Archangelsky, S. and D. W. Brett. 1963. Studies on Triassic fossil plants from Argentina. 2. *Michelilloa waltonii* nov. gen. et sp. from the Ischigualasto Formation. *Ann. Bot.* 27:147–154.

Ash, S. 1985. A short thick cycad stem from the Upper Triassic of Petrified Forest National Park, Arizona and vicinity. In E. H. Colbert and R. R. Johnson, eds., *The Petrified Forest Through the Ages. Mus. North. Ariz. Bull,* ser. 54:17–32.

Audran, J. C. 1981. Pollen and tapetum development in *Ceratozamia mexicana*

(Cycadaceae): Sporal origin of the exinic sporopollenin in cycads. *Rev. Palaeobot. Palynol* 33:315–346.

Audran, J. C. and E. Masure. 1977. Contributions à la connaissance de la composition des sporodermes chez les Cycadales (Prespermaphytes). Étude en microscopie electronique à transmission (M.E.T.) et à balayage (M.E.B.) *Palaeontographica* 162B:115–158.

Baldoni, A. M. and T. N. Taylor. 1982. The ultrastructure of *Trisaccites* pollen from the Cretaceous of southern Argentina. *Rev. Palaeobot. Palynol* 38:23–33.

Benson, M. 1904. *Telangium scotti*, a new species of *Telangium* (Calymmatotheca) showing structure. *Ann. Bot.* 69:161–176.

—— 1908. On the contents of the pollen chamber of a specimen of *Lagenostoma ovoides*. *Bot. Gaz.* 45:409–412.

Bertrand, E. C. 1899. Rémarques sur la structure des grains de pollen de *Cordaites*. *Assoc. Franc. pour l'Avancem. des Sci. C. R. de la 27. Session, Nantes (1898), 2 partie*. Paris.

Bose, M. N., P. K. Pal, and T. M. Harris. 1985. The *Pentoxylon* plant. *Phil. Trans. Roy. Soc. London* 310B:77–108.

Crane, P. R. 1985. Phylogenetic anaylsis of seed plants and the origin of angiosperms. *Ann. Missouri Bot. Gard.* 72:716–793.

Crepet, W. L. 1974. Investigations of North American cycadeoids: The reproductive biology of *Cycadeoidea*. *Palaeontographica* 148B:144–169.

Daghlian, C. P. and T. N. Taylor. 1979. A new structurally preserved Pennsylvanian cordaitean pollen organ. *Amer. J. Bot.* 66:290–300.

Delevoryas, T. 1953. A new cordaitean fructification from the Kansas Carboniferous. *Amer. J. Bot.* 40:144–150.

—— 1963. Investigations of North American cycadeoids: Cones of *Cycadeoidea*. *Amer. J. Bot.* 50:45–52.

—— 1964. A probable pteridosperm microsporangiate fructification from the Pennsylvanian of Illinois. *Palaeontology* 7:60–63.

—— 1965. Investigations of North American cycadeoids: Microsporangiate structures and phylogenetic implications. *Palaeobotanist* 14:89–93.

—— 1968a. Investigations of North American cycadeoids: Structure, ontogeny, and phylogenetic considerations of the cones of *Cycadeoidea*. *Palaeontographica* 121B:122–133.

—— 1968b. Some aspects of cycadeoid evolution. *J. Linn. Soc. Bot.* 61:137–146.

Delevoryas, T. and R. C. Hope. 1971. A new Triassic cycad and its phyletic implications. *Postilla* 150:1–21.

—— 1973. Fertile coniferophyte remains from the late Triassic Deep River Basin, North Carolina. *Amer. J. Bot.* 60:810–818.

—— 1976. More evidence for a slender growth habit in Mesozoic cycadophytes. *Rev. Palaeobot. Palynol* 21:93–100.

Dennis, R. L. and D. A. Eggert. 1978. *Parasporotheca*, gen. nov. and its bearing on the interpretation of the morphology of permineralized medullosan pollen organs. *Bot. Gaz.* 139:117–139.

Dilcher, D. L. 1979. Early angiosperm reproduction: An introductory report. *Rev. Palaeobot. Palynol* 27:291–328.

Doyle, J. A. 1978. Origin of angiosperms. *Ann. Rev. Ecol. Syst.* 9:365–392.

Dufek, D. and B. M. Stidd. 1981. The vascular system of *Dolerotheca* and its phylogenetic significance. *Amer. J. Bot.* 68:897–907.

Eggert, D. A. and R. W. Kryder. 1969. A new species of *Aulacotheca* (Pteridospermales) from the Middle Pennsylvanian of Iowa. *Palaeontology* 12:414–419.

Eggert, D. A. and G. W. Rothwell. 1979. *Stewartiotheca* gen. n. and the nature and origin of complex permineralized medullosan pollen organs. *Amer. J. Bot.* 66:851–866.

Eggert, D. A. and T. N. Taylor. 1971. *Telangiopsis* gen. nov., in Upper Mississippian pollen organ from Arkansas. *Bot. Gaz.* 132:30–37.

Florin, R. 1936. On the structure of the pollen-grains in the Cordaitales. *Svensk Bot. Tidskr.* 30:624–651.

—— 1937. On the morphology of the pollen-grains in some Palaeozoic pteridosperms. *Svensk Bot. Tidskr.* 31:305–338.

—— 1951. Evolution in cordaites and conifers. *Acta Horti Berg.* 15:285–388.

Gould, R. E. 1971. *Lyssoxylon grigsbyi,* a cycad trunk from the Upper Triassic of Arizona and New Mexico. *Amer. J. Bot.* 58:239–248.

Gould, R. E. and T. Delevoryas. 1977. The biology of *Glossopteris:* Evidence from petrified seed-bearing and pollen-bearing organs. *Alcheringa* 1:387–399.

Grauvogel-Stamm, L. 1969. Nouveaux types d'organes reproducteurs males de Coniféres du Grès à *Voltzia* (Trias Inférieur) des Vosges. *Bull. Serv. Carte géol. Als. Lorr.* 22:93–120.

—— 1972. Revision de cones males du "Keuper Inférieur" du Worcestershire (Angleterre) attribués à *Masculostrobus willsi* Townrow. *Palaeontographica* 140B:1–26.

Hall, J. W. and B. M. Stidd. 1971. Ontogeny of *Vesicaspora,* a late Pennsylvanian pollen grain. *Palaeontology* 14:431–436.

Halle, T. G. 1933. The structure of certain fossil spore-bearing organs believed to belong to the pteridosperms. *K. Svenska Vetenskaps-Akad. Handl.,* Ser. 3, 12.

Harris, T. M. 1941. *Caytonanthus,* the microsporophyll of *Caytonia. Ann. Bot.,* n.s. 5:47–58.

—— 1964. *The Yorkshire Jurassic Flora,* vol. 2. *Caytoniales, Cycadales and Pteridosperms.* London: British Museum (Natural History).

—— 1969. *The Yorkshire Jurassic Flora,* vol. 3. *Bennettitales.* British Museum (Natural History).

Harris, T. M. and W. Millington. 1974. *The Yorkshire Jurassic Flora,* vol. 4, part 1. *Ginkgoales.* London: British Museum (Natural History).

Harris, T. M. and J. Miller 1974. *The Yorkshire Jurassic Flora,* vol. 4, part 2. *Czekanowskiales.* London: British Museum (Natural History).

Kerp, J. H. F. 1982. Aspects of Permian palaeobotany and palynology. 2. On the presence of the ovuliferous organ *Autunia milleryensis* (Renault) Krasser

(Peltaspermaceae) in the Lower Permian of the Nahe area (F.G.R.) and its relationship to *Callipteris conferta* (Sternberg) Brongniart. *Rev. Palaeobot. Palynol.* 31:417–427.

Kidston, R. and W. J. Jongmans. 1911. Sur la Fructification de *Neuropteris obliqua* Bgt. *Ext. Arch. Neerl. Sci. Exa. Nat.*, ser. 3B, 1:25–26.

Krassilov, V. A. 1977a. Contribution to the knowledge of the Caytoniales. *Rev. Palaeobot. Palynol.* 24:155–178.

—— 1977b. The origin of angiosperms. *Bot. Rev.* 43:143–176.

Kurmann, M. H. and T. N. Taylor. 1984. The ultrastructure of *Boulaya fertilis* (Medullosales) pollen. *Pollen et Spores* 26:109–116.

Lacey, W. S., D. E. van Dijk, and K. D. Gordon-Gray. 1975. Fossil plants from the Upper Permian in the Mooi River district of Natal, South Africa. *Ann. Natal. Mus.* 22:349–420.

Leisman, G. A. and J. S. Peters. 1970. A new pteridosperm male fructification from the Middle Pennsylvanian of Illinois. *Amer. J. Bot.* 57:867–873.

Long, A. G. 1961. Some pteridosperm seeds from the Calciferous Sandstone Series of Berwickshire. *Trans. Roy. Soc. Edinburgh* 64:401–419.

—— 1966. Some Lower Carboniferous fructifications from Berwickshire, together with a theoretical account of the evolution of ovules, cupules, and carpels. *Trans. Roy. Soc. Edinburgh* 66:345–375.

Mamay, S. H. 1973. *Archaeocycas* and *Phasmatocycas*—new genera of Permian cycads. *J Res. U.S. Geol. Surv.* 1:687–689.

—— 1976. Paleozoic origin of the cycads. *U.S. Geol. Surv. Prof. Paper* 934:1–48.

Mapes, G. 1981. *Halletheca conica*, medullosan synangia from the Wewoka Formation in Oklahoma. *Bot. Gaz.* 143:125–133.

Mapes, G. and G. W. Rothwell. 1984. Permineralized ovulate cones of *Lebachia* from late Paleozoic limestones of Kansas. *Palaeontology* 27:69–94.

Mapes, G. and J. T. Schabilion. 1981. *Vallitheca valentia* gen. et sp. nov., permineralized synangia from the Middle Pennsylvanian of Oklahoma. *Amer. J. Bot.* 68:1231–1239.

Medus, J. 1977. The ultrastructure of some *Circumpolles*. *Grana* 16:23–28.

Meyer-Berthaud, B. and J. Galtier. 1986. Studies on a Lower Carboniferous flora from Kingswood near Pettycur, Scotland. 2. *Phacelotheca*, a new synangiate fructification of pteridospermous affinity. *Rev. Palaeobot. Palynol.* 48:181–198.

Mickle, J. M. and R. L. Leary. 1984. *Aulacotheca collicola* n. sp. (Medullosaceae) from the early Pennsylvanian of the Illinois Basin. *Rev. Palaeobot. Palynol.* 43:343–357.

Millay, M. A. and D. A. Eggert. 1970. *Idanothekion* gen. n., a synangiate pollen organ with saccate pollen from the Middle Pennsylvanian of Illinois. *Amer. J. Bot.* 57:50–61.

—— 1974. Microgametophyte development in the Paleozoic seed fern family Callistophytaceae. *Amer. J. Bot.* 61:1067–1075.

Millay, M. A., D. A. Eggert, and R. L. Dennis. 1978. Morphology and ultrastructure of four Pennsylvanian prepollen types. *Micropaleontology* 24:303–315.

Millay, M. A. and T. N. Taylor. 1970. Studies of living and fossil saccate pollen. *Micropaleontology* 16:463–470.

—— 1974. Morphological studies of Paleozoic saccate pollen. *Palaeontographica* 147B:75–99.

—— 1976. Evolutionary trends in fossil gymnosperm pollen. *Rev. Palaeobot. Palynol.* 21:65–91.

—— 1977. *Feraxotheca* gen. n., a lyginopterid pollen organ from the Pennsylvanian of North America. *Amer. J. Bot.* 64:177–185.

—— 1978. Fertile and sterile frond segments of the lyginopterid seed fern *Feraxotheca. Rev. Palaeobot. Palynol.* 25:151–162.

—— 1979. Paleozoic pollen organs. *Bot. Rev.* 45:301–375.

Miller, C. N. 1977. Mesozoic conifers. *Bot. Rev.* 43:217–280.

Nathorst, A. G. 1909. Paläobotanische Mitteilungen, 8. *K. Svenska Vetenskaps-Akad. Handl.* 45:1–37.

Oliver, F. W. and D. H. Scott. 1904. On the structure of the Palaeozoic seed *Lagenostoma lomaxi,* with a statement of the evidence upon which it is referred to *Lyginodendron. Phil. Trans. Roy. Soc. London* 197B:193–247.

Pant, D. D. and N. Basu. 1979. Some further remains of fructifications from the Triassic of Nidpur, India. *Palaeontographica* 168B:129–146.

Pant, D. D. and D. D. Nautiyal. 1960. Some seeds and sporangia of *Glossopteris* flora from Raniganj Coalfield, India. *Palaeontographica* 107B:41–64.

Pedersen, K. R. and E. M. Friis. 1987. *Caytonanthus* pollen from the Lower and Middle Jurassic. In J. T. Moller, ed., *25 Years of Geology in Aarhus.* Aarhus: Geological Institute, Aarhus University.

Pigg, K. B. and T. N. Taylor. 1985. Anatomically preserved *Glossopteris* from the Beardmore Glacier area of Antarctica. *Ant. J. of the U.S.* 20:5–7.

Pocknall, D. T. 1981. Pollen morphology of the New Zealand species of *Dacrydium* Solander, *Podocarpus* L'Heritier, and *Dacrycarpus* Endlicher (Podocarpaceae). *New Zealand J. Bot.* 19:67–95.

Ramanujam, C. G. K., G. W. Rothwell, and W. N. Stewart. 1974. Probable attachment of the *Dolerotheca* campanulum to a *Myeloxylon–Alethopteris* type frond. *Amer. J. Bot.* 61:1057–1066.

Renault, B. 1896. *Bassin houiller et Permien d'Autun et d'Epinac.* Fasc. 4. *Flore fossile, part 2.* Études Gîtes Mineraux de la France. Ministère des Travaux Publics, Paris.

—— 1902. Sur quelques pollen fossiles, prothalles male, tubes polliniques, etc., du terrain houiller. *C. R. Acad. Sci. Paris* 135:229–234.

Reymanowna, M. 1973. The Jurassic flora from Grojec near Krakow in Poland. Part 2. Caytoniales and anatomy of *Caytonia. Acta Palaeobot.* 14:46–87.

Rothwell, G. W. 1972. Evidence of pollen tubes in Paleozoic pteridosperms. *Science* 175:772–774.

—— 1975. The Callistophytaceae (Pteridospermopsida). 1. Vegetative structures. *Palaeontographica* 151B:171–196.

—— 1977. The primary vasculature of *Cordaianthus concinnus. Amer. J. Bot.* 64:1235–1241.

—— 1980. The Callistophytaceae (Pteridospermopsida). 2. Reproductive features. *Palaeontographica* 173B:85–106.

—— 1981. The Callistophytales (Pteridospermopsida): Reproductively sophisticated Paleozoic gymnosperms. *Rev. Palaeobot. Palynol.* 32:103–121.

—— 1982. *Cordaianthus duquesnensis* sp. nov., anatomically preserved ovulate cones from the Upper Pennsylvanian of Ohio. *Amer. J. Bot.* 69:239–247.

Rothwell, G. W. and D. A. Eggert. 1986. A monograph of *Dolerotheca* Halle, and related complex permineralised medullosan pollen organs. *Trans. Roy. Soc. Edinburgh: Earth Sciences* 77:47–79.

Rothwell, G. W. and J. E. Mickle. 1982. *Rhetinotheca patens* n. sp., a medullosan pollen organ from the Upper Pennsylvanian of North America. *Rev. Palaeobot. Palynol.* 36:361–374.

Rothwell, G. W. and D. A. Eggert. 1982. What is the vascular architecture of complex medullosan pollen organs? *Amer. J. Bot.* 69:641–643.

Rothwell, G. W. and S. Warner. 1984. *Cordaixylon dumusum* n. sp. (Cordaitales). I. Vegetative structures. *Bot. Gaz.* 145:275–291.

Scheuring, B. W. 1974. "Protosaccate" Strukturen, ein weitverbreitetes Pollenmerkmal zur frühen und mittleren Gymnospermenzeit. *Geol. Paläont. Mitt. Innsbruck* 4:1–30.

Schopf, J. M. 1948. Pteridosperm male fructifications: American species of *Dolerotheca*, with notes regarding certain allied forms. *J. Paleont.* 22:681–724.

Scott, A. C. and T. N. Taylor. 1983. Plant/animal interactions during the Upper Carboniferous. *Bot. Rev.* 49:259–307.

Sellards, E. H. 1903. *Codonotheca*, a new type of spore-bearing organ from the Coal-Measures. *Amer. J. Sci.* 16:87–92.

—— 1907. Notes on the spore-bearing organ *Codonotheca* and its relationship with the Cycadofilices. *New Phytol.* 6:175–178.

Smoot, E. L., T. N. Taylor, and T. Delevoryas. 1986. Structurally preserved fossil plants from Antarctica. I. *Antarcticycas*, gen. nov., a Triassic cycad stem from the Beardmore Glacier area. *Amer. J. Bot.* 72:1410–1423.

Srivastava, S. K. 1976. The fossil pollen genus *Classopollis*. *Lethaia* 9:437–457.

Sterling, C. 1963. Structure of the male gametophyte in gymnosperms. *Biol. Rev.* 38:167–203.

Stewart, W. N. 1951. A new *Pachytesta* from the Berryville locality of southeastern Illinois. *Amer. Midl. Nat.* 46:717–742.

Stidd, B. M. 1978a. An anatomically preserved *Potoniea* with in situ spores from the Pennsylvanian of Illinois. *Amer. J. Bot.* 65:677–683.

Stidd, B. M. 1978b. The synangiate nature of *Dolerotheca*. *Amer. J. Bot.* 65:243–245.

—— 1980. The neotenous origin of the pollen organ of the gymnosperm *Cycadeoidea* and implications for the origin of higher taxa. *Paleobiology* 6:161–167.

—— 1981. The current status of the medullosan seed ferns. *Rev. Palaeobot. Palynol.* 32:63–101.

Stidd, B. M. and J. W. Hall. 1970. *Callandrium callistophytoides*, gen. et sp. nov., the probable pollen-bearing organ of the seed fern, *Callistophyton*. *Amer. J. Bot.* 57:394–403.

Stidd, B. M., G. A. Leisman, and T. L. Phillips. 1977. *Sullitheca dactylifera* gen. et sp. nov., a new medullosan pollen organ and its evolutionary significance. *Amer. J. Bot.* 64:994–1002.

Stidd, B. M., M. O. Rischbieter, and T. L. Phillips. 1985. A new lyginopterid pollen organ with alveolate pollen exines. *Amer. J. Bot.* 72:501–508.

Stubblefield, S. P., T. N. Taylor, and C. P. Daghlian. 1982. Compressed plants from the Lower Pennsylvanian of Kentucky (U.S.A.). 1. *Crossotheca kentuckiensis* n. sp. *Rev. Palaeobot. Palynol.* 36:197–204.

Surange, K. R. and S. Chandra. 1974a. Some male fructifications of Glossopteridales. *Palaeobotanist* 21:255–266.

—— 1974b. Further observations on *Glossotheca* Surange and Maheshwari: A male fructification of Glossopteridales. *Palaeobotanist* 21:248–254.

Taylor, T. N. 1965. Paleozoic seed studies: A Monograph of the American species of *Pachytesta*. *Palaeontographica* 117B:1–46.

—— 1970. *Lasiostrobus* gen. n., a staminate strobilus of gymnospermous affinity from the Pennsylvanian of North America. *Amer. J. Bot.* 57:670–690.

—— 1971. *Halletheca reticulatus* gen. et sp. n.: A synangiate Pennsylvanian pteridosperm pollen organ. *Amer. J. Bot.* 58:300–308.

—— 1973. A consideration of the morphology, ultrastructure and multicellular microgametophyte of *Cycadeoidea dacotensis* pollen. *Rev. Palaeobot. Palynol.* 16:157–164.

—— 1976. The ultrastructure of *Schopfipollenites:* Orbicules and tapetal membranes. *Amer. J. Bot.* 63:857–862.

—— 1978. The ultrastructure and reproductive significance of *Monoletes* (Pteridospermales) pollen. *Can. J. Bot.* 56:3105–3118.

—— 1980. Ultrastructural studies of pteridosperm pollen: *Nanoxanthiopollenites* Clendening and Nygreen. *Rev. Palaeobot. Palynol.* 29:15–21.

—— 1982. Ultrastructural studies of Paleozoic seed fern pollen: Sporoderm development. *Rev. Palaeobot. Palynol.* 37:29–53.

Taylor, T. N. and K. L. Alvin. 1984. Ultrastructure and development of Mesozoic pollen: *Classopollis*. *Amer. J. Bot.* 71:575–587.

Taylor, T. N., M. A. Cichan, and A. M. Baldoni. 1984. The ultrastructure of Mesozoic pollen: *Pteruchus dubius* (Thomas) Townrow. *Rev. Palaeobot. Palynol.* 41:319–327.

Taylor, T. N. and C. P. Daghlian. 1980. The morphology and ultrastructure of *Gothania* (Cordaitales) pollen. *Rev. Palaeobot. Palynol.* 29:1–14.

Taylor, T. N., T. Delevoryas, and R. C. Hope. 1987. Pollen cones from the late Triassic of North America and implications on conifer evolution. *Rev. Palaeobot. Palynol.* 53:141–149.

Taylor, T. N. and M. H. Kurmann. 1985. *Boulayatheca*, the new name for the seed fern pollen organ *Boulaya* Carpentier. *Taxon* 34:666–667.

Taylor, T. N. and M. A. Millay. 1977. The ultrastructure and reproductive significance of *Lasiostrobus* microspores. *Rev. Palaeobot. Palynol.* 23:129–137.

—— 1979. Pollination biology and reproduction in early seed plants. *Rev. Palaeobot. Palynol.* 27:329–355.

—— 1981a. Additional information on the pollen organ *Halletheca* (Medullosales). *Amer. J. Bot.* 68:1403–1407.

——. 1981b. Morphologic variability of Pennsylvanian lyginopterid seed ferns. *Rev. Palaeobot. Palynol.* 32:27–62.

Taylor, T. N. and G. W. Rothwell. 1982. Studies of seed fern pollen: Development of the exine in *Monoletes* (Medullosales). *Amer. J. Bot.*, pp. 570–578.

Taylor, T. N. and E. L. Smoot. (In press). The ultrastructure of fossil gymnosperm pollen. *Bull. Soc. Bot. France.*

Taylor, T. N. and M. S. Zavada. 1986. Developmental and functional aspects of fossil pollen. In S. Blackmore and I. K. Ferguson, eds., *Pollen and Spores: Form and Function*, pp. 165–178. Linnean Society, London: Academic Press.

Thomas, H. H. 1925. The Caytoniales, a new group of angiospermous plants from the Jurassic rocks of Yorkshire. *Phil. Trans. Roy. Soc. London* 213B:299–363.

—— 1933. On some pteridospermous plants from the Mesozoic rocks of South Africa. *Phil. Trans. Roy. Soc. London* 222B:193–265.

Townrow, J. A. 1960. The Peltaspermaceae, a pteridosperm family of Permian and Triassic age. *Palaeontology* 3:333–361.

—— 1962. On *Pteruchus*, a microsporophyll of the Corystospermaceae. *Bull. British Museum (Natural History) Geol.* 6:287–320.

—— 1967. On *Rissikia* and *Mataia* podocarpaceous conifers from the Lower Mesozoic of southern lands. *Papers and Proc. Roy. Soc. Tasmania.* 101:103–136.

Trivett, M. L. and G. W. Rothwell. 1985. Morphology, systematics, and paleoecology of Paleozoic fossil plants: *Mesoxylon priapi*, sp. nov. (Cordaitales). *Syst. Bot.* 10:205–223.

Van Konijnenburg-Van Cittert, J. H. A. 1971. In situ gymnosperm pollen from the Middle Jurassic of Yorkshire. *Acta Bot. Neerl.* 20.

Watson, J. 1982. The Cheirolepidiaceae: A short review. *Phyta*, Pant Comm. vol. 265–273.

White, M. E. 1978. Reproductive structures of the Glossopteridales in the plant fossil collection of the Australian museum. *Rec. Austral. Mus.* 31:473–505.

Wieland, G. R. 1906. *American Fossil Cycads: Structure.* Washington: Carnegie Institution of Washington.

Zavada, M. S. and W. L. Crepet. 1985. Pollen wall ultrastructure of the type material of *Pteruchus africanus, P. dubius* and *P. papillatus. Pollen et Spores* 27:271–276.

—— 1986. Pollen wall structure of *Caytonanthus arberi. (Caytoniales) Pl. Syst. Evol.* 153:259–264.

Zavada, M. S. and T. N. Taylor. 1986. The role of self-incompatibility and sexual selection in the gymnosperm-angiosperm transition: A hypothesis. *Amer. Nat.* 128:538–550.

Zhu Jia-Nan and Du Xian-Ming. 1981. A new cycad—*Primocycas chinensis* gen. et sp. nov. from the Lower Permian in Shanxi, China and its significance. *Acta Bot. Sin.* 23:401–404.

5

Major Clades and Relationships in the "Higher" Gymnosperms

PETER R. CRANE

Since the mid-1980s there has been a marked renewal of interest in the phylogeny and classification of seed plants (Meyen 1984; Crane 1985a, 1985b; Doyle and Donoghue 1986a, 1986b, 1987a, 1987b). Several factors seem to have stimulated much of the recent activity. One has been the explicit realization that attempts to resolve the phylogenetic position of flowering plants are inextricably linked to the broader problems of defining the major groups of gymnosperms and resolving their interrelationships (Crane 1984, 1985a). Another has been increased knowledge of several critical groups of fossil gymnosperms (e.g., Rothwell 1981) that has raised questions concerning the systematic homogeneity and relationships of the pteridosperms (seed ferns) as they have been traditionally delimited (Meyen 1984; Crane 1985a).

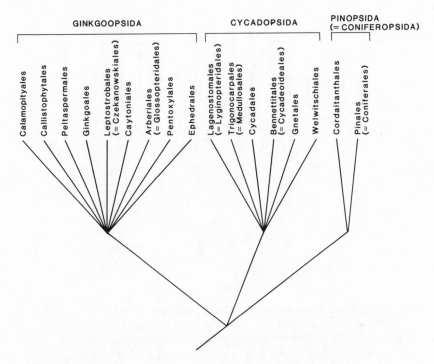

Figure 5.1. Diagrammatic summary of the classification of gymnosperms presented by Meyen (1984:88–90).

Recent attempts to clarify phylogenetic relationships within seed plants have adopted two different methodologies. One approach (e.g., Meyen 1984; see figure 5.1) has been to apply techniques that group organisms and therefore infer relationships on the basis of overall similarity (compare Meyen's discussion of "congregational analysis" with Davis and Heywood's discussion of phenetic methods; Davis and Heywood 1973:110–112; Meyen 1984:5–8). The other has been to apply cladistic techniques that attempt to group organisms only on the basis of shared derived similarities (Hennig 1966; Eldredge and Cracraft 1980; Wiley 1981; see Crane 1985a:718–721 for an outline of cladistic techniques). The two recent studies that have applied cladistic techniques to resolving the phylogenetic relationships of seed plants (Crane 1985a; Doyle and Donoghue 1986b) differ in the exact placement of certain taxa, but nevertheless share significant similarities, particularly with respect to several phylogenetic questions that have previously been controversial (figure 5.2). First, the three extant

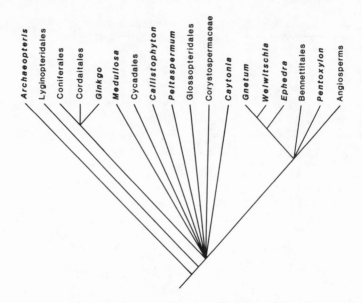

Figure 5.2. Diagrammatic summary of patterns of relationships among major groups of seed plants. This strict consensus representation incorporates only features common to the cladistic analyses of Crane (1985a: figures 20, 21) and Doyle and Donoghue (1986b: figures 4, 5). Note particularly the agreement on relationships within the Gnetales and on the gymnosperm groups most closely related to flowering plants.

genera of Gnetales are placed as more closely related to each other than to any other major group of gymnosperms. The interpretation of the Gnetales as a monophyletic group supports traditional views (Coulter and Chamberlain 1917; Bierhorst 1971) but conflicts with the ideas of Eames (1952) and Meyen (1984), who suggest that *Ephedra* is more closely related to cordaites and the Ginkgoopsida, respectively (e.g., figure 5.1). Second, within the Gnetales, *Gnetum* and *Welwitschia* are placed as more closely related to each other than either is to *Ephedra* (see also Coulter and Chamberlain 1917:403). Third, the Bennettitales (+ *Pentoxylon)* and the Gnetales are resolved as more closely related to each other than they are to other gymnosperms. This is consistent with several earlier interpretations (Arber and Parkin 1907, 1908; Thoday 1911; Takhtajan 1969; Ehrendorfer 1971, 1976; Martens 1971) but conflicts with the view that the Bennettitales are closely related to cycads (Wieland 1906; Meyen 1984) and that Gnetales are derived coniferopsids, whereas the Bennettitales and cycads are part of an independent cycadopsid line of gymnosperm evolution (Chamberlain 1935; Doyle 1978; Stewart 1983; Walker and Walker 1984). Fourth, the angiosperms are placed as more closely related to the Bennettitales *(+Pentoxylon)* and the Gnetales than to other groups of gymnosperms. This has been suggested by several authors (Arber and Parkin 1908; Just 1948; Ehrendorfer 1971, 1976) but conflicts with the widespread view that the Gnetales and angiosperms are only distantly related (Cronquist 1968; Doyle 1978; Walker and Walker 1984). Fifth, certain late Palaeozoic and Mesozoic "pteridosperms," including *Callistophyton, Caytonia,* corystosperms, peltasperms, and glossopterids are resolved as more closely related to the clade formed by angiosperms, Gnetales, and Bennettitales *(+Pentoxylon)* than to other pteridosperm taxa such as the lyginopterids and perhaps the medullosans. This is in broad agreement with the suggestion that certain "Mesozoic seed ferns" are closely related to angiosperms (Gaussen 1946; Thomas 1955; Andrews 1961; Stebbins 1974; Doyle 1978; Retallack and Dilcher 1981; Stewart 1983) but conflicts with the traditional use of the "pteridosperms" as a meaningful group for phylogenetic purposes (e.g., Knoll and Rothwell 1981; Stewart 1983).

While the extent of agreement between the two independent cladistic studies is encouraging, it is important to recognize that the taxa and characters employed were very similar, and although the most recent analyses (Doyle and Donoghue 1986b, 1987a) incorporate significant improvements over earlier work, both in character coding and character definition, knowledge of many characters remains unsatisfactory. Furthermore, all of the analyses completed to date have high levels of homoplasy (convergence and reversals) and data sets from which a large number of almost equally parsimonious cladograms can be derived. It is therefore clear that the pattern of phylogenetic relationships within seed plants is far from finally resolved.

In cladistic analyses the criterion of parsimony (sometimes termed the principle of simplicity or economy; Wiley 1981:20) is used to choose between conflicting hierarchical character patterns in a data matrix. Under this criterion the simplest hierarchical summary of character distributions (the "shortest tree") is preferred, and in evolutionary terms this preferred phylogenetic hypothesis effectively minimizes the amount of homoplasy (parallelism, convergence, and reversal) required by a given cladogram. Changes in the way characters or polarity are scored in the data matrix may affect parsimony and hence the phylogenetic hypothesis preferred. For example, the traditional view that the compound ovulate cone is primitive within the conifers (Florin 1951; Crane 1985a; Doyle and Donoghue 1986b) presupposes that the terminal ovules of extant *Taxus* and the stalked, individually scattered ovules of the Permian conifer *Buriadia* (Pant 1977b) are derived *within* the conifer clade. If this hypothesis were refuted, it would have serious consequences for the accepted homologies and relationships between conifers and cordaites (Florin 1951; Crane 1985a; Doyle and Donoghue 1986b). Similarly, in the context of fossil and extant seed plants, a distal germinal aperture has apparently arisen in the pollen of at least four different clades: conifers, cordaites, Ginkgoales, and other seed plants (Crane 1985a). If extinct seed plants are excluded from consideration, the presence of a distal germinal aperture may be erroneously interpreted as a synapomorphy of all extant seed plants (Hill and Crane 1982). It is therefore clear that reinterpretations of characters that arise either from a more refined understanding of relationships in certain groups, or from the incorporation of additional living or fossil plants, may have a major effect on character patterns and thus on the perceived pattern of relationships among higher taxa (see Doyle and Donoghue 1987b for further examples). In this paper I consider these problems with respect to the fossil history and patterns of relationship in four groups of relatively derived "higher" gymnosperms: the corystosperms, cycads, Bennettitales, and Gnetales. The implications for our current understanding of relationships within seed plants are discussed.

CORYSTOSPERMS

The family Corystospermaceae (Umkomasiaceae *sensu* Meyen 1984) was first established by Thomas (1933) for leaves, pollen organs, and seed organs from the Middle Triassic Molteno Formation of the Upper Umkomaas Valley, Natal (Anderson and Anderson 1983). Leaves were assigned to *Dicroidium*, pollen organs to *Pteruchus*, and seed-bearing organs to three very similar genera, *Pilophorosperma*, *Spermatocodon*, and *Umkomasia*. Although probable fossil leaves of the Corystospermaceae (e.g., *Dicroidium, Pachypteris, Xylopteris;* Townrow 1965) are common and diverse during the Mesozoic, particularly in the southern hemisphere, relatively little reproductive material has been described (Petriella 1979, 1980, 1981, 1983; see Crane 1985a for a brief review).

The corystosperms are generally interpreted as an isolated group of "pteridosperms," and the similarities between *Pteruchus* (figure 5.3A) and the probable lyginopterid pollen organ *Crossotheca* have been frequently cited (Thomas 1933; Townrow 1962). Meyen (1984) places the Corystospermaceae in his order Peltaspermales (class Ginkgoopsida, figure 5.1) along with the Cardiolepidaceae, Peltaspermaceae, and Trichopityaceae. In recent cladistic analyses corystosperms have been placed as the sister taxon to the clade composed of Bennettitales, *Pentoxylon*, Gnetales, and angiosperms (Crane 1985a: figure 22) or the sister group to these taxa with the addition of cycads, peltasperms, *Caytonia*, and glossopterids (Doyle and Donoghue 1987a: figure 4).

Interpretations of the phylogenetic position of the corystosperms hinge to a large extent on the morphological interpretation of their reproductive structures. Thomas (1933) regarded the pollen and ovule-bearing organs as complex branching structures with "bracts" apparently subtending fertile "branches." This is supported in part by the occurrence of a spiral arrangement of "branches" in some species (Pant and Basu 1973, 1979). In contrast, Crane (1985a) and Doyle and Donoghue (1986b) follow Townrow (1962) and, on the basis of dorsiventrality in "branching" and cuticular differentiation in several species, interpret corystosperm reproductive structures as conventional microsporophylls and megasporophylls. Under this interpretation the "bracts" mentioned by Thomas (1933) are viewed as sterile rachial pinnules (figure 5.3C), and in many specimens of both *Pteruchus* and *Umkomasia* (e.g., Anderson and Anderson 1985: plates 195, 196) these structures are apparently absent. The microsporophyll and megasporophyll interpretations are also more consistent with the morphology of the putative corystosperm reproductive structures *Ktalenia circularis*, *Nidiostrobus harrisiana*, and *Kachchhia naviculata*, reviewed below.

Ktalenia circularis *Ktalenia* (figure 3.5B) was originally established to accommodate isolated cupules containing either one or two seeds, from the Lower Cretaceous of Ticó, Santa Cruz Province, Argentina (Archangelsky 1963). Additional material has since been described (Taylor and Archangelsky 1985) that was closely associated or attached to leaves of *Ruflorinia sierra* (Archangelsky 1963). As reconstructed by Taylor and Archangelsky, certain *Ruflorinia* fronds included both a fertile and sterile region, the fertile portion being composed of a central rachis bearing several cupules. Clusters of linear pinnules occur along the rachis interspersed between the cupules in a fashion similar to the sterile rachial pinnules in *Pilophorosperma*, *Spermatocodon*, and *Umkomasia*. These rachial pinnules, together with the occurrence of recurved cupules containing one or two ovules, suggest a close relationship to the Corystospermaceae; if this interpretation is correct, *Ktalenia* provides further support for the

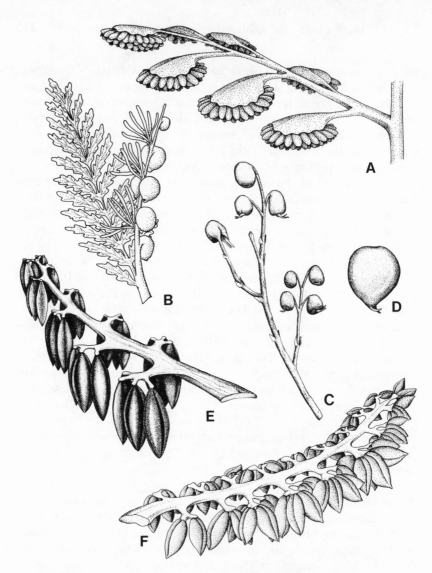

Figure 5.3. Morphology of selected Mesozoic pteridosperms. **A** *Pteruchus africanus*, based on Townrow (1962: figures 1A–D, 2D–G); from Crane (1985a) by permission. × 3.5. **B** Frond system of *Ruflorinia* bearing *Ktalenia* cupules, redrawn from Taylor and Archangelsky (1985: figure 25). × 2. **C** *Umkomasia macleanii*, redrawn from Thomas (1933: figure 1, plate 26, figure 56); from Crane (1985a) by permission. × 1. **D** Corystosperm ovule, based on Thomas (1933: figure 33c); from Crane (1985a) by permission. × 2. **E** *Caytonanthus arberi*, based on Harris (1941b: plate 2, figure 3), redrawn from Crane (1985a). × 6. **F** *Kachchhia navicula*, based on Bose and Banerjee (1984: text figure 67a, d). × 5.

suggestion that the ovulate structures in the group are modified leaves (i.e., megasporophylls). *Ktalenia* is also somewhat similar to the reproductive organs of the glossopterids, in which a fertile structure (variously interpreted as a fertile shoot or leaf; see Retallack and Dilcher 1981) is apparently borne adaxially on an otherwise vegetative leaf (Pant 1977a). It is also interesting that the dehisced cupules of some corystosperms (e.g., *Umkomasia speciosa [Zuberia zuberi]*, Frenguelli 1944: figure 11; *Umkomasia* sp. E, Retallack 1980: figure 21.9F; *Karibacarpon*, Lacey 1976; Holmes and Ash 1979) resemble the cupules of the glossopterid *Lidgettonia africana* (Thomas 1958; Lacey, van Dijk, and Gordon-Gray 1975) and the possible glossopterid *Mooia lidgettonioides* (Lacey, van Dijk, and Gordon-Gray 1975), and that some glossopterids apparently also have uniovulate cupules (e.g., *Denkania indica*, Surange and Chandra 1975).

Nidiostrobus harrisiana *Nidiostrobus* (figure 5.4A, B) is a pollen cone from the Lower Triassic of Nidpur, Madhya Pradesh, India (Bose and Srivastava 1973a, 1973b). Based on association evidence and cuticular similarities, *N. harrisiana* is thought to have been produced by the same plant as leaves of *Dicroidium nipurdensis* (Bose and Srivastava 1971) and ovulate cones of *Nidia ovalis* (Bose and Srivastava 1973b). The pollen cone is about twice as long as wide and may be over 15 cm in length. It consists of a central axis bearing numerous spirally arranged, fan-shaped scales (presumed microsporophylls) with seven to nine elongated pollen sacs on their adaxial surface (Bose and Srivastava 1973a, 1973b; see Pant and Basu 1979:130; Meyen 1984:54 for alternative interpretations). The pollen is bisaccate and nonstriate. The ovulate cone is compact, up to 4 cm long, and bears helically arranged megasporophylls that resemble those of some cycads in having a narrow stalk and an expanded distal head bearing two ovules. Few details of seed cuticles or anatomy are known.

In characters of morphology, venation, and cuticular structure, *D. nipurdensis* is similar to other fossil leaves assigned to *Dicroidium* (Townrow 1965; Bose and Srivastava 1971; Retallack 1977), but the reproductive structures are very different from those associated with other *Dicroidium* species. Bose and Srivastava (1973b) compare the ovulate cones *(Nidia)* with those of *Zamia*, while Meyen (1984) emphasizes the similarities to *Peltaspermum thomasii* Harris (1932a). Equally, the megasporophylls could be compared to a pair of reflexed corystosperm cupules (see Frenguelli 1944: figure 10). Until more complete morphological information or cuticular details of the seeds are available, it will be impossible to resolve these different alternatives.

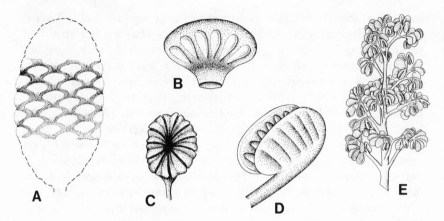

Figure 5.4. Problematic Mesozoic plant fossils. **A** *Nidiostrobus harrisiana* morphology of pollen cone, redrawn from Bose and Srivastava (1973b: text figure 5A). × 0.5. **B** *N. harrisiana*, detail of presumed microsporophyll from adaxial surface, based on Bose and Srivastava (1973b: text figure 5B, C: plate 9). × 2. **C** *Pteroma thomasii*, abaxial view of synangium, redrawn from Harris (1964: fig. 66B); from Crane (1985a) by permission. × 2.5. **D** *Harrisothecium marsilioides*, detail of single synangium, based on Harris (1932b: figure 52b). × 5. **E** *Harrisothecium marsilioides*, based on Harris (1932b: figure 52a, d–f: plate 9). × 0.75.

The microsporophylls and pollen of *Nidiostrobus* are unlike those of either cycads or peltasperms *sensu stricto*. Bose and Srivastava (1973b) have emphasized the similarities with *Harrisothecium marsilioides*, a complex three-dimensionally branched structure bearing bivalved synangia from the Upper Triassic of Greenland (figure 5.4D, E; Harris 1932b; Lundblad 1950, 1961). Each synangium of *Harrisothecium* consists of approximately seven pairs of pollen sacs containing bisaccate pollen. On the basis of association evidence and similarities in cuticular structure, *Harrisothecium* is thought to have been borne on the same plant as leaves of *Ptilozamites nilssonii* (Harris 1932b). The arrangement of pollen sacs in *Nidiostrobus* and *Harrisothecium* is also somewhat similar to that in *Pteroma thomasii* from the Middle Jurassic of Yorkshire (Harris 1964). *Pteroma* is a pinnately organized, two-dimensional structure bearing oval or round fertile laminae (figure 5.4C). Each of these laminae has a ring of about 10 pollen sacs embedded in its undersurface, which contain bisaccate pollen. *Pteroma* is thought to be the pollen organ of the plant that bore leaves of *Pachyp-*

teris papillosa (Harris 1964) and is generally included in Corystospermaceae (Townrow 1965).

Kachchhia navicula *Kachchhia* (figure 5.3F) is a microsporophyll from the Middle–Upper Jurassic flora of Kachchh (Kutch), in northwestern India (Bose and Banerjee 1984). Based on association evidence and similarity of cuticular structure, *Kachchhia* is thought to have been borne on the plant that produced leaves of *Pachypteris specifica*. *Kachchhia* is dorsiventrally flattened with alternate or opposite, simple or forked laterals bearing bilocular synangia on the presumed abaxial surface. Pollen preserved in the sporangia is immature, but the abundant grains attached to the cuticles suggest that the mature pollen was bisaccate. In several respects *Kachchhia* resembles *Antevsia* (pollen organ of the *Peltaspermum* plant; Harris 1932a; Crane 1985a) and *Pteruchus* (e.g., Townrow 1962; see figure 5.3A) but differs from both taxa in having synangia containing two pollen sacs. In the morphology of the microsporophyll, and in details of the synangia and saccate pollen, *Kachchhia* is more like *Caytonanthus* (figure 5.3E), the microsporophylls of the *Caytonia* plant (Harris 1964).

Discussion *Ktalenia*, *Nidiostrobus*, and *Kachchhia* all support the interpretation of the reproductive structures of corystosperms as microsporophylls and megasporophylls, but they also add significantly to the range of morphological diversity in the group and could be interpreted as narrowing the gap between relatively well-known taxa, based on *Dicroidium*, *Pteruchus*, and *Umkomasia*, and other gymnosperms such as *Caytonia*, cycads, peltasperms, and perhaps glossopterids. *Kachchhia* (figure 5.3F), for example, could be viewed as intermediate between "pteridosperms" with unilocular synangia (e.g., *Antevsia*, *Pteruchus*; figure 5.3A), and the tetrasporangiate condition in *Caytonanthus* (figure 5.3E) or angiosperms. Alternatively, *Kachchhia* could be merely a relatively derived form of microsporophyll that evolved within the corystosperm clade. Choosing between these two alternatives and therefore assessing the potential phylogenetic significance of plant organs such as *Ktalenia*, *Nidiostrobus*, and *Kachchhia* is currently impossible without more knowledge of the whole plants on which these organs were borne. Nevertheless, the information presently available highlights the fact that there is currently no clearly defined character or group of characters by which the family Corystospermaceae may be delimited, and that recent considerations of relationships of the corystosperms (Crane 1985a; Doyle and Donoghue 1986b) based largely on the plants described by Thomas (1933) and Townrow (1965) may not necessarily apply to other taxa traditionally included within the family.

CYCADS

Extant cycads comprise approximately 100 species of dioecious pachycaul plants distributed between 10 genera: *Bowenia, Ceratozamia, Cycas, Dioon, Encephalartos, Lepidozamia, Macrozamia, Microcycas, Stangeria,* and *Zamia* (Greguss 1968). They are a predominantly tropical group with their greatest systematic diversity in Mexico, South Africa, and northeastern Australia (Schuster 1932). Traditionally, cycads have been considered a natural group (Chamberlain 1935; Bierhorst 1971; Sporne 1971) distinct from, but perhaps closely related to, the Bennettitales (Cycadeoidales; e.g., Wieland 1906). Meyen (1984) places the Cycadales as an order in his class Cycadopsida (figure 5.1) along with the Lagenostomales (Lyginopteridales), Trigonocarpales (Medullosales), Bennettitales, Gnetales, and Welwitschiales. It has also been suggested that cycads are closely related to lyginopterid pteridosperms (Scott 1923; Arnold 1953) or to medullosans (Wordsell 1906; Delevoryas 1955; Stewart 1983). Recent cladistic approaches (Crane 1985a; Doyle and Donoghue 1986b) have treated cycads as a monophyletic group based on several characters that are apparently unique within seed plants. My own analyses placed the cycads either as the sister group to medullosans (Crane 1985a; cladogram 1) or as equally closely related to medullosans and platyspermic and derivative taxa (Crane 1985a; cladogram 2). Doyle and Donoghue (1986b) also varied in their placement of cycads, resolving them as the sister group of medullosans, the sister group of platyspermic and derivative taxa, or as the sister group to the glossopterid, *Caytonia,* Bennettitales, Gnetales, and angiosperm clade. In other equally parsimonious arrangements cycads are placed *within* the platyspermic clade as the sister group to peltasperms.

Several classifications of the ten extant cycad genera have been proposed (figure 5.5). Usually *Cycas* has been recognized as particularly distinctive and separated from the other genera. It has been variously assigned to its own tribe (Hutchinson 1924), subfamily (Pilger 1926; Schuster 1932), or family (Johnson 1959; see figure 5.5A), although other genera (especially *Stangeria* or *Bowenia*) have sometimes been considered equally isolated within the group. Recently Stevenson (1985; see figure 5.5B) has presented a preliminary outline of a new classification of the cycads. He recognizes two suborders,

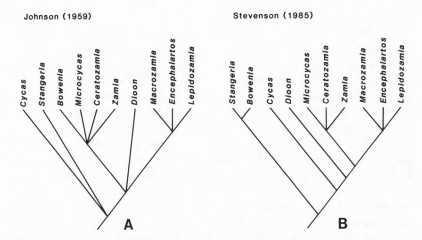

Figure 5.5. Diagrammatic summary of two classifications of extant cycads. **A** Johnson (1959). **B** Stevenson (1985).

Stangerineae *(Bowenia, Stangeria)* and Cycadineae, and, within the Cycadineae, two families, the Cycadaceae *(Cycas)* and Zamiaceae. The Zamiaceae in turn are divided into the subfamilies Diooideae *(Dioon)* and Zamioideae (comprising Zamieae and Encephalarteae).

The only existing cladistic analysis of cycads is that of Petriella and Crisci (1977: figure 1B), in which *Cycas* is placed as the sister taxon to a group in which *Dioon* forms the sister taxon of all other genera. The Encephalarteae *(Encephalartos, Lepidozamia, Macrozamia)* are recognized as a monophyletic sister group to *Bowenia, Ceratozamia, Microcycas, Stangeria,* and *Zamia. Ceratozamia* and *Microcycas* form the sister group to *Zamia* and *Bowenia + Stangeria.* Although several elements in Petriella and Crisci's classification are similar to that of Stevenson (1985), they do not resolve the Zamieae (*sensu* Stevenson 1985; i.e., *Ceratozamia, Microcycas,* and *Zamia*) as a monophyletic group, and they separate the two genera *Microcycas* and *Zamia,* which are generally thought to be closely related (Eckenwalder 1980:714). In addition, there is little discussion of the characters employed, outgroups, and character polarity. The preliminary cladistic analysis presented below and based on characters summarized in table 5.1 incorporates information from the work of Johnson (1959), Petriella and Crisci (1975, 1977), and Stevenson (1985) and provides a preliminary hypothesis of phylogenetic relationships within cycads to be tested with future research.

Characters(see table 5.1)

1.1 Presence of cycasin. All extant genera of cycads produce the methylazoxymethanol glycoside cycasin, but this compound is not known to occur in any other gymnosperm (DeLuca et al. 1980; Moretti, Sabato, and Siniscalo-Gigliano 1983). A similar compound, macrozamin, is also known in many cycads and may be a potential synapomorphy of the group (Moretti, Sabato, and Siniscalo-Gigliano 1983).

1.2 Girdling leaf traces. At least some leaf traces in all cycads arise from the stele on the side of the stem opposite to the leaf that they supply (Chamberlain 1935). This feature is unique in gymnosperms and is useful in determining anatomically preserved fossil stems (e.g., Smoot, Taylor, and Delevoryas 1985). Girdling leaf traces are not present in some cycad seedlings (Wordsell 1906) or any cone axes (Matte 1904), including the portion of the stem-bearing megasporophylls in *Cycas* (D. W. Stevenson, personal communication). Tangential or occasionally radial traces are the norm in the leaves of other fossil and living gymnosperms, although less pronounced girdling traces occur in some angiosperms (Chamberlain 1935).

1.3 Sympodial growth. Growth of the stem in cycads is usually interpreted as sympodial. Cones are generally produced singly at the stem apex and appear to terminate the activity of the apical meristem. Vegetative growth is continued by another meristem at the base of the cone stalk. This meristem does not appear to be in the axil of a leaf (axillary branching is unknown in cycads; Stevenson 1980), but it is uncertain whether it arises *de novo* or from the apical meristem. The one exception to the ubiquitous presence of this form of sympodial growth occurs in ovulate plants of *Cycas*. In these individuals, growth is monopodial and the production of a cluster of megasporophylls does not terminate the activity of the apical meristem. Superficially, monopodial growth occurs in adult plants of *Ceratozamia*, *Encephalartos, Lepidozamia, Macrozamia,* and *Zamia*, which have lateral cones (Chamberlain 1913, 1935; Johnson 1959:66–67; D. W. Stevenson, personal communication; see character 1.19). The unusual pattern of growth seen in cycads is not known to occur in any other group of gymnosperms.

1.4 Simple ovulate cone. All extant cycads produce simple, clearly defined clusters of megasporophylls (cones) and in all genera except *Cycas* the cones are determinate. Doyle and Donoghue (1986b:429) imply that the *Cycas* condition may be primitive in cycads, but it seems more likely that ovulate *Cycas* plants are relatively advanced in this respect (see also Harris 1976). The unexpanded crown of spo-

Table 5.1. Data matrix for characters of extant cycads. Apomorphic characters are marked +, plesiomorphic characters are marked −, and uncertain or missing characters are marked ?.

		Other seed plants	Cycas	Bowenia	Stangeria	Ceratozamia	Zamia	Microcycas	Dioon	Encephalartos	Macrozamia	Lepidozamia
1.1	Presence of cycasin	−	+	+	+	+	+	+	+	+	+	+
1.2	Girdling leaf traces	−	+	+	+	+	+	+	+	+	+	+
1.3	Sympodial growth	−	+	+	+	+	+	+	+	+	+	+
1.4	Simple ovulate cone	−	+	+	+	+	+	+	+	+	+	
1.5	Primary thickening meristem	−	+	+	+	+	+	+	+	+	+	+?
1.6	Simple megasporophylls	−	−	+	+	+	+	+	+	+	+	+
1.7	Biovulate megasporophylls	−	−	+	+	+	+	+	+	+	+	+
1.8	Ovules with micropyles oriented proximally	−	−	+	+	+	+	+	+	+	+	+
1.9	Radiospermic seeds	−	−	+	+	+	+	+	+	+	+	+
1.10	Colored trichomes	−	−	+	+	+	+	+	+	+	+	+
1.11	Specialized alveolate exine structure	−	−	+?	+	+	+	+?	+	+	+	+?
1.12	Stipules on leaves	−	−	+	+	+	+	+	−	−	−	−
1.13	Vascularized stipules	−	−	+	+	−	−	−	−	−	−	−
1.14	Primary rachis of leaf with circular arrangement of vascular bundles	−	−	+	+	−	−	−	−	−	−	−
1.15	Sporophyll vasculature in more than one plane	−	−	+	−	+	+	+	−	+	+	+
1.16	Developing pinnae flat	−	−	−	−	+	+	+	+	+	+	+
1.17	Articulate pinnae	−	−	−	−	+	+	+	−	−	−	−
1.18	Polyxylic stele	−	+	−	−	−	−	−	−	+	+	+
1.19	Lateral cones	−	−	−	−	+	+	−	−	+	+	+
1.20	Cone peduncles with cataphylls	−	−	−	−	−	−	−	+	+	+	+
1.21	Short, curved, colored trichomes	−	−	+	−	−	−	−	−	−	+	+
1.22	Branched trichomes	−	+	−	−	−	+	+	−	−	−	−
1.23	Equally branched trichomes	−	−	−	−	−	+	+	−	−	−	−

rophylls in *Cycas* is very similar to the loose cones of *Dioon* (Chamberlain 1935: compare figures 96, 98), and the developmental difference between the determinate and indeterminate condition is minor, involving only the persistence of the apical meristem. In *Cycas*, pollination occurs prior to expansion, while the cluster of sporophylls most resembles the cones of other species (Niklas and Norstog 1984) and the portion of the axis that bears megasporophylls resembles the cones of other genera in lacking girdling leaf traces. Fossil cycads are also known (new genus 2, Hill 1986) in which *Cycas*-like megasporophylls occur in a definite cone. Doyle and Donoghue (1986b:403) raise the possibility of homology between the ovulate cones of cycads and the fertile secondary shoots of cordaites and conifers. However, these groups differ from cycads in having uniovulate sporophylls that are interspersed among sterile scales (Florin 1954; Mapes and Rothwell 1984) or leaves (e.g., *Buriadia;* Pant and Nautiyal 1967, Pant 1977b). In cycads, the sporophylls are usually biovulate or multiovulate, are more substantial, and are not interspersed among sterile scales. In *Ginkgo* the ovule-bearing structure occurs in the axil of a leaf, and is most easily interpreted as a short shoot bearing two fused fertile leaves (G. W. Rothwell, personal communication). Although the basic arrangement of megasporophylls in cycads may be similar to *Ginkgo*, both the form and the number of megasporophylls borne on the fertile shoot are very different. For these reasons I regard simple ovulate cones as a synapomorphy of all cycads.

1.5 Primary thickening meristem. In all cycads that have been investigated (all genera except *Lepidozamia*) radial growth occurs through the activity of a primary thickening meristem that mainly produces derivatives centrifugally (Stevenson 1980; personal communication). It is this primary thickening meristem rather than the massive shoot apex that is largely responsible for the pachycaul habit in cycads (Stevenson 1980). It is also present in cycads with slender stems (D. W. Stevenson, personal communication) and therefore may have occurred in fossil cycads with a slender habit (Harris 1961; Delevoryas and Hope 1971; Kimura and Sekido 1975; Zhang and Mo 1981; Delevoryas 1982). A similar primary thickening meristem occurs in some monocotyledons (e.g., many palms), but in these taxa the derivatives are mainly produced centripetally. A primary thickening meristem closely resembling that of cycads is not known to occur in any other fossil or living seed plant.

1.6 Simple megasporophylls. In all cycads except *Cycas*, the megasporophylls are simple and nonpinnate. In *Cycas* (figure 5.6A, E) the megasporophyll distal to the seeds is either serrate or pinnately divided, and, by comparison with the pinnately organized sporophylls of primitive seed plants (e.g., lyginopterids, medullosans, *Callistophyton*, and other "pteridosperms"), simple nonpinnate megasporophylls are interpreted as a synapomorphy of all cycads except *Cycas*. Under this interpretation the pinnate megasporophylls of the cycad *Primocycas* (figure 5.6F) from the Permian of China would be considered primitive, whereas the simple megasporophylls of *Phasmatocycas* (figure 5.6H) from the Permian of North America would be considered relatively derived (see below). The interpretation of a single megasporophyll of *Cycas* as homologous to a complete cone of other cycads (Meeuse 1963) is not followed here.

1.7 Biovulate megasporophylls. In all cycads except *Cycas* the megasporophylls typically bear two ovules, although this may vary between one and three (Johnson 1959). In *Cycas* the number of ovules per megasporophyll is generally three or more, although megasporophylls with one or two ovules occasionally occur (Schuster 1932; see also figure 5.6E). By comparison with the more complex fertile leaves typical of primitive seed plants (e.g., figures 5.6G, H and other Palaeozoic "pteridosperms" that presumably bore numerous ovules), biovulate megasporophylls are considered derived within the cycads.

1.8 Ovules with micropyles oriented proximally. On the megasporophylls of all extant cycads except *Cycas*, the ovules are attached parallel to the megasporophyll stalk with their micropyles oriented toward the cone axis. Usually the ovules are borne directly on the broad distal portion of the megasporophyll, but they also may be stalked (*Dioon;* Sabato and DeLuca 1985) or borne on the megasporophyll stalk (e.g., *Bowenia;* Schuster 1932: figure 15M, N; fossil genus 3, Hill 1986). In *Cycas* the ovules are attached along the sides of the megasporophyll with their micropyles oriented either laterally or toward the apex of the megasporophyll, typically at angles of 50°–80° (figure 5.6A, E). The "reflexed" orientation of the ovules in all genera except *Cycas* is interpreted as relatively derived compared to the less specialized ovule arrangement seen in most primitive Palaeozoic seed plants. In this context the ovule orientation in *Cycas* may also be specialized but in a manner different to that in other genera.

1.9 Radiospermic seeds. The ovules of cycads are generally interpreted as radiospermic, although there is evidence of platyspermy in the group, particularly in *Cycas* and the earliest unequivocal fossil cycad *Primocycas* (Zhu and Du 1981). Mature *Cycas* seeds are frequently bilaterally symmetrical (figure 5.6B, D) or more or less two-angled in cross section (Schuster 1932: figure 12S, T), although occasional trigonous forms also occur. Young *Cycas* ovules are also distinctly flat-

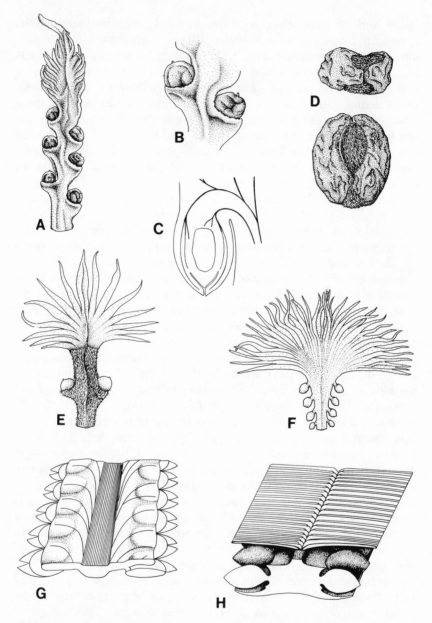

Figure 5.6. Extant and fossil cycads. **A** Megasporophyll of *Cycas media*. × 1. **B** Detail of seeds from specimen illustrated in A. × 2. **C** Diagrammatic longitudinal section through a seed of *Ceratozamia robusta*, redrawn from Stopes (1904: figure 24). **D** Seed of *Cycas revo-*

tened throughout their development (D. W. Stevenson, personal communication), often with two lobes at the apex on either side of the micropyle (Stopes 1903: figure 15; Schuster 1932: figure 12D, P, Y, Z). In addition, the testa splits regularly into two valves on germination (D. W. Stevenson, personal communication; cf. seeds of *Peltaspermum rotula* Harris 1932a:69). The vasculature of cycad seeds is also bilaterally arranged. The seed is supplied either by a single vascular bundle that bifurcates in the base of the seed, or by two distinct vascular bundles that may diverge well within the megasporophyll (Matte 1904, e.g., figures 158, 177; Stopes 1903: figures 1, 18, 20, 22, 23, 24; Stopes 1905; Kershaw 1912). Whichever of these patterns is present, each of the two vascular bundles divides irregularly to send branches to the tissues *both* internal and external to the sclerotesta. Thus most bundles of the inner vascular supply are merely branches from bundles present in the outer system, and the inner and outer vascular supplies are not distinct (figure 5.6C). This contrasts with the superficially similar situation in medullosan ovules with which cycads have often been compared (e.g., Meyen 1984; Crane 1985a), in which the inner and outer vascular systems arise separately from a single bundle at two discrete levels (Stewart 1954; Taylor 1965). It is therefore unlikely that the double vascular systems in the ovules and seeds of medullosans and cycads are homologous (Crane 1985a; character 9.7). Bilateral seed symmetry is most marked in *Cycas,* and in this paper I follow Seward (1917:25) and interpret the seeds of this genus as platyspermic. Contrary to earlier interpretations (Crane 1985a; Doyle and Donoghue 1986b), radiospermic seeds are interpreted here as derived within the cycads and to define a group consisting of all other genera. Meyen (1984:6) adopts the opposite view, that platyspermy is secondarily derived from radiospermy within cycads.

1.10 Colored trichomes. Colored trichomes have been reported in all extant genera of cycads except *Cycas* (Stevenson 1981) and are interpreted here as a relatively derived feature of these taxa compared

---

luta in distal (above) and anterior–posterior (below) views showing 180° rotational ("bilateral") symmetry. × 2. **E** Megasporophyll of *Cycas tonkinensis.* × 1. **F** Megasporophyll of fossil *Primocycas chinensis,* based on Zhu and Du (1981: figure 1, plate 1, figures 4, 4a). × 2. **G** Reconstruction of portion of megasporophyll of fossil *Spermopteris coriacea,* redrawn from Cridland and Morris (1960: figure 10). × 2.5. **H** Reconstruction of portion of megasporophyll of fossil *Phasmatocycas,* redrawn from Gillespie and Pfefferkorn (1986: figure 1). × 5.

with transparent trichomes, which occur in all cycads and which are widely distributed in seed plants.

1.11 Specialized alveolate exine structure. The pollen wall of *Cycas* has a distinct spongy layer in the inner portion of the ektexine, whereas all other cycads examined (all genera except *Bowenia, Lepidozamia,* and *Microcycas;* Audran and Masure 1976, 1977) have a specialized alveolate structure throughout the exine distinct from that in all other seed plants. The pollen wall in *Cycas* is more similar to the poorly organized alveolate ektexine of coniferophytes and "pteridosperms" (Doyle, Van Campo, and Lugardon 1975; Millay and Taylor 1976; Stidd, Rischbieter, and Phillips 1985), and the condition in all other cycads is treated here as relatively derived.

1.12 Stipules on leaves. Stipules of various kinds occur on the foliage leaves of *Bowenia, Ceratozamia, Microcycas, Stangeria,* and *Zamia* (Stevenson 1981). In *Bowenia* and *Stangeria* they are large (see character 1.13), but in the other three genera they are small, simple, and unvascularized. In *Microcycas,* stipules are present only in young plants (D. W. Stevenson, personal communication).

1.13 Vascularized stipules. Among extant cycads, large vascularized stipules on the foliage leaves occur only in *Bowenia* and *Stangeria.* In *Bowenia* the stipules are large and fleshy (Stevenson 1985; personal communication), while in *Stangeria* they are modified into a fused hood-like structure (Stevenson 1981: figure 4). In both genera the stipules function as cataphylls and protect the leaf during its early development (Stevenson 1985). Within cycads this is a unique feature and probably correlates with the absence of true cataphylls on vegetative shoots of *Bowenia* and *Stangeria* (Stevenson 1985).

1.14 Primary rachis of leaf with circular arrangement of vascular bundles. Matte (1904) describes the vascular structure of leaves, pinnae, and pinnules in cycads. All genera examined, with the exception of *Bowenia* and *Stangeria,* have an omega-shaped arrangement of vascular bundles in the primary rachis of the leaf, or some modification of the omega arrangement. In *Stangeria,* however, the bundles are arranged in a ring, while in *Bowenia* up to three concentric rings of vascular bundles may be present. The circular arrangement is interpreted as relatively derived compared to the simple or shallow U-shaped traces of most seed plants.

1.15 Sporophyll vasculature in more than one plane. Sporophyll vasculature in cycads is dichotomous (Stevenson 1982). In *Cycas, Dioon,* and *Stangeria* the dichotomies all occur in a single plane, whereas in *Ceratozamia, Encephalartos, Lepidozamia, Macrozamia, Microcycas,* and *Zamia* the second dichotomy and all subsequent dichotomies are at right angles to the first. *Bowenia* appears to be uniquely specialized in having a series of dichotomies at successive right angles to each other (Stevenson 1982). The occurrence of sporophyll vasculature in more than one plane is interpreted here as specialized relative to the leaf-borne seeds and microsporangia of Palaeozoic "pteridosperms."

1.16 Developing pinnae flat. In *Bowenia*, *Cycas*, and *Stangeria* the pinnae of the developing leaf exhibit either circinnate or conduplicate ptyxis, although in the bipinnate leaves of *Bowenia* the pinnules are flat during development. In all other cycad genera the pinnae are flat throughout their development and borne in a succubous arrangement (Stevenson 1981).

1.17 Articulate pinnae. The presence of a distinct articulation at the base of the pinnae in *Ceratozamia*, *Microcycas*, and *Zamia* is unique within cycads (Johnson 1959). It is interpreted as relatively derived compared to the leaves of other cycads and "pteridosperms," in which pinnae bases are unmodified.

1.18 Polyxylic stele. The occurrence of polyxylic steles has been reported in *Cycas*, *Encephalartos*, *Lepidozamia*, and *Macrozamia* (Greguss 1968:26) but is not always present in every species of these genera. Polyxylic steles are interpreted as a derived feature with respect to the monoxylic steles of other cycads and most other living and fossil seed plants.

1.19 Lateral cones. The production of several cones at a time, which leads to the appearance of laterally borne cones, is characteristic of adult plants of the genera *Ceratozamia*, *Encephalartos*, *Lepidozamia*, *Macrozamia*, and *Zamia* (Johnson 1959:66–67; Stevenson 1981:1105). Juvenile plants of all these genera usually produce only one cone at a time (Stevenson 1981).

1.20 Cone peduncles with cataphylls. The cones of *Dioon*, *Encephalartos*, *Lepidozamia*, and *Macrozamia* are unusual in having cataphylls borne directly on the peduncles. This feature does not occur in any other genus of cycads (D. W. Stevenson, personal communication).

1.21 Short, curved, colored trichomes. The occurrence of short, curved, colored trichomes is characteristic of the developing leaves of *Bowenia*, *Lepidozamia*, and *Macrozamia*, and these hairs do not occur in other cycad genera (Stevenson 1981).

1.22 Branched trichomes. Although the young leaves of all cycad genera bear unbranched trichomes, only *Cycas*, *Microcycas*, and *Zamia* also have trichomes that are branched (Stevenson 1981).

1.23 Equally branched trichomes. In the three cycad genera known to have branched foliar trichomes, the branching is distinctly unequal (e.g., Stevenson 1981: figure 34). Both unequally and equally branched forms are present only in *Microcycas* and *Zamia* (Stevenson 1981: table 1).

Discussion   The characters described above are summarized in table 5.1, and the resulting cladogram is given in figure 5.7. The pattern of relationships is similar to that outlined by Stevenson (1985) except that *Stangeria* and *Bowenia* are placed as the sister group to the Diooideae and Zamioideae rather than as the sister group to all other cycads. The cladogram is also similar to the classification proposed by Johnson (1959) in that *Ceratozamia*, *Microcycas* and *Zamia* are regarded as closely related, and the genera *Encephalartos*, *Lepidozamia*, and *Macrozamia* are placed together as a natural group. With respect to the phylogenetic position of cycads within seed plants, the most critical feature is that *Cycas* is placed as the sister taxon to all other genera, but this conclusion hinges on several critical characters about which opinions have differed (table 5.2). The interpretation of the pinnate multiovulate megasporophylls of *Cycas* as primitive compared to the megasporophylls of all other genera seems relatively secure, but the interpretation of other characters such as the platyspermy and indeterminate ovulate cone in *Cycas* is less certain. It is important to establish more securely whether these features are merely unique specializations (autapomorphies) of *Cycas* or primitive features of the group that have been retained in one living genus. In particular, if the traditional interpretation of cycads as fundamentally radiospermic is adopted (Meyen 1984; Crane 1985a; Doyle and Donoghue 1986b), then seed symmetry sets cycads apart from a major group of relatively derived seed plants (Crane 1985a). However, if the platyspermic condition in *Cycas* is interpreted as primitive for the group as a whole, then it may instead suggest a closer relationship between cycads and other platyspermic taxa (Doyle and Donoghue 1986b:355).

Table 5.2. Different interpretations of characters in *Cycas*. Apomorphies are marked +, while plesiomorphies are marked −.

Authors	*Pinnate Megasporophylls*	*Platyspermic Seeds*	*Indeterminate Cones*
Meyen 1984	−[1]	+	+[1]
Crane 1985a	not considered	+	+
Doyle and Donoghue 1986b	−	+	−
Crane, this paper	−	−	+

[1] Based on citation of Harris (1976); no other interpretation mentioned.

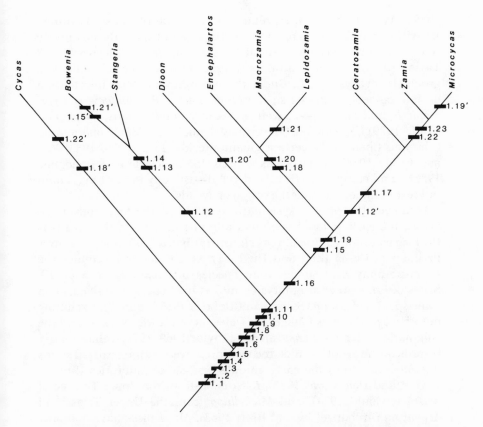

Figure 5.7. Preliminary cladogram summarizing relationships within extant cycads, based on the data in table 5.1. Autapomorphies of genera are not included. Prime marks indicate homoplasy in characters 1.12, 1.15, 1.18, 1.19, 1.20, 1.21, and 1.22. Homoplasy in character 1.20 could be interpreted either as convergence (as plotted) or as reversal in *Ceratozamia, Microcycas,* and *Zamia.*

The possibility that cycads are primitively platyspermic is of considerable interest in view of the seeds in the putative Palaeozoic cycads, *Spermopteris coriacea* (figure 5.6G; Cridland and Morris 1960), and *Phasmatocycas kansana* (figure 5.6H, Mamay 1969, 1976; Gillespie and Pfefferkorn 1986; see also *Sobernheimia* Kerp 1983). Despite recent suggestions to the contrary (Gillespie and Pfefferkorn 1986), the seeds of both taxa appear to be platyspermic (Delevoryas 1982; Meyen 1984; Crane 1985a). I had previously considered seed symmetry as evidence against a cycad relationship for these taxa (Crane

1985a). With a revised interpretation of the evolution of seed symmetry within cycads, this is no longer necessary and opens the possibility that other platyspermic taxa such as *Eremopteris* (Delevoryas and Taylor 1969) and *Tinsleya* (Mamay 1966) may also be related to the group (Delevoryas 1982). Under this hypothesis further information on late Palaeozoic plants such as *Eremopteris, Phasmatocycas, Spermopteris,* and *Tinsleya*—as well as possibly *Emplecopteris, Nystroemia* (Halle 1929), *Dicksonites* (Meyen and Lemoigne 1986), *Fimbriotheca* (Zhu and Chen 1981), certain gigantopterids (Li and Yao 1983; Yao and Crane 1986), peltasperms (Harris 1932a; Doyle and Donoghue 1986b), and callipterids (Haubold 1980)—is likely to be of maximum interest in understanding the origin of cycads.

Whatever the precise systematic position of the taxa considered above, the earliest fossil unequivocally attributable to the cycads is the megasporophyll *Primocycas chinensis* (figure 5.6F) from the Lower Permian of China (Zhu and Du 1981) which closely resembles the megasporophylls of some extant species of *Cycas* (figure 5.6A, E). Seeds of *Primocycas* are platyspermic, as in *Cycas,* and stalked, as in some species of *Dioon* (Sabato and DeLuca 1985: figure 2). Excluding putative cycadaceous foliage (see Delevoryas 1982 for review) and the enigmatic pollen cone *Lasiostrobus* (Taylor 1969, 1970), reliable early records of the group include the well-preserved permineralized stems *Antarcticycas,* from the early–middle Triassic of Antarctica (Smoot, Taylor, and Delevoryas 1985), *Lyssoxylon* from the Upper Triassic of Arizona (Gould 1971), and *Michelliloa* from the Upper Triassic of Argentina (Archangelsky and Brett 1963). All of these have anatomy similar to stems of extant cycads, as well as at least some diagnostic girdling leaf traces. In the Mesozoic, several cycad reproductive structures have been described, although, as noted by Harris (1961), definite seeds are not known on the putative Triassic megasporophylls *Cycadospadix* (Seward 1917; see also Harris 1932b), *Dioonitocarpidium* (Lilienstern 1928; Kräusel 1949; 1953), and *Palaeocycas* (Florin 1933). The Middle Jurassic flora of Yorkshire includes several kinds of unequivocal megasporophylls, some of which have been conclusively attributed to leaves and pollen organs (Thomas and Harris 1960; Harris 1941a, 1961, 1964; Hill 1986). It is clear that, during the Mesozoic and early Tertiary, cycads were widespread in both hemispheres (Arnold 1953; Cookson 1953; Jain 1962, 1971; Harris 1964; Petriella 1969, 1972, 1978; Archangelsky, Petriella, and Romero 1969; Archangelsky and Petriella 1971; Hill 1978). The phylogenetic position of most of these fossils within the cycads remains to be clarified, but at least some of these taxa appear to have been more specialized in some

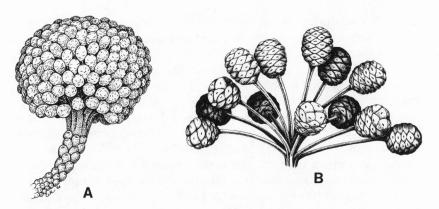

Figure 5.8. Ovulate reproductive structures of the *Pentoxylon* plant. **A** *Carnoconites compactus,* redrawn from Bose, Pal, and Harris (1985: figure 9). × 0.25. **B** *Carnoconites cranwelliae,* based on Harris (1962: text figure 2B, figure 1); from Crane (1985a) by permission. × 1.

respects than their living counterparts. In *Dirhopalostachys* from the late Jurassic–early Cretaceous of Siberia (Krassilov 1975), each ovule of the biovulate megasporophyll was apparently enclosed in a "capsule" (Krassilov 1975). A similar tendency toward enclosure of individual ovules occasionally occurs in extant cycads ("Wucherungen," *Macrozamia;* Schuster 1932: figure 13L; *Stangeria,* Stopes 1903: figure 29).

PENTOXYLON AND THE BENNETTITALES

Pentoxylon is known from the Jurassic and Cretaceous of Australia, India, and New Zealand (Sahni 1948; Harris 1962; Bose, Pal, and Harris 1984, 1985; Drinnan and Chambers 1985, 1986). Its most characteristic features are the unusual arrangement of five vascular segments in the stems *(Pentoxylon),* the flower-like aggregation of the three-dimensionally branched microsporophylls *(Sahnia),* and the spherical–ellipsoidal ovulate heads *(Carnoconites,* figure 5.8A, B), consisting of a mass of sessile ovules (see Crane 1985a for a brief discussion of characters). The relationships of the *Pentoxylon* plant to other gymnosperms are usually regarded as enigmatic (Andrews 1961; Stewart 1983; Bose, Pal, and Harris 1985), although recent cladistic analyses (figure 5.2; Crane 1985a; Doyle and Donoghue 1986a) both support earlier suggestions (Ehrendorfer 1971, 1976) that the group is closely related to the Bennettitales (Crane 1985a) or to the Bennettitales and Gnetales (Doyle and Donoghue 1986b).

The Bennettitales are an ecologically important group of extinct gymnosperms that are widespread and abundant in Triassic, Jurassic, and Cretaceous floras (Harris 1932b, 1969; Crane 1985a, 1987). The cycad-like leaves of the Bennettitales differ from those of cycads and other gymnosperms in the possession of characteristically cutinized paracytic stomata (Harris 1932a), in which the guard cells and subsidiary cells are usually interpreted as having developed ontogenetically by successive divisions of a single mother cell (syndetocheilic, Thomas and Bancroft 1913; Harris 1932a, 1976; Florin 1933). The distinctive stomata combined with the characteristic interseminal scales of the ovulate reproductive structures (Harris 1932b, 1969; Crane 1985a, 1986; see below) has led to the widespread acceptance of the Bennettitales as a natural group (e.g., Harris 1976), but there is less consensus about their relationships to other seed plants. The Bennettitales have often been interpreted as closely related to cycads (Wieland 1906, 1916; Chamberlain 1913), based mainly on their cycad-like foliage and the pachycaul habit of some genera (e.g., *Cycadeoidea*). Other authors have suggested that the two groups are only superficially similar and are probably only distantly related (Arber and Parkin 1907; Scott 1923; Arnold 1953; Stewart 1983). An alternative suggestion is that the Bennettitales are closely related to the Gnetales (Arber and Parkin 1907, 1908; Thoday 1911; Martens 1971; Takhtajan 1969) on the basis of similar details of the seed and micropyle (Berridge 1911; Thoday 1911), the presence of paracytic stomata (Martens 1971), the possible occurrence of vessels in some Bennettitales (Krassilov 1984; Crane 1985a), and the possible equivalent of a "feeder" in *Cycadeoidea* embryos (Crepet 1974:158; Crane 1985a). It has also been suggested that the microsporangiate reproductive structures of *Welwitschia* (see figure 5.12B, below) could represent a highly reduced bisexual bennettitalean "flower" (Arber and Parkin 1908; Crane 1985a; Doyle and Donoghue 1986b). Recent cladistic analyses also support a close relationship between the Bennettitales and Gnetales (Crane 1985a; Doyle and Donoghue 1986b).

Within the Bennettitales Arnold (1947), Taylor (1981), and Stewart (1983) recognize two families, the Cycadeoidaceae and the Williamsoniaceae, while Sporne (1971) also recognizes a third family, the Wielandiellaceae. A preliminary phylogenetic analysis was attempted by Crane (1985a) and is extended here to represent more adequately the morphological diversity known in the group by including four additional taxa, *Leguminanthus siliquosus*, *Bennetticarpus wettsteinii*, *Sturianthus langeri*, and *Amarjolia dactylota* (see table 5.3). Characters

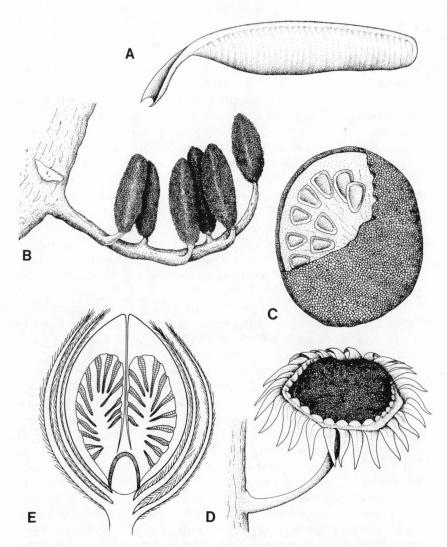

Figure 5.9. Fossil Bennettitales. **A** Reconstruction of *Leguminanthus siliquosus* microsporophyll, based on Kräusel and Schaarschmidt (1966: plates 11, 14, 15; Crane 1986: figures 11.5, 11.6). × 0.75. **B** Reconstruction of *Westersheimia pramelreuthensis* ovulate reproductive structure, based on Kräusel (1949: text figure 14, plate 16, figures 3, 4). × 1. **C** Reconstruction of *Bennetticarpus wettsteinii* ovulate reproductive structure, based on Kräusel (1949: text figure 9A, plate 11, figures 5, 6); "perianth" of *Cycadolepis wettsteinii* bracts not shown. × 0.6. **D** Reconstruction of *Sturianthus langeri* ovulate reproductive structure, based on Kräusel (1948: figures 1, 2, 4, 6). × 6. **E** *Amarjolia dactylota* diagrammatic longitudinal section, based on Bose, Banerjee, and Pal (1984: text figures 3, 4). × 2.5.

and provenance of all other taxa considered are reviewed in Crane (1985a).

Leguminanthus siliquosus (figure 5.9A) is an unusual bennettitalean microsporophyll from the Upper Triassic of Neuewelt, Switzerland, and Lunz, Austria (Kräusel and Schaarschmidt 1966; Crane 1986). It resembles a longitudinally folded taeniopterid leaf with numerous small ellipsoidal pollen sacs on the inner surface. The microsporophyll is folded toward the presumed adaxial surface, but the whole structure is twisted at the base so that the open edge of the microsporophyll is oriented abaxially. Each microsporophyll has a broad sheathing base that would seem to preclude a whorled arrangement, although the microsporophylls may have been aggregated together in a helix (Crane 1986). The most common bennettitalean leaves associated with *Leguminanthus* are referable to *Pterophyllum* (Kräusel 1949; Kräusel and Schaarschmidt 1966).

Westersheimia pramelreuthensis (figure 5.9B) from the Upper Triassic of Lunz, Austria, is a pinnately organized ovulate reproductive organ with five to seven lateral structures each consisting of a typical bennettitalean "gynoecium" composed of interseminal scales and ovules (Kräusel 1949). None of the "gynoecia" show any evidence of having been subtended by a "perianth" or microsporophylls. One specimen (Kräusel 1949: text figure 14) clearly shows that *Westersheimia* was borne laterally on a vegetative axis, but it is unclear whether *Westersheimia* is a branch with several lateral branches, a branch with several megasporophylls, or a single megasporophyll with the "gynoecia" borne on lateral pinnae. The leaves of *Westersheimia* are unknown.

Bennetticarpus wettsteinii (figure 5.9C) is a bennettitalean ovulate reproductive structure from the Upper Triassic of Lunz and Schrambach, Austria (Kräusel 1949). It consists of a large spherical head typically 6–7 cm, but ranging from 3.5 to 13 cm in diameter. The outer surface of the head probably had a leathery texture and is formed from a large number (25,000–30,000) of small interseminal scales (Kräusel 1949). Externally there is no indication of ovules, and it is uncertain how they were pollinated. Internally there are typically 25–30 seeds that were presumably embedded in relatively soft tissue. Each seed was flattened and relatively large compared to those of many Bennettitales. The *B. wettsteinii* ovulate head is thought to have been unisexual and surrounded by a "perianth" composed of *Cycadolepis wettsteinii* scales (Kräusel 1949). The corresponding microsporangiate "flowers" may have been composed of *Haitingeria krasseri* microsporophylls that are associated with *B. wettsteinii* at the Lunz locality (Kräusel 1949; Crane 1986). *Haitingeria* is a broadly elliptical microsporophyll with about seven pairs of distal pinna-like lobes. Each lobe has up to six pairs of pollen-bearing structures. These are not typical bivalved bennettitalean synangia, but whether they are synangia or sporangia cannot be unequivocally determined (Kräusel 1949). The leaves of *B. wettsteinii* are unknown.

Sturianthus langeri (figure 5.9D) is based on a single specimen from the Upper Triassic of Lunz, Austria (Kräusel 1948). It consists of an axis bearing approximately 20 flower-like structures on lateral branches. Each "flower" consists of a flattened, subhemispherical mass of interseminal scales (presumably with embedded ovules) surrounded by a distinct "perianth" of 25–30 bracts that are fused proximally but free distally. Some bracts have a distinct bulge on their inner surface that Kräusel (1949) interpreted as a pollen sac on a microsporophyll. However, no pollen was isolated, and it is equally likely that these bulges are nectaries or some other structure (see Harris 1972). There is therefore no conclusive evidence that *Sturianthus* is bisexual, and it is interpreted here as unisexual. The leaves of *Sturianthus* are unknown.

Amarjolia dactylota (figure 5.9E) is a bisexual bennettitalean "flower" from the Upper Jurassic (or possibly Lower Cretaceous; Varma and Ramanujam 1984) of the Rajmahal Hills, India (Bose, Banerjee, and Pal 1984). The ovules and microsporophylls are surrounded by a "perianth" of helically arranged hairy bracts. There are approximately 12 microsporophylls, and they evidently matured before the ovules (protandrous). As in *Cycadeoidea* (Wieland 1906; Crepet 1974), each microsporophyll is involute and laterally flattened and has a solid distal sterile portion. Adaxially, along each margin each microsporophyll bears a large number of lateral pinnae that project toward the center of the "flower." Each pinna consists almost entirely of two synangia, each composed of two alternating rows of pollen sacs. The center of the "flower" consists of an ellipsoidal mass of immature interseminal scales and ovules. The layer of interseminal scales was apparently continuous over the apex of the "gynoecium," and a sterile "corona" such as occurs in *Williamsoniella* and some *Williamsonia* species (Harris 1969; Crane 1985a) is not present. The leaves of *Amarjolia* are unknown.

Characters (see table 5.3)
3.1 Megasporophylls consisting of a single unicupulate ovule and clustered to form heads. The most straightforward interpretation of the ovulate heads of the Bennettitales and *Pentoxylon* is that they consist

Table 5.3. Data matrix for characters of *Pentoxylon* and selected Bennettitales. Apomorphic characters are marked +, plesiomorphic characters are marked − , and uncertain or missing characters are marked ?.

	Cycadeoidea	*Amariolia dactylota*	*Williamsoniella lignieri*	*Williamsoniella coronata*	*Williamsonia gigas*	*Williamsonia leckenbyi*	*Williamsonia hildae*	*Wielandiella angustifolia*	*Monanthesia magnifica*	*Williamsonia harrisiana*	*Williamsonia sewardiana*	*Sturianthus langeri*	*Bennetticarpus wettsteinii*	*Vardekloeftia sulcata*	*Leguminanthus siliquosus*	*Westersheimia pramelreuthensis*	*Pentoxylon*	*Other seed plants*
3.1 Megasporophylls consisting of a single unicupulate ovule and clustered to form heads	+	+	+	+	+	+	+	+	+	+	+	+	+	+	+?	+	−	−
3.2 Interseminal scales	+	+	+	+	+	+	+	+	+	+	+	+	+	+	+	+	−	−
3.3 Guard cells paracytic, strongly cutinized	+	+	+	+	+	+	+	+	+?	+?	+?	+?	+	+	+	+	−	+/−
3.4 Ovulate heads with numerous ovules (or ovule homologues)	+	+	+	+	+	+	+	+	+	+	+	+	+	+	+?	+	−	−

3.5 "Perianth" of helically arranged bracts	+/−	−	−?	−?	−?	+	+	+	+	+	+	+	+	+	+
3.6 Bivalved synangia	+/−	−	−?	−?	−?	+?	+?	+?	+	+	+	+	+	+	+
3.7 Microsporophylls fused proximally	+/−	+	−?	−?	−?	+?	+?	+?	+	+	+	+	+	+	+
3.8 "Corona"	−	−	−?	−	−	−	−	−	+	+	+	+	+	−	−
3.9 Microsporophylls laterally flattened	−	−	−?·	+	−?	−?	−?	−?	+	+	+	+	+	+	+
3.10 Bisexual "flowers"	+/−	−	−?	−?	−	−	−	−	−?	−	−	+	+	+	+
3.11 Microsporophylls thick, sterile distally	+/−	−	−?	−	−?	−?	−?	−	−?	−	−	+	+	+	+
3.12 Microsporophylls with 3–4 pairs of bivalved synangia	−	−	−?	−	−	−?	−?	−	−?	−	+	+	+	−	−
3.13 Nilssoniopteris foliage	+/−	−	−?	−?	−?	−	−?	−	−?	−	+	+	+	−?	−?
3.14 Scale leaves of the Cycadolepis hypene/ C. nitens type	−	−	−?	−?	−	−	−	+	−	+	−	−	−	−	−

of a mass of highly reduced megasporophylls (Chamberlain 1935; Crane 1985a, 1986; Doyle and Donoghue 1986b) each consisting of a single unitegmic ovule within a "cupule" (Crane 1985a, 1986). This is unique in seed plants, and, irrespective of whether the clustering of megasporophylls is homologous to the condition of the multicarpellary flowers of angiosperms or the cones of conifers and cordaites (Doyle and Donoghue 1986b:429), the structure of the megasporophylls is quite different and unlikely to be homologous.

3.2 Interseminal scales. Interseminal scales, interpreted as sterile ovules or, more strictly, sterile cupules (Harris 1932b; Crane 1985a), are known to occur in all the Bennettitales considered here (table 5.3) except *Leguminanthus*, for which ovulate reproductive structures are unknown. On the basis of the condition in all other known bennettitalean "gynoecia," interseminal scales and unicupulate ovules (character 3.1) are scored as present in the *Leguminanthus* plant. Interseminal scales are unique to the Bennettitales, with the possible exception of the two sterile lobes at the base of the aborted ovule in male "flowers" of *Welwitschia* (figure 5.12B), which may be homologous (Doyle and Donoghue 1986b:416).

3.3 Guard cells paracytic and strongly cutinized. Among seed plants, paracytic stomata are restricted to the Bennettitales, *Gnetum, Welwitschia*, some conifers (Scott and Chaloner 1983), and many angiosperms (Crane 1985a; Doyle and Donoghue 1986b). Although bennettitalean stomata may be distinguished by their characteristic cuticular thickening and orientation perpendicular to the venation (Florin 1933), the possibility that stomata of this type are plesiomorphic with respect to the condition in *Gnetum, Welwitschia*, and angiosperms cannot be excluded. The occurrence of bennettitalean stomata has not yet been demonstrated in *Cycadeoidea, Monanthesia, Sturianthus, Williamsonia harrisiana*, and *W. sewardiana*, but they are interpreted as present in this analysis on the basis of the clear relationship of all of these plants to the Bennettitales.

3.4 Ovulate heads with numerous ovules (or ovule homologues). All of the Bennettitales included here, with the exception of *Leguminanthus*, for which the ovulate reproductive structures are unknown, have heads with very numerous (more than 500) ovules or ovule homologues (interseminal scales). The massive duplication of these structures is unique to the Bennettitales and clearly specialized relative to the much smaller number of ovules in *Pentoxylon* (figure 5.8).

3.5 "Perianth" of helically arranged bracts. All of the Bennettitales considered in this study, with the exception of *Leguminanthus, Var-*

dekloeftia, and *Westersheimia,* have ovulate heads surrounded by a "perianth" of helically arranged bracts. In *Westersheimia* there is no evidence of a "perianth" at the base of the "gynoecia," but in *Vardekloeftia* (Harris 1932b) and *Leguminanthus* the available material is inadequate to judge. *Vardekloeftia* and *Leguminanthus* are scored as lacking a "perianth" in this preliminary analysis. The possibility that the bennettitalean "perianth" is homologous either to the bracts surrounding the "flowers" of Gnetales or the perianth of angiosperms needs detailed consideration but seems unlikely in view of the absence of a "perianth" in *Pentoxylon* and *Westersheimia.*

3.6 Bivalved synangia. In most Bennettitales the microsporangia are borne in distinctive bivalved synangia. However, in several Triassic representatives of the group, such as *Leguminanthus, B. wettsteinii (Haitingeria),* and other taxa (Crane 1986), bivalved synangia do not occur and the sporangia are borne singly. In the relatively derived Bennettitales *Williamsonia sewardiana, W. harrisiana,* and *Monanthesia magnifica,* the corresponding microsporangiate organs are unknown but are taken here to be similar to those of other *Williamsonia* species and *Cycadeoidea* (also characters 3.7, 3.9, 3.11, 3.12). The microsporophyll of *Vardekloeftia* is unknown, but the "gynoecia" occur associated with two species of *Bennettistemon (B. amblum, B. bursigerum;* Harris 1932b), which both lack bivalved synangia (Raunsgaard Pedersen, personal communication). In this analysis the microsporophyll of *Vardekloeftia* is assumed to be similar to these taxa. The microsporophylls of the other Triassic taxa *Sturianthus* and *Westersheimia* are also unknown but assumed to be plesiomorphic in the arrangement of pollen sacs.

3.7 Microsporophylls fused proximally. The only taxa considered here in which the microsporophylls are apparently free and not fused into a cup at the base are *Leguminanthus, Bennetticarpus wettsteinii, (Haitingeria)* and possibly *Vardekloeftia* (based on *B. amblum* and *B. bursigerum*). The possibility that this is homologous with the condition in Gnetales and perhaps certain angiosperms needs further study, particularly in view of the occurrence of proximally fused microsporophylls in *Pentoxylon* (Bose, Pal, and Harris 1985; Drinnan and Chambers 1986: figure 25A).

3.8 "Corona." In several bennettitalean "gynoecia" there is a sterile distal extension of the floral axis on which ovules and interseminal scales fail to develop. Of the taxa considered in this analysis, a "corona" occurs in *Wielandiella angustifolia,* several species of *Williamsonia (W. gigas, W. hildae, W. leckenbyi),* and both species of *William-*

soniella. The "gynoecium" of *Leguminanthus* is unknown but is assumed to be similar to that in the contemporaneous taxon *B. wettsteinii,* and thus to lack a "corona."

3.9 Microsporophylls laterally flattened. In the Bennettitales considered in this paper, the microsporophylls of *Amarjolia, Cycadeoidea,* and *Williamsoniella* are all similar in being laterally flattened with the pinnae directed adaxially toward the center of the "flower." Contrary to my earlier interpretation (Crane 1985a), a similar arrangement occurs in *Williamsonia gigas (Weltrichia sol),* in which the "fertile appendages" (Harris 1969) are most straightforwardly interpreted as inwardly directed pinnae (Crane 1985a). Although *Leguminanthus* is not pinnate, it is laterally flattened and adaxially folded as in *Amarjolia, Cycadeoidea,* and *Williamsoniella.*

3.10 Bisexual "flowers." In the Bennettitales considered in this paper, bisexual "flowers" are known to occur only in *Amarjolia, Cycadeoidea,* and *Williamsoniella.* Contrary to the accepted view (Kräusel and Schaarschmidt 1966; Doyle and Donoghue 1986b), there is no direct evidence that the "flowers" of *Sturianthus* were bisexual (see above), and they are scored as unisexual. The relative disposition of microsporophylls and ovules in *B. wettsteinii, Leguminanthus,* and *Vardekloeftia* is uncertain, but there is nothing in the information currently available to suggest that these reproductive structures were bisexual. Based on comparison with *Pentoxylon* and all other gymnosperms, except possibly the Gnetales, the bisexual condition is considered derived.

3.11 Microsporophylls thick, sterile distally. In the Bennettitales considered in this paper the microsporophylls of *Amarjolia, Cycadeoidea,* and *Williamsoniella* all have a distinct thick, sterile portion distally that is specialized compared to the distally unmodified microsporophylls of other taxa.

3.12 Microsporophylls with 3–4 pairs of bivalved synangia. The microsporophylls of *Williamsoniella coronata* and *W. lignieri* are very similar, with three to four pairs of bivalved synangia. In other synangiate bennettitalean microsporophylls, the number of synangia is much larger.

3.13 Nilssoniopteris foliage. Williamsoniella coronata and *W. lignieri* are thought to have borne foliage of *Nilssoniopteris vittata* and *N. major,* respectively (Harris 1969). The similarities between these two leaf species in morphology, venation, and cuticle, combined with characters of the "flowers," strongly suggest that the two species of *Williamsoniella* are closely related (Harris 1969).

3.14 Scale leaves of the Cycadolepis hypene/C. nitens type. These two species of scale leaves are thought to have been borne around the "gynoecia" of *Williamsonia hildae* and *Williamsonia leckenbyi*, respectively (Harris 1969). They are closely similar in shape and cuticular structure (Harris 1969) and, together with other characters of the "flowers," they strongly suggest that these two species of *Williamsonia* are closely related.

Discussion The characters described above are summarized in table 5.3, and the resulting cladogram is given in figure 5.10. The results are obviously preliminary, and the extent to which they accurately reflect the phylogeny of the Bennettitales will depend largely on the validity of the assumptions concerning missing characters. None of the characters are known in all the taxa, and about 20 percent of the data matrix is based on extrapolation rather than direct observation. The taxa included do, however, adequately represent the range of known morphological diversity in the Bennettitales, and the cladogram highlights important areas of ignorance that need to be addressed. In addition, even with the presently crude evaluation of relationships, the cladogram raises several interesting points concerning the congruence between the cladogram and stratigraphy, the extent to which *Cycadeoidea* is representative of the group, and the possibilities for future research.

The earliest unequivocal Bennettitales are of Upper Triassic age. The five most plesiomorphic Bennettitales are from the Upper Triassic, and the five most apomorphic are Middle Jurassic to Lower Cretaceous. Between these extremes, however, *Williamsonia* is resolved as paraphyletic at a similar grade of organization to *Wielandiella* and *Monanthesia*. These three genera, ranging in age from Upper Triassic *(Wielandiella)* to Upper Cretaceous *(Monanthesia)*, are grouped at three nodes and do not display any clear stratigraphic pattern. One of the youngest taxa included, *Cycadeoidea*, is resolved as relatively derived, and is therefore unlikely to be typical of the group. In addition, the bisexual "flowers" that have figured prominently in comparisons of the Bennettitales with angiosperms and Gnetales are almost certainly an independent development within the Bennettitalean clade, occurring only in the three most highly derived genera. Neither of the reports of possible bisexual bennettitalean "flowers" in the Upper Triassic *(Sturianthus, Wielandiella)* are reliable. Furthermore, while most bennettitalean "flowers" have a "perianth" and pollen sacs borne in characteristic synangia, these may also be relatively derived features that developed *within* the bennettitalean clade. Thus the similarities between the synangia of *Harrisothecium* (figure 5.4D, E) and

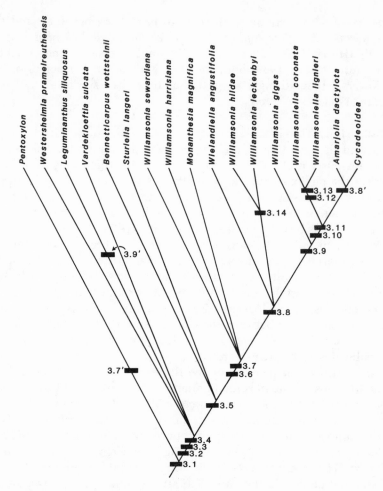

Figure 5.10. Preliminary cladogram summarizing relationships within *Pentoxylon* and the Bennettitales, based on the data in table 5.3. Autapomorphies are not included. Prime marks indicate homoplasy in characters 3.7, 3.8, and 3.9.

some Bennettitales are probably the result of convergence, and at least for microsporophylls the primitive condition in the Bennettitales is not very different from the relatively generalized condition in other seed plants, such as *Callistophyton*, peltasperms, and corystosperms. The same arguments may also apply to the similarities between the perianth of angiosperms and the scale leaves of some bennettitaleans.

The ovulate heads of the Bennettitales are more difficult to interpret. Both Crane (1985a) and Doyle and Donoghue (1986b) interpret
the "floral axis" bearing uniovulate cupules as a stem bearing numerous, highly reduced megasporophylls. This interpretation is straightforward but results in an substantial morphological hiatus between
the Bennettitales and *Pentoxylon*, and other gymnosperms. One possible alternative is to interpret the "flower" as a leaf-derived structure
that has assumed a terminal position. While this breaks the "rules" of
comparative morphology (Harris 1976), it could account for the unusual structure of *Westersheimia* (Crane 1986).

As discussed above, it is unclear whether *Westersheimia* is a branch
with several lateral branches, a branch with several lateral megasporophylls, or a single megasporophyll with the "gynoecia" borne on
lateral pinnae. The first interpretation creates no difficulties in regarding the uniovulate cupules as reduced megasporophylls. The second and third interpretations, however, eliminate this possibility but
more easily reconcile the "gynoecia" of Bennettitales with those of
the *Pentoxylon* plant. In *Pentoxylon (Carnoconites)* the fertile shoot
bears six to eight *(C. compactus;* figure 5.8A) or about twelve (*C.
cranwelliae;* figure 5.8B) "peduncles" that are fused to form a cup-like
structure at the base. In *C. compactus* each "peduncle" gives rise to
several "pedicels" that bear clusters of ovules at their apex, but in *C.
cranwelliae* each "peduncle" bears only one "gynoecium." Instead of
interpreting the peduncles in *C. compactus* as a whorl of branches
(implicit in Crane 1985a; Doyle and Donoghue 1986b), they could also
be interpreted as a whorl of megasporophylls bearing "gynoecia" on
lateral pinnae. Under this interpretation the flower-like *Sahnia* pollen
organs and the *Carnoconites* ovulate structures of *Pentoxylon* would
both consist of a cluster of fertile leaves, and the situation in *Carnoconites* would be compatible with a megasporophyll interpretation of
the *Westersheimia* organ. The two major difficulties with this interpretation are the formation of a radially symmetrical "gynoecium" from
a dorsiventrally organized megasporophyll, and the required shift
from a lateral to terminal position in the Bennettitales. Both morphological transitions are unusual, but there are precedents for both—for
example, the terminal ovule of *Taxus* and terminal ovaries in angiosperms (Harris 1976:131). If future work supports the megasporophyll
interpretation of the "gynoecia" in *Pentoxylon* and the Bennettitales,
then it will have relatively little effect on relationships outlined in
figure 5.10, but could have a major effect on the relationships of the
Bennettitales within seed plants as a whole. Clearly, further work on
Triassic Bennettitales or reproductive structures similar to *Wester-*

sheimia is most likely to enhance our knowledge of the morphology and relationships of the group.

GNETALES

Although strikingly different in habit, the three extant genera of Gnetales, *Ephedra*, *Welwitschia*, and *Gnetum*, are united by a range of vegetative and reproductive features (figures 5.11, 5.12A). Recent cladistic analyses (Crane 1985a; Doyle and Donoghue 1986b) have resolved the Gnetales as a monophyletic group more closely related to angiosperms, Bennettitales, and *Pentoxylon* than to other seed plants (figure 5.2). Within the group *Welwitschia* and *Gnetum* are more closely related to each other than either is to *Ephedra* (figure 5.12A), and this suggests that at least some of the obvious similarities between *Gnetum* and angiosperms, such as dicotyledon-like leaves, are more likely a result of covergence than inheritance from a common ancestor (Crane 1985a; Doyle and Donoghue 1986b).

Striate-ribbed ("ephedroid") pollen similar to that of *Ephedra* and *Welwitschia* occurs as early as the Middle Permian in North America (Wilson 1959). This kind of pollen is probably primitive within the Gnetales relative to the spinulose inaperturate grains of *Gnetum* (Crane 1985a; Doyle and Donoghue 1986b). During the mid-Cretaceous (Barremian–Cenomanian) ephedroid pollen is particularly abundant in the low to middle paleolatitude floras of Northern Gondwana and Southern Laurasia (Brenner 1976; Doyle, Jardiné, and Doerenkamp 1982; Muller 1984; Crane 1987). Much of this fossil pollen closely resembles that of extant Gnetales (Trevisan 1980), but a variety of more specialized forms (e.g., *Galeacornea*, *Elaterocolpites*) with horn-like or elater-like projections (figure 5.13A) also occur in these floras (Muller 1984; Crane 1987) and add to the evidence of insect pollination in the group (Bino, Dafni, and Meeuse 1984; Bino, Devente, and Meeuse 1984). In some mid-Cretaceous sequences local fluctuations in the abundance of ephedroid pollen coincide with fluctuations in the abundance of angiosperm grains (Doyle, Jardiné, and Doerenkamp 1982), and this is consistent with limited macrofossil evidence suggesting that some mid-Cretaceous angiosperms and Gnetales may have been autecologically similar as early successional herbs or shrubs of disturbed environments (Crane and Upchurch 1987).

Cretaceous Macrofossils   In contrast to the palynological record, the macrofossil record of the Gnetales is poor, and the two most securely identified fossils are *Drewria potomacensis* from zone I (probable Ap-

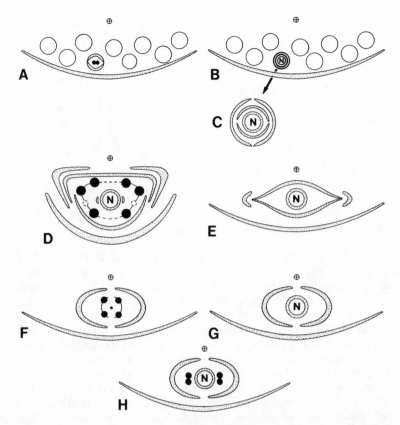

Figure 5.11. Comparison of microsporangiate and ovulate floral diagrams in the Gnetales; from Crane (1985a) by permission, but F, G, H have been modified on the basis of Doyle and Donoghue (1986b), Martens (1971), and Takaso (1985) to indicate that the outer "integument" in *Ephedra* is probably derived from a pair of lateral rather than anterior–posterior bracteoles. **A** *Gnetum* microsporangiate "flower." **B** *Gnetum* ovulate "flower." **C** *Gnetum* ovulate "flower" interpretation. **D** *Welwitschia* microsporangiate "flower." **E** *Welwitschia* ovulate "flower." **F** *Ephedra* microsporangiate "flower." **G** *Ephedra* ovulate "flower." **H** *Ephedra* abnormal ovulate flower with microsporophylls.

tian; Brenner 1963; Doyle 1969) of the Potomac Group, eastern North America (Upchurch and Crane 1985; Crane and Upchurch 1987), and *Eoanthus zherikhinii* from the Lower Cretaceous (Barremian–Aptian) of the Lake Baikal area, Mongolia (Krassilov 1986).

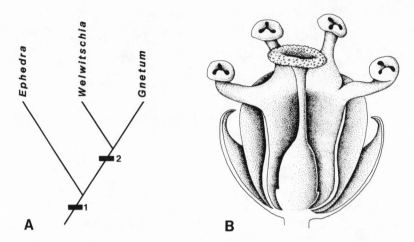

Figure 5.12. Extant Gnetales. **A** Cladogram summarizing relationships among extant Gnetales. Characters at 1 include opposite and decussate phyllotaxy in leaves, bracts, and bracteoles, multiple axillary buds, circular bordered pits in the protoxylem, vessels (derived independently from those of angiosperms), compound strobili, and a single terminal ovule with a second integument formed from a pair of opposite bracteoles (see figure 5.11). Characters at 2 include anastomoses in foliar venation, four nucleate male gametophyte, tetrasporic megagametophyte development, absence of archegonia, cellular embryogeny, and the presence of a "feeder" in the embryo. See Crane (1985a) and Doyle and Donoghue (1986b) for details. **B** Microsporangiate "flower" of *Welwitschia* in longitudinal section showing aborted ovule (not sectioned) surrounded by a whorl of microsporophylls, and associated bracteoles (compare with figure 5.11D), based on Martens (1971: figure 70). × 10.

Drewria (figure 5.13I) is a herb or small shrub with slender stems, axillary monopodial branching, and opposite decussate leaves borne at swollen nodes. The leaves are oblong with sheathing bases and a distinctive pattern of reticulate venation closely resembling that of *Welwitschia* cotyledons. Attached reproductive structures consist of dichasia of three short, loose spikes that are borne either terminally or in the axil of a leaf. Seeds on the two lateral spikes of a dichasium are surrounded by small "bracts," but details of their arrangement are uncertain. Seeds do not occur on the central spike and the "bracts" are always smaller, suggesting either that the ovules aborted or that the central spike bore pollen organs. Pollen sacs containing *Welwitschia*-like pollen occur in the same thin bed as the *Drewria* material.

Eoanthus (figure 5.13H) is based on only two impressions that are difficult to interpret. The specimens consist of four ovule-bearing structures in whorls around a central axis. Immediately below this whorl is a further whorl consisting of sterile bracts. Each of the four ovulate structures has been interpreted as an open follicle with a single ovule but could equally represent an ovule with associated bracteoles (Krassilov 1986). Macerations have demonstrated *Ephedripites* pollen in the micropyles of the ovules.

In addition to *Drewria* and *Eoanthus* several similar macrofossils that may be referable to the Gnetales have been described from the mid-Cretaceous. These include some specimens of *Conospermites hakeaefolius* from the Cenomanian of Czechoslovakia (Velenovský and Viniklář 1926; plate 3, figure 5, lefthand specimen; see figure 5.13J) that closely resemble leaves of *Drewria* in pattern of venation, and *Cyperacites potomacensis* (Berry 1911) and *Casuarina covilli* (Ward 1895; see figure 5.13K) from the Potomac Group, which have features resembling *Drewria* and *Eoanthus*, respectively. From the early Cretaceous of Mongolia other possible Gnetales include *Gurvanella dictyoptera*, *Cyperacites* sp. (figure 5.13L), "*Potamogeton*-like spike" (Krassilov 1982), and *Baisia hirsuta* (Krassilov and Bugdaeva 1982).

Triassic Macrofossils Putative gnetalean macrofossils from the early Mesozoic are even less common than in the early Cretaceous. The anatomy of *Hexagonocaulon minutum* stems from the Triassic of the South Shetland Islands (Lacey and Lucas 1981) has been compared with stems of *Ephedra;* details of rays in permineralized stems of *Schilderia* from the Upper Triassic flora of Arizona have been compared with *Gnetum* (Daugherty 1941); and the remains described as *Sanmiguelia lewisi* from the late Triassic Trujillo Formation of northwest Texas (Ash 1976) also appear to exhibit specialized features that occur in extant Gnetales (Cornet 1986; see Brown 1956; Becker 1972; Tidwell, Simper, and Thayn 1977 for further discussion of this genus). The most interesting of putative Triassic Gnetales is *Dechellyia gormanii* from the Upper Triassic Chinle Formation of northeastern Arizona (Ash 1972). *Dechellyia* consists of shoots with attached leaves and seeds (figure 5.13C, D). The leaves are opposite and decussate but flattened into a single plane as in some extant conifers (e.g., *Metasequoia*). The seeds, however, have a narrowly elliptical wing distally, and are borne in one or two pairs rather than in cones. Pollen cones associated with *Dechellyia* (*Masculostrobus clathratus*, figure 5.13F, G) contain *Equisetosporites chinleanus* (figure 5.13B, E), which resembles pollen of *Ephedra* and *Welwitschia* (Scott 1960), although details of

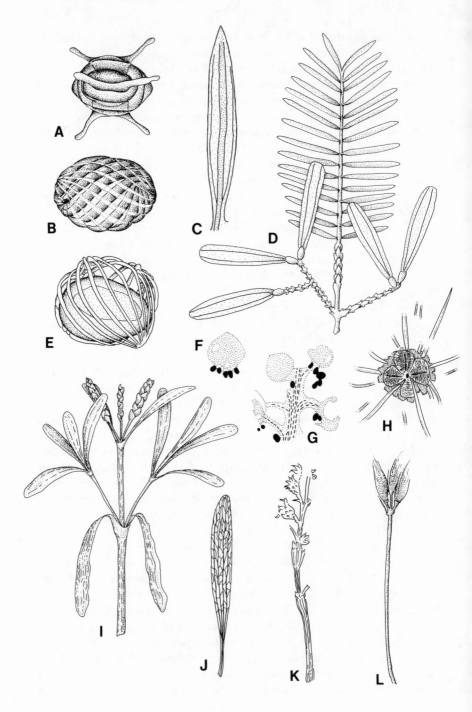

pollen wall ultrastructure differ (Zavada 1984). Several features of *Dechellyia* are consistent with a gnetalean relationship including the opposite–decussate phyllotaxy, details of seed wing and leaf venation, and characters of the pollen. However, other features are more difficult to interpret in terms of extant Gnetales, particularly the winged seed and the "cones" of leaf-like microsporophylls. Until these features can be accounted for either by additional information or in terms of related plants, the gnetalean affinity of *Dechellyia* will remain uncertain.

Discussion Taken together, the microfossil data supported by the limited macrofossil evidence suggest that the Gnetales were diverse and widespread during the mid-Cretaceous. Based on fossil pollen, the group may extend as far back as the Middle Permian, but palynological evidence is not so far supported by unequivocal gnetalean macrofossils in the latest Palaeozoic and early Mesozoic. In the mid-Cretaceous, *Drewria* and *Eoanthus* indicate that the apparently sparse macrofossil record of the Gnetales may be due both to problems of distinguishing gnetalean remains from those of contemporary angiosperms and to the underrepresentation of herbs or small shrubs as macrofossils (Crane and Upchurch 1987). The gnetalean features of *Dechellyia*, and perhaps related genera, also raise the possibility that early representatives of the Gnetales may have been confused with

Figure 5.13. Morphology of fossil Gnetales and putative related taxa. **A** *Elaterosporites* pollen grain. × 450. **B** *Equisetosporites* pollen grain, based on Chaloner (1969: plate 4.2, figure 7). × 1100. **C** Leaf of *Dechellyia gormanii* showing probable venation, based on Ash (1972: text figure 6a–c). × 4. **D** Shoots and seeds of *Dechellyia gormanii*, redrawn from Ash (1972: text figure 5b). × 0.75. **E** Pollen grain isolated from *Masculostrobus clathratus*, redrawn from Ash (1972: text figure 6i). × 500. **F** Microsporophyll of *Masculostrobus clathratus* showing pollen sacs, redrawn from Ash (1972: text figure 6d). × 10. **G** Cone fragment of *Masculostrobus clathratus*, redrawn from Ash (1972: text figure 6h). × 10. **H** *Eoanthus zherikhinii*, redrawn from Krassilov (1986: figure 1). × 3. **I** Reconstruction of vegetative and reproductive parts of *Drewria potomacensis*, based on Crane and Upchurch (1987). × 1.25. **J** *Conospermites hakeaefolius*, redrawn from Velenovský and Viniklář (1926: plate 3, figure 5, lefthand specimen). × 0.5. **K** *Casuarina covilli*, redrawn from Ward (1895: plate 3, figure 2). × 1. **L** *Cyperacites* sp., based on Krassilov (1982: plate 19, figures 240–243). × 1.5.

conifers. It seems likely that future work will add to our knowledge of the Gnetales not only through new discoveries but also through rein- terpretation of previously described fossil "angiosperm" and "conifer" material. As a more extensive record of the group emerges and as more inclusive phylogenetic analyses are attempted, the most parsi- monious explanation of the relationships of the Gnetales as a whole may also change. In recent cladistic analyses (Crane 1985a; Doyle and Donoghue 1986b) the Gnetales are firmly linked with the Bennetti- tales, *Pentoxylon*, and angiosperms. Any attempt to place the Gnetales close to conifers and cordiates, as they are often interpreted, requires that the Bennettitales, *Pentoxylon*, and angiosperms also be placed as relatively derived coniferopsids: an arrangement that seems intu- itively implausible (Doyle and Donoghue 1986b:373). However, if rel- atively plesiomorphic extinct Gnetales lacked many of the anthophyte synapomorphies, it could significantly reduce the coherence of the anthophytes as a whole. In particular, the discovery of securely deter- mined fossil Gnetales with significant coniferopsid similarities (per- haps like *Dechellyia*) could allow the Gnetales to be repositioned as coniferophytes without affecting the relationships of Bennettitales, *Pentoxylon*, and angiosperms.

GENERAL DISCUSSION

From the brief outlines presented above, it is clear that our under- standing of the morphology and diversity of different groups of seed plants is highly uneven. Some, such as the Bennettitales and cycads, are clearly enough defined for us to begin to assess primitive and derived character states, resolve the interrelationships of the constit- uent taxa, and identify specific questions that need to be addressed by future research. Others, however, are less straightforward to define and currently consist of just a few well-known taxa with a halo of potentially relevant but incompletely understood fossils (e.g., corys- tosperms). These difficulties in defining monophyletic groups are par- ticularly acute in the late Palaeozoic and Mesozoic pteridosperms, where a paucity of anatomical information is combined with rela- tively generalized morphology. In the corystosperms even the "core taxa" remain surprisingly poorly known (Thomas 1933; Townrow 1965). The situation in the Gnetales is better, in that the extant taxa are relatively well understood, but the abundance and diversity of fossil gnetalean pollen, as well as the putative macrofossils reviewed above, raise the possibility that we currently have access to an unre- presentative sample of the diversity of group as a whole. As additional information accrues on the Gnetales and other taxa, it will be "necessary that classification schemes be revised frequently and kept abreast of modern knowledge. Obsolete classifications retard progress and become misleading" (Arnold 1948:12).

One feature of seed plant evolution that is clear from both stratigraphy and recent cladistic analyses is that none of the five groups of extant seed plants (cycads, conifers, *Ginkgo*, Gnetales, and angiosperms) were present in the earliest phases of spermatophyte evolution. All these groups, with the possible exception of angiosperms (Crane 1985a; Doyle and Donoghue 1986b), were products of subsequent diversification between the early Pennsylvanian and late Triassic. This radiation corresponds broadly to the prominent polychotomy in figure 5.2; and of the taxa placed at or above this node, only the medullosans and cordaites have a fossil history that extends back into the Mississippian. To a large extent our ability to clarify the pattern of this radiation will hinge on our ability to learn more about the morphology and biology of late Palaeozoic and Mesozoic seed ferns. It is already clear that the plants traditionally grouped as "pteridosperms" are of diverse relationships (Meyen 1984; Crane 1985a; Doyle and Donoghue 1986b), and it seems inevitable that this conclusion will be strengthened by future work. Just as progress in resolving the phylogeny of early land plants necessitated dismantling the "catchall group," the Psilophyta (Banks 1975;401), progress in understanding phylogenetic relationships within seed plants will require major revision of current concepts of "pteridosperms."

Within the "anthophytes" the pattern of relationship between the Bennettitales, Gnetales, and angiosperms remains unclear, and the principle gap in our knowledge concerns the fossil Gnetales. Information on additional diversity in Gnetales and Bennettitales, particularly of relatively plesiomorphic taxa, would be of considerable value in further clarifying relationships within anthophytes. It seems likely that the Gnetales have a substantial macrofossil record and may indeed have already been described in the literature as conifers (e.g., *Dechellyia*) or perhaps angiosperms (e.g., *Conospermites*). Cases such as this, and others in the history of paleobotany (e.g., progymnosperms, pteridosperms; Oliver and Scott 1904; Beck 1960), tend to support Meyen's conviction (1984:7) that "we have already encountered, but have failed to identify, members of most of the existing families and orders of higher plants." Knowledge of this overlooked botanical diversity is something that only paleobotany can contribute and will be crucial to developing a more detailed understanding of phylogeny and evolution in fossil and living seed plants.

ACKNOWLEDGMENTS

I am grateful for very helpful discussions with C. B. Beck, J. A. Doyle, C. R. Hill, K. Raunsgaard Pedersen, W. E. Stein, D. W. Stevenson, D. C. White, and Z. Yao during the preparation of this paper. Demetrios Betinis and Clara Richardson drew the illustrations, and Elaine Zeiger typed the manuscript. I also thank Dr. George Rogers, editor, *Annals of the Missouri Botanical Garden*, for permission to reproduce several illustrations from Crane (1985a). This work was partially supported by National Science Foundation grant BSR 83 14592 to the author.

LITERATURE CITED

Anderson, J. M. and H. M. Anderson. 1983. *Palaeoflora of Southern Africa: Molteno Formation (Triassic).* Rotterdam: Balkema.
—— 1985. *Palaeoflora of Southern Africa: Prodromus of South African Megafloras Devonian to Lower Cretaceous.* Rotterdam: Balkema.
Andrews, H. N. 1961. *Studies in Paleobotany.* New York: Wiley.
Arber, E. A. N. and J. Parkin. 1907. On the origin of angiosperms. *J. Linn. Soc., Bot.* 38:29–80.
—— 1908. Studies on the evolution of angiosperms: The relationship of the angiosperms to the Gnetales. *Ann. Bot. (London)* 22:489–515.
Archangelsky, S. 1963. A new Mesozoic flora from Ticó, Santa Cruz Province, Argentina. *Bull. British Museum (Natural History), Geol.* 8:45–92.
Archangelsky, S. and D. W. Brett. 1963. Studies on Triassic fossil plants from Argentina. 2. *Michelilloa waltonii* nov. gen. et sp. from the Ischigualiasto Formation. *Ann. Bot. (London)* 27:147–154.
Archangelsky, S. and B. Petriella. 1971. Notas sobre la flora fosile de la zona de Ticó, Provincia de Santa Cruz. 9. Nuevos datos acerca de la morfologia foliar de *Mesodescolea plicata* Arch. (Cycadales, Stangeriaceae). *Bol. Soc. Argent. Bot.* 14:88–94.
Archangelsky, S., B. Petriella, and E. Romero. 1969. Nota sobre el bosque petrificado del Cerro Bororó (Terciaro Inferior), Provincia de Chubut. *Ameghiniana* 6:119–126.
Arnold, C. A. 1947 *An Introduction to Paleobotany.* New York: McGraw-Hill.
——1948. Classification of gymnosperms from the viewpoint of paleobotany. *Bot. Gaz.* 110:2–12.
—— 1953. Origin and relationships of the cycads. *Phytomorphology* 3:51–65.
Ash, S. R. 1972. Late Triassic plants from the Chinle Formation in northeastern Arizona. *Palaeontology* 15:598–618.

——1976. Occurrence of the controversial plant fossil *Sanmiguelia* in the Upper Triassic of Texas. *J. Paleont.* 50:799–804.

Audran, J. C. and E. Masure. 1976. Précisions sur l'infrastructure de l'exine chez les Cycadales (Prespermaphytes). *Pollen et Spores* 18:5–26.

—— 1977. Contributions à la connaissance de la composition des sporodermes chez les Cycadales (Prespermaphytes). Étude en microscopie electronique à transmission (M.E.T) et à balayage (M.E.B.) *Palaeontographica, Abt. B* 162:115–158.

Banks, H. P. 1975. Reclassification of Psilophyta. *Taxon* 24:401–413.

Beck, C. B. 1960. Connection between *Archaeopteris* and *Callixylon*. *Science* 131:1524–1525.

Becker, H. F. 1972. *Sanmiguelia*, an enigma compounded. *Palaeontographica, Abt. B* 138:181–185.

Berridge, E. M. 1911. On some points of resemblance between gnetalean and bennettitean seeds. *New Phytol.* 10:140–144.

Berry, E. W. 1911. Systematic paleontology: Pteridophytae, Cycadophytae, Gymnospermae, Monocotyledonae, Dicotyledonae. In W. B. Clark, ed., *Lower Cretaceous*, pp. 214–597. Baltimore: Maryland Geological Survey.

Bierhorst, D. W. 1971. *Morphology of Vascular Plants.* New York: Macmillan.

Bino, R. J., A. Dafni, and A. D. J. Meeuse. 1984. Entomophily in the dioecious gymnosperm *Ephedra aphylla* Forsk. (= *E. alte* C. A. Mey.), with some notes on *E. campylopoda* C. A. Mey. 1. Aspects of the entomophilous syndrome. *Proc. Kon. Ned. Akad. Wet.* 87:1–13.

Bino, R. J., N. Devente, and A. D. J. Meeuse. 1984. Entomophily in the dioecious gymnosperm *Ephedra aphylla* Forsk. (= *E. alte* C. A. Mey.), with some notes on *E. campylopoda* C. A. Mey. 2. Pollination droplets, nectaries and nectarial secretion in *Ephedra*. *Proc. Kon. Ned. Akad. Wet.* 87:15–24.

Bose, M. N. and J. Banerjee. 1984. The fossil floras of Kachchh. 1. Mesozoic megafossils. *Palaeobotanist* 33:1–189.

Bose, M. N., J. Banerjee, and P. K. Pal. 1984. *Amarjolia dactylota* (Bose) comb. nov., bennettitalean bisexual flower from the Rajmahal Hills, India. *Palaeobotanist* 32:217–229.

Bose, M. N., P. K. Pal, and T. M. Harris. 1984. *Carnoconites rajmahalensis* (Wieland) comb. nov. from the Jurassic of Rajmahal Hills, India. *Palaeobotanist* 32:368–369.

—— 1985. The *Pentoxylon* plant. *Phil. Trans. Roy. Soc. London* 310B:77–108.

Bose, M. N. and S. C. Srivastava. 1971. The genus *Dicroidium* from the Triassic of Nidpur, Madhya Pradesh, India. *Palaeobotanist* 19:41–51.

—— 1973a. *Nidiostrobus* gen.nov., a pollen-bearing fructification from the Lower Triassic of Godpad River Valley, Nidpur. *Geophytology* 2:211–212.

—— 1973b. Some micro- and megastrobili from the Lower Triassic of Gopad River Valley, Nidpur. *Geophytology* 3:69–80.

Brenner, G. J. 1963. The spores and pollen of the Potomac Group of Maryland. *Maryland Dept. Geol. Mines Water Res. Bull.* 27:1–215.

—— 1976. Middle Cretaceous floral provinces and early migrations of angiosperms. In C. B. Beck, ed., *Origin and Early Evolution of Angiosperms*, pp. 23–47. New York: Columbia University Press.

Brown, R. W. 1956. Palmlike plants from the Dolores Formation (Triassic), southwestern Colorado. *U.S. Geol. Surv. Prof. Paper* 274-H:205–209.

Chaloner, W. G. 1969. Triassic spores and pollen. In R. H. Tschudy and R. A. Scott, eds., *Aspects of Palynology*, pp. 291–309. New York: Wiley.

Chamberlain, C. J. 1913. *Macrozamia moorei*, a connecting link between living and fossil cycads. *Bot. Gaz.* 55:141–154.

—— 1935. *Gymnosperms, Structure and Evolution*. Chicago: University of Chicago Press.

Cookson, I. C. 1953. On *Macrozamia hopeites*—an early Tertiary cycad from Australia. *Phytomorphology* 3:306–312.

Cornet, B. 1986. The leaf venation and reproductive structures of a late Triassic angiosperm, *Sanmiguelia lewisii*. *Evol. Theor.* 7:231–309.

Coulter, J. M. and C. J. Chamberlain. 1917. *Morphology of Gymnosperms*. Chicago: University of Chicago Press.

Crane, P. R. 1984. Misplaced pessimism and misguided optimism: A reply to Mabberley. *Taxon* 33:79–82.

—— 1985a. Phylogenetic analysis of seed plants and the origin of angiosperms. *Ann. Missouri Bot. Gard.* 72:716–793.

——1985b. Phylogenetic relationships in seed plants. *Cladistics* 1:329–348.

—— 1986. The morphology and relationships of the Bennettitales. In B. A. Thomas and R. A. Spicer, eds., *Systematic and Taxonomic Approaches in Palaeobotany*, pp. 163–175. Oxford: Oxford University Press.

—— 1987. Vegetational consequences of the angiosperm diversification. In E. M. Friis, W. G. Chaloner, and P. R. Crane, eds, *The Origins of Angiosperms and Their Biological Consequences*, pp. 107–144. Cambridge: Cambridge University Press.

Crane, P. R. and G. R. Upchurch. 1987. *Drewria potomacensis* gen. et sp.nov., an early Cretaceous member of Gnetales from the Potomac Group of Virginia. *Amer. J. Bot.* 74:1722–1736.

Crepet, W. L. 1974. Investigations of North American cycadeoids: The reproductive biology of *Cycadeoidea*. *Palaeontographica, Abt. B.* 148:144–169.

Cridland, A. A. and J. E. Morris. 1960. *Spermopteris*, a new genus of pteridosperms from the Upper Pennsylvanian Series of Kansas. *Amer. J. Bot.* 47:855–859.

Cronquist, A. 1968. *The Evolution and Classification of Flowering Plants*. New York: Houghton Mifflin.

Daugherty, L. H. 1941. *The Upper Triassic Flora of Arizona*. Washington: Carnegie Institution of Washington. Vol. 526.

Davis, P. H. and V. H. Heywood. 1973. *Principles of Angiosperm Taxonomy*. New York: Krieger.

Delevoryas, T. 1955. The Medullosae—structure and relationships. *Palaeontographica, Abt. B.* 97:114–167.

—— 1982. Perspectives on the origin of cycads and cycadeoids. *Rev. Palaeobot. Palynol.* 37:115–132.

Delevoryas, T. and R. C. Hope. 1971. A new Triassic cycad and its phyletic implications. *Postilla* 150:1–21.

Delevoryas, T. and T. N. Taylor. 1969. A probable pteridosperm with eremopteroid foliage from the Allegheny Group of northern Pennsylvania. *Postilla* 133:1–14.

DeLuca, P., A. Moretti, S. Sabato, and G. Siniscalco-Gigliano. 1980. The ubiquity of cycasin in cycads. *Phytochemistry* 19:2230–2231.

Doyle, J. A. 1969. Cretaceous angiosperm pollen of the Atlantic coastal plain and its evolutionary significance. *J. Arn. Arbor.* 50:1–35.

—— 1978. Origin of angiosperms. *Ann. Rev. Ecol. Syst.* 9:365–392.

Doyle, J. A. and M. J. Donoghue. 1986a. Relationships of angiosperms and Gnetales: A numerical cladistic analysis. In B. A. Thomas and R. A. Spicer, eds., *Systematic and Taxonomic Approaches in Palaeobotany*, pp. 177–198. Oxford: Oxford University Press.

—— 1986b. Seed plant phylogeny and the origin of angiosperms: An experimental cladistic approach. *Bot. Rev.* 52:321–431.

—— 1987a. The origin of angiosperms: A cladistic approach. In E. M. Friis, W. G. Chaloner, and P. R. Crane, eds., *The Origins of Angiosperms and Their Biological Consequences*, pp. 17–49. Cambridge: Cambridge University Press.

—— 1987b. The importance of fossils in elucidating seed plant phylogeny and macroevolution. *Rev. Palaeobot. Palynol.* 50:63–95.

Doyle, J. A., S. Jardiné, and S. Doerenkamp. 1982. *Afropollis*, a new genus of early angiosperm pollen, with notes on the Cretaceous palynostratigraphy and paleoenvironments of northern Gondwana. *Bull. Centres Rech. Explor.-Prod. Elf-Aquitaine* 6:39–117.

Doyle, J. A., M. Van Campo, and B. Lugardon. 1975. Observations on exine structure of *Eucommiidites* and Lower Cretaceous angiosperm pollen. *Pollen et Spores* 17:429–486.

Drinnan, A. N. and T. C. Chambers. 1985. A reassessment of *Taeniopteris daintreei* from the Victorian Early Cretaceous: A member of the Pentoxylales and a significant Gondwanaland plant. *Austral. J. Bot.* 33:89–100.

—— 1986. Flora of the Lower Cretaceous Koonwarra Fossil Bed (Korumburra Group), South Gippsland, Victoria. *Mem. Assoc. Australas, Palaeontol.* 3:1–77.

Eames, A. J. 1952. Relationships of the Ephedrales. *Phytomorphology* 2:79–100.

Eckenwalder, J. E. 1980. Taxonomy of the West Indian cycads. *J. Arn. Arbor.* 61:701–722.

Ehrendorfer, F. 1971. *Systematik und Evolution: Spermatophyta, Samenpflanzen. Lehrbuch der Botanik für Hochschulen.* Stuttgart: Fisher.

—— 1976. Evolutionary significance of chromosomal differentiation patterns in gymnosperms and primitive angiosperms. In C. B. Beck ed., *Origin and Early Evolution of Angiosperms*, pp. 220–240. New York: Columbia University Press.

Eldredge, N., and J. Cracraft. 1980. *Phylogenetic Patterns and the Evolutionary Process.* New York: Columbia University Press.

Florin, R. 1933. Studien über die Cycadales des Mesozoikums, nebst erörternungen über die spatöffnungsapparate der Bennettitales. *K. Svenska Vetenskaps-Akad. Handl.* 12:1–134.

—— 1951. Evolution in cordaites and conifers. *Acta Horti Berg.* 15:285–388.

—— 1954. The female reproductive organs of conifers and taxads. *Biol. Rev. Cambridge Philos. Soc.* 29:367–389.

Frenguelli, J. 1944. Las especies del género "*Zuberia*" en la Argentina. *An. Mus. La Plata Paleont. B, Paleobot.* 2:1–30.

Gaussen, H. 1946. Les Gymnospermes, actuelles et fossiles. *Trav. Lab. Forest. Toulouse, tome 2. Étud. Dendrol.*, sect. 1, 1(3, 5):1–26.

Gillespie, W. H., and H. W. Pfefferkorn. 1986. Taeniopterid lamina on *Phasmatocycas* megasporophylls (Cycadales) from the Lower Permian of Kansas, U.S.A. *Rev. Palaeobot. Palynol.* 49:99–116.

Gould, R. E. 1971. *Lyssoxylon grigsbyi*, a cycad trunk from the Upper Triassic of Arizona and New Mexico. *Amer. J. Bot.* 58:239–248.

Greguss, P. 1968. *Xylotomy of the Living Cycads, with a Description of Their Leaves and Epidermis.* Budapest: Akadémiai Kiadó.

Halle, T. G. 1929. Some seed-bearing pteridosperms from the Permian of China. *K. Svenska Vetenskaps-Akad. Handl.*, ser. 3, 6 (8):1–24.

Harris, T. M. 1932a. The fossil flora of Scoresby Sound East Greenland. Part 2. Description of seed plants *incertae sedis* together with a discussion of certain cycadophyte cuticles. *Meddel. Grønland* 85(3):1–114.

—— 1932b. The fossil flora of Scoresby Sound East Greenland. Part 3. Caytoniales and Bennettitales. *Meddel. Grønland* 85(5):1–133.

—— 1941a. Cones of extinct Cycadales from the Jurassic rocks of Yorkshire. *Phil. Trans. Roy. Soc. London* 231B:75–98.

—— 1941b. *Caytonanthus*, the microsporophyll of *Caytonia*. *Ann. Bot. (London)* 5:47–58.

—— 1961. The fossil cycads. *Palaeontology* 4:313–323.

—— 1962. The occurrence of the fructification *Carnoconites* in New Zealand. *Trans. Roy. Soc. New Zealand, Geol.* 4:17–27.

—— 1964. *The Yorkshire Jurassic Flora. 2. Caytoniales, Cycadales and Pteridosperms.* London: British Museum (Natural History).

—— 1969. *The Yorkshire Jurassic Flora. 3. Bennettitales.* London: British Museum (Natural History).

—— 1973. *The strange Bennettitales.* Nineteenth Sir Albert Charles Seward Memorial Lecture, 1971. Lucknow: Birbal Sahni Institute of Palaeobotany.

—— 1976. The Mesozoic gymnosperms. *Rev. Palaeobot. Palynol.* 21:119–134.

Haubold, H. 1980. Zur Gattung *Callipteris*. Brongniart. *Z. geol. Wiss. Berlin* 6:747–767.

Hennig, W. 1966. *Phylogenetic Systematics.* Urbana: University of Illinois Press.

Hill, C. R. 1986. Reproductive biology and diversity among fossil cycads. *Amer. J. Bot.* 73:700–701. (Abstract.)

Hill, C. R. and P. R. Crane. 1982. Evolutionary cladistics and the origin of angiosperms. In K. A. Joysey and A. E. Friday, eds., *Problems of Phylogenetic Reconstruction,* pp. 269–361. New York: Academic Press.

Hill, R. S. 1978. Two new species of *Bowenia* Hook. ex Hook.f. from the Eocene of eastern Australia. *Austral. J. Bot.* 26:837–846.

Holmes, W. B. K. and S. R. Ash. 1979. An early Triassic megafossil flora from the Lorne Basin, New South Wales. *Proc. Linn. Soc. N.S.W.* 103:47–70.

Hutchinson, J. 1924. Contributions toward a phylogenetic classification of flowering plants, 3. The genera of gymnosperms. *Kew Bull.* 1924:49–66.

Jain, K. P. 1962. *Fascivarioxylon mehtae* gen. et sp. nov., a new petrified cycadean wood from the Rajmahal Hills, Bihar, India. *Palaeobotanist* 11:138–143.

—— 1971. Taxonomic observations on the genus *Sewardioxylon* Gupta—a junior synonym of *Fascisvarioxylon* Jain. *Palaeobotanist* 19:251–252.

Johnson, L. A. S. 1959. The families of cycads and the Zamiaceae of Australia. *Proc. Linn. Soc. N.S.W.* 84:64–117.

Just, T. K. 1948. Gymnosperms and the origin of angiosperms. *Bot. Gaz.* 110:91–103.

Kerp, J. H. F. 1983. Aspects of Permian palaeobotany and palynology. I. *Sobernheimia jonkeri* nov. gen., nov. sp.: A new fossil plant of cycadalean affinity from the Waderner Gruppe of Sobernheim. *Rev. Palaeobot. Palynol.* 38:173–183.

Kershaw, E. M. 1912. Structure and development of the ovule of *Bowenia spectabilis. Ann. Bot. (London).* 26:625–646.

Kimura, T. and S. Sekido. 1975. *Nilssoniocladus* n. gen. (Nilssoniaceae, n. fam.) newly found from the early Lower Cretaceous of Japan. *Palaeontographica, Abt. B* 153:111–118.

Knoll, A. H. and G. W. Rothwell. 1981. Paleobotany: Perspectives in 1980. *Paleobiology* 7:7–35.

Krassilov, V. A. 1975. *Dirhopalostachyaceae*—a new family of proangiosperms and its bearing on the problem of angiosperm ancestry. *Palaeontographica, Abt. B* 153:100–110.

—— 1982. Early Cretaceous flora of Mongolia. *Palaeontographica, Abt. B* 181:1–43.

—— 1984. New paleobotanical data on origin and early evolution of angiospermy. *Ann. Missouri Bot. Gard.* 71:577–592.

—— 1986. New floral structure from the Lower Cretaceous of Lake Baikal area. *Rev. Palaeobot. Palynol.* 47:9–16.

Krassilov, V. A., and E. V. Bugdaeva. 1982. Achene-like fossils from the Lower Cretaceous of the Lake Baikal area. *Rev. Palaeobot. Palynol.* 36:279–295.

Kräusel, R. 1948. *Sturiella langeri* nov. gen., nov. sp., eine Bennettitee aus der Trias von Lunz (Nieder-Österreich). *Senckenbergiana* 29:141–149.

—— 1949. Koniferen und andere Gymnospermen aus der Trias von Lunz, Nieder-Österreich. *Palaeontographica, Abt. B* 84:35–82.

—— 1953. Ein neues *Dioonitocarpidium* aus der Trias von Lunz. *Senckenbergiana* 34:105–108.

Kräusel, R. and Schaarschmidt. 1966. Die Keuperflora von Neuewelt bei Basel. 4. Pterophyllen und Taeniopteriden. *Schweiz. Paläontol. Abh.* 84:1–64.

Lacey, W. S. 1976. Further observations on the Molteno flora of Rhodesia. *Arnoldia Rhod.* 7:1–14.

Lacey, W. S. and R. C. Lucas. 1981. The Triassic flora of Livingston Island, South Shetland Islands. *Brit. Anarct. Surv. Bull.* 53:157–173.

Lacey, W. S., D. E. van Dijk, and K. D. Gordon-Gray. 1975. Fossil plants from the Upper Permian in the Mooi River district of Natal, South Africa. *Ann. Natal. Mus.* 22:349–420.

Li, X., and Yao, Z. 1983. Fructifications of gigantopterids from South China. *Palaeontographica, Abt. B* 185:11–26.

Lilienstern, H. R. 1928. *Dioonites pennaeformis* Schenk, eine fertile Cycadee aus der Lettenkohle. *Paläont. Zeitschr.* 10:91–107.

Lundblad, A. B. 1950. Studies in the Rhaeto-Liassic floras of Sweden. 1. Pteridophyta, Pteridospermae and Cycadophyta from the Mining District of NW Scania. *K. Svenska Vetenskaps-Akad. Handl.*, ser. 4,1 (8):1–82.

—— 1961. *Harrisothecium* nomen novum. *Taxon* 10:23–24.

Mamay, S. H. 1966. *Tinsleya*, a new genus of seed-bearing callipterid plants from the Permian of Texas. *U.S. Geol. Surv. Prof. Paper* 523-E:1–15.

—— 1969. Cycads: Fossil evidence of late Paleozoic origin. *Science* 164:295–296.

—— 1976. Paleozoic origin of the cycads. *U.S. Geol. Surv. Prof. Paper* 934:1–48.

Mapes, G. and G. W. Rothwell. 1984. Permineralized ovulate cones of *Lebachia* from late Palaeozoic limestones of Kansas. *Palaeontology* 27:69–94.

Martens, P. 1971. Les Gnétophytes. *Handb. Pflanzenanat.* 12(2):1–295.

Matte, H. 1904. Recherches sur l'appareil libéro-ligneux des Cycadacées. *Mém. Soc. Linn. Normandie* 22:1–233.

Meeuse, A. D. J. 1963. The so-called "megasporophyll" of *Cycas*—a morphological misconception: Its bearing on the phylogeny and the classification of the Cycadophyta. *Acta Bot. Neerl.* 12:119–128.

Meyen, S. V. 1984. Basic features of gymnosperm systematics and phylogeny as evidenced by the fossil record. *Bot. Rev.* 50:1–111.

Meyen, S. V. and Y. Lemoigne. 1986. *Dicksonites pluckenetii* (Schlotheim) Sterzel and its affinity with Callistophytales. *Geobios* 19:87–97.

Millay, M. A. and T. N. Taylor. 1976. Evolutionary trends in fossil gymnosperm pollen. *Rev. Palaeobot. Palynol.* 21:65–91.

Moretti, A., S. Sabato, and G. Siniscalo-Gigliano. 1983. Taxonomic significance of methylazoxymethanol glycosides in the cycads. *Phytochemistry* 22:115–117.

Muller, J. 1984. Significance of fossil pollen for angiosperm history. *Ann. Missouri Bot. Gard.* 71:419–443.

Niklas, K. J. and K. Norstog. 1984. Aerodynamics and pollen grain depositional patterns on cycad megastrobili: Implications on the reproduction of three cycad genera *(Cycad, Dioon* and *Zamia). Bot. Gaz.* 145:92–104.

Oliver, F. W. and D. H. Scott. 1904. On the structure of the Palaeozoic seed

Lagenostoma lomaxi, with a statement of the evidence upon which it is referred to *Lyginodendron. Phil. Trans Roy. Soc. London* 197B:193–247.

Pant, D. D. 1977a. The plant of *Glossopteris. J. Indian Bot. Soc.* 56:1–23.

—— 1977b. Early conifers and conifer allies. *J. Indian Bot. Soc.* 56:23–37.

Pant, D. D. and N. Basu. 1973. *Pteruchus indicus* sp. nov. from the Triassic of Nidpur, India. *Palaeontographica, Abt. B* 144:11–24.

—— 1979. Some further remains of fructifications from the Triassic of Nidpur, India. *Palaeontographica, Abt. B.* 168:129–146.

Pant, D. D. and D. D. Nautiyal. 1967. On the structure of *Buriadia heterophylla* (Feistmantel) Seward & Sahni and its fructification. *Phil. Trans. Roy. Soc. London* 252B:27–48.

Petriella, B. 1969. *Menucoa cazaui* nov. gen. et sp. tronco petrificado de Cycadales, Provincia de Río Negro, Argentina. *Ameghiniana* 6:291–302.

—— 1972. Estudio de maderas petrificadas del Terciario inferior del área central de Chubut (Cerro Bororó). *Rev. Mus. La Plata*, n.s. 6:159–254.

—— 1978. Nuevos Hallazgos de Cycadales fosiles en Patagonia. *Bol. Asoc. Lat. Paleobot. Paly.* 5:13–16.

—— 1979. Sinopsis de las Corystospermaceae (Corystospermales, Pteridospermophyta) de Argentina. 1. Hojas. *Ameghiniana* 16:81–102.

—— 1980. Sinopsis de las Corystospermaceae (Corystospermales, Pteridospermophyta) de Argentina. 2. Estructuras fertiles. *Ameghiniana* 17:168–180.

—— 1981. Sistematica y vinculaciones de las Corystospermaceae H. Thomas. *Ameghiniana* 18:221–234.

—— 1983. Sinopsis de las Corystospermaceae (Corystospermales, Pteridospermophyta) de la Argentina. 3. Troncos y cronoestratigrafia. *Ameghiniana* 20:41–46.

Petriella, B. and J. V. Crisci. 1975. Estudios numericos en Cycadales. 1. Cycadales actuales: Sistemica. *Bol. Soc. Argent. Bot.* 16:231–247.

—— 1977. Estudios numericos en Cycadales. 2. Cycadales actuales: Simulacion de arboles evolutivos. *Obra Mus. La Plata* 3:151–159.

Pilger, R. 1926. Cycadaceae. In A. Engler and K. Prantl, eds., *Die naturlichen Pflanzenfamilien*, vol. 13:44–82. Leipzig: Engelmann.

Retallack, G. J. 1977. Reconstructing Triassic vegetation of eastern Australasia: A new approach for the biostratigraphy of Gondwanaland. *Alcheringa* 1:247–278.

—— 1980. Late Carboniferous to Middle Triassic megafossil floras from the Sydney Basin. *Geol. Surv. N.S.W. Bull.* 26:384–430.

Retallack, G. J. and D. L. Dilcher. 1981. Arguments for a glossopterid ancestry of angiosperms. *Paleobiology* 7:54–67.

Rothwell, G. W. 1981. The Callistophytales (Pteridospermopsida): Reproductively sophisticated Paleozoic gymnosperms. *Rev. Palaeobot. Palynol.* 32:103–121.

Sabato, S. and P. DeLuca. 1985. Evolutionary trends in *Dion* (Zamiaceae). *Amer. J. Bot.* 72:1353–1363.

Sahni, B. 1948. The Pentoxyleae: A new group of Jurassic gymnosperms from the Rajmahal Hills of India. *Bot. Gaz.* 110:47–80.

Schuster, J. 1932. Cycadaceae. In A. Engler, ed., *Das Pflanzenreich* 99 (IV.1):1–168. Leipzig: Engelmann.

Scott, A. C. and W. G. Chaloner. 1983. The earliest fossil conifer from the Westphalian B of Yorkshire. *Proc. Roy. Soc. London* 220B:163–182.

Scott, D. H. 1923. *Studies in Fossil Botany. 2. Spermophyta.* London: Black.

Scott, R. A. 1960. Pollen of *Ephedra* from the Chinle Formation (Upper Triassic) and the genus *Equisetosporites. Micropaleontology* 6:271–276.

Seward, A. C. 1917. *Fossil Plants. 3. Pteridospermae, Cycadofilices, Cordaitales and Cycadophyta.* Cambridge: Cambridge University Press.

Smoot, E. L., T. N. Taylor, and T. Delevoryas. 1985. Structurally preserved fossil plants from Antarctica. 1. *Antarcticycas*, gen. nov., a Triassic cycad stem from the Beardmore Glacier area. *Amer. J. Bot.* 72:1410–1423.

Sporne, K. R. 1971. *Morphology of Gymnosperms.* London: Hutchinson.

Stebbins, G. L. 1974. *Flowering Plants: Evolution Above the Species Level.* Cambridge: Harvard University Press.

Stevenson, D. W. 1980. Radial growth in the Cycadales. *Amer. J. Bot.* 67:465–475.

—— 1981. Observations on ptyxis, phenology and trichomes in the Cycadales and their systematic implications. *Amer. J. Bot.* 68:1104–1114.

—— 1982. Sporophyll vasculature in the Cycadales. *Bot. Soc. Amer. Misc. Pub.* 162:25.

—— 1985. A proposed classification of the Cycadales. *Amer. J. Bot.* 72(6):971–972. (Abstract.)

Stewart, W. N. 1954. Structure and affinities of *Pachytesta illinoense* comb. nov. *Amer. J. Bot.* 41:500–508.

—— 1983. *Paleobotany and the Evolution of Plants.* New York: Cambridge University Press.

Stidd, B. M., M. O. Rischbieter, and T. L. Phillips. 1985. A new lyginopterid pollen organ with alveolate pollen exines. *Amer. J. Bot.* 72:501–508.

Stopes, M. C. 1903. Beiträge zur Kenntnis der Fortpflanzungsorgane der Cycadeen. *Flora* 93:435–482.

—— 1905. On the double nature of the cycadean integument. *Ann. Bot.* (London). 19:561–566.

Surange, K. R. and S. Chandra. 1975. Morphology of the gymnospermous fructifications of the *Glossopteris* flora and their relationships. *Palaeontographica, Abt. B* 149:153–180.

Takaso, T. 1985. A developmental study of the integument in gymnosperms. 3. *Ephedra distachya* L. and *E. equisetina* Bge. *Acta Bot. Neerl.* 34:33–48.

Takhatajan, A. 1969. *Flowering Plants: Origin and Dispersal.* Edinburgh: Oliver and Boyd.

Taylor, T. N. 1965. Paleozoic seed studies: A monograph of the American species of *Pachytesta. Palaeontographica, Abt. B* 117:1–46.

—— 1969. Cycads: Evidence from the Upper Pennsylvanian. *Science* 164:294–295.

—— 1970. *Lasiostrobus* gen. n., a staminate strobilus of gymnospermous affinity from the Pennsylvanian of North America. *Amer. J. Bot.* 57:670–690.

—— 1981. *Paleobotany: An Introduction to Fossil Plant Biology.* New York: McGraw-Hill.

Taylor, T. N. and S. Archangelsky. 1985. The Cretaceous pteridosperms *Ruflorinia* and *Ktalenia* and implications on cupule and carpel evolution. *Amer. J. Bot.* 72:1842–1853.

Thoday. M. G. 1911. The female inflorescence and ovules of *Gnetum africanum,* with notes on *Gnetum scandens. Ann. Bot. (London)* 25:1101–1135.

Thomas, H. H. 1933. On some Pteridospermous plants from the Mesozoic rocks of South Africa. *Phil. Trans. Roy. Soc. London* 222B:193–265.

—— 1955. Mesozoic pteridosperms. *Phytomorphology* 5:177–185.

—— 1958. *Lidgettonia,* a new type of fertile *Glossopteris. Bull. British Museum (Natural History), Geol.* 3:177–189.

Thomas, H. H. and N. Bancroft. 1913. On the cuticles of some recent and fossil cycadean fronds. *Trans. Linn. Soc. London, Bot.* 8:155–204.

Thomas, H. H. and T. M. Harris. 1960. Cycadean cones of the Yorkshire Jurassic. *Senck. leth.* 41:139–161.

Tidwell, W. D., A. D. Simper, and G. F. Thayn. 1977. Additional information concerning the controversial Triassic plant: *Sanmiguelia. Palaeontographica, Abt. B* 163:143–151.

Townrow, J. A. 1962. On *Pteruchus,* a microsporophyll of the Corystospermaceae. *Bull. British Museum (Natural History), Geol.* 6:287–320.

—— 1965. A new member of the Corystospermaceae Thomas. *Ann. Bot.* (London) 29:495–511.

Trevisan, L. 1980. Ultrastructural notes and considerations on *Ephedripites, Eucommiidites* and *Monoculcites* pollen grains from Lower Cretaceous sediments of southern Tuscany (Italy). *Pollen et Spores* 22:85–132.

Upchurch, G. R. and P. R. Crane. 1985. Probable gnetalean megafossils from the Lower Cretaceous Potomac Group of Virginia. *Amer. J. Bot.* 72(6):903. (Abstract.)

Varma, Y. S. R. and C. A. K. Ramanujam. 1984. Palynology of some Upper Gondwana deposits of Palar Basin, Tamil Nadu, India. *Palaeontographica, Abt. B.* 190:37–86.

Velenovský, J. and L. Viniklář. 1926. Flora Cretacea Bohemiae, 1. *Roz. Stát. Geol. Úst. Česk. Rep.* 1:1–57.

Walker, J. W. and A. G. Walker. 1984. Ultrastructure of Lower Cretaceous angiosperm pollen and the origin and early evolution of flowering plants. *Ann. Missouri Bot. Gard.* 71:464–521.

Ward, L. F. 1895. The Potomac Formation. *U.S. Geol. Surv. Ann. Rept.* 15:307–397.

Wieland, G. R. 1906. *American Fossil Cycads: Structure.* Vol. 34, no. 1. Washington: Carnegie Institution of Washington.

—— 1916. *American Fossil Cycads: Taxonomy.* Vol. 2. Washington: Carnegie Institution of Washington.

Wiley, E. O. 1981. *Phylogenetics: The Theory and Practice of Phylogenetic Systematics.* New York: Wiley.

Wilson, L. R. 1959. Geological history of the Gnetales. *Okla. Geol. Notes* 19:35–40.

Wordsell, W. C. 1906. The structure and origin of the Cycadaceae. *Ann. Bot.* (London) 20:129–159.

Yao, Z. and P. R. Crane. 1986. Gigantopterid leaves with cuticles from the Lower Permian of China. *Amer. J. Bot.* 73:715–716.

Zavada, M. W. 1984. Angiosperm origins and evolution based on dispersed fossil pollen ultrastructure. *Ann Missouri Bot. Gard.* 71:444–463.

Zhang, S. Z. and Mo, Z. C. 1981. On the occurrence of cycadophytes with slender growth habit in the Permian of China. *Geol. Soc. Amer. Spec. Paper* 187:237–242.

Zhu Jia-Nan and Chen Gong-xin. 1981. *Fimbriotheca tomentosa* Zhu et Chen—A new genus and species from Permian of China and its systematic position. *Acta Bot. Sin.* 23:487–491.

Zhu Jia-Nan and Du Xian-Ming. 1981. A new cycad—*Primocycas chinensis* gen. et sp. nov. from the Lower Permian in Shanxi, China and its significance. *Acta Bot. Sin.* 23:401–404.

6

Cordaitales

GAR W. ROTHWELL

::::

Cordaitales are an extinct order of gymnosperms with a well-docu-
mented geological range of Namurian to Permian. Members of the
group constitute one of the most prominent components of the late
Paleozoic terrestrial vegetation, and are considered to be closely related, or potentially ancestral, to conifers. Some species were small
shrubs and possibly mangroves, while others were trees. The order is
represented by abundant fossils preserved by cellular permineralization in peat, and as coalified compression/impressions and stump
casts in clastic sediments.

Putative cordaitean remains have been reported from around the
world in beds that range from Devonian to Cretaceous, but the order
is best represented in Euramerian Carboniferous (or equivalent) deposits (Grand'Eury 1877; Seward 1917; Crookall 1970; Trivett and
Rothwell 1985; Costanza 1985). Additional compression/impression
fossils from the Angara floras of eastern Asia (Middle Carboniferous-
Upper Permian; Meyen 1984) are assigned to the Cordaitales, but they
differ from the largely older and better-known Euramerian forms,

and their exact systematic relationships remain equivocal. Leaf and stem fragments and compressed fertile structures from southern hemisphere Gondwana deposits (Pant 1982), from Devonian and Lower Carboniferous sediments, and from post-Permian strata (Seward 1917) may have been produced by cordaiteans. However, such remains also could represent several other groups of vascular plants, and, in the absence of diagnostic cordaitean reproductive structures, they do not present adequate evidence for extending the geographic or stratigraphic ranges of the order (Pant 1982; Rothwell 1982a; Meyen 1984). A detailed investigation of the late Paleozoic, South American putative cordaitean fossils recently reviewed by Archangelsky (1986) may provide such evidence in the near future.

BUILDING A CONCEPT

Fossils that we now assign to the Cordaitales have been the subject of numerous investigations since the beginning of the nineteenth century, but our understanding of the plants represented by such fossils dates from the pioneering work of Grand'Eury (1877). Among treatments of several groups of Carboniferous tracheophytes from Europe, Grand'Eury assembled associations of organs to characterize a new order of gymnosperms, the Cordaitales (Cordaitées of Grand'Eury 1877). He discovered interconnections among some organs, and recognized transitions in the features of other organs that allowed him to identify several genera as parts of the same types of plants. Grand'Eury also offered several elegant reconstructions to characterize the general form of cordaitean plants (e.g., figure 6.1A), but he was careful to separate evidence of attachment from hypotheses of relationships (note discontinuities in the cordaitean reconstructions of Grand'Eury 1877, figure 6.1A). Among more recent authors similar caution has not always been expressed, and some restorations of cordaitean plants appear to be based more upon conjecture than evidence from the fossils (e.g., figure 6.1B; Cridland 1964; Costanza 1985).

Among the structural transitions and interconnections that Grand'Eury recognized are stems that show a surface pattern (viz., *Cordaicladus*, figure 6.2C) in some places and a septate pith (viz., *Artisia*, figure 6.2D) in other places, stem compressions with attached leaves and cones, and stumps with evidence of septate pith. Through loans of permineralized material from his contemporary Bernard Renault (Jean Galtier, personal communication), he also was able to determine some features of anatomy for the leaves and stems, and to

A

B

C

Figure 6.1. Reconstructions of Euramerian cordaitean plants. **A** Cordaitean trees with *Eu-Cordaites* leaves (at **a**), *Dory-Cordaites* leaves (at **b**), and *Poa-Cordaites* leaves (at **c**); from Grand'Eury (1877). **B** Putative mangrove cordaite; from Cridland (1964), by permission of *Palaeontology.* **C** Shrub-like *Cordaixylon dumusum;* from Rothwell and Warner (1984), by permission of the University of Chicago Press.

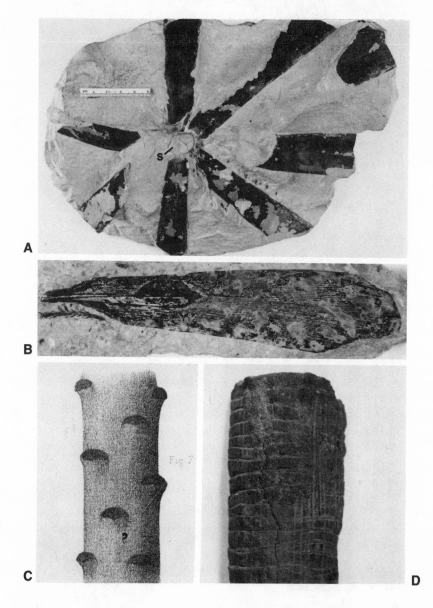

Figure 6.2. Morphology of Euramerian cordaitean organs. **A** *Cordaites* leaves attached to mold/cast of stem (at **s**), from Middle–Upper Pennsylvanian (Lansing-Douglas Groups) near Ottawa, Kansas. × 0.3. (Courtesy Dr. A. T. Cross.) **B** *Cordaites* leaf from Upper Pennsylvanian deposits near Hamilton, Kansas. × 0.7. (Courtesy Dr. G. Mapes.) **C** *Cordaicladus* stem surface from Grand'Eury (1877). **D** *Artisia* pith cast from Pennsylvanian deposits of Michigan. × 0.6. (Courtesy Dr. A. T. Cross.)

demonstrate that cordaiteans were one of the groups of gymnosperms that produced wood assignable to *Dadoxylon* (or *Araucarioxylon*).

GENERAL FEATURES OF EURAMERIAN CORDAITALES

As a result of continuing studies, the general features of Carboniferous Euramerian cordaiteans are becoming increasingly well characterized, and an understanding of diversity within the group is beginning to emerge. There now is good evidence that large, much-branched trees as well as smaller trees (figure 6.1A) inhabited extra-basinal lowlands, relatively stable parts of flood plains, and possibly more upland environments (Grand'Eury 1877; Scott 1977; Trivett and Rothwell 1985), while putative mangrove forms (figure 6.1B) and shrubs (figure 6.1C) lived in the peat-forming swamps (Cridland 1964; Raymond and Phillips 1983; Rothwell and Warner 1984).

Cordaiteans have eustelic stems that produce abundant wood (figure 6.3A) and bear helically arranged strap-shaped leaves (figure 6.2A). Many species also display a septate pith (figure 6.3A). The reproductive organs are compound, monosporangiate strobili (figures 6.4A, B) in which either platyspermic ovules or saccate prepollen (or true pollen) is produced (figures 6.4D; 6.5A). Strobili occur either as axillary branches (figures 6.4C; Baxter 1959; Trivett and Rothwell 1985, in press) or as branches with no definite relationship to the leaves (Grand'Eury 1877; Rothwell and Warner 1984).

TAXONOMY AND NOMENCLATURE

Both the taxonomy and the nomenclature of organs assignable to cordaites have been the subject of considerable confusion and controversy (e.g., Arnold 1967; Meyen 1984; Rothwell and Warner 1984; Costanza 1985; Trivett and Rothwell 1985), with disagreement continuing among recent workers. The order Cordaitales is based upon the leaf genus *Cordaites* Unger (1850, figure 5), the type species of which, *C. borassifolia* (Sternberg) Unger (1850), has as its type the same specimen as *Flabellaria borassifolia* Sternberg (1825) and *Pychnophyllum borassifolia* (Sternberg) Brongniart (1849). However, both *Flabellaria* Sternberg and *Pychnophyllum* Brongniart have been rejected as latter homonyms (Arnold 1967), leaving *Cordaites* Unger as the legitimate name. As most broadly employed, *Cordaites* includes strap-shaped leaves produced by a wide variety of plants that transcend familial and possibly even ordinal boundaries (Meyen 1984; Rothwell and Warner 1984). For this reason the presence of *Cordaites*-type leaves

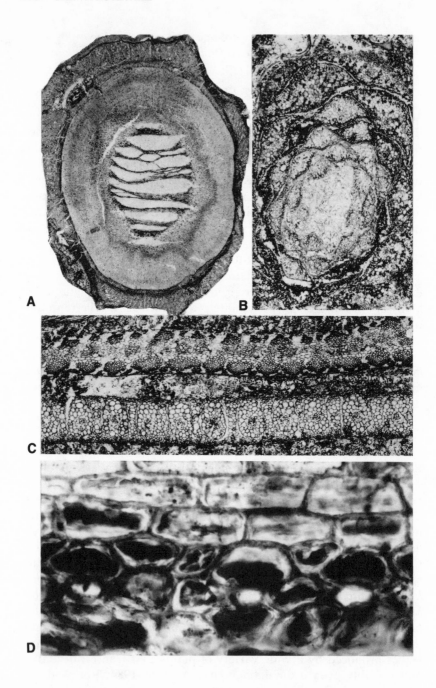

alone is not used in this treatment to establish or extend the stratigraphic or geographic ranges of the Cordaitales, and reports of cordaiteans that are based only on fragments of leaves or stem patterns are not included.

A list of the most commonly accepted genera of cordaitean organs is presented in table 6.1. Compressed (or mold/cast) cordaitean stems with attached leaves (figure 6.2A) are assigned to *Cordaites*, while those that display characteristic leaf scar patterns are assigned to *Cordaicladus* Grand'Eury (1877; figure 6.2C). Compressions or mold/ cast specimens that reflect the presence of a septate pith are referred to *Artisia* Sternberg (1838; figure 6.2D). Permineralized cordaitean stems (figure 6.3A) have been described as species of *Cordaixylon* Grand'Eury (1877), *Cordaioxylon* Felix (1882), *Cordaites* (e.g., Cohen and Delevoryas 1959), *Mesoxylon* Scott and Maslen (1910), *Mesoxylopsis* Scott (1919), *Mesoxyloides* Maslen (1930), and *Pennsylvanioxylon* Vogellehner (1965), but the systematics of these genera is confused and recent work suggests that genera of stems may not reflect distinct genera of whole plants (Trivett and Rothwell 1988). Older treatments also included *Callixylon*, *Pitus*, and *Poroxylon* as cordaitean plants, but all of these genera have since been placed among other groups of vascular plants (Beck 1960; Long 1979; Rothwell 1982a).

Cordaianthus Grand'Eury (1877) and *Cordaitanthus* Feistmantel (1876) are used by different workers for the same types of cordaitean strobili (figures 6.4A, B; Rothwell 1977; Meyen 1984). *Cordaianthus* traditionally has been favored by North American authors and is a *nomina conservanda* (Voss et al. 1983:426). However, *Cordaitanthus* Feistmantel (1876) has an earlier date of effective publication and is not included in the list of names over which *Cordaianthus* Grand'Eury (1877) is conserved. *Cordaitanthus* Fiestmantel must, therefore, be regarded as the legitimate name (Meyen 1984). Isolated cordaitean fructifications of slightly different morphology and internal anatomy are assigned to *Gothania* Hirmer (1933).

Figure 6.3. Anatomy of Euramerian cordaitean organs. **A** *Mesoxylon thompsonii* stem in cross section, from Middle Pennsylvanian sediments of Iowa. × 3.6. **B** Cross section of *Cordaixylon dumusum* apex showing needle/scale-like leaves of some cordaitean branches, from the Upper Pennsylvanian of Ohio. × 26. **C** Cross section of two *Cordaites filicis*-type leaves produced by *Mesoxylon priapi*, from the Upper Pennsylvanian of Ohio. × 16. **D** Abaxial epidermis of *Cordaixylon dumusum* leaf from the Upper Pennsylvanian of Ohio. × 580.

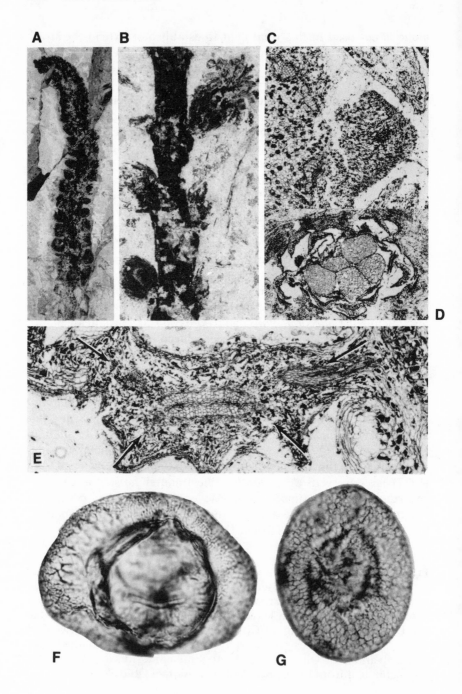

Ovules of the types that were produced by cordaitean plants are known as *Cardiocarpon* Brongniart (1928), *Cordaicarpon* Geinitz (1862; figure 6.5B) (= *Cordaicarpus*) or *Samaropsis* Goeppert (1864) when compressed, and as *Cardiocarpus* Brongniart (1874; figures 6.5C, E), *Rhabdocarpus* Brongniart (1874), *Diplotesta* Brongniart (1874), *Sarcotaxus* Brongniart (1874), *Leptocaryum* Brongniart (1874), *Taxospermum* Brongniart (1874), *Mitrospermum* Arber (1910; figures 6.5 D, F–H) and *Nucellangium* Andrews (1949) when anatomically preserved. Additional ovules of varying modes of preservation have been assigned to *Diplotesta* Brongniart (1874) and *Carpolithus* L. (Seward 1917). However, when used in this way, the generic concepts implied by such names are broadly circumscribed. Like isolated *Cordaites* leaves, the genera of such ovules are not always systematically diagnostic. Saccate grains produced by cordaitean fructifications are assignable to *Florinites* (figure 6.4G), *Felixipollenites*, or *Sullisaccites* (figure 6.4F) when found dispersed (Millay and Taylor 1974; Trivett and Rothwell 1985).

STRUCTURE OF EURAMERIAN CORDAITEAN ORGANS

Leaves Most *Cordaites* leaves (figures 6.1A, 6.2A, B) are roughly spatulate, with veins that dichotomize infrequently and appear to be parallel. Grand'Eury (1877) recognized three subgenera for the mor-

Figure 6.4. Cordaitean fructifications from Euramerian Pennsylvanian sediments. **A** Compressed, immature *Cordaitanthus*-type strobilus from Pennsylvanian sediments of central Kansas. × 0.7. **B** Compressed, mature pollen-producing strobilus from Pennsylvanian sediments of Kansas. Note numerous sporangia protruding from apex of secondary shoot at upper right. × 1.4. **C** Transverse section near apex of *Mesoxylon birame* fertile stem, from the Middle Pennsylvanian of Kansas. Note parenchymatous pith at this level (at left) and base of *Cordaitanthus concinnus*-type cone diverging at right. × 20. **D** Transverse section of secondary shoot of *Cordaitanthus*-type *Cordaixylon dumusum* pollen cone showing five sporangia of a single fertile scale with mature grains. × 16. **E** Transverse section of primary axis of *Cordaitanthus concinnus*, from Middle Pennsylvanian deposits of Kansas, showing four ranks of appendage divergence (arrows). × 30. **F** *Sullisaccites*-type grain from pollen sac of *Gothania*-type pollen cone of *Mesoxylon priapi*. Note monolete suture on proximal surface. × 450. **G** *Florinites*-type grain from *Cordaitanthus concinnus* pollen cone. Note absence of suture from proximal surface. × 600.

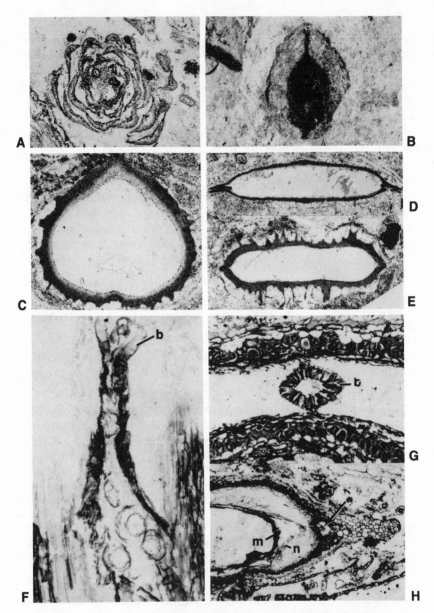

Figure 6.5. Ovulate structures of Euramerian cordaitaleans. **A** Transverse section of ovulate secondary shoot of *Cordaixylon dumusum* (viz., *Cordaitanthus duquesnensis*), with attached ovules (at center). × 16. **B** *Cardiocarpon*-type compressed ovule found associated with *Eu-*

phologies of compressed cordaitean leaves produced by different types of cordaitean plants. These are *Eu-Cordaites* for large spatulate leaves with rounded tips (**a** in figures 6.1A, 6.2B), *Dory-Cordaites* for slightly smaller leaves with pointed tips (**b** in figure 6.1A), and *Poa-Cordaites* for narrow, grass-like cordaitean leaves (**c** in figure 6.2A). More recently, heterophylly has been recognized among cordaitean plants (Rothwell 1982a), and scale or needle-like leaves (figure 6.3B) are now known to be produced around buds, at the base of branches of several taxa (Rothwell 1982; Meyen 1984; Costanza 1985), and along the entire length of some branches of *Cordaixylon dumusum* (Rothwell and Warner 1984).

The surface of most large cordaitean leaves is characterized by longitudinal ridges that correspond to the positions of veins and interspersed sclerenchyma bundles, and furrows that lie adjacent to mesophyll tissue (figures 6.2B; 6.3C). Some leaves are amphistomatic, while others are hypostomatic. Epidermal cells are rectangular and occur in longitudinally oriented rows (figure 6.3D). Stomata are arranged in longitudinal rows positioned above the furrows, and consist of sunken guard cells flanked by two lateral and two terminal subsidiary cells (figure 6.3D).

Stems   As is characteristic of most gymnospermous taxa, cordaitean stems are eustelic, with secondary vascular tissue production from a bifacial vascular cambium and periderm production from a phellogen. In many stems the pith consists of diaphragms of tissue that are separated by hollow chambers (figure 6.3A), but a few species pro-

---

Cordaites leaves (figure 6.2B) and *Cordaitanthus*-type strobili (figure 6.4A) in Upper Pennsylvanian sediments near Hamilton, Kansas. Note prominent wing and micropyle (vertical line at apex). × 3. **C** Near-median longitudinal section (primary plane) of *Cardiocarpus spinatus*, from Middle Pennsylvanian sediments of Kansas. × 3. **D** Transverse section of *Mitrospermum leeanum* from the Middle Pennsylvanian of Iowa. Note lateral wings. × 3. **E** Transverse section of *Cardiocarpus spinatus*, from Middle Pennsylvanian sediments of Kansas. × 3. **F** Longitudinal section of pollen chamber of *Mitrospermum vinculum*, from the Upper Pennsylvanian of Ohio. Note numerous *Sullisaccites*-type grains in pollen chamber and nucellar beak (**b**) at apex. × 81. **G** Transverse section through nucellar beak (**b**) of *Mitrospermum vinculum*. × 89. **H** Transverse section at mid-level of *Mitrospermum vinculum* showing megaspore membrane (**m**), nucellus (**n**), vascular bundle in integument (at arrow), and sarcotestal wing at right. × 22.

Table 6.1. Genera of organs from Euramerian and Angara deposits that are attributed to Cordaitales. V = Vojnovskyaceae, R = Rufloriaceae.

Organ	Permineralization	Compression/Impression Mold/Cast
Euramerian Cordaitales		
Stem	*Cordaixylon* *Mesoxylon*	*Cordaicladus*
Pith		*Artisia*
Leaf	*Cordaites*	*Cordaites*
Strobili	*Cordaitanthus* *Gothania*	*Cordaitanthus*
Ovule	*Cardiocarpus* *Mitrospermum* *Nucellangium* *Cyclospermum* *Rhabdocarpus* *Diplotesta* *Sarcotaxus* *Leptocaryum* *Gutbieri* *Taxospermum*	*Cardiocarpon* *Cordaicarpon* *Samaropsis* *Carpolithus*
Pollen/ Prepollen	*Florinites* *Felixipollinites* *Sullisaccites*	*Florinites* *Felixipollenites?* *Sullisaccites?*
Root	*Amyelon* *Stelastellara*	
Angaran Cardaitales		
Leaf		*Cordaites* (V) *Rufloria* (R) *Sparsistomites* *Papillophyllites*
Bract		*Nephropsis* (V) *Lepeophyllum* (R) *Crassinerva* (R)
Fructifications		*Vojnovskya* (V) *Krylovia* (R) *Gaussia* (R)

Organ	Permineralization	Compression/Impression Mold/Cast
		Bardocarpus (R)
		Suchoviella (R)
		Cladostrobus (R)
		Pechorostrobus (R)
		Kuznetskia (V)
		Niazonaria
Seed		*Samaropsis* (V,R)
		Tungussocarpus
		Sylvella (V)
Pollen		*Cladaitina* (R)
		Cordaitinia (R)

duced a solid parenchymatous pith. It is the chambered anatomy that results in the *Artisia* configuration in compressed and mold/cast specimens (figure 6.2D). At the apices of all stems the pith is parenchymatous (figure 6.4C, at left), with the separation of plates and the shrinkage of cells producing the diaphragms of mature specimens.

Xylem maturation is endarch in the cauline bundles, but varies from endarch to mesarch in the leaf traces. Some stems have a typical sympodial vascular architecture like that of extant conifers (e.g., *Cordaixylon dumusum*), while others have the sympodia interconnected by reparatory strands like those of some seed ferns (Beck 1970; Rothwell 1976; Trivett and Rothwell 1988). In still other species the leaf traces are not interconnected to form sympodia—i.e., *Mesoxylon sutcliffii*, *M. platypodium*, *M. priapi*, and possibly *M. thompsonii* (Whiteside 1974; Trivett and Rothwell 1985, 1988). In the latter forms the leaf traces diminish in size when followed down the stem until they no longer can be identified at the margin of the pith (Whiteside 1974; Trivett and Rothwell 1985, 1988). Therefore, there are no cauline bundles in the latter species.

Leaf traces diverge from the margin of the pith as single bundles in some species (e.g., *M. birame*), as paired bundles in other species (e.g., *M. priapi*), and as both single and paired bundles in still other species (e.g., *C. dumusum*). The bundles divide several times within the stem cortex until there are about 8 to 12 bundles at the level of leaf divergence.

Branching typically is axillary, with either one or two branches in the axil of some leaves. Vegetative branches either occur singly (Rothwell and Warner 1984) or are paired with a cone in the axil of a leaf (Trivett and Rothwell 1985). The cones of some species occur either paired or singly (Baxter 1959; Trivett and Rothwell in press), while in other species only a single cone has been found in the axil of a leaf (Trivett and Rothwell 1985). In still other species the cones do not appear to be associated with leaves at all (Grand'Eury 1877; Rothwell and Warner 1984). Because it is not known whether the latter fructifications were derived from the apical meristem or from adventitious buds, they have been termed epicormic branches.

The cortex of young stems either is characterized by alternating strands of sclerenchyma and parenchyma, like the *Sparganum* or *Dictyoxylon* cortex of seed ferns and aneurophytes, (figure 6.3A) or is relatively thin with prominent bulging leaf bases. A cork cambium was produced beneath the epidermis in several species (figure 6.3A, at top), and at proximal levels the primary cortex is replaced by a zone of periderm.

Secondary vascular tissues were produced by a bifacial vascular cambium, and a thick zone of wood developed at proximal levels of the shoots and roots. In some species the secondary xylem consists of narrow tracheids with uni- or biseriate pits and low uniseriate rays, and is quite similar to that of conifers (e.g., *C. dumusum;* see Vogellehner 1965). In other species the tracheids are somewhat larger, with multiseriate pitting; the rays are numerous and biseriate (e.g., *M. multirame*); and the wood is reminiscent of some seed ferns (e.g., *Callistophyton boyssetii*).

Roots Isolated cordaitean roots are assignable to the genus *Amyelon* Williamson (1874) but *Amyelon*-type roots are also found in strata that yield no other cordaitean remains (e.g., *A. bovius* Barnard 1962). Therefore, either *A. bovius* and similar species must be excluded from the genus (Cridland 1964) or else *Amyelon* must be treated as a form genus. Specimens of *Amyelon* are characterized by a small actinostele with two to five protoxylem strands, and there is abundant dense wood at proximal levels. At some levels there is a parenchymatous center in the roots, and some specimens of this type have been named *Stelastellara parvula* Baxter (1965). In many specimens the primary cortex is aerenchymatous. The latter features are quite variable, and do not support interpretations (e.g., Costanza 1985) that species of *Amyelon* are diagnostic of species of cordaitean plants (Cridland 1964).

Fructifications Cordaitean strboili range up to about 25 cm long, but most *Cordaitanthus* specimens do not exceed 10 cm. Strobili have a flattened primary axis upon which bracts arise from opposite sides.

When compressed, the strobili are easily recognized because none of the appendages overlie the primary axis, and they appear to be borne in a decussate arrangement (figures 6.4A, B). However, permineralized specimens demonstrate that the bracts are four-ranked (figure 6.4E; Rothwell 1977; Daghlian and Taylor 1979). In the axil of each bract is a radial secondary axis (figure 6.4B) that bore helically arranged scale-like leaves (viz., scales; figure 6.5A). Most of the scales are vegetative with a pointed tip, but some bore either terminal pollen sacs (figure 6.4B) or ovules (figure 6.5A). In *Cordaitanthus* the pollen sacs are arranged in a ring at the tip of a scale, and only a few bore mature pollen at any given time (figure 6.4D; Delevoryas 1953; Rothwell and Warner 1984). Specimens of *Gothania* typically are larger than *Cordaitanthus* and have the pollen sacs arranged in a linear file at the tip of the fertile scale. They were characterized by synchronous development of the pollen sacs such that all of the grains appeared mature at the same time (Daghlian and Taylor 1979; Trivett and Rothwell 1985).

Ovules Permineralized cordaitean ovules conform to the form-order Cardiocarpales (Seward 1917). They are symmetrically flattened (viz., have 180° symmetry; Rothwell 1986; figure 6.5), often with a cordate base and pointed apex (figures 6.5B, C). Ovules typically have a narrow micropyle (figure 6.5B) and a nucellus that is fused to the integument at the base of the ovule and free distally (figure 6.5H). The ovules are vascularized by a disc of tracheids at the base of the nucellus, and by two integumentary bundles (figure 6.5H). Species of *Cardiocarpus* and *Nucellangium* are typically oval in transverse sections (figure 6.5E), while in *Mitrospermum* there are two narrow wings at the margin of the integument (figures 6.5D, H). In *Nucellangium* (Stidd and Cosentino 1976) the nucellar tracheids extend distally to the midlevel, while in *Mitrospermum compressum* the integumentary bundles divide to form several strands within each wing (Taylor and Stewart 1964).

The pollen chamber of cordaitean ovules consists of a conical area at the tip of the nucellus that is distal to the megaspore membrane. It is surrounded by a uniseriate wall (figure 6.5F) at the apex of which the cells may be differentiated to form a nucellar beak (figure 6.5G). There is no pollen chamber floor such as that found in ovules of primitive gymnosperms with hydrasperman reproduction (Rothwell 1986).

Pollen and Prepollen The microgametophytes of cordaiteans are monosaccate (figures 6.4F, G). They had either proximal germination (i.e., trilete or monolete) and are considered to be prepollen (figure 6.4F), or they have no definite haptotypic mark (or germinal aperture) and may represent true pollen (figure 6.4G). At present, prepollen of the *Felixipollenites*-type is known from the pollen sacs of *Gothania lesliana* (Daghlian and Taylor 1979), and prepollen of the *Sullisaccites*-type has been recovered from the *Gothania*-type cones of *Mesoxylon priapi* (Trivett and Rothwell 1985). In contrast, *Florinites* grains with no haptotypic mark occur within the microsporangia of *Cordaitanthus concinnus* (Delevoryas 1953; figure 6.4G) and in the *Cordaitanthus*-type cones of *Cordaixylon dumusum* (Rothwell and Warner 1984).

REPRODUCTIVE BIOLOGY

Cordaitean fructifications are monosporangiate, like those of living conifers, but the number of cones thus far found in attachment to stems is too small to provide evidence whether the plants were monoecious or dioecious. If extant conifers may be used as an analogue, then we can expect that most cordaiteans were monoecious, whereas a few were dioecious. The large and long-lived vegetative body of cordaitean plants suggest that they were polycarpic, an interpretation that is supported by the discovery of cones of *Cordaixylon dumusum* that are attached both at the branch tips and to stems with abundant secondary tissues (Rothwell and Warner 1984). Simultaneous maturation of the pollen sacs in *Gothania lesliana* (Daghlian and Taylor 1979) and in the *Gothania*-type cones of *Mesoxylon priapi* suggests that reproduction in some plants occurred in a flush, whereas the apparently sequential maturation of pollen sacs in *Cordaitanthus concinnus* and the *Cordaitanthus*-type cones of *Cordaixylon dumusum* suggests that reproduction in other plants may have been more or less continuous.

With the exception of exine structure, microgametophytes of cordaiteans are poorly known. Cordaitean pollen (viz., *Florinites*) with internal cells was originally interpreted to have a jacket layer like the microgametophytes of heterosporous pteridophytes (Florin 1936). However, reinterpretation of the material reveals that the peripheral features are the result of folds in the exine, and that the microgametophytes produced an axial row of cells like those of modern conifers (Taylor and Millay 1979). Whether the prepollen of cordaiteans (i.e., *Felixipollinites* and *Sullisaccites*) also developed more like the microgametophytes of extant conifers than heterosporous pteridophytes remains to be discovered.

Early ovule development in cordaitean plants is known best from a combination of the *Cordaitanthus duquesnensis* cones (Rothwell 1982b) of *Cordaixylon dumusum* and the more mature, isolated ovules that are assignable to *Cardiocarpus oviformis* (Rothwell and Warner 1984). All of these organs presumably were produced by the *Cordaixylon dumusum* plant (Rothwell and Warner 1984). Later ovule development and some facets of early embryogeny have been described for *Nucellangium* (Stidd and Cosentino 1976).

The ovule originated as an ellipsoidal bulge at the tip of a fertile scale. As the scale matured and elongated, the bulge increased in size and two lateral bulges appeared. These were the primordia of the integument. Continued growth led to the integument overarching the nucellus and forming a micropyle.

Two integumentary tips protrude from the apex of small specimens with an otherwise fully formed micropyle, and are nearly identical to the micropylar arms that characterize the pollination-drop mechanism recently described for several living species of conifers (Owens, Simpson, and Molder 1981; Singh and Owens 1981; Owens and Blake 1983). In the extant species the pollination drop is suspended between the micropylar arms, and in post-pollination stages the arms either turn inward to close the micropyle or shrivel. The occurrence of similar structures on pre-pollination ovules of *Cardiocarpus oviformis* (Rothwell 1982b) and their absence from immediate post-pollination stages suggest a similar pollination mechanism in *Cordaixylon dumusum*. Also, like extant conifers, the micropyle of *Cardiocarpus oviformis* was sealed shortly after pollination.

Subsequent ovule development is comparable to that of the pteridospermous cardiocarpalean ovules *Callospermarion pusillum* and *C. undulatum* (Rothwell 1971, 1980). The ovules increased in size, and the sarcotesta of the integument became recognizable. Although only tiny ovules have thus far been found in attachment to *Cordaitanthus* scales (e.g., figure 6.5A), functional ovules must have remained attached until they had reached full size and had acquired the nutritional reserves to nourish a growing embryo (Rothwell 1982b). During this period the megagametophyte increased in size to occupy nearly the entire seed cavity, and the nucellus was reduced to a papery-thin membrane (figure 6.5H). After the ovules reached full size, fibrous cells of the sclerotesta differentiated and completed integument development (figures 6.5G, H). In the most mature ovules the megagametophyte became cellular and two ovoid archegonia were differentiated at the micropylar end. Radial alignment of certain cells of the megametophytes in *Cardiocarpus spinatus* and *C. oviformis* (Andrews and Felix 1952) demonstrates that some cordatiean megagametophytes had centripetal cellularization like those of most extant gymnosperms (Rothwell 1971, 1980; Singh 1978).

Although embryos are almost never preserved within the seeds of Paleozoic gymnosperms, the presence of a young embryo within an archegonium of *Nucellangium* (Stidd and Cosentino 1976) provides some evidence of embryogeny in cordaiteans. The embryo is ellipsoidal and cellular, and 0.2 mm long. This indicates that either a free-nuclear (coenocytic) stage like that of most extant gymnosperms was absent from cordaitean embryogeny, or that cordaitean embryos became cellular at a much earlier stage than in most other gymnosperms. Also, there is no evidence that a suspensor like that of most gymnospermous embryos was produced by *Nucellangium* The extreme rarity of cordaitean seeds with demonstrable embryos supports the interpretation that there was no period of dormancy before germination (Rothwell 1982b). Seedlings and young cordaitean sporophytes have yet to be discovered.

CORDAITEAN FOSSILS FROM THE ANGARA FLORA

Among the compressed plant fossils from Middle Carboniferous–Upper Permian Angara deposits of eastern Asia are abundant cordaitean-type leaves assigned to *Cordaites*, *Sparsistomites*, *Papillophyllites*, and *Rufloria* (figure 6.6A), as well as several genera of reproductive remains that are considered to be of cordaitean origin (table 6.1; Meyen 1982). Although placed in the Cordaitales by Meyen, the spatulate leaves are relatively small (viz., up to about 10 cm) with morphological features that are as similar to those of the extant conifers *Agathis* and *Podocarpus* as they are to Euramerian *Cordaites*. Also, features of the microsporangiate structures and most of the megasporangiate fructifications are not comparable to those of the Euramerian cordaites, and this author considers the affinities of much of the Angaran material to remain equivocal.

The fossils consist of isolated organs interpreted as parts of the same types of plants (Meyen 1984), based primarily upon their common occurrence at several localities. However, some remains have characteristic cuticular features that aid in their identification (Meyen 1982). Meyen (personal communication) considers the evidence for such associations to be strengthened if the organs that he interprets as parts of the same type of plant are structurally similar to equivalent combinations of organs from other plants. In his 1982 and 1984 treatments, Meyen recognized two families of such remains: the Vojnovskyaceae and the Rufloriaceae (table 6.1).

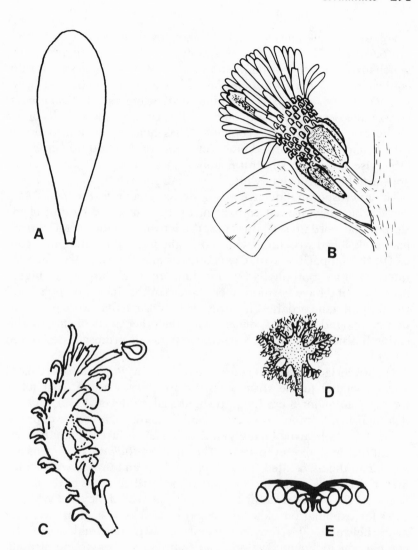

Figure 6.6. Putative cordaitean remains from late Paleozoic sediments of the Angara flora. **A** Outline of *Rufloria* leaf (Rufloriaceae); redrawn from Meyen (1982). Natural size. **B** Reconstruction of *Vojnovskya* (Vojnovskyaceae) showing secondary ovulate shoot attached to primary axis distal to a *Nephropsis* bract. (Courtesy S. V. Meyen.) **C** Line diagram of *Krylovia* (Rufloriaceae) showing ovulate sporophylls (?) diverging from axis; redrawn from Meyen (1982). **D** *Kuznetskia*, microsporangiate fertile structure (sporophyll?), assigned to Vojnovskyaceae by Meyen; redrawn from Meyen (1982). **E** Reconstruction of *Gaussia* ovulate fructification (Rufloriaceae) in mid-longitudinal section; redrawn from Meyen (1982).

Vojnovskyaceae consists of *Vojnovskya* ovulate fructifications (figure 6.6B) and *Kuznetskia* pollen-producing structures (figure 6.6D), which are associated with *Cordaites* leaves. *Vojnovskya* is structurally equivalent to *Cordaitanthus* in that it has a primary axis that bears fertile axes above the axils of *Nephropsis*-type bracts, and *Samaropsis*-type ovules occur terminally on helically arranged sporophylls of the fertile axes (figure 6.6B). *Vojnovskya* has helically arranged bracts rather than the four ranked bracts of *Cordaitanthus* and *Gothania*, but otherwise the basic structural homologies among all three genera appear to be appropriately interpreted (Meyen 1982).

The microsporangiate structures are assignable to the genus *Kuznetskia*, but they are not equivalent to the compound strobili of the Euramerian cordaites. Rather, they ("microsporoclads" of Meyen) are palmately lobed structures with sporangia borne at the margin (figure 6.6D). By comparison with *Cordaitanthus* and *Gothania* they may be interpreted as sporophylls (?), but they are much larger, are highly branched, and have far more sporangia than the linear sporophylls of the Euramerian cordaites. If *Kuznetskia* "sporophylls" were produced in a compound strobilus, evidence for the other parts of the fructification has not been found. *Kuznetzskia* pollen is protosaccate (Meyen 1982).

The Rufloriaceae consists of the spatulate leaf *Rufloria* (figure 6.6A) and associated fertile structures. *Rufloria* leaves grade from foliar forms up to about 8 cm long, to small scales (Meyen 1982), and are distinguished from Angaran *Cordaites* leaves by the presence of prominent stomata-bearing furrows on the abaxial surface ("stomatiferous dorsal furrows" of Meyen 1982). The smallest *Rufloria* specimens are similar to the associated scales *Lepeophyllum* and *Crassinerva*, which Meyen describes as having subtended the fertile regions of the fructifications. The ovulate fructifications ("polysperms" of Meyen; see Meyen 1984 for explanation of terminology) consist of a central axis that bears helically (?) arranged appendages with terminal ovules. By comparison to *Vojnovskya* and the Euramerian compound strobili, ovulate structures of Rufloriaceae can be interpreted as simple cones ("polysperms" of Meyen) consisting of a central axis to which sporophylls with terminal ovules (viz., *Samaropsis*) were attached. Compound strobili are not known for the Rufloriaceae. *Krylovia* (Figure 6.6C) has elongated sporophylls arranged along an elongated axis, while the axis of *Gaussia* (figure 6.6E) is much shorter and the sporophylls radiate outward in a single plane to give the entire fructification a peltate shape. In a third genus, *Bardocarpus*, the helically arranged sporophylls are extremely short, and the fructification appears to consist of a long axis that bears sessile ovules (Meyen 1982). Like Vojnovskyaceae, isolated Rufloriaceae ovules are assignable to *Samaropsis*.

Microsporangiate fructifications of the Rufloriaceae are equivalent to simple cones, consisting of a central axis and helically arranged microsporophylls ("microsporoclads" of Meyen). In *Pechorostrobus* a single ellipsoidal sporangium terminates each short sporophyll, but in *Cladostrobus* the sporophylls have a narrow stalk and a distal lamina. The sporangia are attached in clusters to the adaxial (and possibly lateral) surface of the stalk region. Although assigned to the Cordaitales by Meyen (1982, 1984), *Pechorostrobus* and *Cladostrobus* are simple strobili that possibly could have been produced by primitive conifers. This latter supposition is supported by the fact that *Cladaitina* microsporophylls are nearly identical to those of the most ancient walchian conifer cones (Mapes 1983) and to the Triassic voltzialean *Darneya* (Grauvogel-Stamm 1978). Pollen of the Rufloriaceae is monosaccate and assignable to *Cladaitina* and *Cordaitina*.

SYSTEMATICS AND EVOLUTION

Cordaites and conifers long have been suspected to be closely related taxa, but the closeness of their relationship was convincingly established through the exhaustive studies of Florin (1951). More recent treatments have continued to support and strengthen this interpretation (e.g., Rothwell 1982b, 1986; Crane 1985; Doyle and Donoghue 1986), but the exact nature of the relationship and possible progenitors of both are the subject of several contrasting hypotheses. Florin (1951) interpreted cordaites as possibly ancestral to conifers, while more recent treatments have either left the exact nature of the relationship equivocal or have interpreted them to be derived from a common ancestor (Rothwell 1977). In cladistic terminology cordaites and conifers may be interpreted to be closely related sister groups (Crane 1985; Doyle and Donoghue 1986).

Following the recognition of Progymnospermopsida as the most probable ancestor to gymnosperms (Beck 1960), the Archaeopteridales were interpreted to be the group from which conifers (and, by implication, cordaites) most likely were derived (Beck 1970, 1976, 1981; Meyen 1984). However, potentially significant differences in both structure and reproductive biology have been recognized between the Archaeopteridales and cordaites/conifers (Rothwell 1976, 1982b), and it now appears plausible that seed ferns may have been the ancestral group. In the latter case, the features of primitive cordaite reproductive biology are hypothesized as having arisen among the seed ferns, and the cordaites/conifers are hypothesized as having arisen through developmental changes that had a dramatic impact on the adult morphology of the vegetative and fertile shoot systems (Rothwell 1982a).

The Cordaitales clearly were one of the more significant components of the late Paleozoic land flora. Regardless of which (if any) of the hypotheses of cordaitean relationships ultimately is proven to be most accurate, the group occupies a crucial position in the systematics and phylogeny of gymnosperms. Of perhaps equal importance, cordaiteans undoubtedly played an important role in establishing and developing complex plant community structure in a variety of habitats (e.g., Phillips 1980; Raymond and Phillips 1983; Phillips, Peppers, and DiMichele 1985), and thereby served a crucial role in the evolution of terrestrial ecosystems.

LITERATURE CITED

Andrews, H. N. 1949. *Nucellangium*, a new genus of fossil seeds previously assigned to *Lepidocarpon*. *Ann. Missouri Bot. Gard.* 36:479–505.

Andrews, H. N. and C. N. Felix. 1952. The gametophyte of *Cardiocarpus spinatus* Graham. *Ann. Missouri Bot. Gard.* 39:127–135.

Arber, A. 1910. On the structure of the Paleozoic seed *Mitrospermum compressum* (Will.). *Ann. Bot.* 24:491–510.

Archangelsky, S. 1986. Late Paleozoic fossil plant assemblages of the southern hemisphere: Distribution, composition, paleoecology. In T. W. Broadhead, ed., *Land Plants: Notes for a Short Course*, organized by R. A. Gastaldo. University of Tennessee, Knoxville, Department of Geological Sciences. *Studies in Geol.* 15:128–142.

Arnold, C. A. 1967. The proper designations of the foliage and stems of the Cordaitales. *Phytomorphology* 17:346–350.

Barnard, P. D. W. 1962. Revision of the genus *Amyelon* Williamson. *Palaentology* 5:213–24.

Baxter, R. W. 1959. A new cordaitean stem with paired axillary branches. *Amer. J. Bot.* 46:163–169.

—— 1965. *Stelastellara parvula*, a new genus of unknown affinity from the American Carboniferous. *Univ. Kansas Sci. Bull.* 11:1119–1139.

Beck, C. B. 1960. The identity of *Archaeopteris* and *Callixylon*. *Brittonia* 12:351–368.

—— 1970. The appearance of gymnospermous structure. *Biol. Rev.* 45:379–400.

—— 1976. Current status of the Progymnospermopsida. *Rev. Palaeobot. Palynol.* 21:5–23.

—— 1981. *Archaeopteris* and its role in vascular plant evolution. In K. J. Niklas, eds., *Paleobotany, Paleoecology, and Evolution,* vol. 1, pp. 193–229. New York: Praeger.

Brongniart, A. 1828–1836. *Histoire des Végétaux Fossiles on Recherches Botaniques et Géologiques sur les Végétaux Renfermés dans les Diverses Couches du Globe,* vol 1. Paris: G. Durfour and E. d'Ocagne.

—— 1849. Tableau des genres de végéteaux fossiles considérés sous le point de vue de leur classification botanique et de leur distribution géologique. *Dictionnaire Univ. Histoire Nat.* 13:1–127.

—— 1874. Les graines fossiles trouvées a l'état silicifié dans le terrain houiller de Saint-Étienne. *Ann. Sci. Nat., Botanique* 20:234–265.

—— 1881. *Recherches sur les graines fossiles silicifées.* Paris: G. Masson.

Cohen, L. M and T. Delevoryas. 1959. An occurrence of *Cordaites* in the Upper Pennsylvanian of Illinois. *Amer. J. Bot.* 46:545–549.

Costanza, S. H. 1985. *Pennsylvanioxylon* of the Middle and Upper Pennsylvanian coals from the Illinois basin and its comparison with *Mesoxylon. Palaeontographica* 197B:81–121.

Crane, P. R. 1985. Phylogenetic analysis of seed plants and the origin of angiosperms. *Ann. Missouri Bot. Gard.* 72:716–793.

Cridland, A. A. 1964. *Amyelon* in American coal balls. *Palaeontology* 7:189–209.

Crookall, R. 1970. Fossil plants of the Carboniferous rocks of Great Britain. *Mem. Geol. Surv. Great Britain. Palaeontology* 4:793–840.

Daghlian, C. P. and T. N. Taylor. 1979. A new structurally preserved Pennsylvanian cordaitean pollen organ. *Amer. J. Bot.* 66:290–300.

Delevoryas, T. 1953. A new male cordaitean fructification from the Kansas Carboniferous. *Amer. J. Bot.* 40:144–150.

Doyle, J. A. and M. J. Donoghue. 1986. Relationships of angiosperms and Gnetales: A numerical cladistic analysis. In B. A. Thomas and R. A. Spicer, eds., *Systematic and Taxonomic Approaches in Palaeobotany,* pp. 177–198. Oxford: Oxford University Press.

Feistmantel, O. 1876. Versteinerungen der böhmischen Kohlenablagerungen. *Palaeontographica* 23:223–316.

Felix, G. 1882. Über die Versteinerten Hölzer von Frankenberg. *Sitzengsberichte naturforschenden Ges. Leipzig Neunte Jahrgang* 1882:5–9.

Florin, R. 1936. On the structure of the pollen-grains in the Cordaitales. *Svensk Bot. Tidskr.* 30:624–651.

—— 1951. Evolution in cordaites and conifers. *Acta Horti Bergiani* 15:285–389.

Geinitz, H. B. 1962. Dyas oder die Zechsteinformation und das Rothliegende. *Die Pflanzen der Dyas und Geologisches,* Band 2. Leipzig.

Goeppert, H. R. 1964. Die fossil Flora der permischen Formation. *Palaeontographica* 12:1–224.

Grand'Eury, F. C 1877. Flore carbonifère du Département de la Loire et du centre de la France. *Acad. sci. Inst. France Mém.* 24:1–624.

Grauvogel-Stamm, L. 1978. La flore du Grès à Voltzia (Buntsandstein Supérieur) des Vosges du Nord (France): Morphologie, anatomie, interprétations phylogénetique et paléogéographique. *Université Louis Pasteur de Strasbourg, Institut de Géologie, Memoire.* 50:1–225.

Hirmer, M. 1933. Zur Kenntnis dur strukturbietenden Pflanzenreste des jüngreren Paläozoikums. *Palaeontographica* 77B:121–140.

Long, A. G. 1979. Observations on the Lower Carboniferous genus *Pitus* Witham. *Trans. Roy. Soc. Edinburgh* 70:111–127.

Mapes, G. 1983. Permineralized *Lebachia* pollen cones. *Amer. J. Bot.* 70:74. (Abstracts.)

Maslen, A. J. 1930. The structure of *Mesoxylon platypodium* and *Mesoxyloides. Ann. Bot.* 44:503–533.

Meyen, S. V. 1982. Fructifications of Upper Paleozoic Cordaitanthales from the Angara Region. *Paleont. Jour.* 2:107–119.

—— 1984. Basic features of gymnosperm systematics and phylogeny as evidenced by the fossil record. *Bot. Rev.* 50:1–112.

Millay, M. A., and T. N. Taylor. 1974. Morphological studies of Paleozoic saccate pollen. *Palaeontographica* 147B:75–99.

Owens, J. N., and M. D. Blake. 1983. Pollen morphology and development of the pollination mechanism in *Tsuga heterophylla* and *T. mertensiana. Can. J. Bot.* 12:3041–3048.

Owens, J. N., S. J. Simpson, and M. Molder. 1981. Sexual reproduction of *Pinus contorta.* 1. Pollen development, the pollination mechanism, and early ovule development. *Can. J. Bot.* 59:1828–1843.

Pant, D. D. 1982. The Lower Gondwana gymnosperms and their relationships. *Rev. Palaeobot. Palynol.* 37:55–70.

Phillips, T. L. 1980. Stratigraphic and geographic occurrences of permineralized coal-swamp plants—Upper Carboniferous of North America and Europe. In D. L. Dilcher and T. N. Taylor, eds., *Biostratigraphy of Fossil Plants,* pp. 25–92. Stroudsburg, Pa.: Dowden, Hutchinson & Ross.

Phillips, T. L., R. A. Peppers, and W. A. DiMichele. 1985. Stratigraphic and interregional changes in Pennsylvanian coal-swamp vegetation: Environmental inferences. *International J. Coal Geol.* 5:43–109.

Raymond, A. and T. L. Phillips. 1983. Evidence for an Upper Carboniferous mangrove community. In H. J. Teas, ed., *Tasks for Vegetation Science,* pp. 19–30. The Hague: W. Junk.

Rothwell, G. W. 1971. Ontogeny of the Paleozoic ovule, *Callospermarion pusillum. Amer. J. Bot.* 58:706–715.

—— 1976. Primary vasculature and gymnosperm systematics. *Rev. Palaeobot. Palynol.* 22:193–206.

—— 1977. The primary vasculature of *Cordaianthus concinnus. Amer. J. Bot.* 64:1235–1241.

Rothwell, G. W. 1980. The Callistophytaceae (Pteridospermopsida). 2. Reproductive features. *Palaeontographica* 173B:85–106.

—— 1982a. New interpretations of the earliest conifers. *Rev. Palaeontol. Palynol.* 37:7–28.

—— 1982b. *Cordaianthus duquesnensis* sp. nov., anatomicaly preserved ovulate cones from the Upper Pennsylvanian of Ohio. *Amer. J. Bot.* 69:239–247.

—— 1986. Classifying the earliest gymnosperms. In B. A. Thomas and R. A. Spicer, eds., *Systematic and Taxonomic Approaches in Palaeobotany*. pp. 137–161. Oxford: Oxford University Press.

Rothwell, G. W. and S. Warner, 1984. *Cordaixylon dumusum* n. sp. (Coradaitales). 1. Vegetative structures. *Bot. Gaz.* 145:275–291.

Scott, A. C. 1977. A review of the ecology of Upper Carboniferous plant assemblages, with new data from Strathclyde. *Palaeontology* 20:447–473.

Scott, D. H. 1919. On the fertile shoots of *Mesoxylon* and an allied genus. *Ann. Bot.* 33:1–21.

Scott, D. H. and A. J. Maslin. 1910. On *Mesoxylon*, a new genus of Cordaitales —preliminary note. *Ann. Bot.* 24:236–239.

Seward, A. C. 1917. *Fossil Plants*, vol. 3. Cambridge: Cambridge University Press.

Singh, H. 1978. *Embryology of gymnosperms*. Berlin: Gebruder Borntraeger.

Singh, H. and J. N. Owens. 1981. Sexual reproduction of Engelmann spruce *(Picea engelmannii)*. *Can. J. Bot.* 59:793–810.

Sternberg, G. K. 1825. *Versuch einer geognostischen botanischen Darstellung der Flora der Vorwelt*, vol. 1, part 4: Leipzig and Prague.

—— 1838. *Versuch einer geognostischen botanischen Darstellung der Flora der Vorwelt*, vol. 1, parts 7, 8:81–220. Leipzig and Prague.

Stidd, B. M. and K. Cosentino. 1976. *Nucellangium:* Gametophytic structure and relationship to *Cordaites. Bot. Gaz.* 137:242–249.

Taylor, T. N. and M. A. Millay. 1979. Pollination biology and reproduction in early seed plants. *Rev. Palaeobot. Palynol.* 27:329–355.

Taylor, T. N. and W. N. Stewart. 1964. The Paleozoic seed *Mitrospermum* in American coal balls. *Palaeontographica* 115B:51–58.

Trivett, M. L. and G. W. Rothwell. 1985. Morphology, systematics and paleoecology of Paleozoic fossil plants: *Mesoxylon priapi*, sp. nov. (Cordaitales). *Syst. Bot.* 10:205–223.

—— 1988. Diversity among Paleozoic Cordaitales: The vascular architecture of *Mesoxylon birame* Baxter. *Bot. Gaz.* 149:116–125.

Unger, F. 1950. *Genera et Species Planatarium Fossilium*. Vienna.

Vogellehner, D. 1965. Untersuchungen zur Anatomie und Systematik der verkieselten Kölzer aus dem frankischen und sudthüringischen Keuper. *Erlanger geol. Abh.* 59:1–76.

Voss, E. G. et al. 1983. *International Code of Botanical Nomenclature*. Utrecht: Bohn, Scheltema and Holkema.

Whiteside, K. L. 1974. *Petrified cordaitean stems from North America*. Ph.D. dissertation, University of Iowa.

7

Morphology and Phylogeny of Paleozoic Conifers

JOHANNA A. CLEMENT-WESTERHOF

⊞

$\mathbf{A}$mong the gymnosperms, the conifers represent a puzzling and much discussed group, especially with respect to the structure of the reproductive organs. Whereas the pollinferous conifer cone is generally regarded as a simple strobilus, the organization of the ovuliferous cones has given rise to controversies: the conifer ovuliferous cone has been interpreted as a flower (e.g., Pilger 1926) or an inflorescence (e.g., Celakovsky 1890; Hagerup 1933). However, Florin's detailed study on Paleozoic conifers (1938–1945) demonstrated that the ovuliferous cones of most fossil and recent conifers can be derived from a compound strobilus, consisting of a cone axis bearing spirally arranged bract/fertile dwarf-shoot complexes. These were well exemplified in the Walchiaceae (Lebachiaceae; cf. Florin). As in the Cordaitales, Florin interpreted the walchiaceous ovules as terminally borne on stalk-like fertile scales. The polliniferous cones in the earliest conifers were found to be nearly modern in appearance.

Florin considered the ovuliferous dwarf-shoot of the Walchiaceae to be structurally homologous with that of the Cordaitales, and consequently interpreted the ovuliferous fructifications of this family as transitional between those of the Cordaitales and younger conifers (which also were interpreted to have "stalked ovules").

Florin's concept was generally accepted and his interpretation of the coniferous ovuliferous dwarf-shoots was not challenged until Schweitzer (1963) proved that at least in *Pseudovoltzia*—regarded by Florin as transitional between the Walchiaceae and Triassic conifers —the ovules were directly attached on foliaceous fertile scales and not terminally borne on stalk-like megasporophylls. Schweitzer rejected the concept of "stalked ovules" also for younger conifers, but not with respect to the Walchiaceae.

Although a few other specialists (e.g., Harris 1976) mildly questioned his observations, until now Florin's concepts have found acceptance in most botanical and paleobotanical textbooks.

Especially during the last few years, several fundamental investigations of Paleozoic conifers and conifer-like taxa have been initiated. Apart from new anatomical studies, the study of cuticles has significantly contributed to better insight into the organization of ovuliferous dwarf-shoots. Some authors still follow Florin's views with respect to the Walchiaceae; others have minor or major objections. In any case the current studies reveal a much greater diversity in reproductive morphology of Paleozoic conifers (and conifer-like taxa) than envisaged by Florin. In most phylogenetic analyses of conifers, the Walchiaceae, in Florin's concept, still play an important role. However, the concept of this family has recently been radically revised (Clement-Westerhof 1984; Visscher et al. 1986; Kerp et al., in press) by emphasizing the presence of an abaxial, rather than terminal, ovule attachment.

The objectives of the present paper are the following.

(1) To present a survey of coniferous and conifer-like taxa from Euramerian, Angara, and Gondwana floras, with special reference to their ovuliferous fructifications. Useful information about Cathysian conifers was not available. The following taxa have been selected: the families Walchiaceae, Majonicaceae, Ullmanniaceae, Buriadiaceae, and Feru gliocladaceae, the natural genera *Sashinia* and *Kungurodendron*, and the species *"Lebachia" lockardii*, as well as the form-genera *Thu ringiostrobus* and *Moyliostrobus*. The descriptions of these taxa are based on original (some not yet formally published) diagnoses or (re)descriptions. Each taxon will be briefly discussed. (2) To provide a historical survey of the Walchiaceae with respect to the different

morphological, taxonomic, nomenclatural, and phylogenetic interpretations of the family and its constituents. (3) To offer a comparison of the Cordaitales and other conifer and conifer-like taxa with the Walchiaceae in order to estimate their relationship with this family in its revised concept. (4) To provide a general discussion of conifer phylogeny based on the presently available information on ovuliferous fructifications of Paleozoic conifers, together with an attempt to define the term "conifer" in such a way that both extinct and extant representatives can be included.

TERMINOLOGY

Since there is great variety in the descriptive terminology, the present author decided to apply the following terminology in all descriptions in order to achieve a measure of uniformity.

Terminology of elements of the ovuliferous cone:

Bract: A foliar appendage of the cone axis.

Ovuliferous dwarf-shoot: A lateral branch of limited growth, arising in the bract axil.

Scale: A foliar appendage of an ovuliferous dwarf-shoot; scales can be divided into (a) sterile scales and (b) fertile scales (ovule-bearing foliar appendages).

Ovule/seed: Often it is not possible to distinguish with certainty between ovules and seeds. For practical purposes it is considered that specimens found dispersed represent seeds. Another problem with respect to the seeds is the angiosperm-centered terminology. Although coniferous affinity of the seeds described is sometimes disputable, they certainly do not belong to angiosperms. Therefore, terms such as *anatropous* (often erroneously used for *inverted*) and *atropous* cannot be applied. The present author suggests the following terminology for the gymnospermous seed:

Seed axis: Line between abcission area, chalaza, and micropyle, when this line coincides with the longitudinal axis of the nucellus, the seed axis is *straight;* otherwise the seed axis is *bent.* These terms refer to the seed itself. Related to the position of the ovule/seed are the terms *erect,* used when the micropyle is directed away from the cone axis, and *inverted,* used when the micropyle is pointing toward the cone axis.

Terminology of cuticle structures:

Stoma: Stomatal aperture together with guard cells.

Subsidiary cells: Cells enclosing the stoma.

Stomatal complex: Stoma together with subsidiary cells.

Dicyclic stomatal complex: Stoma enclosed by two rings of cells; the inner ring is formed by the subsidiary cells, the outer ring by the encircling cells.

EURAMERIAN TAXA

Family Walchiaceae (Goppert 1865) Schimper 1870

Type-genus: *Walchia* Sternberg 1825

Description (based on the emended diagnosis of Kerp et al., in press): Presumed main axis leafy when young. Pinnately branched lateral shoot systems, consisting of penultimate branches with two lateral series of parallel ultimate branches, situated in one plane. Leaves bifacial, spirally arranged; heterophylly may occur. Ovuliferous cone compound; cone axis bearing spirally arranged bracts with ovuliferous dwarf-shoots freely arising in their axils. Ovuliferous dwarf-shoots bilaterally symmetrical, slightly flattened, provided with a number of sterile scales (on the order of 10–30): number of fertile scales reduced (1–3) and more or less adaxially situated. Fertile scales straight, bearing abaxially (in reference to dwarf-shoot axis) a single inverted ovule; position of attachment of ovules ranging from subapical to central. Fertile and sterile scales only slightly connate. Ovule/seed bilaterally symmetrical, platyspermic, seed axis bent (figure 7.1 c, d). Nucellus free to its base from the single integument. Pollen/archegonial chamber present; megaspor membrane cutinized. Polliniferous cone simple, cone axis bearing spirally arranged hyposporangiate microsporophylls. Prepollen circular to elliptical in polar view; monosaccoid. Epidermis of leaves, bracts, scales, and microsporophylls amphistomatic; stomata situated in longitudinal bands or rows. Stomatal complexes mono- to incompletely tricyclic; 4–8 subsidiary cells, sometimes bearing papillae. Hair bases present. Papillae on epidermal cells may occur.

Remarks: Natural genera included in the Walchiaceae are *Walchia*, *Ernestiodendron*, *Otovicia*, and *Ortiseia* (for descriptions supplementary to the family description, see below). Form-genera within the family are *Culmitzschia* for vegetative material with epidermal structure, *Walchiostrobus* for ovuliferous fructifications with 1–3 fertile scales per dwarf-shoot (figure 7.2d; see Kerp and Clement-Westerhof, in press), and *Walchianthus* for polliniferous cones with epidermal structure (see Visscher et al. 1986).

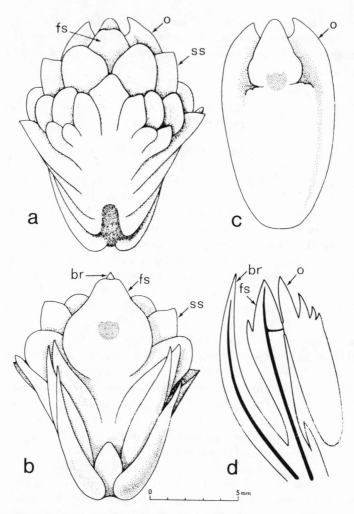

Figure 7.1. *Ortiseia jonkeri:* Reconstruction of bract, ovuliferous dwarf-shoot, and ovule. Here and in all subsequent figures, br=bract, fs=fertile scale, ss=sterile scale, o=ovule. **a** Dwarf-shoot with ovule, abaxial view. **b** Bract and dwarf-shoot, adaxial view. **c** Ovule, obverse view. **d** Tentative longitudinal section of bract, dwarf-shoot and ovule, showing vascularization. (**a, b, c** reproduced from Clement-Westerhof 1984.)

Figure 7.2. a–c: *Otovicia hypnoides.* **a** Lateral shoot system. × 1. **b** Ovuliferous dwarf-shoot, adaxial view. × 6. **c** Seed. × 5. **d** *Walchiostrobus* sp. Ovuliferous dwarf-shoot, adaxial view; note bilobate apex of fertile scale. × 3.7. (**a, b, c** reproduced from Kerp et al., in press.)

Walchia Sternberg 1825

    Status: Natural genus within the Walchiaceae

    Type-species: *Walchia piniformis* Sternberg 1825, emend. Clement-Westerhof 1984; Upper Carboniferous–Lower Permian, Europe. Also recorded from Canada and China. Other species recognized: *W. garnettensis*, *W. goeppertiana*. For detailed descriptions of the species, see Forin (1938, 1939).

    Supplementary description: Lateral shoot systems arranged in false

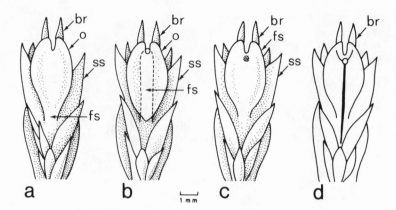

Figure 7.3. *Walchia piniformis:* Reconstruction of bract, ovuliferous dwarf-shoot, and ovule, adaxial view. **a** Interpretation of Florin (1938). **b** Interpretation of Mapes and Rothwell (1984). **c** Interpretation of Clement-Westerhof (1984). **d** Tentative reconstruction of vascularization of fertile scale. (**a** and **c** after Clement-Westerhof 1984.)

whorls on the main axis. Leaves narrow triangular or lanceolate, decurrent on the axis. Heterophylly may occur; leaves on main axis and penultimate axes sometimes bifurcate, on ultimate axes entire, with acute apex. Ovuliferous cones approximately 7 cm long, 1.5 cm wide; cone axis bearing bifurcate bracts; ovuliferous dwarf-shoots (figure 7.3c) about 8 mm long, not reaching apex of subtending bracts, provided with a large number (+ 14 in *W. piniformis*) of sterile scales and a single fertile scale. Sterile scales varying from lanceolate to narrow subtriangular, with acute apex. Fertile scale obovate with bilobate apex, projecting beyond the sterile scales and bearing a single inverted ovule subapically on its abaxial surface. Detailed organization of ovule unknown. Prepollen probably assigned to *Potonieisporites* Bhardwaj 1954. Leaves and bracts unequally amphistomatic, with most stomata on adaxial surface; stomata arranged in longitudinal bands on both sides of a stomata-free median zone. Stomatal complexes mono- to incompletely dicyclic. Hair bases abundant.

Remarks: Florin's diagnosis of *Walchia* (1938) has been emended with respect to the oviliferous cones and pollen grains (Clement-Westerhof 1984). The bilobate structure in the ovuliferous dwarf-shoot, formerly interpreted as a terminal ovule by Florin (figure 7.3a), represents a fertile scale (figure 7.3c, d). Also the radial arrangement of the scales as described by Florin cannot be convincingly corroborated from his unretouched illustrations.

Mapes and Rothwell (1984) also reinterpreted the ovuliferous dwarf-shoot of *W. piniformis*. They believe the bilobate structure represents an inverted ovule, terminally attached to a fertile scale (figure 7.3b). Such an organization would require a protruding fertile scale, visible in abáxial view of the dwarf-shoot; this has not so far been observed. However, the assumption of a bilateral symmetry rather than a radial symmetry with respect to the dwarf-shoot of *Walchia* is shared by the present author.

Ernestiodendron Florin 1934

Status: Natural genus within the Walchiaceae

Type-species: *Ernestiodendron filiciforme* (Sternberg) Florin 1934, emend. Clement-Westerhof 1984: Upper Carboniferous–Lower Permian, Europe. Further mentioned from the Untied States. For detailed description of the species, see Florin (1939).

Supplementary description: Leaves entire, falcate, narrow triangular or lanceolate, not decurrent on the axis, apex acute. Ovuliferous cones lax, approximately 10 cm long, 2.5 cm wide; cone axis bearing entire bracts; ovuliferous dwarf-shoots projecting beyond apex of subtending bracts, approximately 14 mm long, provided with a number of sterile scales and a single fertile scale. Sterile scales narrow, subtriangular, with acute apex; fertile scales obovate with obtuse apex, projecting far beyond the sterile scales and bearing a single inverted ovule subapically on the abaxial surface. Detailed organization of ovule unknown. Prepollen probably assignable to *Potonieisporites* Bhardwaj 1954. Leaves and bracts amphistomatic; stomata arranged in single or, sometimes over a short distance, double longitudinal rows. Stomatal complexes mono- to incompletely dicyclic. Hair bases abundant.

Remarks: As in *Walchia*, the author suggests the presence in *Ernestiodendron* of a fertile scale with abaxial, subapical ovule attachment instead of a "stalked" ovule. Moreover, the author believes there is only a single ovule per dwarf-shoot, in contrast to Florin's interpretation of 3–5 ovules.

Otovicia Kerp et al., in press

Status: Natural genus within the Walchiaceae

Type-species: *Otovicia hypnoides* (Brongniart) Kerp et al. (in press); Upper Carboniferous–Lower Permian, Europe. For detailed descriptions, see Kerp et al. (in press).

Figure 7.4. a–b: *Ortiseia jonkeri.* **a** Ovuliferous dwarf-shoot, abaxial view. × 8. **b** Ovule, reverse view. × 7. **c–e:** *Majonica alpina.* **c** Shoot. × 4. **d** Dwarf-shoot with subtending bract, abaxial view. × 6. **e** Seed, lateral view. × 3. **f** *Pseudovoltzia liebeana.* Dwarf-shoot, adaxial view; note slit-shaped sites of ovule attachment on lateral fertile scales. × 4.

Supplementary description: Leaves narrow, subtriangular, oblong or lanceolate, decurrent on the axis. Heterophylly may occur; leaves on main axis and penultimate axis sometimes bifurcate, on ultimate axes entire with acute to slightly obtuse apex (figure 7.2a). Ovuliferous cones approximately 7 cm long, 1.5 cm wide; cone axis bearing bifurcate bracts; ovuliferous dwarf-shoots (figure 7.2b) approximately 8 mm long, not reaching apex of subtending bracts, provided with a great number (approximately 12 in *O. hypnoides*) of sterile scales and 2 fertile scales. Sterile scales lanceolate, narrow, subtriangular, with acute apex. Fertile scales resembling sterile scales. Ovule/seed ovate, up to 7 mm long; integument extending into wing-like seam, broad, sometimes cordate at base, tapering to micropylar area and there forming two narrow extensions. Micropylar area acute (figure 7.2c). Prepollen assignable to *Potonieisporites* Bhardwaj 1954. Leaves and bracts unequally amphistomatic, with most stomata on adaxial surface; stomata in longitudinal bands on both sides of a stomata-free median zone. Stomatal complexes mono- to incompletely dicyclic. Hair bases abundant.

Remarks: The establishment of new genus *Otovicia* is based on a reinterpretation of Florin's material, and new coniferous remains collected from Oberhausen (southwest Germany). The vegetative remains agree with the material formerly ascribed to *Walchia hypnoides*. The ovuliferous fructifications are characterized by two fertile scales per dwarf-shoot, whereas *Walchia* has only a single fertile scale.

Ortiseia Florin 1964

Status: Natural genus within the Walchiaceae

Type-species: *Ortiseia leonardii* Florin 1964, emend. Clement-Westerhof 1984; Upper Permian, Italy. Other species recognized: *O. visscheri*, *O. jonkeri*. For detailed descriptions, see Clement-Westerhof (1984).

Supplementary description: Leaves entire, triangular, ovate, or lanceolate, not decurrent on the axis; apex acute or obtuse. Ovuliferous cones approximately 6 cm long, 2 cm wide; cone axis bearing entire bracts with acute apex. Ovuliferous dwarf-shoots (figures 7.1, 7.4a) not reaching apex of subtending bracts or equaling their length, approximately 15 mm long, provided with a large number of sterile scales (20–30) and a single fertile scale. Sterile scales ranging from lanceolate to obovate; apices acute to obtuse. Fertile scale rhomboid, with an obtuse apex (in *O. jonkeri* sometimes slightly bilobate) slightly projecting beyond the sterile scales and bearing a single

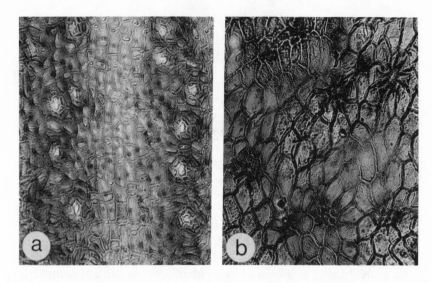

Figure 7.5. Epidermal structure of abaxial leaf surface. **a** *Ortiseia leonardii;* note hair bases. × 150. **b** *Dolomitia cittertiae.* × 120.

inverted ovule in the middle of its abaxial surface. Ovule/seed (figures 7.1, 7.4b) ovate, sometimes elliptical, up to 15 mm long. Basal part of integument extending into two lateral horn-shaped projections and one triangular or rotund differentiation, bordering the abcission area; on the reverse side occurs a transverse line of small protrusions. Micropylar area obtuse. Prepollen assignable to *Nusckoisporites* Potonie et Klaus 1954. Leaves (figure 7.5a) and bracts amphistomatic; stomata generally arranged in longitudinal rows, sometimes doubled over some distance. Stomatal complexes di- to incompletely tricyclic; subsidiary cells surrounded by a similar number of encircling cells. Hair bases abundant.

Remarks: Especially *O. jonkeri* shows an overall similarity to *Walchia* and *Ernestiodendron*. Vegetative remains from the Alpine Permian, generally identified as *Walchia*, probably belong to this species of *Ortiseia*. The epidermal structure of *Ortiseia* resembles that of *Ernestiodendron* in the arrangement of the stomata.

Discussion of the Walchiaceae.   A historical survey is presented here of the different morphological, taxonomic, nomenclatural, and phylogenetic interpretations of the Walchiaceae, with emphasis on the ovuliferous fructifications.

(1) Florin (1945) established the family Lebachiaceae, comprising

the natural genera *Lebachia* Florin 1938 and *Ernestiodendron* Florin 1934. He interpreted *Lebachia* to have ovuliferous dwarf-shoots characterized by many sterile scales and a single fertile scale bearing terminally an erect ovule, and *Ernestiodendron* to possess dwarf-shoots characterized by a strongly varying number (0–30) of sterile scales and 3–5 fertile scales, each bearing terminally an erect of inverted ovule. The form-genus *Walchiostrobus* Florin 1940 was established in order to accommodate ovuliferous fructifications agreeing with those of *Lebachia* of *Ernestiodendron* but not assignable on a species level. The family was thought to represent a transitional stage between Cordaitales and late Permian–Triassic conifers. Conifers exhibiting an "ovuliferous scale" (consisting of more or less fused sterile and fertile scales) could be derived from the *Lebachia* type and the Cephalotaxaceae from the *Ernestiodendron* type.

(a) Authors who follow Florin's taxonomic, nomenclatural, morphological, and phylogenetic interpretations are, e.g., Wilde (1944), Darrah (1960), and Taylor (1981). Others agree in general with Florin's interpretations but prefer acceptance of a close phylogenetic relationship between Cordaitales and Lebachiaceae, rather than regarding the Cordaitales as ancestral stock, e.g., Rothwell (1982), Mapes and Rothwell (1984), and Crane (1985).

(b) Authors agreeing with Florin's interpretations and following his hypothesis with the exception of a close phylogenetic relationship with the Cordaitales are Pant and Nautiyal (1967) and Beck (1981). Pant and Nautiyal suggest a *Buriadia*-like ancestor for the Lebachiaceae. Beck derives both the Cordaitales and the Lebachiaceae from a common *Archeopteris*-like ancestor; this cannot be considered as a close phylogenetic relationship.

(c) Authors agreeing with Florin's interpretations and following his hypothesis with the exception of the presumed lebachiaceous ancestry of younger conifers are Schweitzer (1963) and Krassilov (1971). Schweitzer rejects the presence of "stalked" ovules in *Pseudovoltzia* and derives most conifers via *Pseudovoltzia* from the Lycopsida, rather than via the Lebachiaceae from the Cordaitales; only the Cephalotaxaceae could be derived from the *Ernestiodendron* type, as suggested by Florin. Krassilov considers the Lebachiaceae as a specialized group, closer to the Cordaitales than to the conifers, and questions a Lebachiacean ancestry of conifers.

(2) Meyen (1984) accepts Florin's taxonomic concepts, nomenclature, and interpretation of the *Lebachia* type. However, he challenges the presence of "stalked" ovules in the *Ernestiodendron* type and interprets these structures as fertile scales, showing places of ovule attach-

ment on their abaxial surface. He regards such an organization as representing a transitional stage between the organization of dwarf-shoots of *Sashinia*-like ancestor and that of younger conifers such as *Pseudovoltzia*. He believes *Lebachia* could be derived from the Cordaitales and could occupy a peripheral position in conifer evolution.

(3) Clement-Westerhof (1984) does not agree with Florin with respect to taxonomic, nomenclatural, morphological, and phylogenetic interpretations. She showed that the name *Lebachia* is invalid, and reintroduced the name *Walchia* Sternberg for remains formerly ascribed to *Lebachia*. Consequently the family name Walchiaceae (Goppert) Schimper was reintroduced. Zimmermann (1959) also does not accept Florin's nomenclatoral approach. Instead of Lebachiaceae he introduced the name Walchiaceae Zimmermann, based on the erroneous view that the name *Ernestiodendron* represents a synonym of *Walchia;* Zimmermann's view is followed by Remy and Remy (1977). As a consequence of her detailed study of the morphology of *Ortiseia*, Clement-Westerhof interprets both *Walchia* and *Ernestiodendron* to have a single fertile scale per dwarf-shoot, characterized by abaxial ovule attachment. The family Walchiaceae is emended to comprise only genera with a single fertile scale per dwarf-shoot: *Walchia*, *Ernestiodendron*, and *Ortiseia*. Dwarf-shoots with several ovules (Florin's *Ernestiodendron* type) are also interpreted as having fertile scales characterized by abaxial ovule attachment. However, they are not considered to be assignable to the Walchiaceae.

Cordaitalean ancestry of the Walchiaceae is improbable because of the marked difference in ovule attachment in these groups. With respect to possible descendants from the Walchiaceae, it is suggested that no taxon can be directly derived from this family.

The diagnosis of *Walchiostrobus* Florin 1940 is emended by Visscher et al. (1986) to constitute a form-genus for fructifications showing walchiaceous affinity and provided with a single fertile scale per dwarf-shoot.

(4) Kerp et al. (in press) have recently demonstrated an association of walchiaceous vegetative material with ovuliferous fructifications, characterized by two fertile scales per dwarf-shoot *(Otovicia)*, thus necessitating a revision of the diagnosis of the family. The diagnosis of *Walchiostrobus* (Visscher et al. 1986) is emended in order to accommodate ovuliferous fructifications with 1–3 fertile scales and several scales (see Kerp and Clement-Westerhof, in press). Ovuliferous fructifications characterized by at least four fertile scales (as represented by Florin's *Ernestiodendron* type within his concept of *Walchiostrobus*) are assigned to *Thuringiostrobus* Kerp et al. (in press), a new form-

genus for ovuliferous fructifications not assignable to a natural family (for description, see below, and figure 7.10).

Family Majonicaceae Clement-Westerhof 1987
Type-genus: *Majonica* Clement-Westerhof 1987
Description: Shoots bearing spirally arranged, entire, bifacial leaves; heterophylly may occur. Ovuliferous cones compound; cone axis bearing spirally arranged, entire bracts with ovuliferous dwarf-shoots arising in bract axils. Dwarf-shoots free from or partially fused with bracts. Ovuliferous dwarf-shoots bilaterally symmetrical and provided with 1–15 sterile scales and 2–3 fertile scales, each bearing basally a single inverted ovule on its abaxial surface (i.e., abaxial in reference to dwarf-shoot axis). Two fertile scales lateral, emerging, and recurved; the third, if present, emerging medially on the adaxial side of the dwarf-shoot and often provided with a pro-tuberance (figures 7.7b, 7.8a, b) distally bordering the position of ovule attachment. Ovule attachment on lateral fertile scales explic-itly laterally situated or more or less shifted to the adaxial side of the dwarf-shoot. Fertile and sterile scales moderately to consider-ably connate. Basal part of dwarf-shoot stalk-like. Ovule/seed bilat-erally symmetrical; seed axis straight (figures 7.6d, 7.8b, c). Nucel-lus ovate, (nearly) free to its base from the integument. Pollen/archegonial chamber present; megaspore membrane cutinized. Pol-liniferous cone simple; cone axis bearing spirally arranged micros-porophylls. Pollen bisaccoid. Epidermis of leaves, bracts, scales, and microsporophylls amphistomatic; stomata arranged in more or less conspicuous rows or scattered. Stomatal complexes mono-cyclic, provided with 5–10 papillate subsidiary cells. Papillae may occur on epidermal cells.
Remarks: Natural genera included in the Majonicaceae are *Majon-ica*, *Dolomitia*, and *Pseudovoltzia* (for descriptions supplementary to the family description, see below).

Majonica Clement-Westerhof 1987
Status: Natural genus with the Majonicaceae
Type-species: *Majonica palina* Clement-Westerhof 1987; Upper Per-mian, Europe. For detailed descriptions, see Clement-Westerhof (1987).
Supplementary description: Leaves ovate or lanceolate, with mostly obtuse apices; leaf bases decurrent on the axis; leaves varying in shape and dimensions, indicating heterophylly (figure 7.4c). Bracts with acute apices and obviously sinuate margins. Ovuliferous dwarf-

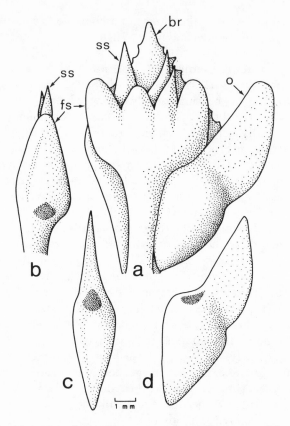

Figure 7.6. *Majonica alpina:* Reconstruction of bract, ovuliferous dwarf-shoot, and ovule. **a** Bract, ovuliferous dwarf-shoot, and ovule, adaxial view. **b** Dwarf-shoot, lateral view. **c** Ovule, obverse view. **d** Ovule obverse/oblique view. (Reproduced from Clement-Westerhof 1987.)

shoots (figures 7.4d, 7.6) arising in bract-axils, almost free from bracts, mostly projecting beyond the bracts, approximately 10 mm long, and showing two planes of symmetry. Dwarf-shoots bearing 1–5 sterile scales forming median part, and two laterally emerging fertile scales forming the blanks. Sterile scales narrow, subtriangular, with acuminate apices; fertile scales ovate with obtuse apices. Sites of ovule attachment distinctly lateral; rarely a third central site of ovule attachment is present on the adaxial side of the dwarf-shoot. Fertile and sterile scales extensively connate. Ovule/seed (figures 7.4e, 7.6) up to 15 mm long, triangular in cross section. Basal part of integument extending into a flattened wing-like differentia-

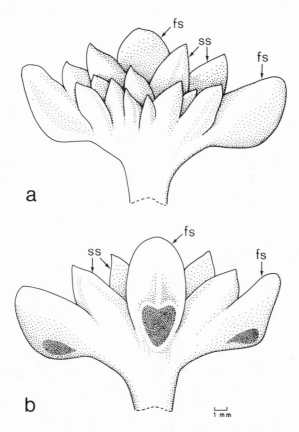

Figure 7.7. *Dolomitia cittertiae:* Reconstruction of ovuliferous dwarf-shoot. **a** Abaxial view. **b** Adaxial view. (Reproduced from Clement-Westerhof 1987.)

tion perpendicular to and in direct line with the abscission area. Micropylar area acute. Specimens mostly laterally compressed, with a boomerang-like outline. Microsporophylls with distal part showing sinuate margin; pollen bisaccoid and bitaeniate. Stomata arranged in more or less irregular rows.

Remarks: So far as is known, the occurrence of *Majonica* is limited to the Alpine Upper Permian, where this genus represents an important component of the floras.

Dolomitia Clement-Westerhof 1987
Status: Natural genus within the Majonicaceae
Type-species: *Dolomitia cittertiae* Clement-Westerhof 1987; Upper

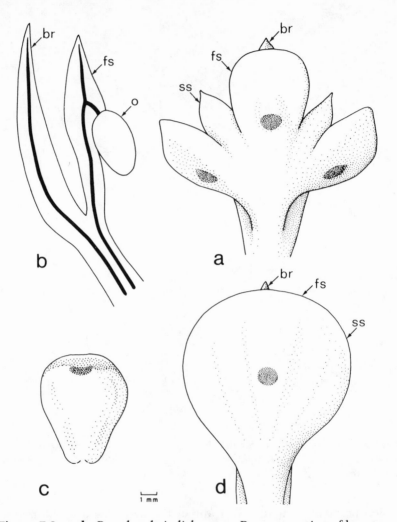

Figure 7.8. a–b: *Pseudovoltzia liebeana.* **a** Reconstruction of bract and ovuliferous dwarf-shoot, adaxial view. **b** Longitudinal section of bract, dwarf-shoot, and ovule, showing vascularization. **c** *Pseudovoltzia sjerpii:* Seed. **d** *Ullmannia:* Reconstruction of bract and dwarf-shoot. (**b** after Schweitzer 1963; **d** reinterpreted after Florin 1944.)

Permian, Italy. For detailed descriptions, see Clement-Westerhof (1987).

Supplementary description: leaves oblong or lanceolate, with obtuse apex, decurrent on the axis; variation in shape and dimension of leaves indicating heterophylly. Shape and dimensions of bracts and ovules unknown. Ovuliferous dwarf-shoot (figure 7.7) bearing approximately 15 sterile scales and 3 fertile scales. Sterile scales subtriangular, with acuminate or acute apices. Fertile scales oval, with obtuse apices. Sites of ovule attachment on lateral fertile scales barely on the adaxial side of the dwarf-shoot. Sterile and fertile scales moderately connate. Basal part of dwarf-shoot relatively short indicating partial fusion with bract. Stomata arranged in more or less irregular rows or scattered (figure 7.5b).

Remarks: As known at present, *Dolomitia* is limited to the Alpine Upper Permian, where it represents a minor component.

Pseudovoltzia Florin 1927
    Status: Natural genus within the Majonicaceae
    Type-species: *Pseudovoltzia liebeana* (Geinitz) Florin 1927; Upper Permian, Europe. Other species recognized: *P. sjerpii* Clement-Westerhof 1987. For detailed description of the species, see Schweitzer (1962, 1963) and Clement-Westerhof (1987).
    Supplementary description: Leaves ovate to lanceolate, slightly decurrent on the axis, apex obtuse to acute. Heterophylly occurs. Ovuliferous cones unlignified, approximately 7 cm long, 2 cm wide. Bracts with acute apices. Ovuliferous dwarf-shoots (figures 7.4f, 7.8a, b) partially fused with subtending bracts, their vascular strands originating separately with their xylem regions facing; dwarf-shoots equaling bracts in length, approximately 20 mm long, bearing two sterile and three fertile scales, the median fertile scale clearly showing in longitudinal sections the protuberance distal to position of ovule attachment (Schweitzer 1963: plates V2, 3; figure 7.8b). Sterile scales subtriangular, with acute apices; fertile scales oval, with obtuse apices. Ovule/seed (figure 7.8c) ovate, up to 7 mm long, micropylar area obtuse. Stomatal complexes sunken and arranged in longitudinal rows.
    Remarks: *P. liebeana* is well known from the European Upper Permian. The occurrences of *P. sjerpii*, however, as so far observed, is limited to the Vicentinian Alps (Alpine Upper Permian).

Discussion of the Majonicaceae. Schweitzer's anatomical studies (1963) of the ovuliferous cone of *Pseudovoltzia liebeana* represented a milestone in conifer research. In Florin's opinion the ovuliferous dwarf-shoot in *Pseudovoltzia* was free from the bract and bore five sterile scales and three stalk-like, separate fertile scales, bearing terminally inverted ovules. Schweitzer, however, demonstrated that (1) bract and dwarf-shoot are partially fused; (2) the ovules are directly inserted on foliaceous fertile scales (already suggested by Walton 1928). Schweitzer did not interpret the ovules as abaxial on fertile scales. However, an abaxial insertion in *Pseudovoltzia* has been independently suggested by Meyen (1984) and the present author (1984) as a result of their investigations on *Sashinia* and *Ortiseia*, respectively.

The anatomical studies provide no information on the arrangement of the fertile scales. According to Schweitzer, they are arranged in one plane. However, specimens figured by Florin (1944: plate CLXXIX/CLXXX, 9–10), Schweitzer (1963: plate I, 5–6), and the present author (figure 7.4f) consistently show a longitudinal fold in the lateral fertile scales. Combined with the slit-shaped sites of ovule attachment (compare figure 7.4f), this indicates that in *P. liebeana* the positions of ovule attachment are not situated solely in a single plane. In another species, *P. sjerpii* (Clement-Westerhof 1987), however, such a condition (i.e., planation) is more or less attained.

It is here considered that the Majonicaceae differ from most known extinct and extant conifers in one distinct aspect: the fertile scales (and consequently the ovules are not arranged in one plane (with the exception of *Pseudovoltzia sjerpii*). Meyen (1981, 1984) described a radial arrangement of fertile and sterile scales in *Sashinia* (see description below). Yet this fructification is quite different from those ascribed to the Majonicaceae. Thus the position of the fertile scales in the Majonicaceae could well be derived from an originally radial arrangement of fertile and sterile scales—which, compared with the arrangement of microsporophylls in the polliniferous cones, is by no means surprising. The organization observed in *Dolomitia*, moreover, suggests descendance from an ancestral stock characterized by at least three fertile scales and a large number of sterile scales.

Family Ullmanniaceae Zimmermann 1959
Type-genus: *Ullmannia* Goppert 1850
Type-species: *Ullmannia bronnii* Goppert 1850; Upper Permian, Europe. Further mentioned from China. Other species recognized: *U. frumentaria* Goppert 1850. For detailed descriptions of the species, see Florin (1944); Schweitzer (1962, 1963).
Description: Shoots bearing entire bifacial leaves, ovate or lanceolate, decurrent on the axis, and with obtuse or acute apices. Ovuliferous cones compound, approximately 6 cm long, 2.5 cm wide; cone axis bearing spirally arranged, entire bracts, with acute apices. Ovuliferous dwarf-shoots (figure 7.8d) axillary, almost free from bracts, approximately 12 mm long, bilaterally symmetrical,

more or less globular, approximately 9 mm long, and provided with a single integument. Micropylar area obtuse; nucellus ovate, free to its base; pollen/archegonial chamber present; megaspore membrane cutinized. Polliniferous cone simple; cone axis bearing spirally arranged hypopeltate microsporophylls. Pollen bisaccoid. Epidermis of leaves amphistomatic; stomata situated in more or less irregular rows. Stomatal complexes monocyclic, with 4–10 (12) papillate subsidiary cells. Epidermal papillae may occur.

Discussion of the Ullmanniaceae.   Florin and Schweitzer have different opinions concerning both the morphology of the ovuliferous fructifications and their classification at the species level. Florin (1944) described the ovule as terminal and inverted on a free, emergent stalklike fertile scale; he assigned the fructifications to *U. bronnii*. Schweitzer (1962, 1963) interpreted the ovule as directly inserted basally on the fertile scale (forming the median part of the dwarf-shoot), although the exact place of ovule attachment is uncertain; he assigned the fructifications to *U. frumentaria*.

Compared with coeval taxa, the ovuliferous dwarf-shoots ascribed to *Ullmannia* exhibit a "modern" organization: the fusion of scales is so complete that they are hardly recognizable. However, there is no fusion with the subtending bract.

The author believes that the natural status of the genus *Ullmannia* (and consequently the Ullmanniaceae) is by no means ascertained. Further comments on the family must await a restudy of the material concerned.

Other Taxa
Moyliostrobus Miller and Brown 1973.
   Status: Form-genus
   Type-species: *Moyliostrobus texanum* Miller and Brown 1973; Lower Permian, United States. For detailed descriptions, see Miller and Brown (1973).
   Description: Shoots and epidermal structure unknown. Ovuliferous cone compound, approximately 2 cm wide; cone axis bearing spirally arranged, entire bracts, with acute apices. Ovuliferous dwarf-shoots axillary, free from bracts, up to 12 mm long, flattened, equaling length of subtending bracts, bearing 20–50 spirally arranged sterile scales and a single ovule. Sterile scales absent opposite ovule and at apex of dwarf-shoot. Ovule erect, platyspermic, attached at base of dwarf-shoot (figure 7.9d), approximately 12 mm long, provided with two integumentary extensions, abaxially and adaxially

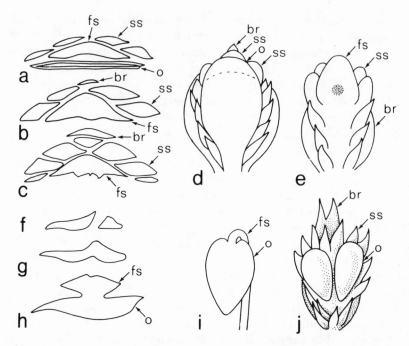

Figure 7.9. a–e: *Moyliostrobus texanum.* **a–c:** Details derived from a single transverse cone section. The three dwarf-shoots figured here reflect a sequence through the apical part of a dwarf-shoot. **a** Most apical; note integumentary extensions of an ovule. **b** Slightly more basal; note apex of bract. **c** Section at +¾ of the dwarf-shoot; note possible site of ovule attachment with scar of vascular strand. **d** Reconstruction of dwarf-shoot and ovule, adaxial view; after Miller and Brown (1973). **e** Tentative reconstruction of dwarf-shoot, adaxial view. **f–j:** *"Lebachia" lockardii.* **f–h:** Sequence of sections through a single fertile scale with ovule. **f** Integumentary extensions of ovule. **g** Basal part of ovule. **h** Position of ovule attachment. **i** Reconstruction of ovule attachment; after Mapes and Rothwell (1984). **j** Tentative reconstruction, adaxial view, with two ovules. (**a–d** after Miller and Brown 1973; **f–i** after Mapes and Rothwell 1984.)

situated on both sides of the micropyle. Seed shows evidence of probable embryo tissue.

Discussion

The embryos in *Moyliostrobus* are the oldest so far recorded in conifers. With respect to the organization of the ovuliferous dwarf-shoots, the following information may be derived from the description of Miller and Brown (1973): (1) There is no evidence that the ovule emerges basally on the dwarf-shoot. (2) No micropyle is shown; no evidence of an archegonial chamber near the presumed micropyle is provided; thus there is no evidence that the integumentary extensions are situated at the micropylar area.

The structure Miller and Brown described as a dwarf-shoot resembles that of *Ortiseia*. Combined with facts (1) and (2), another interpretation may be suggested. As in *Ortiseia*, the ovule in *Moyliostrobus* could well have been attached on the abaxial surface (in reference to the axis of the dwarf-shoot) of a fertile scale (compare figure 7.9a–c, e). Moreover, two sectioned dwarf-shoots (figure 7.9c) show a possible position of ovule attachment with the scar of a vascular strand. In this interpretation *Moyliostrobus* shows a remarkable resemblance to *Ortiseia*.

In the original concept, the diagnosis of *Moyliostrobus* excludes assignment to the Walchiaceae. However, if the interpretation here presented (figure 7.9e) is correct, *Moyliostrobus* agrees with the diagnosis of *Walchiostrobus* (Kerp and Clement-Westerhof, in press). In that case, the cone as described here has to be classified as *Walchiostrobus texanum*.

"Lebachia" lockardii Mapes and Rothwell 1984.
   Status: Conifer species erroneously assigned to *Lebachia;* Upper Carboniferous (to perhaps Lower Permian), Kansas. For detailed descriptions, see Mapes and Rothwell (1984).
   Description: Cone-bearing axes with spirally arranged, simple or bifurcate leaves. Ovuliferous cones compound, approximately 5.5 cm long, 1.5 cm wide; cone axis bearing spirally arranged, bifurcate bracts, with ovuliferous dwarf-shoots arising in their axils, free from bracts; ovuliferous dwarf-shoots up to 7 mm long, bilaterally symmetrical, flattened, bearing 20–30 sterile scales and 1–2 fertile scales (rarely 3–5), the latter emerging on the adaxial side and bearing a terminal inverted ovule. Sterile scales with acute apices, occasionally bilobate. Fertile scales comparable to sterile scales in anatomical and morphological features. Ovule bilaterally symmetrical and winged, up to 2.3 mm wide; basal part rounded or cordate, micropylar area tapering. Nucellus free to its base; pollen/archegonial chamber present. Megaspore membrane cutinized. Epidermis of leaves, bracts, and scales unequally amphistomatic, with most stomata on adaxial surfaces; stomata arranged in longitudinal bands. Stomatal complex monocyclic, provided with 5–9 subsidiary cells bearing papillae. Hair bases present. Epidermal papillae

may occur. Associated prepollen monosaccoid and assignable to *Potonieisporites* Bhardwaj 1954.

Discussion

The description of the ovuliferous cones is based largely on pre-served anatomical structures. The bilateral symmetry of the dwarf-shoot and the organization of the ovule have been clearly demon-strated. There is reason to assume that two ovules per dwarf-shoot regularly occur in *"Lebachia" lockardii* (see tentative reconstruction in figure 7.9j). This is in contradiction to the diagnosis of *Lebachia* (now *Walchia*), which emphasizes the presence of a single ovule per dwarf-shoot.

A questionable point to the interpretation of the material is the ovule attachment in *"L." lockardii*. The presumed terminal attach-ment in this conifer is not sufficiently shown by means of figures. The transverse sections could well indicate an abaxial ovule attachment. In any case, these sections clearly show that the seed axis is not straight (figure 7.9f–h), as is suggested in the reconstructions (Mapes and Rothwell 1984; figures 7.1, 7.9i).

If an interpretation of the dwarf-shoots with ovules borne abaxially on fertile scales were found to be correct, there would exist an overall resemblance to *Otovicia*, especially when *"L." lockardii* is interpreted with two fertile scales per dwarf-shoot (compare figures 7.2b, 7.9j). Kerp et al. (in press) noticed a similarity not only in form of the ovuliferous dwarf-shoots but also in epidermal structure. It is not unlikely, therefore, that this North American conifer is comparable to the European genus *Otovicia* (see below).

Thuringiostrobus Kerp and Clement-Westerhof in press
   Status: Form-genus
   Type-species: *Thuringiostrobus meyenii;* Upper Carboniferous–Lower Permian, Europe. Other species recognized: *T. florinii* Kerp et al.; Upper Carboniferous–Lower Permian. For detailed descriptions of the species, see Kerp and Clement-Westerhof (in press).
   Description: Cone axis bearing spirally arranged, radially to bilat-erally symmetrical, axillary ovuliferous dwarf-shoots, free from subtending bracts. Dwarf-shoots bearing 0–30 sterile scales and at least 4 straight fertile scales, showing, subapically on the abaxial surface of each, a site of ovule attachment. Fertile and sterile scales free or only slightly fused.
   Supplementary descriptions of species: *T. meyenii* characterized by bifurcate bracts and slightly flattened, more or less bilaterally sym-

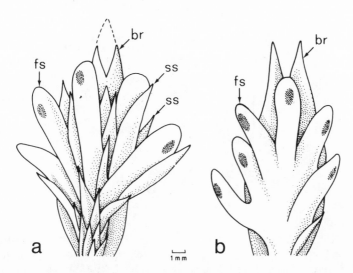

Figure 7.10. *Thuringiostrobus:* Reconstructions of bract and ovuliferous dwarf-shoot, adaxial view. **a** *T. florinii.* **b** *T. meyenii.* (Reinterpreted after Florin 1944.)

metrical, ovuliferous dwarf-shoots, approximately 12 mm long and provided with none or a few sterile scales (figure 7.10b). *T. florinii* characterized by probably radially symmetrical ovuliferous dwarf-shoots, approximately 12 mm long, bearing many (up to 30) sterile scales (figure 7.10a).

Remarks: The form-genus *Thuringiostrobus* has been established in order to accommodate ovuliferous fructifications with at least four fertile scales, formerly included in *Walchiostrobus* (as *"Ernestiodendron"* type). Florin interpreted these fructifications to have stalked, erect or inverted ovules. Both Meyen (1984) and the present author (1984) interpreted the "stalked ovules" as fertile scales with abaxial ovule attachment. These ovuliferous fructifications have not been found to be associated with vegetative remains. Nevertheless, they are here considered to resemble those of the Walchiaceae, especially *T. florinii*, which represents the more primitive form. According to Florin (1951), its fertile and sterile scales are spirally arranged.

ANGARA TAXA

Sashinia Meyen 1978, associated with *Quadrocladus* Madler 1957 (vegetative remains) and *Dvinostrobus* Meyen, in press (polliniferous cones).

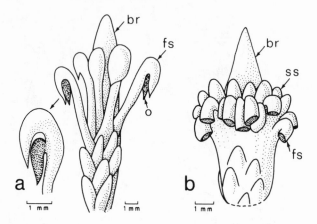

Figure 7.11. a *Sashinia aristovensis:* Reconstruction of bract, ovuliferous dwarf-shoot, and ovules, adaxial view. **b** *Kungurodendron sharovii:* Reconstruction of bract and dwarf-shoot, adaxial view. (**a** after Meyen 1984; **b** after Meyen, in press.)

Status: See remarks
Type-species: *Sashinia aristovensis* Meyen 1978, associated with *Quadrocladus dvinensis* Meyen 1978; Upper Permian, Soviet Union. Other species recognized: *Sashinia borealis* Meyen 1981. For detailed descriptions of the species, see Meyen (1978, 1981, 1984).
Description: Shoots with spirally arranged bifacial, lanceolate to obovate, entire leaves, with obtuse apices, and decurrent on the axis; assignable to *Quadrocladus.* Ovuliferous cones lax, compound, approximately 4 cm long, 1.5 cm wide; cone axis bearing spirally arranged bracts, with ovuliferous dwarf-shoots arising free in their axils. Ovuliferous dwarf-shoots (figure 7.11a) about 15 mm long, with radially arranged sterile and fertile scales. Fertile scales apically situated, bearing single, inverted ovules on their abaxial surfaces, covered by the recurved apices of the fertile scales. Bracts, sterile and fertile scales resembling in morphology and epidermal structure vegetative leaves of *Quadrocladus* type. Ovules approximately 5 mm long, megaspore membrane cutinized. Polliniferous cone lax, assigned to *Dvinostrobus* Meyen, in press. Cone axis bearing spirally arranged peltate microsporophylls bearing clusters of sporangia in the middle of the proximal stalk-like part. Pollen assignable to *Lueckisporites* Pontonie and Klaus 1954. Epidermis of leaves, bracts, and scales amphistomatic stomata scattered; stoma-

tal complexes mono- to incompletely dicyclic, with 4–6 subsidiary cells.

Remarks: Meyen (1984) advocates the constitution of assemblage genera—i.e., providing vegetative and fertile parts with different generic and specific names. The present author follows his nomenclature in the description of his material. Further, Meyen (1981) suggests that the *Sashinia* plant might comprise a family of its own, although he did not assign a family rank to this conifer in his proposed system (Meyen 1984). He suggests the possibility that the uniformity of foliage leaves, bracts, and sterile and fertile scales (seed stalks in his terminology) in *Sashinia* is the result of dedifferentiation. It is remarkable that the male cone (*Dvinostrobus* Meyen, in press) has not been involved in this process of dedifferentiation. In 1984 he presented a phylogenetic series from a *Sashinia*-like ancestor via the taxon here referred to as *Thuringiostrobus meyenii*. Although shoots of the *Quadrocladus* type (Madler 1957; Schweitzer 1962; Ullrich 1964) and pollen grains assignable to *Lueckisporites* (Visscher 1971) form an essential component of western European and Alpine Permian mega- and Palynofloras, respectively, *Sashinia*-like fructifications have not been recorded from these regions.

Kungurodendron Meyen, in press, associated with *Cyparissidium* Meyen in press (vegetative material)

Status: see remarks.

Type-species: *Kungurodendron sharovii* Meyen, in press, associated with *Cyparissidium appressum* (Zalessky) Meyen, in press; Lower Permian, Soviet Union.

Description: Pinnately branched shoot systems bearing spirally arranged entire leaves, with apices characterized by a long mucro. Shoots assigned to *Cyparissidium appressum*. Ovuliferous cone compound; cone axis bearing spirally arranged, long, leaf-like bracts (nearly identical to the sterile leaves) and dwarf-shoots, not reaching apex of subtending bracts. Dwarf-shoots flattened, provided basally with reduced sterile scales and subapically with numerous (approximately 10) fertile scales (seed stalks) emerging on the adaxial side of the dwarf-shoot; sterile "fertile" scales apically inserted on the abaxial side of the dwarf-shoot. Fertile scales narrow and hooked, apically showing a site of ovule attachment (figure 7.11b). Polliniferous cones simple; cone axis bearing spirally arranged microsporophylls with leaf-like distal parts. Pollen probably asaccate and resembling *Inaperturopollenites* Pflug and Thomson 1953. Epidermis of leaves epistomatic; stomata mostly arranged into two papillose bands on the adaxial surface. Hair bases occasionally present.

Remarks: Meyen (in press) proposes a separate subfamily for conifers with apically attached ovules, the Kungurodendroideae. Besides *Kungurodendron* this taxon includes *Timanostrobus* Meyen, in press and *"Lebachia" lockardii* Mapes and Rothwell 1984.

GONDWANA TAXA

Family Buriadiaceae *sensu* Pant 1977 (see discussion).
Type-genus: *Buriadia* Seward and Sahni 1920
Status: Natural genus within the Buriadiaceae
Type species: *Buriadia heterophylla* (Feistmantel) Seward and Sahni, emend. Pant and Nautiyal 1967; Lower Gondwana, Karharbari beds (Permo-Carboniferous), India. Further recorded from Brazil and Korea. For detailed descriptions, see Pant and Nautiyal (1967).
Description: Shoots irregularly branched, bearing spirally arranged leaves; leaves entire, bifurcate or multifid, bifacial or trifacial. Fertile parts of (not always ultimate) shoots bearing laterally attached, solitary, stalked, inverted ovules (figure 7.12a), replacing a leaf in the phyllotactic spiral. Ovule platyspermic, up to 4.5 mm long, tapering to narrow micropylar region, and carinate, one carina continuing as a stalk, the other extending into a little horn; micropyle situated between stalk and horn. Integument showing tough outer and delicate inner cuticle. Nucellus free to its base, with touch cuticle; megaspore membrane not cutinized. Epidermis of leaves epistomatic; stomata arranged in wide longitudinal bands on adaxial surface(s) (if trifacial, also a narrow band on the third surface); epidermal cells bearing hollow papillae, and near margin bulging out to form cutinized hairs. Seeds with few stomata and papillate epidermal cells. Stomatal complexes mono- to incompletely dicyclic, each with 5–7 papillate subsidiary cells. Associated pollen grains asaccate, monocolpate, and resembling *Ginkgocycadophytus* Samoilovich 1953.

Discussion of the Buriadiaceae.   Pant (1977) proposed the name Buriadiales as a suprageneric category for *Buriadia*. The name Buriadiaceae has not been formally introduced but is applied by Meyen (1984) in his proposed Gymnosperm System.

With respect to the relationships of *Buriadia*, Pant and Nautiyal (1967) suggest several possibilities: (1) *Buriadia* is a primitive conifer,

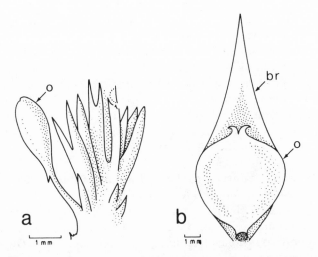

Figure 7.12. a *Buriadia heterophylla:* Shoot with ovule. **b** *Ferugliocladus riojanum:* Reconstruction of bract and ovule. (**a** after Pant and Nautiyal 1967; **b** interpreted after Archangelsky and Cuneo 1987.)

possibly ancestral to the Walchiaceae. They observe a resemblance in foliage and epidermal structure and present a phylogenetic series leading from a *Buriadia*-like ancestor via hypothetical stages of reduction and concentration of fertile shoots to a compound cone of, e.g., *Walchia piniformis* (see also Pant 1977, 1982). (2) *Buriadia* represents a completely new group of plants, unclassifiable among the presently known groups of gymnosperms.

Meyen (1968) suggested that *Buriadia* could represent an independently evolving group of plants in which, as in the Walchiaceae, the ovuliferous shoots were reduced and transformed into ovuliferous dwarf-shoots arising in bract axils. This case of parallelism would suggest heterogeneity in the conifers.

However, the major differences between *Buriadia* and a typical conifer are the following: (1) Absence in *Buriadia* of bract/dwarf-shoot complexes concentrated in a compound cone. (2) The seeming absence of a preformed abscission area in the *Buriadia* seed; dispersed seeds show a broken stalk. In any case, the abscission area is situated near the micropylar end and not near the chalaza. Thus, the seed axis is arched. (3) Absence of a cutinized megaspore membrane. This characteristic is especially peculiar: the combination of a thick nucellar cuticle and a noncutinized megaspore membrane represents one of the important characteristics of Caytoniales and Bennettitales (Harris

1953), but is unusual in conifers. (4) The pollen grains represent a type so far unknown in conifers.

Generally, authors attribute *Buriadia* to conifers—e.g., Beck (1981), Meyen (1984), and Miller (1985).

Remarks: Zimina (1983) described *Paraburiadia* from the Upper Permian of southern Primor'e. This genus resembles *Buriadia* but shows simple leaves and erect ovules with bifid apices. Details of epidermal structure and internal structure of ovules are unknown. The genus is not classified on a suprageneric level.

Family Ferugliocladaceae Archangelsky and Cuneo 1987

Type-genus: *Ferugliocladus* Archangelsky and Cuneo 1987

Description: Branching irregular and radial; shoots bearing spirally arranged, small, uniform, lanceolate leaves. Compact ovuliferous and polliniferous cones terminally situated on shoots. Ovuliferous cone consisting of a cone axis with bracts and apparently or actually free, erect ovules that are spirally arranged. Seeds platyspermic, reaching their greatest size while still in cones (figure 7.12b). Polliniferous cones consisting of a cone axis bearing spirally arranged, proximally reflexed microsporophylls. Pollen monosaccoid.

Remarks: Natural genera included in the Ferugliocladaceae are *Ferugliocladus* and *Ugartecladus* (see supplementary descriptions). A form-genus for dispersed seeds within the Ferugliocladaceae is *Eucerospermum* Feruglio.

Ferugliocladus Archangelsky and Cuneo 1987

Status: Natural genus within the Ferugliocladaceae

Type-species: *Ferugliocladus riojanum* Archangelsky and Cuneo 1987; Lower Gondwana, Gangamopteris zone, Argentina. Other species recognized: *F. patagonicus* (Feruglio) Archangelsky and Cuneo. For detailed descriptions of the species, see Archangelsky and Cuneo (1987).

Supplementary description: Branching to fifth order. Ovuliferous cone approximately 2 cm long; cone axis bearing lanceolate, entire bracts and free ovules or ovuliferous complexes. Presumed ovules axillary to subaxillary with respect to the bracts, subcircular, approximately 6 mm long, each erect with bifid apex; seeds, each with micropylar tube and lateral wings, about 7 mm long; nucellus cuticle present, megaspore membrane cutinized. Epidermal structure scarcely known; leaves probably amphistomatic, stomata scattered; each stomatal complex provided with approximately 6 subsidiary cells. Associated pollen grains resembling *Cannanoropollis* Potonie and Sah.

Ugartecladus Archangelsky and Cuneo 1987
   Status: Natural genus within the Ferugliocladaceae
   Type-species: *Ugartecladus genoensis* Archangelsky and Cuneo 1987;
   Lower Gondwana, Gangamopteris zone, Argentina. For detailed descriptions, see Archangelsky and Cuneo (1987).
   Supplementary description: Branching to fourth order. Ovuliferous cone approximately 4 cm long; cone axis bearing lanceolate, entire bracts, and free ovules or ovuliferous complexes. Presumed ovules sessile and subaxillary with respect to the bracts, subcircular, approximately 5 mm long, with simple nucellar apices; seeds with single mucronate apices and short lateral wings.

Discussion of the Ferugliocladaceae.   The authors propose two interpretations of the organization of the ovuliferous cones of the Ferugliocladaceae: (1) There is an ovuliferous complex, consisting of an erect ovule, entirely fused with one or more scales. This would suggest a highly advanced stage of conifer evolution; hypothetical evolutionary steps from a *Walchia*-like ancestor to the Ferugliocladaceae are presented. (2) The ovules represent unitary structures; therefore the cones can be regarded as simple strobili. As noted by the authors, it remains a crucial point if such taxa with seemingly simple strobili may be regarded as true conifers. Possible relationship with the Buriadiaceae is discussed and a common ancestry is suggested.

   Remarks: There is a certain resemblance between *Paraburiadia* (Zimina 1983) and *Ferugliocladus* with respect to the erect ovules with bifid apices.

GENERAL DISCUSSION

Phylogenetic considerations of conifers are invariably based on the interpretation of the reproductive organs, especially the ovuliferous fructifications. For a survey of the various theories on the organization of ovuliferous cones, one is referred to the work of Pilger (1926) and Florin (1954).

   As early as the late nineteenth century, some botanists considered the "ovuliferous scale" (e.g., in Pinaceae) to represent axillary dwarf-shoots (e.g., Celakovsky 1890). However, this concept was not generally accepted until Florin (1938–1945) published his detailed studies on Paleozoic conifers. Florin suggested that the cones of most modern conifers can be derived from the unambiguously compound strobili of the Walchiaceae ("Lebachiaceae").

Consequently, Walchiaceae occupies a central position in most phylogenetic considerations. The family is generally thought to represent a link between the Cordaitales and younger (late Permian–recent) conifers. However, some authors (e.g., Schweitzer 1963) reject the Walchiaceae (or only *Walchia;* Meyen 1984) as ancestral to younger conifers, and they do not accept Florin's interpretation of the coniferous fertile scale as stalk-like with a terminal erect or recurved ovule.

The recent revision of the Walchiaceae (Clement-Westerhof 1984; Kerp et al., in press), with its natural genera, *Walchia, Ernestiodendron, Otovicia,* and *Ortiseia,* in which the presence of "stalked" ovules was rejected, necessitates a further reconsideration of the relationship among conifers. In the following paragraphs the most important coniferous and conifer-like taxa will be briefly discussed in order to estimate their degree of relationship to the Walchiaceae. For purposes of discussion four categories may be distinguished:

(1) This category includes conifers represented by the form-genus *Thuringiostrobus* and the families Majonicaceae and Ullmanniaceae, as well as the Angara genus *Sashinia.* The organization of their ovuliferous dwarf-shoots is characterized by abaxial ovule attachment and can therefore readily be compared to that in the Walchiaceae. If interpreted as radially symmetrical *Thuringiostrobus* (notably *T. florinii*) represents the most primitive condition in this category. The organization of the ovuliferous dwarf-shoots of the Walchiaceae could well be derived from an organization resembling, in general terms, that in *T. florinii. Sashinia* also represents a rather primitive taxon as reflected in the radial symmetry of the dwarf-shoot, subapical ovule attachment, and probably bent seed axis (figure 7.11a).

Within the Majonicaceae one observes a bilateral symmetry and a tendency toward an arrangement of fertile and sterile scales in one plane, resulting in flattening of the dwarf-shoot; the seed axis is straight. The organization of dwarf-shoot appendages, however, suggests an ancestral stock characterized by radial arrangement of fertile and sterile scales, as assumed in *T. florinii.* The Ullmanniaceae would represent the most derived condition because of the flattened dwarf-shoot and total fusion of the single fertile scale with the sterile scales.

(2) This category comprises the North American taxa *"Lebachia" lockardii* and *Moyliostrobus,* together with the Angara genus *Kungurodendron.* These taxa have been described, by the authors concerned, as a possessing terminal (erect or recurved) ovules. This would imply

a crucial difference from the abaxial ovule attachment in the Walchi-aceae. Therefore, this category, when interpreted with terminal ovules, cannot be directly associated with a walchiaceous organization, un-less the terminal attachment were to represent a condition derived from an abaxial ovule attachment. However, the North American taxa show, in other characteristics, a great resemblance to the Walchi-aceae. In addition, *Moyliostrobus* may be alternatively interpreted as having an abaxial ovule attachment (figure 7.9a–e). In *"Lebachia" lockardii* the bent seed axis (figure 7.9f–i) is difficult to reconcile with a terminal attachment; the comparison of these American specimens with the walchiaceous genus *Otovicia*, by Kerp et al. (in press), should be seriously considered.

A comparison with *Otovicia* would imply that the slight resem-blance between the dwarf-shoots of *"L." lockardii* and *Kunguroden-dron* (Meyen, personal communication) cannot be related to the na-ture of ovule attachment.

(3) This category, characterized by the presence of simple cones, includes the Ferugliocladaceae. This family can hardly be associated with the Walchiaceae. Such association would imply that the ovule has to be interpreted as the only remnant of an ovuliferous dwarf-shoot, and even then the erect, scarcely stalked or sessile ovule strongly differs from the abaxially attached ovule in the Walchiaceae.

(4) The last category recognized is represented by the Buriadiaceae. The solitary ovules of *Buriadia* can hardly be associated with com-pound structures. Moreover, the organization of the ovule itself is totally different from that in the Walchiaceae and all other Paleozoic conifers and conifer-like taxa.

Summarizing, it is considered that, on the basis of organization of the ovuliferous fructifications, a relatively close relationship with the Walchiaceae can be demonstrated only for category (1) *(Thuringiostro-bus,* Majonicaceae, Ullmanniaceae, *Sashinia).* Such a relationship for category (2) is still uncertain. For categories (3) and (4) there are no arguments for a close relationship with the Walchiaceae.

It may even be suggested that a coniferous group characterized by *Thuringiostrobus* (notably *T. florinii*) was ancestral to the Walchi-aceae, *Sashinia,* and the Majonicaceae (see figure 7.13).

In order to estimate which catagories can be seen as the ancestors of the Mesosoic and Cenozoic conifers, it is necessary to discuss briefly the Triassic record. In general terms it can be stated that Triassic conifers clearly display the presence of compound ovuliferous cones. This would imply that categories (3) and (4)—all Gondwana taxa— represent unlikely ancestors of the post-Paleozoic coniferous stock.

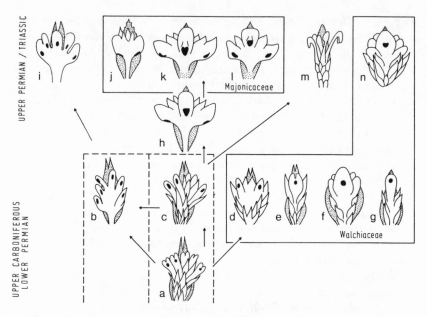

Figure 7.13. Some major phylogenetic trends in the organization of ovuliferous dwarf-shoots of early conifers: **a** Hypothetical stage. **b** *Thuringiostrobus meyenii.* **c** *T. florinii.* **d** *Otovicia hypnoides.* **e** *Walchia piniformis.* **f** *Moyliostrobus texanum* (tentative interpretation) **g** *Ernestiodendron filiciforme.* **h** Hypothetical stage. **i** *Voltziopsis africana.* **j** *Majonica alpina.* **k** *Dolomitia cittertiae.* **l** *Pseudovoltzia liebeana.* **m** *Sashinia aristovensis.* **n** *Ortiseia jonkeri.*

The suggestion of a possible phylogenetic relationship between the Ferugliocladaceae and the Araucariaceae (Archangelsky 1985) seems to be unjustified because of the compound nature of the araucariaceous cone (e.g., Stockey 1982).

The most apparent ancestors of post-Paleozoic conifers have to be sought in the Walchiaceae and related groups—category (1) or (2)—and one's conclusions will be based largely on the interpretation of ovule attachment as abaxial or terminal. Florin interpreted the ovules in Triassic conifers to be recurved and attached terminally on stalk-like fertile scales that varied from being free from *(Glyptolepis)*, to totally adnate to (fructifications ascribed to *Voltzia*), sterile scales. This interpretation has been followed by several authors—e.g., Miller (1982). Townrow (1967) and Grauvogel-Stamm (1978) also interpreted the dwarf-shoots of, respectively, *Voltziopsis* and *Aethophyllum* to be

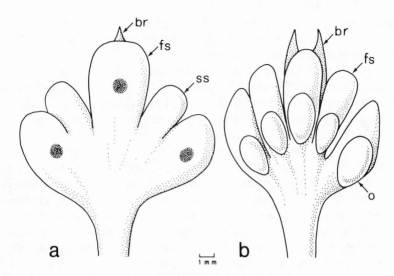

Figure 7.14. a Ovuliferous dwarf-shoot, ascribed to *Voltzia:* Reconstruction of bract and dwarf-shoot, adaxial view. **b** *Voltziopsis africana:* Reconstruction of bract, dwarf-shoot, and ovules, adaxial view. (Reinterpreted after Florin 1944.)

characterized by totally adnate fertile scales. Florin considered the Late Permian genus *Pseudovoltzia* to be transitional between the Walchiaceae (in his concept) and Triassic conifers.

Schweitzer (1963), however, unambiguously proved that in *Pseudovoltzia* the ovules are directly inserted on foliaceous fertile scales and supplied by a vascular strand, branching from that of the fertile scale. Schweitzer also suggested such an ovule attachment for fructifications of post-Paleozoic conifers (e.g., *Voltziopsis*, and the presumed cones of *Voltzia;* see figure 7.14) and consequently considered them to be derived from cones with a structural organization like those of *Pseudovoltzia*. Subsequently, Meyen (1984) and Clement-Westerhof (1984) interpreted this ovule attachment as abaxial on fertile scales of the dwarf-shoots. Schweitzer's concept is supported by the organization in *Majonica* and *Dolomitia*.

It is beyond the scope of this contribution to discuss the ovuliferous fructifications of all Triassic conifers. However, this author considers that at least some of them—e.g., the cones ascribed to *Voltzia* (figure 7.14b) and even those of the family Cheirolepidiaceae—can be derived from a cone structure like that observed in the Majonicaceae.

Other genera, characterized by about five fertile scales and an absence of sterile scales, can perhaps be derived directly from cones like those of *Thuringiostrobus meyenii* (figure 7.10).

Summarizing, it is suggested, on the basis of the structure of ovuliferous cones, that a group characterized by the form-genus *Thuringiostrobus*, rather than the Walchiaceae as generally understood, should be regarded as the ancestral stock for most conifers. Some taxa can be derived via the Majonicaceae.

It is apparent that the very incompletely known group represented by the form-genus *Thuringiostrobus* (especially *T. florinii*) should be regarded as the most primitive known conifers. This group might be interpreted as the ancestral stock of the extinct families Walchiaceae (as redefined), Majonicaceae, and Cheirolepidiaceae, as well as all other fossil and extant conifers in which it can be demonstrated or reasonably suggested on the basis of phylogenetic and ontogenetic considerations that (1) the ovuliferous cone represents a compound structure consisting of a cone axis bearing spirally arranged bract/dwarf-shoot complexes, and (2) the ovule is abaxial (in reference to the dwarf-shoot axis) on a fertile scale. In effect, this combination of characters seems to provide an appropriate means of diagnosing conifers among the various extinct and extant gymnosperms.

With regard to the Paleozoic taxa under consideration, such a definition would imply that the Buriadiaceae and the Ferugliocladaceae are difficult to interpret as true conifers. Also the concept of the terminal ("stalked") ovules cannot be reconciled with this diagnosis, unless such terminal ovules are regarded as derived from an originally abaxial attachment. In figure 7.13 the author's view of major phylogenetic trends in the organization of ovuliferous dwarf-shoots of early conifers has been visualized.

The direct ancestry of the earliest conifers still remains highly obscure. According to Florin (1951b), the Walchiaceae ("Lebachiaceae") could well be derived from a cordaitalean organization (see figure 7.15a). Indeed, the ovuliferous fructifications of the Cordaitales are compound, although the bract/dwarf-shoot complexes are not spirally arranged as in conifers. However, the terminal, erect ovule attachment in *Cordaianthus* is in crucial conflict with the definition of a conifer presented here. In the view of the author, the abaxial ovule attachment in the conifers cannot be derived from the terminal attachment in the Cordaitales.

The bilobate shape of the fertile scale in some Walchiaceae (*Walchia*, figure 7.3c; *Walchiostrobus*, figure 7.2d; *Ortiseia*, figure 7.1) may

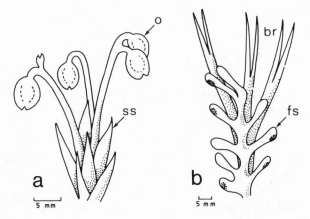

Figure 7.15. a *Cordaianthus pseudofluitans:* Reconstruction of ovuliferous dwarf-shoot and ovules, adaxial view (bract invisible). **b** *Trichopitys heteromorpha:* Tentative reconstruction of bract and dwarf-shoot, adaxial view. (**a** after Florin 1944; **b** after Florin 1951a.)

reflect a primitive condition, similar to the bifurcate bracts in *Walchia, Otovicia,* and *Thuringiostrobus (T. meyenii)*. This is relevant to the author's argument because the tips of the bifurcate bracts are thought to be vascularized (Florin 1944). There is, thus, no reason to reject the possibility of vascularization of the lobes of the fertile scales (figure 7.3d). It is possible, therefore, that, as in *Pseudovoltzia* (Schweitzer 1963), the part of the fertile scale distal to the position of ovule attachment was vascularized; if true, this would exclude an originally terminal attachment. Moreover, the bent seed axis in the earliest conifers represents a fundamental difference from the straight seed axis in the Cordaitales. The bent seed axis in more advanced conifers can be regarded as a derived condition caused by recurving of the fertile scales.

Another important difference between the conifers and the Cordaitales is the long recognized difference in the organization of the polliniferous organs. Apart from their compound structure, the cordaitalean polliniferous organs display terminally attached, erect pollen sacs. Although the simple condition of the conifer pollen cone might be derived from the compound cordaitalean structure (compare Wilde 1944), the generally accepted abaxial attachment of pollen sacs in conifers (Florin 1951b) cannot be reconciled with the terminal position in the Cordaitales.

Recapitulating, the organization of both ovuliferous and polliniferous organs makes it difficult to interpret the Cordaitales as the direct
ancestors of the conifers.

Zimmermann (1959) and Krassilov (1971) suggest the possibility
that the organization of ovuliferous fructifications as observed in *Trichopitys* (figure 7.15b) could be ancestral to that of the conifers. These
fructifications consist of an axis bearing repeatedly forked bracts subtending dwarf-shoots provided with only fertile scales. According to
Florin (1951a), the bracts are spirally arranged and the ovules recurved and terminally attached on more or less spirally arranged
fertile scales. Meyen (1984) described the bracts as arranged in two
rows, the dwarf-shoots as strongly flattened and the ovules as subapically attached. Both authors considered *Trichopitys* a primitive Ginkgophyte. Some ovuliferous fructifications figured by Florin (1939: plate
CXIX/CL, 1–3), here referred to as *Thuringiostrobus meyenii*, resemble
the fructifications of *Trichopitys*. It is evident that this interesting
taxon needs to be restudied in order to assess whether it should be
regarded as a conifer or as a more or less related, nonconiferous
gymnosperm.

It may be concluded that, at present, there is no candidate that can
be considered as the direct ancestor of the coniferous stock. In the
concept presented here the ovuliferous dwarf-shoot of a possible
ancestor might correspond to the hypothetical form included in figure
7.13a.

ACKNOWLEDGMENTS

The author is greatly indebted to Dr. S. V. Meyen for providing descriptions and illustrations of Angara conifers. Many thanks are due
to Dr. S. Archangelsky and Dr. R. Cuneo for permission to use information from their manuscript. Grateful appreciation is rendered to
Dr. J. H. F. Kerp, Mr. H. A. J. M. Swinkels, and Mr. R. Verwer for their
descriptions and illustrations of Early Permian conifers. I am most
grateful to Dr. H. Visscher for his essential support in modeling the
rough material. My special gratitude goes to Mr. M. Clement for his
everlasting patience and support during my investigations. Many
thanks are also due to Dr. G. K. Mapes and Dr. J. H. A. Van Konijnenburg-Van Cittert for the fruitful discussions on the earliest conifers. I
am most grateful to Mr. H. Rypkema, who made the drawings, and
Mr. J. Elsendoorn for preparing the photographs.

LITERATURE CITED

Archangelsky, S. 1985. Aspectos evolutivos de las coniferas gondwanicas del Paleozoico. *Bull. Sect. Sci. Com. Trav. hist. scient.* 8:115–124.

Archangelsky, S. and R. Cuneo. 1987. Ferugliocladaceae, a new conifer family from the Permian of Gondwana. *Rev. Palaeobot. Palynol.* 51:3–30.

Beck, C. B. 1981. *Archeopteris* and its role in vascular plant evolution. In K. J. Niklas, ed., *Paleobotany, Paleoecology and Evolution*, vol. 1, pp. 193–223. New York: Praeger.

Celakovsky, L. 1980. Die Gymnospermen: Eine morphologisch–phylogenetische Studie. *Abh. Bohm. Ges. Wiss.* (Math.–Nat.) 7(4):1–148.

Clement-Westerhof, J. A. 1984. Aspects of Permian palaeobotany and palynology. 4. The conifer *Ortiseia* Florin from the Val Gardena Formation of the Dolomites and the Vicentinian Alps (Italy) with special reference to a revised concept of the Walchiaceae (Goppert) Schimper. *Rev. Palaeobot. Palynol.* 41:51–66.

—— 1987. Aspects of Permian palaeobotany and palynology. 7. The Majonicaceae, a new family of Late Permian conifers. *Rev. Palaeobot. Palynol.* 52:375–402.

Crane, P. R. 1985. Phylogenetic analysis of seed plants and the origin of angiosperms. *Ann. Missouri Bot. Gard.* 72:716–793.

Darrah, W. C. 1960. *Principles of Paleobotany*, 2nd ed. New York: Macmillan.

Florin, R. 1927. Preliminary descriptions of some Paleozoic genera of Coniferae. *Ark. Bot.* 21A (13):1–7.

—— 1938–1945. Die Koniferen des Oberkarbons und des unteren Perms. 1–8. *Palaeontographica* 85B:1–729.

—— 1951a. The morphology of *Trichopitys heteromorpha* Saporta, a seed-plant of Palaeozoic age, and the evolution of the female flowers in the Ginkgoinae. *Acta Hort. Berg.* 15:79–109.

—— 1951b. Evolution of cordaites and conifers. *Acta. Hort. Berg.* 15:285–388.

—— 1954. The female reproductive organs of conifers and taxads. *Biol. Rev.* 29:367–388.

—— 1964. Uber *Ortiseia leonardii* n. gen. et sp., eine Konifere aus den Grodener Schichten in Alto Adige (Sudtirol). *Mem Geopalaeontol. Univ. Ferrara* 1(1):3–11.

Goppert, H. R. 1850. *Monographie der fossilen Coniferen.* Leiden: Arnz.

Grauvogel-Stamm, L. 1978. La flore du Grès à *Voltzia* (Buntsandstein Supérieur) des Voages du Nord (France): Morphologie, anatomie, interprétations phylogenetique et paleogeographique. Univ. Louis Pasteur de Srasbourg, Inst Géol. Mém. 50:1–225.

Hagerup, O. 1933. Zur Organogenie und Phylogeny der Knoiferen-Zapfen. *Biol. Medd.* 10(7):1–82.

Harris, T. M. 1953. Mesozoic seed cuticles. *Svensk. Bot. Tidskr.* 48(2):281–291.

—— 1976. The Mesozoic gymnosperms. *Rev. Palaeobot. Palynol.* 21:119–135.

Kerp, J. H. F. and J. A. Clement-Westerhof. (In press). Aspects of Permian palaeobotany and palynology: The form genus *Walchiostrobus* Florin reconsidered.

Kerp, J. H. F., H. A. J. M. Swinkels, and R. Verwer. (In press). Aspects of Permian palaeobotany and palynology. A conifer-dominated flora from the Rotliegendes of Oberhausen (?Upper Carboniferous–Lower Permian; SW-Germany) with special reference to an emendation of the Walchiaceae. *Rev. Paleobot. Palynol.*

Krassilov, V. A. 1971. Evolution and systematics of conifers (critical review). *Paleontol. Zh.* 1:7–20.

Madler, K. 1957. *Ullmannia*-blatter und andere Koniferenreste aus dem Zechstein der Bohrung Friedrich Heinrich 57. *Geol. Jb.* 73:75–90.

Mapes, G. and G. W. Rothwell. 1984. Permineralized ovulate cones of *Lebachia* from Late Palaeozoic limestones of Kansas. *Palaeontology* 27(1):69–94.

Meyen, S. V. 1968. Some general questions concerning the systematics and evolution of conifers in connection with the open ovule of *Buriadia. Paleontol. Zh.* 4:28–31.

—— 1978. Permian conifers of the West Angaraland and new puzzles in the coniferalean phylogeny. *Palaeobotanist* 25:298–313.

—— 1981. Some true and alleged Permo-Triassic conifers of Siberia and Russian Platform and their alliance. *Palaeobotanist* 28–29:161–176.

—— 1984. Basic features of gymnosperm systematics and phylogeny as evidenced by the fossil record. *Bot. Rev.* 50:1–111.

—— (In press). Permian conifers of Western Angaraland. *Palaeontographica.*

Miller, C. N. 1982. Current status of Paleozoic and Mesozoic conifers. *Rev. Palaeobot. Palynol.* 37:99–144.

—— 1985. A critical review of S. V. Meyen's "Basic features of gymnosperm systematics and phylogeny as evidenced by the fossil record." *Bot. Rev.* 51:295–318.

Miller, C. N. and J. T. Brown. 1973. A new voltzialean cone bearing seeds with embryos from the Permian of Texas. *Am. J. Bot.* 60(6):561–569.

Pant, D. D. 1977. Early conifers and conifer allies. *J. Indian Bot. Soc.* 56:23–37.

—— 1982. The Lower Gondwana gymnosperms and their relationships. *Rev. Palaeobot. Palynol.* 37:55–70.

Pant, D. D. and D. D. Nautiyal. 1967. On the structure of *Buriadia heterophylla* (Feistmantel) Seward & Sahni and its fructifications. *Philos. Trans. Roy. Soc. London* 252:27–48.

Pilger, R. 1926. Gymnospermae 6: Klasse Coniferae. In A. Engler and K. Prantl, eds., *Die naturlichen Pflanzenfamilien*, vol. 13, pp. 121–403. Leipzig: Englemann.

Remy, W. and R. Remy. 1977. *Die Floren des Erdaltertums.* Essen: Gluckauf.

Rothwell, G. W. 1982. New interpretations of the earliest conifers. *Rev. Palaeobot. Palynol.* 37:7–28.

Schweitzer, H. J. 1962. Die Makroflora des niederrheinischen Zechsteins. *Fortschr. Geol. Rheinl. West.* 6:331–377.

—— 1963. Der weibliche Zapfen von *Pseudovoltzia liebeana* und seine Bedeutung fur die Phylogenie der Koniferen. *Palaeontographica* 113B:1–29.

Seward, A. C. and Sahni, B. 1920. Indian Gondwana plants: A revision. *Palaeont. indica*, n.s. 7(1):1–41.

Sternberg, K. 1825. *Versuch einer geognostisch-botanischen Darstellung der Vorwelt.* 1(4). Regensburg: Brenck.

Stockey, R. A. 1982. The Araucariaceae: An evolutionary perspective. *Rev. Palaeobot. Palynol.* 37:133–154.

Taylor, T. N. 1981. *Paleobotany: An Introduction to Fossil Plant Biology.* New York: McGraw-Hill.

Townrow, J. A. 1967. On *Voltziopsis*, a southern conifer of Lower Triassic age. *Pap. Proc. Roy. Soc. Tasmania* 101:173–188.

Ullrich, H. 1964. *Zur Stratigraphie und Palaontologie der marin beeinflussten Randfazies des Zechsteinsbeckens in Ostthuringen und Sachsen. Freib. Forschungsh.* C 169:1–163.

Visscher, H. 1971. The Permian and Triassic of the Kingscourt outlier, Ireland —a palynological investigation related to regional stratigraphical problems in the Permian and Triassic of western Europe. *Geol. Surv. Irel. Spec. Pap.* 1:1–114.

Visscher, H., J. H. F. Kerp, and J. A. Clement-Westerhof. 1986. Aspects of Permian palaeobotany and palynology. 6. Towards a flexible system of naming palaeozoic conifers. *Acta Bot. Neerl.* 34(2):87–99.

Walton, J. 1928. On the structure of a Palaeozoic Cone-Scale and the evidence it furnishes of the primitive nature of the Double Cone Scale in the Conifers. *Mem. & Proc. Manchester Lit. and Phil. Soc.* 73:1–6.

Wilde, M. H. 1944. A new interpretation of coniferous cones. 1. Podocarpaceae *(Podocarpus). Ann. Bot.*, n.s. 8(29):1–41.

Zimina, V. G. 1983. Conifers from the Upper Permian of Southern Primor'e. *Paleontol. Zh.* 1:111–119.

Zimmermann, W. 1959. *Die Phylogenie der Pflanzen*, 2nd ed. Stuttgart: Fischer.

8

Gymnosperms of the Angara Flora

SERGEI V. MEYEN

The term "Angaraland" is currently used for a continent that existed throughout the late Paleozoic within modern northern Asia. The Angara paleofloristic kingdom of the same period also embraced at times eastern parts of Europe, northern parts of central and southwest Asia, and probably northern Greenland and Alaska as well. The epithet "Angara" may be applied also to the Permo-Triassic floras of Siberia and adjoining regions.

The Angara gymnosperms are currently no better understood than European late Paleozoic gymnosperms were a century ago. Hence they have gained little or no attention in paleobotanical textbooks. For instance, Taylor (1981) mentioned only *Cladostrobus, Vojnovskya,* and *Rufloria,* and Stewart (1983) omitted Angara gymnosperms altogether. During the last decade the Angara gymnosperms have been more extensively studied, and presently most of them are given a proper place in the system of suprageneric taxa. The pertinent data have been summarized by the author (Meyen 1982a, 1984a, 1986b, 1987, and in press).

The Angara gymnosperms deserve plant taxonomists' attention not only because they supplement the diversity of gymnosperms known from other floras but also because some of them exhibit a very long-term retention of characteristics of fairly primitive plant types that earlier migrated from the equatorial belt. This phenomenon of extra-equatorial persistence is very characteristic for the Angara flora (Meyen 1984a, 1984b). Therefore, for gymnosperm phylogeny the Angara plants may play the same role as recent tribal peoples play in the reconstruction of prehistoric human societies.

Stratigraphic and Phytogeographic Settings The stratigraphic nomenclature adopted in the Soviet Union is used in this paper. Equivalents of standard series and stages cannot be reliably specified in Siberia, where they are replaced by regional biostratigraphic units called "phytostratigraphic horizons" (figure 8.1; see details in Meyen 1982a).

The territory covered by the Carboniferous Angara flora is termed the Angara area (figure 8.2). Its southwestern part from the Middle Carboniferous onward belonged to the Kazakhstan province, and the northeastern part belonged to the Vilyui-Verkhoyansk district. In the Permian, the independent Angara kingdom can be recognized as consisting of the Angara area (sensu stricto) and Subangara area (western and southern frontier regions of the kingdom). The Angara area is further subdivided into provinces and districts. All these phytochoria have been reviewed elsewhere (Meyen 1982a).

Criteria for Establishing Assemblage Taxa Below, various assemblage taxa (assemblage genera and assemblage species) are used— i.e., sets of form taxa established for different organs (leaves, seeds, and so on) belonging to the same plants. Reconstructions of the assemblage taxa rest on several criteria symbolized as A1, A2, E, M, and OC. These symbols are provided when relevant assemblage taxa are first mentioned, to indicate to the reader the criteria used for reconstruction of lifetime association of permanent dispersed organs (see details in Meyen, 1986c).

A1: Repeated association of dispersed organs in many oligo- or polydominant burials. This criterion is usually supported by criterion E (see below).

A2: Such frequent repetition of association of dispersed organs that

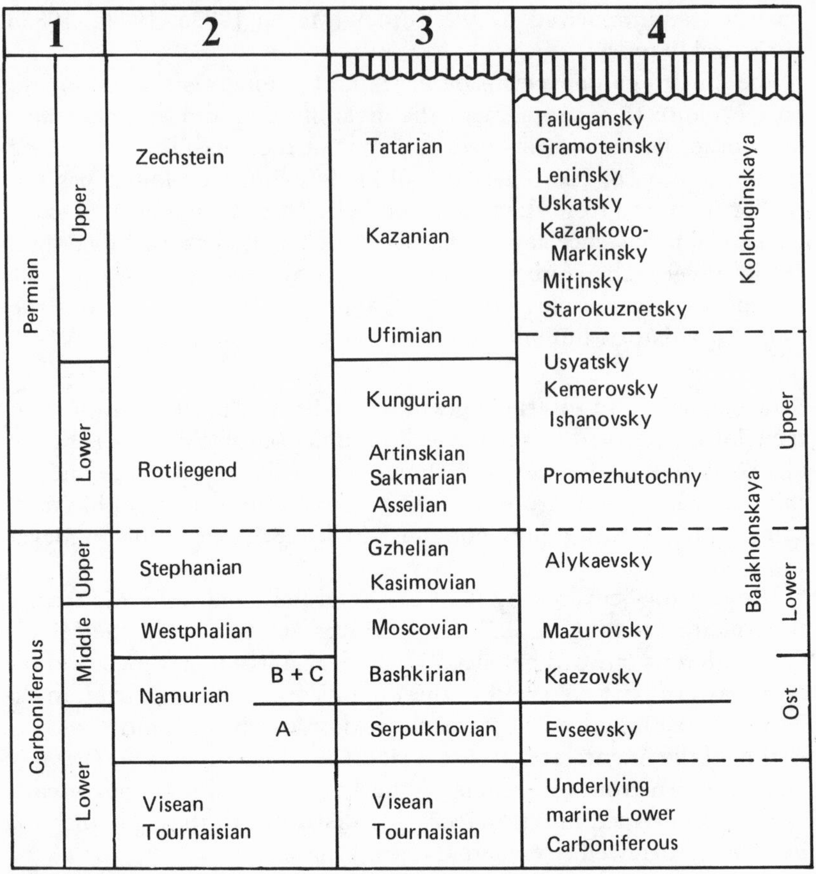

1		2	3	4	
Permian	Upper	Zechstein	Tatarian	Tailugansky Gramoteinsky Leninsky Uskatsky	Kolchuginskaya
			Kazanian	Kazankovo-Markinsky Mitinsky Starokuznetsky	
			Ufimian	Usyatsky	
	Lower	Rotliegend	Kungurian	Kemerovsky Ishanovsky	Upper
			Artinskian Sakmarian Asselian	Promezhutochny	Balakhonskaya
Carboniferous	Upper	Stephanian	Gzhelian Kasimovian	Alykaevsky	Lower
	Middle	Westphalian	Moscovian	Mazurovsky	
		Namurian B + C	Bashkirian	Kaezovsky	Ost
		A	Serpukhovian	Evseevsky	
	Lower	Visean Tournaisian	Visean Tournaisian	Underlying marine Lower Carboniferous	

Figure 8.1. Correlation of Upper Paleozoic phytostratigraphical units of Kuznetsk basin with standard units. **1** Standard systems and series adopted in the USSR. **2** Units adopted in western Europe. **3** Stages adopted in the European part of the USSR. **4** Phytostratigraphical horizons (left column) and series (at right side) of Kuznetsk basin, Ost–Ostrogskaya subseries.

the possibility of accidental cooccurrence of organs belonging to different plants is, for all practical purposes, excluded. These are monodominant burials or burials that are totally devoid of remains not belonging to the assemblage taxon.

E: Typological extrapolation supported by A1 or A2. This criterion can be better explained with an example. Judging from some gross

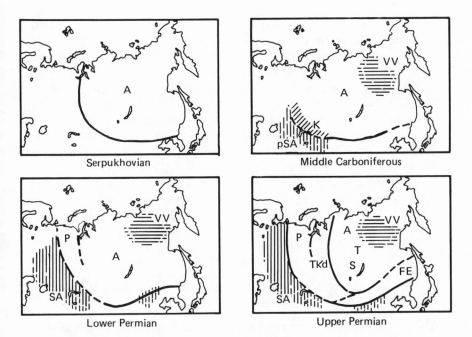

Figure 8.2. Paleophytochoria of Angaraland. (A = Angara area (kingdom); K = Kazakhstan province; pSA = phytochorion preceding Subangara area; VV = Vilyui-Verkhoyansk district; SA = Subangara area; P = Pechora province; TKd = Taimyr-Kuznetsk district; T = Tunguska district; S = Siberian province; FE = Far Eastern province.)

morphological and epidermal features, the genus *Compsopteris* is closely related to *Callipteris* subgenus *Feonia,* and the latter to *Lepidopteris,* known in association with *Peltaspermum*-like fructifications. Because *Compsopteris* and *Peltaspermum*-like fructifications are found together (in the absence of competing possible lifetime associates), we can treat the *Compsopteris-Peltaspermum* set as an assemblage genus.

M: Markers (similar epidermal characters, characteristic resin bodies in the mesophyll, identical epiphyllous fungi, and so on) repeated in different dispersed organs. This criterion is usually supported by A1, A2, and/or E.

OC: Discovery of organs in organic connection.

Morphological Terminology Gymnosperm fructifications are described below in terms previously proposed by the author (Meyen

1982c, 1984a). A seed (or ovule; "seed" is used below to cover both) with its stalk is termed a "monosperm," and any aggregation of monosperms or of sessile seeds is termed a "polysperm." Polysperms may be simple or compound (an aggregation of simple polysperms). "Phyllosperm" means an unmodified seed-bearing leaf, and a modified leaf-like seed-bearing organ is called a "cladosperm." The term "peltoid" is used for peltate polysperms with abaxially attached seeds. Scales or leaf-like organs accompanying polysperms form "circasperms." A branched male fructification that cannot be qualified as a microsporophyll or microstrobilus is termed a "microsporoclad."

Many Angara gymnosperms produced protosaccate pollen. The terms "protosaccate" and "protosaccus" are etymologically irrelevant. They imply that the protosaccate structure precedes the saccate, which is not always so. Because of this, the terms "quasisaccus" and "quasisaccate" are more appropriate.

SYSTEMATICS AND EVOLUTION OF ANGARA GYMNOSPERMS

The Angara gymnosperms fit well the system proposed by the author (Meyen 1984a). This system has elicited a number of critical comments (Beck 1985; Miller 1985; Rothwell 1985), most of which have been answered by the author elsewhere (Meyen 1986c).

Some Angara gymnosperms are conventionally affiliated with certain suprageneric taxa—i.e., are treated as satellite genera of these taxa (on the concept of satellite genus, see Meyen 1984a; Thomas and Brack-Hanes 1984).

Class Ginkgoopsida   *Order Callistophytales.*   For a reliable affiliation of fossils in Callistophytales, one should reveal anatomical characters of axes, seeds, and synangia. Typical Euramerian callistophytons (Rothwell 1981) produced seeds and synangia on unmodified (or weakly modified) leaves. The author (Meyen 1984a) suggested that this order also comprises more primitive genera, having platyspermic seeds and synangia not yet transferred onto the leaf lamina and attached to pinnately divided axes (figure 8.3). Among Euramerian Middle Carboniferous plants of this type is the assemblage genus *Eremopteris-Cornucarpus-Pterispermostrobus* (A2). Seeds similar to *Cornucarpus* are known from the Middle–Upper Carboniferous of Angaraland. These are *Angarocarpus ungensis,* usually accompanying (A1) the fronds *Paragondwanidium sibiricum.* The latter are undoubtedly associated (A1, M) with the fructification *Gondwanotheca sibirica* (Meyen 1984a; Neu-

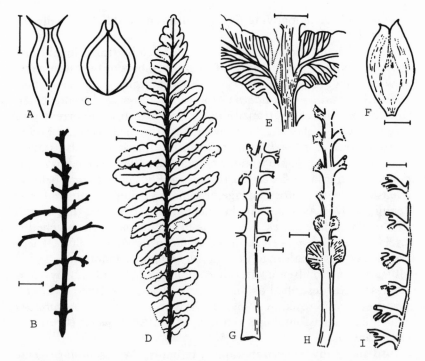

Figure 8.3. Euramerian seeds **(A, C)** and polysperms **(B)** associated with *Eremopteris* fronds, and Angara plants producing *Paragondwanidium sibiricum* fronds **(D, E)**, *Angarocarpus ungensis* seeds **(F)**, and *Gondwanotheca sibirica* polysperms **(G–I)**. Scale bar = 5 mm **(A, C, E–I)** and 10 mm **(B, D).** (After Meyen 1984a.)

burg 1948). The generic name *Gondwanotheca* is unfortunate because these plants have never been recorded in Gondwana. Moreover, the word "theca" implies a male fructification. Neuburg (1948) mistakenly interpreted apical seed scars on lateral branches as synangia. Sterile fronds may be once- or twice-pinnate, and small (young?) fronds are nearly entire, with faintly lobed margins. The fronds are much more reduced than *Eremopteris*.

The Lower Carboniferous of Angaraland does not yield plants from which this assemblage genus may be derived. Among Lower Carboniferous Euramerian plants externally similar to *Paragondwanidium*, the genus *Archaeopteridium* can be noted. *Angarocarpus ungensis* shares common features with *Cornucarpus* and *Samaropsis bicaudata* (Seward 1917). *Gondwanotheca* is very similar to pinnate polysperms (so far unnamed) accompanying *Eremopteris* in the Middle Carboniferous (figure 8.3C).

The *Paragondwanidium* fronds are connected by transitional forms with *Angaridium* (Meyen 1982a). Typical members of the latter genus show denticulate pinnule margins. Such margins have never been recorded in *Paragondwanidium*. More deeply dissected fronds of *Angaridium* are linked by transitional forms with fronds currently referred to *Rhodea javorskyi* (Gorelova 1978) from the lowermost Middle Carboniferous. *R. javorskyi* differs from Euramerian members of the genus in its once-pinnate fronds and may be better referred to *Angaridium*. Euramerian *Rhodea* (correct name *Rhodeopteridium*) belongs, at least partly, to protostelic lagenostomans having *Heterangium* axes (Jennings 1976). As in eustelic lagenostomans, their seeds might have been enclosed in cupules. No cupule-like remains (with attached or abscised seeds) have been recorded so far in the Angara flora. Therefore the lagenostomalean affinity of *Angaridium* is doubtful.

Angaridium fronds are often accompanied (A1 and A2) by small wingless seeds described as *Angarocarpus* (Gorelova 1978) or *Cordaicarpus baranovii* (Sukhov 1969). Beds with dominating *Angaridium* yield miospore assemblages dominated by *Psilohymena (Remysporites) psiloptera*.

Until the study of permineralized remains, the assemblage genus *Paragondwanidium-Gondwanotheca-Angarocarpus* and the genus *Angaridium* may be treated as satellite taxa of the order Callistophytales. The ancestors of these plants had obviously migrated into Angaraland by the end of the early Middle Carboniferous.

Order Peltaspermales.   The Peltaspermales were initially restricted to frontier regions of Angaraland: since the Artinskian, to the Subangara area; and since the early Late Permian, to the Pechora and Far Eastern provinces and the Taimyr-Kuznetsk district of the Angara area. In the Permo-Triassic they moved into central regions of Angaraland (the Siberian platform).

The family Trichopityaceae (figure 8.4) is reliably known only in the Fore-Urals (Subangara area). The characteristics of *Trichopitys* as given by Florin (1949) and widely adopted in the literature are erroneous in key points (Meyen 1984a). The errors are partly explained by the fact that Florin had not been able to inspect the holotype of *T. heteromorpha* (type-species) and other specimens from the type locality stored in the Museum Nationale d'Histoire Naturelle (Paris). These were kindly shown to the present author by Dr. C. Blanc-Louvel. These specimens (figure 8.4A–C) clearly show fronds with mixed pinnate

Figure 8.4. Euramerian *Trichopitys heteromorpha* **(A–G)** and members of Trichopityaceae from Fore-Urals, western Angaraland **(H–M).** **A** Apical part of fertile frond. **B, C** Lateral fertile branch in holotype. **D** Attachment of seed according to Florin. **E** Reconstruction of fertile axillary shoot. **F, G** Seed attachment, abaxial **(F)** and lateral **(G)** views. **H, I** Foliage of *Mauerites*. **J–M** The cladosperm *Biarmopteris pulchra*, general view; seed scars (black, **J, M**), detail of the same cladosperm, seed shown in black **(L).** Scale bar = 1 cm **(A–C, I, J, M)** and 1 mm **(K). (E, F, H–K, M,** after Meyen 1984a.)

and dichotomizing subdivisions. The rachides grade into linear leaf blades and are characteristically longitudinally striated (*Sparganum* cortex?). The axillary polysperms were not trimerous, as Florin thought, but planated, with seeds abaxially attached to apical widenings of the pinnately arranged flat lateral appendages (figure 8.4E, G). Toward the frond apex the pinnate cladosperms gradually decrease in size, then become ternate (figure 8.4B, C) and, finally, turn into solitary flattened nonaxillary seed stalks. Thus the apical portion of the seed-bearing frond (figure 8.4A) is identical to those pinnate lateral polysperms that occupy axillary positions in lower portions of the frond.

Sterile fronds of the Lower Permian genus *Mauerites* (figure 8.4H, I) from the Fore-Urals are very similar to those of *Trichopitys* in having the same intergradations between rachides and leaf blades and the same longitudinal striation of the rachides. Other features of *Mauerites* are epicuticular ribbing and resin bodies in the mesophyll, both features being very characteristic for many Peltaspermales (Meyen 1982a, 1984a). The associated cladosperms, *Biarmopteris* (A1, M), differ from those of *Trichopitys* by having a more dissected lamina and shorter seed-bearing laterals with less pronounced apical widenings (figure 8.4J–M; Meyen 1983, 1984a).

Unlike other members of Peltaspermales, the cladosperms of the Trichopityaceae retain many leafy characters, thus occupying an intermediate position between phyllosperms of Carboniferous Callistophytales *(Dicksonites, Callistophyton)* and strongly modified cladosperms of more advanced Peltaspermales, where leafy characters of fructifications are reduced to a minimum.

The dichotomously divided distal parts of the fronds and the dichotomizing leaves subtending bracts in *Trichopitys* are very similar to leaves of *Ginkgophyllum*, particularly of the type-species *G. grassetii* (the author's own observations in the same museum in Paris). *Ginkgophyllum grassetii* bears numerous resin bodies both in the axis and leaves. Some of the leaves described by Zalessky (1932, 1933, 1937) as *Dicranophyllum* and *Mauerites* from the Lower Permian of the Fore-Urals are very similar to *G. grassetii*.

There are no gross morphological differences between *G. grassetii* and the genus *Sphenobaiera* as defined by Harris et al. (1974). Mesozoic members of *Sphenobaiera* are currently referred to Ginkgoales, but Permian members of the genus may well belong to the Trichopityaceae being referred to *Ginkgophyllum*. If the gross morphological definition of *Sphenobaiera* proposed by Harris et al. (1974) is adopted, this genus must be merged with *Ginkgophyllum* and the name *Sphen-*

obaiera accordingly rejected. Another older synonym of *Sphenobaiera* (and *Ginkgophyllum*) is *Sclerophyllina* (Knobloch 1972).

The family Peltaspermaceae has been initially placed among "Mesozoic pteridosperms," although members were even more widespread in the Permian (Barthel and Haubold 1980; Kerp 1982; Meyen 1970, 1982a, 1983, 1984a). They are dominant plants in the uppermost Permian *Tatarina* flora of western Angaraland. The Angara Peltaspermaceae can be subdivided into four conventional groups.

1. Callipterids (Meyen 1970, 1982a; Meyen and Migdisova 1969). This group comprises fronds forming a continuous series from once-pinnate *(Compsopteris, Comia)* to at least twice-pinnate *(Callipteris,* including subgenus *Feonia)*. The epidermal structure, particularly stomata, is largely the same as in Triassic Peltaspermaceae. These fronds are accompanied (A1, E) by racemose aggregations of *Peltaspermum*-like peltoids (Meyen 1982a; text figure 18) that differ from the *Autunia*-like, bilaterally symmetrical cladosperms associated with Euramerian *Callipteris* (Kerp 1982; Meyen 1984a).

2. *Tatarina* and allied plants. Leaves identical in gross morphology to *Tatarina*, but devoid of cuticular characteristics, belong to the genus *Pursongia*, which was initially affiliated with Gondwana glossopterids. The *Tatarina* leaves may be entire (figure 8.5), lobed, or once-pinnate. In the latter case they are similar to *Compsopteris*, but differ in the absence of a midrib in the pinnules. The epidermal features are roughly the same as in other Peltaspermaceae. The leaf apex may be notched with a central short projection in the notch (figure 8.6A). Simple leaves of *Tatarina* might have been phyllodia. If so, the bipartite apex reflects an originally forked rachis, and the apical projection in the notch is a vestigial leaf lamina (as in phyllodia of some angiosperms). The hypothesis of the phyllodial nature of the *Tatarina* simple leaves is indirectly supported by the presence of numerous hypodermal strands which may be homologous to the *Sparganum*-like cortex in rachides of ancestral forms.

These leaves are accompanied (A2, E, M) by peltoids (figure 8.5) arranged in some species into globose heads and referred by Gomankov (Gomankov and Meyen 1986) to *Peltaspermopsis*. He referred some other compound polysperms and isolated peltoids to the genus *Lopadiangium* (Zhao et al. 1980).

In *Peltaspermopsis* the seed scars are smaller than in *Peltaspermum rotula* (type-species), and their position suggests a radial orientation of the major seed plane. Radial ridges on the outer side of the peltoids may bear a projection identical to the apical projection of *Tatarina*

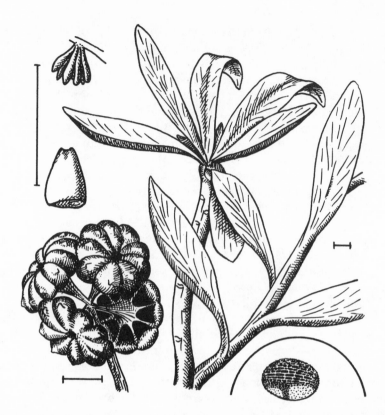

Figure 8.5. Leafy shoots of *Tatarina conspicua* and associated peltoids, *Peltaspermopsis buevichae;* seeds, *Salpingocarpus variabilis;* synangia, *Permotheca striatifera;* and pollen, *Vittatina subsaccata* f. *con-
nectivalis.* Scale bar = 1 cm. (From Gomankov and Meyen 1986.)

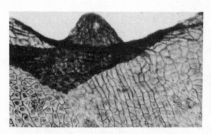

Figure 8.6. A Leaf of *Tatarina conspicua*, projection in apical notch.
× 68. **B** *Peltaspermopsis buevichae*, projection on adaxial surface of
peltoid. × 68.

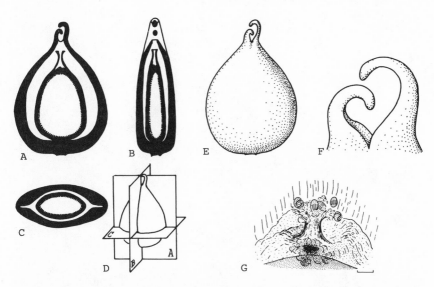

Figure 8.7. *Salpingocarpis bicornutus* seeds. **A** Section in primary plane. **B** Section in secondary plane. **C** Equatorial section. **D** Scheme of sections shown in **A–C**. **E** Reconstruction of general view. **F** Reconstruction of bicornute apex. **G** Apex of nucellus with salpinx and numerous *Protohaploxypinus* pollen grains inside seed, megaspore membrane (below) densely stippled (drawn from photograph, scale bar = 100 μm). (From Gomankov and Meyen 1986.)

leaves (figure 8.6A, B). The seed-bearing disc has a papillose margin comparable to that in valves of *Leptostrobus* (Leptostrobales–Czekanowskiales).

Seeds found in attachment to *Peltaspermopsis* and in sedimentary association with the peltoids and *Tatarina* leaves (A2, M, E, OC) belong to the genus *Salpingocarpus* (figures 8.5, 8.7; Gomankov and Meyen 1986). The perinucellar space forms a slit dividing the integument into two valves along the major plane of the seed. This bivalved structure of the integument, seen both in the arrangement of cutinized membranes and isolated seed stones, is also present in seeds associated with *Glossophyllum-Stiphorus*, *Sporophyllites* (see below), and Mesozoic Ginkgoales. A similar slit, although shorter, is observed in younger seeds of *Callospermarion* (Callistophytales) and young *Ginkgo* ovules.

The salpinx in *Salpingocarpus* is long, sometimes with a funnel-shaped (figure 8.7G) or lacerated apex (Meyen 1984a: figures 17B, C,

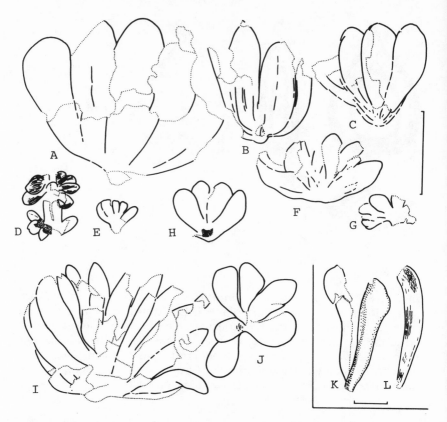

Figure 8.8. *Permotheca.* Synangia associated with *Tatarina* and producing *Protohaploxypinus* pollen **(A–G)**; synangium associated with *Phylladoderma* subgenus *Phylladoderma* and producing *Vesicaspora* pollen **(H)**; synangia associated with *Phylladoderma* subgenus *Aequistomia* and producing the same pollen **(I, J)**; and sporangia yielding *Vittatina* pollen **(K, L)**. Scale bar = 5 mm **(A–J)** and 1 mm **(K, L)**. (From Gomankov and Meyen 1986.)

19A). The seed apex bears two horns (figure 8.7). Among pollen found in the micropyle, the quasidisaccate, striated *Protohaploxypinus* is strikingly dominant. One of the *Peltaspermum* species is associated with striated asaccate pollen of the genus, *Vittatina* (figure 8.5; A2, E). Detached synangia yielding the same two pollen types belong to the genus *Permotheca* (figure 8.8A–G, K, L), also encompassing synangia of the family Cardiolepidaceae (see below). The sporangia are basally fused.

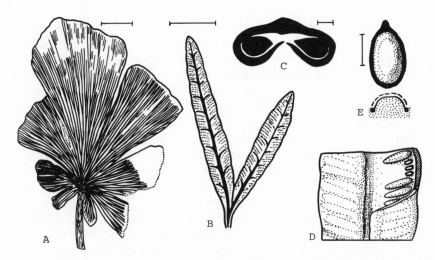

Figure 8.9. *Rhipidopsis ginkgoides* leaf, **(A)**; *Sporophyllites petschorensis*, cladosperm **(B–D)**; and seeds **(E)** probably belonging to same plant. **B** General view of cladosperm in adaxial view. **C** Reconstruction of cladosperm in cross section. **D** Reconstruction of cladosperm in abaxial view, with a portion of revolute margin abolished to show seeds and seed scars. Scale bar = 1 cm **(A, B)** and 1 mm **(C, E)**. **(A,** after Zalessky 1934.)

The systematic affinity of the pollen genera *Protohaploxypinus* and *Vittatina* has been enigmatic for a long time. Such pollen has been affiliated with pteridosperms, glossopterids, conifers, and *Welwitschia*-like plants. No one could have imagined that such pollen was produced by peltaspermaceous plants, because the Triassic members of the family produced monocolpate nonstriated pollen of the *Cycadopites* type (Townrow 1960), as in Ginkgoales. Interestingly, the Gondwana Arberiales (glossopterids) also produced identical pollen (Gould and Delevoryas 1977; Surange and Chandra 1974). *Protohaploxypinus*-like pollen has been also recorded in the Triassic conifer *Rissikia* (Townrow 1967).

3. *Sporophyllites-Rhipidopsis* group (figure 8.9). The genus *Sporophyllites* was initially described as a microsporophyll (Fefilova 1978), an interpretation that was questioned by the present author (Meyen 1982a). E. I. Poletaeva (Syktyvkar) kindly loaned the type material of *S. petschorensis* (type-species). The main axis of the compound polysperm bears numerous helically(?) arranged cladosperms having short stalks and forked seed-bearing laminae with revolute margins (figure

8.9C, D). Numerous small seed scars are arranged in a row on each side of the midrib near the line of lamina folding. Many small seeds (figure 8.9E) are spread over the surrounding rock. The seeds look like detached stones of Mesozoic ginkgoalean seeds (Harris et al. 1974), but are smaller. The macerated nucelli of the seeds found *in situ* were mistaken by Fefilova (1978) for sporangia. The pollen masses observed by her under the folded margins of the cladosperm obviously resulted from pollination (as observed in *Peltaspermopsis*). The seeds appear to have produced much pollination-liquid exudate, thus catching huge amounts of pollen. The pollen is of the *Vitreisporites* type, as in Mesozoic Caytoniales. The paired capsules of *Caytonia* can be easily derived from the *Sporophyllites*-like cladosperms. These cladosperms were associated most likely with the leaves, *Rhipidopsis ginkgoides* (A1, E; figure 8.9A). The putative association between *Sporophyllites* and *R. ginkgoides* is indirectly supported by common features of *Rhipidopsis* and the caytonialean leaves, *Sagenopteris*, both showing a palmate dissection and long petiole, although in *Sagenopteris* the venation is reticulate. In the Pechora Fore-Urals, *Rhipidopsis* is confined to the late Upper Permian, where the miospore assemblages are often dominated by *Vitreisporites*. The suggestion that *Sporophyllites* and *R. ginkgoides* (type-species of the genus) belong to the same plants does not imply that all species of *Rhipidopsis* were associated with similar fructifications. Palmate *Rhipidopsis*-like leaves might have arisen independently in different taxa.

4. *Stiphorus-Kirjamkenia-Glossophyllum* group (figure 8.10). A detailed description of the assemblage genus *Stiphorus-Glossophyllum* is given elsewhere (Gomankov and Meyen 1986; Meyen 1982a, 1983). The genus *Stiphorus* comprises paired, elliptical cladosperms with abaxial, submarginal seed rows (figure 8.10B, D). Judging from seed scars, the major plane of the seed was oriented transversely to the midrib. In the *Tatarina* flora, *S. biseriatus* is associated with leaves, *Glossophyllum permiense* (A2, M; figure 8.10A); and in the Permo-Triassic of Siberia *S. crassus* is associated with leafy shoots, *Kirjamkenia lobata*, producing both entire (*Glossophyllum*-like) and palmately dissected (*Sphenobaiera*-like) leaves (figure 8.10E–G). The seeds (figure 8.10C) accompanying *S. biseriatus* (A2, M), and showing the same resin bodies, are similar to detached stones of Mesozoic ginkgoalean seeds (Harris et al. 1974). The cladosperms, *Stiphorus*, are obviously similar to *Sporophyllites*. They are also similar (if not identical) to Triassic *Leuthardtia*, associating (A1, M) with *Glossophyllum* and *Sphenobaiera*-like leaves (see details in Meyen 1984a; Gomankov and Meyen 1986). Epidermal characters of these plants suggest their

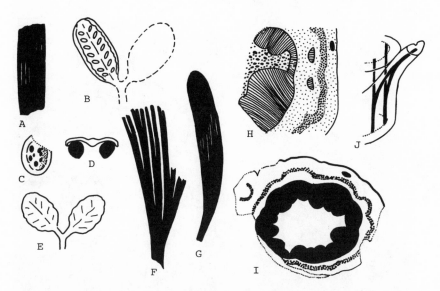

Figure 8.10. Assemblage genus *Stiphorus-Kirjamkenia-Glossophyllum.*
A *Glossophyllum permiense,* leaf fragment. **B** Reconstruction of abaxial view of *Stiphorus biseriatus* showing two rows of seed scars. **C** Seed
associated with preceding plant parts; resin bodies shown in black. **D**
Stiphorus, cladosperm in cross section (scheme); seeds shown in black.
E *S. crassus,* adaxial view. **F, G** *Sphenobaiera*-like **(F)** and *Glossophyllum*-like **(G)** leaves associated with *S. crassus.* **H–J** Anatomical structure of *Kirjamkenia lobata,* double leaf trace in cortex **(H)**; cross section of stem, xylem shown in black **(I)**; scheme of primary bundles
and diverging leaf trace entering petiole **(J)**. Drawings not to same
scale.

close relationship. The cuticle of *Kirjamkenia lobata* (Sadovnikov 1983)
is very similar to that of *Tatarina.*

Previously *Glossophyllum* was placed into Ginkgoales, and the ginkgoalean family Glossophyllaceae has been proposed (Tralau 1968).
This genus resembles *Eretmophyllum* so closely that the difference
between the two is uncertain. The presence of resin bodies in *Eretmophyllum* was regarded as a distinguishing character (Kräusel 1943),
but the present author has found resin bodies in *G. florinii* (type-
species) from the type locality as well as in *G. permiense* of the *Tatarina* flora. In establishing the ginkgoalean affinity of both *Glossophyllum* and *Eretmophyllum,* special attention has been usually paid to
two veins entering the leaf base, in contrast to a single vien in both
Leptostrobales (= Czekanowskiales) and conifers. In *Kirjamkenia lo-*

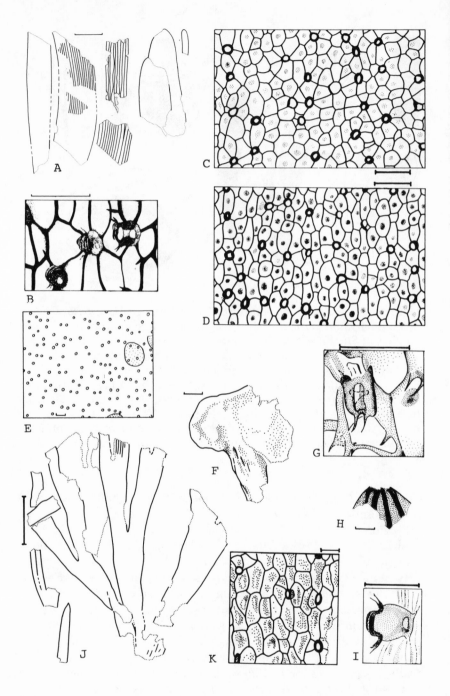

bata (figure 8.10H, J) two leaf traces that diverge independently from adjacent axial primary bundles fuse at the leaf base. The resulting bundle dichotomizes again in the leaf petiole (Sadovnikov 1983). A suppression of the fusion would result in two veins entering the leaf. The group *Stiphorus-Kirjamkenia-Glossophyllum* shows a combination of characters of the orders Peltaspermales, Ginkgoales, and Leptostrobales that suggests a close relationship between these orders (for further details see Gomankov and Meyen 1986; Meyen 1984a).

The family Peltaspermaceae also includes the Upper Permian fronds, *Odontopteris rossica* and *O. tartarica*, described by Zalessky (1927, 1929) from the Russian platform. In several localities they are accompanied (A2) by peltoids arranged into racemose compound polysperms. In venation these fronds resemble *Iniopteris*, which is very similar to *Psygmophyllum expansum*. *Iniopteris* and *Syniopteris* are regarded as synonyms of *Psygmophyllum* (Burago 1982). These plants may also belong to the Peltaspermales. *P. cuneifolium* resembles *Mauerites* and is morphologically closer to Trichopityaceae.

The family Cardiolepidaceae was erroneously placed into Coniferales (Meyen 1977, 1976–1978; see details in Meyen 1982a, 1984a). It includes plants with foliage quite different from that of conifers (figure 8.11). One leaf type *(Phylladoderma)* has lanceolate or linear laminae with parallel venation and one vein entering the base. The leaves are externally identical to the Mesozoic conifer, *Podozamites*. Another type (figure 8.11J, K) has palmately dissected laminae and externally resembles *Sphenobaiera*. The female fructifications, *Cardiolepis* (figure 8.12), found in association with *Phylladoderma* (A2, M) are modified peltoids in which the seeds, arranged around the peltoid stalk, are

Figure 8.11. Leaves of Cardiolepidaceae and their epidermal structure. *Phylladoderma* subgenus *Aequistomia annulata* (**A, H**); *P. (A.) rastorguevii* (**B, I**); *P. (A.)* sp. (**F**); and *Doliostomia pechorica* (**J, K**). **A** Outlines of leaves, and venation seen in naturally macerated compressions. **B** Stomata with laterally displaced epistomatal chambers and guard cells. **C, D** Cuticle of opposite leaf sides. **E** Distribution of stomata (shown by circles); resin bodies stippled. **F** Fragment of leaf with rhomboid lamina. **G** Stoma with cutinized wings of guard cells. **H** Venation in leaf apex as seen in naturally macerated compression. **I** Cuticle of epistomatal chamber, and lateral view of guard cells. **J** Outlines of leaves. **K** Epidermal structure. Scale bar = 1 cm (**A**); 5 mm (**J**); 1 mm (**F, M**); 100 μm (**B–E**); 50 μm (**G, I–K**). (From Gomankov and Meyen 1986.)

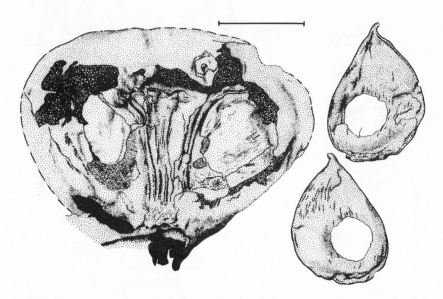

Figure 8.12. Seed-bearing capsule of *Cardiolepis piniformis* in longi-tudinal section (at left) and associated seeds *Nucicarpus piniformis* with projecting micropylar tubes (at right). Scale bar = 1 cm. (From Ignatiev 1983.)

nearly entirely enclosed by the disc which is bent down and embraces the stalk. The pollen found in the seed micropyles and in the associ-ated synangia, *Permotheca* (figure 8.8I, J), is of the *Vesicaspora* type— i.e., quasisaccate and bilaterally symmetrical (Meyen 1984a: figure 18A, B). The synangia are indistinguishable in their gross morphology from those producing the pollen, *Protohaploxypinus*, and belonging to the Peltaspermaceae. All the organs of the Cardiolepidaceae contain resin bodies or ducts. The same ducts are observed in axes accom-panying *Phylladoderma* leaves in the Kazanian of the Kama River region. These axes bear helically arranged appendages also character-ized by resin ducts. Sometimes the appendages are crowned by pel-toids with seed scars. Similar fructifications were described by Vladi-mirovich (1984) as *Quasistrobus ramiflorus*. Some of the fructifications accompanying *Phylladoderma* were probably open peltoids, not semi-closed capsules like *Cardiolepis*. It is also possible that these racemose aggregations of peltoids belong to plants with *Odontopteris rossica* fronds, although in the latter the resin bodies have never been ob-served. In any case, the close relationship between the Cardiolepida-ceae and Peltaspermaceae is evident.

Indubitable members of Cardiolepidaceae are known throughout the Upper Permian.

Other Members of Ginkgoopsida. Upper Permian fronds from western Angaraland referred to *Rhaphidopteris* (Meyen 1979) probably belong to Umkomasiaceae (=Corystospermaceae; Meyen 1984a). If so, one can suggest a northern origin of the family dominating the Triassic floras of Gondwana.

No reliable data are available on the presence of the orders Ginkgoales, Leptostrobales, and Arberiales in the Angara flora. The leaf genera previously affiliated with Ginkgoales (*Mauerites, Rhipidopsis, Glossophyllum, Phylladoderma,* and so on—see above) are either referred to other taxa or remain enigmatic (*Uralobaiera, Baieridium,* and so forth). The *Tatarina* flora contains *Sphenarion*-like leaves (Gomankov and Meyen 1986), but this is insufficient evidence of the presence of the Leptostrobales. *Glossopteris*- and *Gangamopteris*-like leaves occur rarely in Siberia, the far eastern Soviet Union, and Mongolia (Meyen 1982a; Zimina 1977), but associated fructifications have not been found.

Class Cycadopsida Among Paleozoic gymnosperms only Lagenostomales (=Lyginopteridales) and Trigonocarpales (=Medullosales) belong to Cycadopsida as outlined by the author (Meyen 1982c, 1984a). Reliable data on the presence of lagenostomans in the Angara flora are utterly absent. The Upper Devonian of the Minussa basin yields *Moresnetia*-like fructifications. Scheckler (see Rothwell 1985:323) treats the genus as ovulate cupules. But the Upper Devonian flora of the basin (and Siberia in general) does not qualify as Angara flora, because the rise of the independent Angara area occurred later (the very end of the Devonian to beginning of the Carboniferous).

The presence of Trigonocarpales, although suggested by diverse *Neuropteris* fronds (Gorelova 1978; Meyen 1982a; Neuburg 1948, 1961), has not been supported so far by findings of fructifications and permineralized *Medullosa*-like stems. Among Angara *Neuropteris*, imparipinnate fronds produced in the Euramerian flora by more advanced members of the order have not been observed, whereas the paripinnate fronds, unknown in the Euramerian area in post-Westphalian beds, occur until the Upper Permian (near Balkhash Lake and in western Taimyr). This exemplifies the extraequatorial persistence of primitive characters in the Angara flora (see "Discussion").

Class Pinopsida Earlier the author subdivided the Pinopsida into the orders Cordaitanthales and Pinales (Meyen 1984a). More recently, a third order, Dicranophyllales, was recognized (Meyen and Smoller 1986). All three orders appear in Angaraland later than in equatorial phytochoria. Cordaitean leaves appear in the Middle Carboniferous somewhat earlier than *Dicranophyllum*. In the Euramerian flora both Cordaitanthales and Dicranophyllales appear in the Namurian A (Serpukhovian stage of the Lower Carboniferous). The Cordaitanthales disappear first in western Angaraland and then in Siberia (Meyen 1982a), and the Dicranophyllales (Except *Zamiopteris*) disappear first in Siberia (near the end of the Carboniferous) and later in western Angaraland (in the middle Upper Tatarian). The role of Dicranophyllales in late Paleozoic floras has been underestimated. The dicranophylls are the characteristic and, in places, even the dominant plants in the Upper Paleozoic of Kazakhstan and the European part of the Soviet Union.

Conifers appear in central regions of Angaraland near the Permian–Triassic boundary and soon become dominant. They are very rare both in the Pechora basin and far eastern provinces of the Permian, but their shoots become dominant megafossils in western Angaraland from the Artinskian onward. Judging from palynological data, conifers appeared along the west and southwest frontier regions of Angaraland much earlier, in the Middle Carboniferous.

Order Cordaitanthales.   Most paleobotanical textbooks present only a few morphological types of cordaitanthalean fructifications and leaves. A survey of Euramerian taxa (Ignatiev and Meyen, in press), as well as the studies of Angara members of the order during the last decade (Meyen 1982a, 1982b, 1984a), has shown the much wider diversity of the order.

The Angara cordaitanthaleans have been referred to the families Vojnovskyaceae and Rufloriaceae. The presence of Cordaitanthaceae has not been established. In the meantime, the only generic name to accommodate some Angara leaves is *Cordaites*. That is why the family name Cordaitaceae has been rejected, and *Cordaites* is treated as a form-genus, various species of which may belong to different families (Meyen 1982a, 1984a).

The families Vojnovskyaceae (Neuburg 1963) and Rufloriaceae (Ledran 1966) were introduced when data on their fructifications were very meager. Both the size and limits of these families remain uncertain to date. The main defining character of these families is the dorsal stomatiferous furrows on rufloriaceous leaves. Neither pollen nor female fructifications provide clear-cut criteria for recognition of the two families. Quasimonosaccate *Cordaitina*-like pollen was pro-

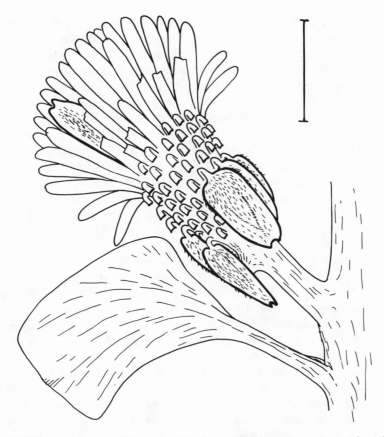

Figure 8.13. Reconstruction of *Vojnovskya paradoxa*, lateral poly-sperm (most seeds removed to show hooked seed stalks), and *Nephropsis*-like bract. Scale bar = 1 cm.

duced by both Rufloriaceae *(Pechorostrobus)* and Vojnovskyaceae (according to criterion A2; Gomankov and Meyen 1980). Other Rufloriaceae *(Cladostrobus)* produced a peculiar pollen of the *Cladaitina* type (Maheshwari and Meyen 1975) having intrareticuloid structure of the saccus, and proximal(?) germination. Both *Rufloria* and *Cordaites* (i.e., leaves with and without dorsal furrows) are associated with scaly bracts described as *Nephropsis*. Such bracts were found in attachment in *Vojnovskya paradoxa* (figure 8.13; Meyen 1982a, 1982b, 1984a; Neuburg 1965). Some *Nephropsis* species possess dorsal furrows, as in accompanying *Rufloria* leaves (A2, M). *Nephropsis* is unknown in the Carboniferous. Such bracts might have appeared independently in the two families in the Permian. Another possibility is that in the Carboniferous the two groups had not attained the family level of divergence. In other words, the independent status of the families is not adequately established and the allotment of genera between the two families often rests on circumstantial evidence alone.

Krassilov and Burago (1981) published a paper on the structure of the female fructification, *Gaussia*, and on the systematic affinity of Angara cordaitanthaleans. They had material of a single species and had not studied firsthand other taxa of Angara cordaitanthalean fructifications, relying on previously published, poor illustrations, and erroneous descriptions. As was shown elsewhere (Meyen 1982a, Addendum; 1984a), both the morphological interpretations and taxonomical conclusions of Krassilov and Burago are untenable, and their view that seeds of *Gaussia* were enclosed in a pistil-like organ is pure fantasy.

The Vojnovskyaceae may illustrate some of the most primitive characters of the Cordaitanthales. The initial interpretation of *Vojnovskya* by Neuburg (1963, 1965) as bisexual organs was much influenced by the discovery of allegedly bisexual fructifications in glossopterids. The obconical supra-axillary polysperms of *Vojnovskya* bear apical rod-shaped appendages among which a single attached seed was found. Neuburg decided that the appendages were microsporophylls, and her view has been adopted in the literature (Andrews 1961; Maekawa 1962; Takhtajan et al. 1963). Harris during his visit to Moscow in 1963 suggested to the author that Neuburg's interpretation was erroneous (for Harris' opinion, see also Mamay 1976–1978:295). Harris compared the apical appendages with the interseminal scales of Bennettitales. The error of Neuburg became evident when the author found indisputable male fructifications of Angara cordaitanthaleans. The attachment of *Nephropsis*-like bracts at a distance below the attachment of lateral polysperms has been confirmed. Paired scars left by the polysperm and subtending bracts are helically arranged on the main axis and can be homologized with bract-axillary complexes of other Pinopsida. Smaller appendages covering the major lower portion of the lateral polysperms were treated (Meyen 1982a, 1982b; Neuburg 1965) as sterile scales. Further studies have shown, however, that the appendages are hooked and apically widened (Meyen 1986b) —i.e., they are externally identical to seed stalks of *Krylovia*, *Gaussia*, and some *Cordaitanthus*. Thus most of the seeds were attached along the polysperm sides rather than at its apex (figure 8.14). Therefore the apical appendages are not interseminal scales but rather are homolo-

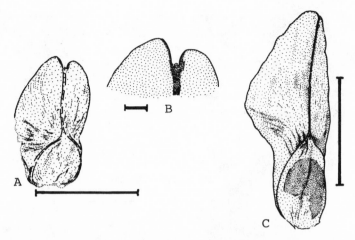

Figure 8.14. *Sylvella brevialata* seeds, probably associated with *Cordaites* leaves. Scale bar = 1 cm **(A, C)** and 1 mm **(B).** (From Ignatiev 1983.)

gous with sterilized seed stalks (funiculodia) of primitive conifers (see *Kungurodendron* below).

Judging from *Vojnovskya paradoxa*, the compound polysperms were very large, but not much larger than the biggest compound polysperms of *Cordaitanthus*. The helical arrangement of bract-axillary complexes of *Vojnovskya* appears more primitive than the four-ranked (in coupled pairs) arrangement of such complexes in most members of *Cordaitanthus*.

Carboniferous beds of Angaraland yield only isolated simple polysperms devoid of apical sterile appendages. *Nephropsis*-like scales are not present (the fructifications might have been subtended by foliage leaves). These polysperms belong to the genera *Krylovia* and *Gaussia*, currently referred to the Rufloriaceae on the basis of sedimentary associations (A1, A2) with *Rufloria* leaves. One should note, however, that the Carboniferous Angara cordaitanthalean polysperms were of the same basic type.

Microsporoclads associated with Angara leaves ascribed to *Cordaites* (devoid of dorsal furrows) belong to *Kuznetskia* (A1, A2). So far only Upper Permian members of the latter have been described (Meyen 1982a, 1982b). These are profusely branching organs, sometimes with a forked main rachis. Branching occurs in one plane (excepting terminal branchlets?). The sporangia are solitary, and the sporangial

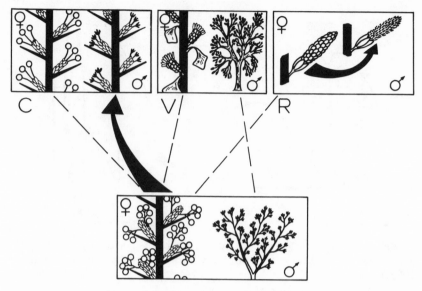

Figure 8.15. Scheme of gamoheterotopic transformation (by means of transfer of characters from one sex to another) of fructifications in Cordaitanthales. Male fructifications of Cordaitanthaceae **(C)** and Rufloriaceae **(R)** are modeled after polysperms; fructifications of Vojnovskyaceae **(V)** retain distinct sexual dimorphism, as in hypothetical ancestral group (shown below), where microsporoclads were similar to lagenostomalean microsporoclads and less reduced than in Vojnovskyaceae. Arrows show gamoheterotopic transfer of characters, and broken lines show usual inheritance of characters. (From Meyen 1986a.)

wall encloses characteristic rod-like bodies. Similar, albeit larger, microsporoclads are known in the Lower Permian, but their planation is less pronounced. *Kuznetskia* and allied fructifications strikingly resemble leafless microsporoclads of Lagenostomales and Calamopityales (Meyen 1984a; Millay and Taylor 1979; Skog and Gensel 1980), but do not show a tendency toward fusion of sporangia into synangia. In microsporoclad structure the Vojnovskyaceae are undoubtedly more primitive than the Cordaitanthaceae, which have strobiloid male fructifications. These Angara plants provide the basis for understanding which are ancestral members of the order Cordaitanthales as a whole (figure 8.15).

The sporangia of *Kuznetskia* yielded pollen similar to *Cordaitina*, but with less pronounced quasisaccae. The quasimonosaccate pollen accompanying Angara *Cordaites* and some species of *Rufloria* is of the

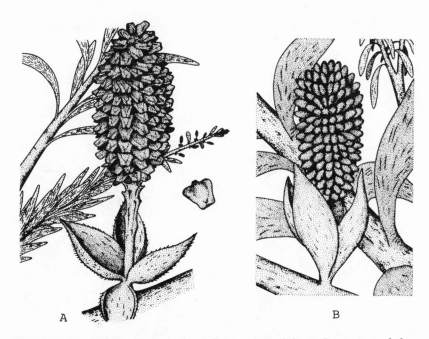

A   B

Figure 8.16. Polysperm, *Suchoviella synensis* **(A)**, and microstrobilus *Pechorostrobus bogovii* **(B)**, associated with *Rufloria synensis* leaves. (Reconstruction prepared by I. A. Ignatiev.)

same general type and never shows an inner corpus, or central body, like that of the pollen genera *Felixipollenites*, *Sullisaccites*, and *Florinites* of Euramerian cordaitanthaleans. The quasisaccate structure is comparable to that known in Archaeopteridales. Once again we observe a retention of a primitive character in Angara plants.

The Vojnovskyaceae may also comprise the seed genus *Sylvella* (figure 8.14), invariably associated (A1, A2) with *Cordaites*-like leaves. These seeds resemble coniferalean seeds in having a long asymmetrical wing. The latter embraces both the nucellar parts of the seed and a very long micropyle.

The Rufloriaceae can be clearly distinguished from the Vojnovskyaceae among the Upper Permian forms. The Upper Permian Rufloriaceae show a uniform structural plan in both male and female fructifications (figures 8.15, 8.16). The simple polysperms *Suchoviella* (designated as "Rs" in Meyen 1984a; the generic name is validated in Ignatiev and Meyen, in press) and the associated (A2, M) microstrobilus *Pechorostrobus* both consist of the main axis bearing an involucre of *Lepeophyllum*-like sterile scales subtending the fertile part of the

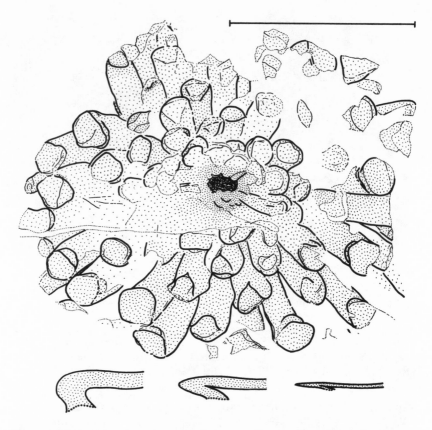

Figure 8.17. *Gaussia cristata* polysperm. Drawn from transfer preparation, abaxial view; below are schemes of seed stalk flattening; seed scars shown by dotted lines. Upper Carboniferous, Minussa basin. Scale bar = 1 cm.

axis (figure 8.16). The fertile part bears crowded solitary sporangia or seeds, both helically arranged. Similar fructifications with seed stalks or sporangia, although without involucres preserved, are known from the Upper Carboniferous.

In the Carboniferous genus *Krylovia*, the seeds may be helically arranged along an elongated axis or concentrated at the apex of a shorter, slightly obconical axis. The two forms are linked by transitional forms within the same sedimentary association and hence are treated as intraspecifc variations (Meyen 1982a, 1982b). Forms with a short axis closely resemble the umbellate polysperm *Gaussia cristata* (figure 8.17). In the Lower and early Upper Permian genus *Bardocar-*

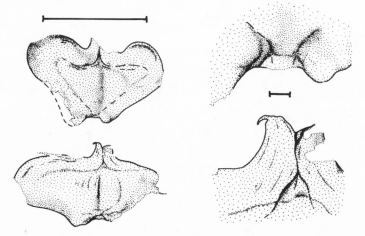

Figure 8.18. *Bardocarpus aliger* seeds. At left, general view (scale bar = 1 cm); at right, seed base and apex (scale bar = 1 mm). (From Ignatiev 1983.)

pus, the axis is long and the seeds are sessile, or nearly so. Detached *Bardocarpus* seeds (figure 8.18) are sometimes accompanied (A1) by scaly leaves that could belong to subtending involucres. The sedimentary associations do not provide reliable hints as to the type of foliage leaves *(Cordaites* or *Rufloria)* produced by these plants. On the basis of the close similarity to *Suchoviella,* the *Bardocarpus* polysperms were affiliated with Rufloriaceae (Meyen 1982a, 1984a). However, the polysperms of *Vojnovskya* also bear crowded seeds along the axis. Therefore the affinity of *Bardocarpus* with the Vojnovskyaceae cannot be ruled out.

Compound polysperms earlier provisionally designated as *Laxostrobus* (Meyen 1966) stand apart. The name *Laxostrobus* is nomenclaturally invalid, and in order to avoid nomenclatural confusions, these polysperms are called below "Rb" (after the associated leaves, *Rufloria brevifolia;* criterion A2). Lateral simple polysperms of Rb are short and strongly flattened, and the seed stalks are decurrent and of different length. The similarity with primitive coniferalean seed scales is striking. However, the coniferalean affinity of Rb is ruled out by the total absence of coniferalean shoots in the Upper Permian flora of Siberia, from which Rb comes. The absence of bracts in Rb is remarkable. This is not a very rare feature, however, as bracts also have not been recorded in the conifer *Timanostrobus* (see below) and some cordaitanthalean fructifications. The latter come from the Pechora basin and were kindly shown to the author by I. A. Ignatiev.

Microstrobili associated (A2) with both Rb and *Rufloria brevifolia* belong to *Cladostrobus* (Maheshwari and Meyen 1975). They are similar to coniferalean microstrobili in having a long microsporophyll stalk crowned by a rhomboid shield. The pollen found *in situ* is of the *Cladaitina* type. The saccus shows a negative intrareticuloid sculpture and covers the central body both equatorially and distally, leaving a proximal unenclosed area. The degree of enclosure by the saccus is variable, some grains appearing asaccate. Both the intact grains and detached bodies and sacci are often folded, and then boat-shaped, simulating monocolpate grains. That is why they were described in the literature as *Ginkgocacydophytus*.

Order Dicranophyllales.   A detailed survey of the Angara dicranophyllalean genera and of the relationship between the Dicranophyllales and other pinopsids is given elsewhere (Meyen and Smoller 1986). The dicranophyllalean affinity of the pertinent Angara taxa rests on epidermal and certain gross-morphological features of leaves. These features are sufficiently peculiar to be estimated as reliable taxonomic markers of the whole group.

The most widespread genus is *Dicranophyllum* (Barthel 1977). Its linear, once- or repeatedly dichotomizing (up to four times), and occasionally simple leaves bear two submarginal, dorsal furrows and marginal microdenticulation. The furrow structure is strikingly similar to that of the older members of *Rufloria*, suggesting a relationship between the two groups. *Mostotchkia* (figure 8.19B, C), known in the Upper Carboniferous of Siberia and Permian of western Angaraland, differs from *Dicranophyllum* in possessing only simple leaves. The Upper Permian *M. gomankovii* shows the same dorsal furrows and stomata, even in minor details, as in *Entsovia*. The leaves of *Entsovia* (figure 8.19F, G) are strap-shaped or spathulate, with numerous paired dorsal furrows. The latter enter lateral leaf margins (in the Upper Carboniferous *E. rara;* Glukhova 1978), or all extend to the leaf apex (remaining species). Numerous and, at times, paired furrows are present in the scale-like leaves of *Slivkovia* (figure 8.19E), which comprise leafy shoots that resemble those of conifers. In stomatal structure and the marginal denticulation, the leaves of *Slivkovia* are closer to *Dicranophyllum* than to *Entsovia*. The upper nonstomatiferous epidermis of all these genera is identical.

The compact stomatal bands, stomata, upper epidermis, and marginal microdenticulation of *Slivkovia* and *Dicranophyllum* are the same as in *Lesleya delafondii* (figure 8.19I; Lower Permian of France) and *Zamiopteris* (as described from the Permian of the Pechora basin;

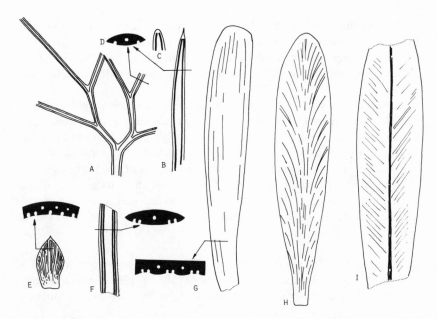

Figure 8.19. Dicranophyllales. **A** *Dicranophyllum effusum,* dorsal furrows shown by submarginal lines. **B** *Mostotchkia longifolia,* dorsal furrows shown by thick lines. **C** *M. gomankovii,* leaf apex with two dorsal furrows. **D** Diagram of cross sections of leaves shown in **A–C.** **E** *Slivkovia petschorensis,* leaf and diagram of its cross section. **F** *Entsovia lorata,* leaf and diagram of its cross section. **G** *E. rarinervis,* leaf and diagram of its cross section. **H** *Zamiopteris.* **I** *Lesleya delafondii.* Drawings not to same scale. (**A–F** from Meyen and Smoller 1986.)

figure 8.19H). The latter two genera differ in venation (pinnate in *Lesleya* and fan-shaped in *Zamiopteris*). In gross morphology and epidermal features, *Zamiopteris* is linked by transitional forms with leaves of the *Cordaites* type. On the other hand, once-forked *Dicranophyllum* leaves appear identical to *Gomphostrobus* leaves of the Walchiaceae, which also have compact stomatal bands and marginal microdenticulation.

The above-mentioned epidermal characters do not appear in the Ginkgoopsida and are not characteristic of the Cycadopsida. Narrow spaces between veins as observed, for instance, in *Paragondwanidium* and *Angaridium* never show such distinct lateral limits or such constant width as the dorsal furrows of the leaves of Dicranophyllales and Rufloriaceae. Among Paleozoic plants, the marginal microdenticulation occurs only in the Pinopsida.

The fructifications of the Angara dicranophyllaleans are unknown. In sedimentary associations (A1) there occur racemose, simple polysperms having nonbranched, thin axes and long seed stalks (as in *Krylovia*). Comparable polysperms occur together with *Lesleya delafondii*.

In Siberia, both *Mostotchkia* and *Dicranophyllum* rarely occur in the Middle–Upper Carboniferous, and *Zamiopteris* is common in the Permian. In the Subangara area, *Dicranophyllum*, *Mostotchkia*, and *Entsovia* are common, whereas *Slivkovia* is rare. *Mostotchkia*, *Entsovia*, and *Slivkovia* are known in several localities of the Pechora basin. *Dicranophyllum* occurs in many Upper Paleozoic localities of Kazakhstan, where K. Z. Salmenova has also recorded solitary *Entsovia* and *Slivkovia*. In the *Tatarina* flora, as well as in the Permo-Triassic assemblages of Siberia, the dicranophyllaleans are absent.

Order Pinales.   Conifers are numerous in the Subangara area, and have been described in detail elsewhere (Meyen 1986b, and in press). Below, only major observations and conclusions are briefly summarized.

When initiating his study of the conifers of western Angaraland at the beginning of the 1970s, the author expected to find intermediates between *Lebachia* and *Pseudovoltzia*, reflecting the phylogeny of conifers proposed by Florin (1938–1945, 1951). Later he concluded tentatively (Meyen 1976–1978) that the Angara conifers were too peculiar to conform to Florin's phylogeny. Both of these suppositions have proved to be only partly justified. Peculiar plants hitherto referred by the author to conifers are presently affiliated with other orders (*Cardiolepis* with Peltaspermales, and *Slivkovia* with Dicranophyllales). True conifers from the Kungurian, Ufimian(?), Kazanian, Tatarian, and the Permo-Triassic have proved to be much more primitive than one would have expected, judging from their stratigraphic position.

Before discussing factual data, it is expedient to emphasize that both the taxonomy and nomenclature of earlier conifers need substantial reform. Presently for a number of key taxa a reliable typification is lacking, and the taxonomy and nomenclature of fertile and sterile shoots have not been properly separated, as in more recent conifers and most other fossil plant taxa. The type material of the most important genera *(Walchia, Ernestiodendron, Ullmannia, Pseudovoltzia, Voltzia)* consists of sterile shoots, the epidermal characters of which cannot be studied.

Considering the necessity of (1) independent generic names for vegetative shoots and fructifications, and (2) a strict following of the principle of typification, the present author (Meyen 1986b, in press) suggested retaining the generic name *Walchia* (= *Lebachia;* see Clement-Westerhof 1984) for sterile shoots alone. The female fructifications referred by Florin (1938) and his followers to *Lebachia* are accordingly assigned to a new genus, *Lebachiella*, with the new type-species, *Lebachiella florinii* (female fructifications hitherto placed into *Lebachia piniformis*). *Walchia* is regarded as a satellite genus of the family Lebachiellaceae. The latter replaces the Walchiaceae in the natural system of conifers. The name Walchiaceae is restricted to a form-group of coniferalean shoots similar to *Walchia. Ernestiodendron* comprises only sterile shoots, and the associated female fructifications are referred to *Walchiostrobus*.

The family Lebachiellaceae consists of two subfamilies. A more primitive subfamily, Kungurodendroideae (figure 8.20J, M, P. Q), comprises the Euramerian Upper Carboniferous species *"Lebachia" lockardii* (Mapes and Rothwell 1984), deserving separation into a new genus, as well as the Subangara genera *Kungurodendron* (Kungurian, Middle Fore-Urals), *Timanostrobus*, and, probably, *Concholepis* (both from the Upper Permian of the Timan Range). In this subfamily, the seeds are apical in position on the seed stalks (as was thought erroneously by Florin to be true of *Lebachia* and *Ernestiodendron;* Clement-Westerhof 1984; Meyen 1984a, 1986). The axillary fertile shoots of *Kungurodendron* (figure 8.21A) are flattened and dorsiventral, and bear numerous hooked seed stalks on the adaxial side. Both the apex and distal part of the adaxial side of the fertile shoots are occupied by funiculodia (a new term introduced for sterilized seed stalks often bearing scars of aborted seeds and showing the same epidermal structure as fertile seed stalks). The proximal part of the axillary fertile shoots are covered by sterile scales that are more similar to underdeveloped foliage leaves than to seed stalks or funiculodia. In contrast to *Lebachiella*, the bracts are not forked and are epistomatic. They differ from foliage leaves of the axis bearing the compound polysperm only in larger size and some other minor details.

The microstrobili of *Kungurodendron* (A2, M) yielded pollen (figure 8.21B–F). The pollen exine was initially thought by the author to be columellar (under the light microscope). A TEM study of the pollen by B. Lugardon (Toulouse), to be published elsewhere, indicates that only the nexine is clearly lamellar, and that the sexine is composed of a thin layer of large alveolae, the residual walls of which simulate columellae. This columellar-like layer embraces the grain both equatorially and distally, leaving a wide proximal area with a small slit having two or three asymmetrical rays. The distal aperture is absent.

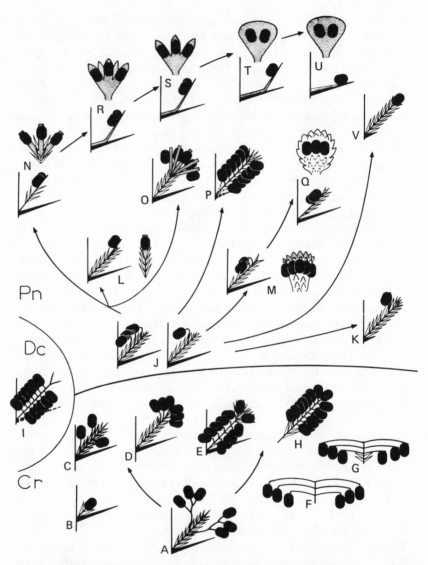

Figure 8.20. Main types of female fructifications in Cordaitanthales (Cr), Dicranophyllales (Dc), and Pinales (Pn). Bracts and seeds in black; seed scales **(N, R–U)**, leaf-like seed stalks **(L, O)** and funiculodia **(M)** stippled; sterile scales shown by narrow white triangles; homologous organs shown identical; spirally arranged organs two-ranked. **A** *Cordaitanthus pseudofluitans.* **B** *C. diversiflorus* (sterile scales problematical). **C** *C. zeilleri.* **D** *C. duquesnensis.* **E** *Vojnovskya.* **F** *Gaussia cristata.* **G** *G. scutellata.* **H** *Suchoviella.* **I** *Dicranophyllum.* **J** *"Lebachia"*

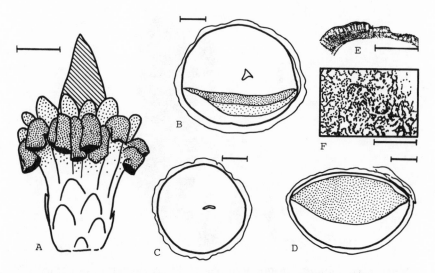

Figure 8.21. *Kungurodendron sharovii.* **A** Reconstruction of axillary fertile shoot and subtending bract (lined) in adaxial view; seed stalks densely stippled; seed scars shown in black; funiculodia sparsely stippled. **B–F** Pollen grains with trilete **(B)** and dilete **(C)** proximal slit; boat-shaped grain **(D)**; lateral optical view of columellar-like elements **(E)**; surface view of columellar-like layer **(F)**. Scale = 2 mm **(A)** and 10 μm **(B–F)**. **(A–D**, from Meyen 1986b.)

The presence of both the proximal slit and columellar-like layer, as well as the globose outline and the absence of the distal aperture, allows one to draw parallels between *Kungurodendron* and the problematical Pennsylvanian genus *Lasiostrobus* (Taylor 1970; Taylor and Millay 1977). The affinity of *Lasiostrobus* with more primitive Lebachiellaceae seems probable.

The genus *Timanostrobus* (figure 8.22) is even more peculiar. Its leafy shoots (A2, M, OC) resemble *Pagiophyllum* and *Brachyphyllum*. As in other primitive conifers, the compound polysperms and microstrobili crown nonmodified leafy shoots. The lateral, simple poly-

---

lockardii. **K** *Buriadia.* **L** *Lebachiella florinii.* **M** *Kungurodendron.* **N** *Walchiostrobus.* **O** *Sashinia.* **P** *Timanostrobus.* **Q** *Concholepis.* **R** *Pseudovoltzia, Voltzia.* **S** *Swedenborgia, Aethophyllum,* and allied Triassic genera. **T, U** More advanced Mesozoic and Cenozoic conifers with partly **(T)** and completely **(U)** fused seed scale and bract. **V** taxads.

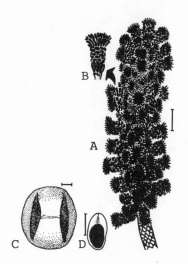

Figure 8.22. *Timanostrobus muravievii.* Reconstruction of compound polysperm (**A**) and simple lateral polysperm (**B**); pollen (**C**); seed with megaspore membrane shown in black (**D**). Scale bar = 1 cm (**A**); 1 mm (**D**); 10 μm (**C**). (From Meyen 1986b.)

sperms are strikingly similar to those of *Vojnovskya*, being nonflat-tened and bearing crowded, helically arranged seeds and apical, rod-shaped sterile scales (funiculodia?) along the elongated, stout axis. Unlike *Vojnovskya*, bracts have not been observed, although they could have been strongly reduced and hence unrecognizable among basal seed stalks of the lateral polysperms. The pollen found both *in situ* and in seed micropyles is globose and often folded into boat-shaped grains. In contrast to *Kungurodendron*, the sexine is not columellar-like and shows dense perforations in the tectum. The proximal area lacks perforations and bears a well-developed monolete to trilete scar.

Polysperms of the genus *Concholepis* (figure 8.20Q) crown leafy shoots indistinguishable from those of *Ullmannia frumentaria*, but the polysperm structure is totally different in the two taxa. That is why the independent nomenclature for sterile and fertile shoots currently applied to *Ullmannia* is necessary. The free bracts of *Concholepis* are long and loosely spaced (bracts in *U. frumentaria* cannot be recog-nized in published photographs). The axillary fertile shoots are strongly flattened and obovate in outline. Both the abaxial side and the margin are covered by short projections externally similar to, but shorter than, apical sterile scales of *Timanostrobus* or funiculodia of *Kungu-*

rodendron. A row of submarginal projections is present on the other-wise smooth adaxial surface. Remains of megaspore membranes were found adpressed to the adaxial surface, but both the number and mode of attachment of the seeds remain uncertain. The axillary, fertile shoot complex of *Concholepis* might have arisen from that of *Kungurodendron.* Accordingly, it is interpreted as a false seed scale, because true seed scales of Voltziaceae and their descendants were formed of fused seed stalks with a minor (and still problematical) participation of sterile members of the axillary fertile shoot (Meyen 1984a, 1986b, in press).

Among Angara conifers a true seed scale appears to have been present only in *Pseudovoltzia? cornuta.* The question mark in this binomial is included because the relevant polysperms show free bracts (unlike a partial fusion between the bract and seed scale in *P. lie-beana*) and both the number and attachment of seeds are still uncertain.

The assemblage genus *Sashinia-Dvinostrobus-Quadrocladus* (A2, M, OC) from the *Tatarina* flora belongs to the subfamily Lebachiellaceae (figure 8.20L, O), as the seeds are abaxially (not apically) attached to their stalks. The compound polysperm *Sashinia* is remarkable in its seemingly primitive habit. Both the bracts and the sterile scales of the fertile shoot are indistinguishable from foliage leaves. Moreover, the seed stalks also resemble foliage leaves, even having a mucronate apex and the same epidermal structure. This unification of the organs is obviously secondary in origin (Meyen 1981, 1984a). Free seed stalks are arranged into a bunch or a compressed (telescoped) helix at the apex of the axillary fertile shoot, so that even faint hints of seed scale formation are totally absent. This character is undoubtedly primarily primitive; hence *Sashinia* may be referred to Lebachielloideae, also devoid of seed scales. *Sashinia* differs from other members of the subfamily *(Lebachiella, Ortiseia),* apart from other characters, in an unusual seed stalk structure. The blunt apex of the stalk is widened into a pad that is recurved abaxially and entirely covers the young ovule.

The associated microstrobilus *Dvinostrobus* (Gomankov and Meyen 1986; Meyen 1981, 1984a), has several primitive traits such as the absence of a heel in the distal shield, and the attachment of micro-sporangia by slender sporangiophores to the middle part of a long microsporophyll stalk. The advanced characters are a thin main axis of the microstrobilus, which may not have continued a leafy shoot axis, and the quasisaccate pollen with three wide taenia *(Scutaspor-*

UPPER DEVONIAN	CARBONIFEROUS			PERMIAN		PERMO-TRIASSIC		
	LOWER	MIDDLE	UPPER	LOWER	UPPER			
								Pycnoxylic woods
								Forked rachis
								Cardiopteroid pinnules
								Paripinnate neuropterid fronds
								Simple leaves
								Needle-shaped leaves
								Taeniopteroid leaves
								Cycadophyte-like leaves
								Reticulate venation
								Dorsal stomatiferous furrows
								Free cupules
								Phyllosperms and clado-sperms
								Coniferalean seed-scale
								Specialized bracts
								Trigonocarpalean seeds
								Telangiopsis-like microsporoclad
								Synangia

ites). *Dvinostrobus* resembles both *Cladostrobus* (see above) and the Triassic putatively voltzialean microstrobili *Sertostrobus* and *Darneya* (Grauvogel-Stamm 1978).

DISCUSSION

Remains of the gymnosperms discussed above are often main components of Angara plant megafossil assemblages from the Middle Carboniferous to the Permo-Triassic. There are also some other gymnosperm remains of uncertain affinities, but their more detailed studies would hardly influence the overall picture.

The following salient features characterize the gymnospermous component of the Angara flora (figures 8.23, 8.24).

1. Gymnosperms are very scarce and nondiversified in the Lower Carboniferous, where only *Angaropteridium–Cardiopteridium*-like plants are prominent in plant megafossil assemblages, any associated cupule-like organs being totally absent. The constant absence of fructifications in most burials of *Angaropteridium* suggests a largely vegative propagation of these plants (semiaquatic or aquatic?).

2. Among more widely spread Angara plants, only *Neuropteris* and *Angaropteridium* may belong to Cycadopsida. These plants are obviously of equatorial origin. *Angaropteridium* is strikingly similar to the Lower Carboniferous Euramerian genus *Cardiopteridium*. In the Permian of the Subangara area and of the inner parts of the Angara area (Tunguska basin, northeastern Soviet Union, northern Mongolia), *Neuropteris* is absent, and indisputable members of the Cycadopsida have never been recorded. Considering that the same pertains to Gondwanaland, one can conclude that in the Late Paleozoic the Cycadopsida were essentially equatorial plants.

3. The class Pinopsida is represented by all three orders, each appearing later than in the Euramerian flora. The most primitive order, Cordaitanthales, was the most common component in inner regions of Angaraland, whereas Dicranophyllales (excepting *Zamiopteris*) and the most advanced order Pinales were more abundant in outer regions (Subangara area). Unlike equatorial phytochoria, the inner regions of Angaraland retained the dominance of cordaitanthaleans until nearly the end of the Permian. On the contrary, conifers penetrated into

Figure 8.23. Stratigraphic ranges of characters of gymnosperms in equatorial phytochoria (thin lines) and Angara flora (thick lines); triangles denote characters showing extraequatorial persistence.

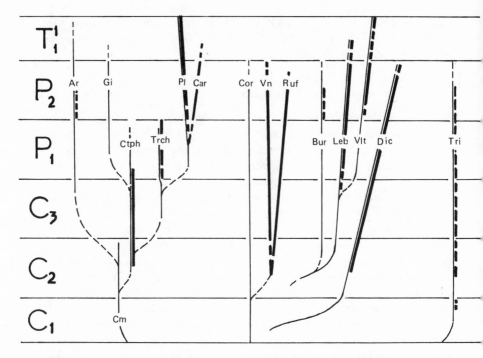

Figure 8.24. Stratigraphic ranges of suprageneric taxa of gymnosperms of Angara flora (thick lines) plotted against general phylogeny of gymnosperms. Cm = Calamopityales; Ar = Arberiales; Gi = Gigantonomiales (gigantopterids of Cathaysia); Ctph = Callistophytales; Trch = Trichopityaceae; Pl = Peltaspermaceae; Car = Cardiolepidaceae; Cor = Cordaitanthaceae; Vn = Vojnovskyaceae; Ruf = Rufloriaceae; Bur = Buriadiaceae; Leb = Lebachiellaceae; Vlt = Voltziaceae; Dic = Dicranophyllales; Tri = Trigonocarpales.

these regions as late as the Permo-Triassic—i.e., when connections between all northern floras suddenly and drastically increased.

Thus current knowledge of the Angara members of the Pinopsida, together with relevant data on Gondwana floras, clearly supports an equatorial or subequatorial origin of the three orders of Pinopsida. More primitive conifers were thermophilic. Cryophilic groups of conifers appeared much later.

4. The gymnosperms of the class Ginkgoopsida are relatively uncommon in inner regions of Angaraland. Their maximum diversification is observed in the Subangara area and along the periphery of the Angara area. Peltaspermaceae of the Subangara area and of the Pe-

chora province exhibit various similarities to Caytoniales, Leptostrobales, and Ginkgoales. These three orders might have originated in the Subangara area or in adjacent subequatorial regions.

5. The available data on Angara gymnosperms fit well the general ideas on the origins of new suprageneric taxa of higher ranks mostly in phytochoria of low latitudes (Meyen 1984a, 1984b). The extraequatorial gymnosperms are represented to a large extent by persistent, archaic morphological types.

6. Entire orders and even classes exhibited preferential geographic restrictions during the Late Paleozoic. There is no explanation why in the Permian of Angaraland, even in warmer southern regions, indisputable members of Cycadopsida (excepting relatively rare occurrences of *Neuropteris*) are absent. This preferential low latitudinal distribution of Cycadopsida continued in later periods. Ginkgoopsida penetrated farther to the north in late Paleozoic, but did not reach high latitudes. On the contrary, primitive Pinopsida (Cordaitanthales, particularly Vojnovskyaceae) were among the dominant plants (together with articulates in the Permian, and *Angaropteridium* in the Carboniferous) at very high latitudes. This distribution of classes among climate zones would be impossible to predict solely on the basis of available morphological characters. Obviously the pattern of geographical distribution of the gymnosperm suprageneric taxa rested solely on unknown physiological (rather than morphological and anatomical) grounds.

ACKNOWLEDGMENTS

I thank Mr. I. A. Ignatiev (Moscow) for providing some illustrations and data on cordaitanthalean fructifications. The help of Mrs. E. I. Poletaeva (Syktyvkar) in obtaining the type-material of *Sporophyllites petschorensis* is gratefully acknowledged. Special thanks are extended to Dr. B. Lugardon (Toulouse) for the TEM study of *Kungurodendron* pollen and comments on its structure. I am very thankful to Professor C. B. Beck for correcting my English.

LITERATURE CITED

Andrews, H. N. 1961. *Studies in Paleobotany*. New York and London: Wiley.
Barthel, M. 1977. Die Gattung *Dicranophyllum* Gr. Eury in den varistischen Innensenken DDR. *Hallesch. Jhb. Geowiss.* 2:73–86.
Barthel, M. and H. Haubold. 1980. Zur Gattung *Callipteris* Brongniart. Teil 1.

Die Ausbildung von *Callipteris conferta* (Sternberg) Brongniart in mitteleuropäischen Rotliegenden. *Schriftener. geol. Wiss. Berlin* 16:49–105.

Beck, C. B. 1985. Gymnosperm phylogeny—a commentary on the views of S. V. Meyen. *Bot. Rev.* 51:273–294.

Burago, V. I. 1982. The morphology of leaves of the genus *Psygmophyllum*. *Paleontol. Zhurn.* 2:128–136. (in Russian).

Clement-Westerhof, J. A. 1984. Aspects of Permian palaeobotany and palynology. 4. The conifer *Ortiseia* Florin from the Val Gardena Formation of the Dolomites and the Vicentinian Alps (Italy) with special reference to a revised concept of the Walchiaceae (Göppert) Schimper. *Rev. Palaeobot. Palynol.* 41:51–166.

Fefilova, L. A. 1978. *Sporophyllites* from the Upper Permian deposits of North Fore-Urals and its taxonomic position. *Trudy Inst. Geol. Komi Fil. AN SSSR* 25:29–41 (in Russian).

Florin, R. 1938–1945. Die Koniferen des Oberkarbons und des untern Perms. 1–8. *Palaeontographica* 85B:1–729.

—— 1949. The morphology of *Trichopitys heteromorpha* Saporta, a seed-plant of Paleozoic age, and the evolution of the female flower in the Ginkgoinae. *Acta Horti Bergiani* 15:80–109.

—— 1951. Evolution in cordaites and conifers. *Acta Horti Bergiani* 15:285–388.

Glukhova, L. V. 1978. On the discovery of *Entsovia* in the Carboniferous of the Tunguska Basin. *Paleontol. Zhurn.* 3:139–140 (in Russian).

Gomankov, A. V. and S. V. Meyen. 1980. On relations between assemblages of plant micro- and megafossils in the Permian of Angaraland. *Paleontol. Zhurn.* 4:114–122 (in Russian).

—— 1986. *Tatarina*-flora (composition, distribution in the Late Permian of Eurasia). *Trudy Geol. Inst. Akad. Nauk SSSR* 401 (in Russian).

Gorelova, S. G. 1978. The flora and stratigraphy of the coal-bearing Carboniferous of middle Siberia. *Palaeontographica* 165B:53–77.

Gould, R. E. and T. Delevoryas. 1977. The biology of *Glossopteris:* Evidence from petrified seed-bearing and pollen-bearing organs. *Alcheringa* 1:387–399.

Grauvogel-Stamm, L. 1978. La flore du grès à *Voltzia* (Buntsandstein Supérieur) des Vosges du Nord (France): Morphologie, anatomie, interprétations phylogénique et paléogéographique. *Univ. L. Pasteur de Strasbourg, Inst. Géol. Mém.* 50:1–225.

Harris, T. M., W. Millington, and J. Miller. 1974. *The Yorkshire Jurassic Flora*, Vol. 4, parts 1 and 2. *Ginkgoales* and *Czekanowskiales*. London: British Museum (Natural History).

Ignatiev, I. A. 1983. Seeds from the Permian of the Pechora Fore-Urals (materials to revision of main taxa). *Moscow VINITI, Dep.* 6126-83:1–61 (in Russian).

Ignatiev, I. A. and S. V. Meyen (In press). *Suchoviella*—a new genus of Rufloriaceae from the Permian of western Angaraland and the main aspects of the taxonomy and phylogeny of Cordaitanthales. *Rev. Palaeobot. Palynol.*

Jennings, J. R. 1976. The morphology and relationships of *Rhodea, Telangium, Telangiopsis,* and *Heterangium. Amer. J. Bot.* 63:1119–1133.

Kerp, J. H. F. 1982. Aspects of Permian palaeobotany and palynology. 2. On the presence of the ovuliferous organ *Autunia milleriensis* (Renault) Krasser (Peltaspermaceae) in the Lower Permian of the Nahe area (F.R.G.) and its relationship to *Callipteris conferta* (Sternberg) Brongniart. *Acta Bot. Neerl.* 31:417–427.

Knobloch, E. 1972. Der Gattungsname *Sphenobaiera* Florin ist illegitim. *Taxon* 21:545–546.

Krassilov, V. A. and V. I. Burago. 1981. New interpretation of *Gaussia* (Vojnovskyales). *Rev. Palaeobot. Palynol.* 32:227–237.

Kräusel, R. 1943. Die Ginkgophyten der Trias von Lunz in Nieder-Österreich und von Neuewelt bei Basel. *Palaeontographica* 87B:59–93.

Ledran, C. 1966. Contributions à l'étude des feuilles de Cordaitales. *Thesis, Fac. Sci. Acad. Reims.*

Maekawa, F. 1962. *Vojnovskya* as a presumable ancestor of angiosperms. *J. Jap. Bot.* 37(5):149–152.

Maheshwari, H. K. and S. V. Meyen. 1975. *Cladostrobus* and the systematics of cordaitalean leaves. *Lethaia* 8:103–123.

Mamay, S. H. 1976–1978. Vojnovskyales in the Lower Permian of North America. *Palaeobotanist* 25:290–297.

Mapes, G. and G. W. Rothwell. 1984. Permineralized ovulate cones of *Lebachia* from Late Palaeozoic limestones of Kansas. *Palaeontology* 27:69–94.

Meyen, S. V. 1966. Cordaiteans of the Upper Palaeozoic of North Eurasia. *Trudy Geol. Inst. Akad. Nauk SSSR* 150:1–184 (in Russian).

—— 1970. Epidermisuntersuchungen an permischen Landpflanzen des Angaragebietes. *Paläontol. Abh.* 3B(3/4):523–552.

—— 1976–1978. Permian conifers of the west Angaraland and new puzzles in the coniferalean phylogeny. *Palaeobotanist* 25:298–313.

—— 1977. Cardiolepidaceae—a new coniferalean family from the Upper Permian of north Eurasia. *Paleontol. Zhurn.* 3:130–140 (in Russian).

—— 1979. Permian predecessors of the Mesozoic pteridosperms in western Angaraland, USSR. *Rev. Palaeobot. Palynol.* 28:191–201.

—— 1981. Some true and alleged Permotriassic confiers of Siberia and Russian Platform and their alliance. *Palaeobotanist* 28–29:161–176.

—— 1982a. The Carboniferous and Permian floras of Angaraland (a synthesis). *Biol. Mem.* 7:1–109.

—— 1982b. Fructifications of Upper Palaeozoic cordaitanthaleans of Angaraland. *Paleontol. Zhurn.* 2:109–120 (in Russian).

—— 1982c. Gymnosperm fructifications and their evolution as evidenced by palaeobotany. *Zhurn. Obshch. Biol.* 43:303–323 (in Russian).

—— 1983. The systematics of peltaspermaceous pteridosperms and their position in the phylogeny of gymnosperms. *Bull. Moskovsk. Obshch. Ispyt. Prirody, otd. Biol.* 88:3–14 (in Russian).

—— 1984a. Basic features of gymnosperm systematics in phylogeny as evidenced by the fossil record. *Bot. Rev.* 50:1–111.

—— 1984b. Phylogeny of higher plants and florogenesis. *Proc. 27th Intern. Geol. Congr.*, vol. 2. pp. 97–109. *Palaeontology*, Utrecht: VNU Science Press.

—— 1986a. A hypothesis of the bennettitalean origin of angiosperms by means of a gamoheterotopy (transfer of characters from one sex to another). *Zhurn. Obshch. Biol.* 47:291–309 (in Russian).

—— 1986b. Permian conifers of western Angaraland. *Moscow VINITI Dep.* pp. 1–140 (in Russian).

—— 1986c. Gymnosperm systematics and phylogeny: A reply to commentaries by C. B. Beck, C. N. Miller and G. W. Rothwell. *Bot. Rev.* 52:300–320.

—— 1987. *Fundamentals of Palaeobotany*. London: Chapman and Hall.

—— (in press). Permian conifers of western Angaraland. *Palaeonotgraphica*.

Meyen, S. V. and A. V. Migdisova. 1969. Epidermal studies of Anagara *Callipteris* and *Compsopteris*. *Trudy Geol. Inst. Akad. Nauk SSSR* 190:59–84 (in Russian).

Meyen, S. V. and H. G. Smoller. 1986. The genus *Mostotchkia* Chachlov (Upper Palaeozoic of Angaraland) and its bearing on the characteristics of the order Dicranophyllales (Pinopsida). *Rev. Palaeobot. Palynol.* 47:205–223.

Millay, M. A. and T. N. Taylor. 1979. Paleozoic seed fern pollen organs. *Bot. Rev.* 51:295–318.

Neuburg, M. F. 1948. Upper Palaeozoic flora of Kuznetsk basin. *Paleontol. SSSR* 12(3, 2):5–342 (in Russian).

—— 1961. Present state of the question on the origin, stratigraphic significance and age of Paleozoic floras of Angaraland. *C. r. 4 Congr. Strat. Géol. Carbonifère Heerlen* 2:443–452.

—— 1963. Family Vojnovskyaceae. In A. L. Takhtajan et al., eds., *Osnovy Paleontologii: Golosemennye i. Pokrytosemennye*, p. 301. Moscow: Gosgeoltekhizdat (in Russian).

—— 1965. Permian flora of the Pechora Basin, part 3. *Trudy Geol. Inst. Akad. Nauk SSSR* 116:1–144 (in Russian).

Rothwell, G. W. 1981. The Callistophytales (Pteridospermopsida): Reproductively sophisticated Paleozoic gymnosperms. *Rev. Palaeobot. Palynol.* 32:103–121.

—— 1985. The role of comparative morphology and anatomy in interpreting the systematics of fossil gymnosperms. *Bot. Rev.* 51:319–327.

Sadovnikov, G. N. 1983. New data on morphology and anatomy of the genus *Kirjamkenia*. *Paleontol. Zhurn.* 4:76–81 (in Russian).

Seward, A. C. 1917. *Fossil Plants*, vol. 3. Cambridge: Cambridge University Press.

Skog, J. E. and P. G. Gensel. 1980. A fertile species of *Triphyllopteris* from the early Carboniferous (Mississippian) of southwestern Virginia. *Amer. J. Bot.* 67:440–451.

Stewart, W. N. 1983. *Paleobotany and the Evolution of Plants*. Cambridge: Cambridge University Press.

Sukhov, S. V. 1969. Seeds of late Palaeozoic plants of middle Siberia. *Trudy Sib. Inst. Geol. Geofiz i Min. Syr.* 64:1–264 (in Russian).

Surange, K. R. and S. Chandra. 1974. Some male fructifications of Glossopteridales. *Palaeobotanist* 21:255–266.

Takhtajan, A. L., V. A. Vakhrameev, and G. P. Radczenko, eds., 1963. *Osnovy Paleontologii: Golosemennye i. Pokrytosemennye.* Moscow: Gosgeoltekhizdat.

Taylor, T. N. 1970. *Lasiostrobus* gen. n., a staminate strobilus of gymnospermous affinity from the Pennsylvanian of North America. *Amer. J. Bot.* 57:670–690.

—— 1981. *Paleobotany: An Introduction to Fossil Plant Biology.* New York: McGraw-Hill.

Taylor, T. N. and M. A. Millay. 1977. The ultrastructure and reproductive significance of *Lasiostrobus* microspores. *Rev. Palaeobot. Palynol.* 23:129–137.

Thomas, B. A., and S. D. Brack-Hanes. 1984. A new approach to family groupings in the Lycophytes. *Taxon* 33:247–255.

Townrow, J. A. 1960. The Peltaspermaceae, a pteridosperm family of Permian and Triassic age. *Palaeontology* 3:333–361.

—— 1967. On *Rissikia* and *Mataia* podocarpaceous conifers from the Lower Mesozoic of southern lands. *Pap. Proc. Roy. Soc. Tasmania* 101:103–136.

Tralau, H. 1968. Evolutionary trends in the genus *Ginkgo. Lethaia* 1:63–101.

Vladimirovich, V. P. 1984. Type Kazanian flora of Kama Embayment. *Moscow VINITI Dep.* 4571–84:1–91 (in Russian).

Zalessky, M. D. 1927. Flore permienne des limites ouraliennes de l'Angaride. *Mém. Com. Géol., n. sér.* 176:1–52.

—— 1929. Sur les débris de nouvelles plantes permiennes. *Bull. Acad. Sci. URSS, cl. sci. phys.-math.* 7:677–689.

—— 1932. On two new species of *Dicranophyllum* from the Artinskian deposits of Fore-Urals. *Bull. Acad. Sci. URSS, cl. sci. math. et natur* 9:1361–1364 (in Russian).

—— 1933. Observations sur les végétaux nouveaux du terrain permien inférieur de l'Oural, *Bull. Acad. Sci. URSS, 7th ser.* 2:283–292.

—— 1934. Observations sur les végétaux permiens du bassin de la Petchora, vol. 1. *Bull. Acad. Sci. URSS, cl. sci. math. et natur* 2–3:241–290.

—— 1937. Sur la distinction de l'étage Bardien dans le Permien de l'Oural et sur sa flore fossile. *Probl. Paleontol.* 2–3:37–101.

Zhao Xiuhu, Mo Zhuangguan, Zhang Sanzhen, and Tao Zhaoqi. 1980. Late Permian flora in western Guizhou and eastern Yunnan. In *Stratigraphy and Palaeontology of the Upper Permian Coal-bearing Formations in Western Guizhou and Eastern Yunnan,* pp. 70–122. Bejing: Science Press.

Zimina, V. G. 1977. *Flora of the Early and Beginning of Late Permian of South Primorie.* Moscow: Nauka (in Russian).

9

The Cheirolepidiaceae

JOAN WATSON

▦

During the past two decades both palynological studies and work on macro-remains have clearly established the Cheirolepidiaceae as an important Mesozoic conifer family of a diversity probably unparalleled in any other conifer family, extinct or living. It is now obvious that members of this family displayed a quite remarkable range of morphology, habit, and habitat, the full extent of which may not yet be fully recognized. To date, the single most reliable character on which to base assignment to this family is possession of the distinctive and unusual pollen of the genus *Classopollis* Pflug. Indeed, it is beginning to look as though it may be the only reliable character, and one of considerable evolutionary significance. The possession of *Classopollis*-bearing male cones, together with whatever female mechanism was involved, may be the only unifying feature of phylogenetic significance. It should be made clear from the outset that our knowledge of the female reproductive structures is quite superficial. The cones or cone scales of very few species are known in any detail, and of these the only clear facts that emerge are that the seeds were enclosed in some way in scales of considerable complexity which yield up to 10 layers of cuticle upon maceration. Of course, *Classopollis* pollen has long excited interest by its tectate-like exine and perforations because of their similarity to angiospermous pollen (Pettitt and Chaloner 1964; Chaloner 1976). Chaloner has suggested that it may have had a parallel function to that of tectate angiosperm pollen (Heslop-Harrison 1976) carrying tapetally derived "recognition" substances concerned with pollen-receptor interaction. Archangelsky (1968) has noted the absence of *Classopollis* pollen from the micropyles of isolated ovules, but of the pollen-receiving tissue and pollination mechanism we know nothing. This important gap in our knowledge of this family has been the most disappointing aspect of what has otherwise been fairly spectacular progress over the past decade or so.

HISTORICAL BACKGROUND

The genera of vegetative shoots that are now well established as members of the family Cheirolepidiaceae were in the past attributed to various living conifer families, mainly the Araucariaceae, Taxodiaceae, and Cupressaceae, largely on the basis of morphological similarity to living genera. Clearly these three families embrace a considerable variety of leaf morphology and phyllotaxis, and this variety is reflected in cheirolepidiaceous foliage genera, the names of some of them indicating early opinions on their affinities. The shoots may broadly be divided into two groups, the first group comprising those of *Brachyphyllum* or *Pagiophyllum* type with spirally arranged leaves. These were generally thought to be of araucariaceous or taxodiaceous affinities. The second group are what we might call the frenelopsids. This group is typified by the genus *Frenelopsis* Schenk, which has whorls of leaves and a jointed appearance, thus resembling members of the Cupressaceae. The individual genera and component species of both groups are dealt with in detail below. One interesting thing about them is that until 1975 any connection between these two groups was entirely unsuspected and all work had proceeded entirely independently, with no hint of any relationship between them other than the fact that they were undoubtedly conifers.

The first group includes *Hirmeriella muensteri* (Schenk), long known to have attached male cones bearing unusual nonsaccate pollen and associated female cone scales of a distinctive form, distally lobed. This species gave the family its name. Apart from nomenclature problems, progress involving this species and other members of this group was steady over a considerable number of years. Initially the structure of the female cone scales, but later also the unusual pollen, gradually led to the realization that they probably belonged to a now extinct family. In more recent years progress has been linked to the increasingly important palynological studies of dispersed *Classopollis* pollen. Hörhammer (1933) described two species of conifers each with female cones, on the basis of which he allotted them to two distinct genera, *Cheirolepis* Schimper and *Hirmeriella* Hörhammer. The *Cheirolepis* species also bore male cones yielding nonsaccate pollen which he figured. Hirmer and Hörhammer (1934) then proposed a special family, the "Cheirolepidaceen," occasioned by the remarkable structure of the female cone scale. Work in the next two decades (e.g., Harris 1957; Chaloner 1962) added information but did not materially alter the status of any of the genera or species involved. In 1963 Takhtajan pointed out that *Cheirolepis* Schimper was a later homonym of *Cheirolepis* Boissier of the Compositae and substituted the name *Cheirolepidium* Takhtajan. At the same time he emended the family name to Cheirolepidiaceae based on his new name. In my opinion the family name was validly established as Cheirolepidiaceae at this point and is unaffected by nomenclatural changes that followed in later years. Jung (1967, 1968) demonstrated that the lobed scale of *Cheirolepidium* Takhtajan was an ovuliferous scale that had been subtended in the axil of a persistent bract and shed as a unit from the ripe cone. The bracts had been named *Hirmeriella* by Hörhammer, and thus the two genera were shown to be synonymous. Under the rules of nomenclature, the appellation *Hirmeriella* Hörhammer was declared by Jung (1968) to have priority. Unfortunately an unnecessary orthographic change was also made, deleting the second *i* to give *Hirmerella*, on which was based the name of a suggested new subfamily "Hirmerelloidae." The confused usage of both generic and family names that ensued from this name change included a number of variants both deliberate and misspelled and has been dealt with in some detail elsewhere (Alvin 1982; Watson 1982). Attention need not be drawn to this again; suffice it to say that recent authors have returned to the use of the generic name *Hirmeriella* Hörhammer and the family name Cheirolepidiaceae.

The frenelopsids, on the other hand, have a long history during which, though consistently attributed to the Cupressaceae, the precise diagnostic characters of the genera were really not known. A number of contributing factors caused this continued lack of understanding, including loss and inaccessibility of type material during the first half of this century. However, the single outstanding reason is that frenelopsids are not present in Jurassic floras, on which historically much more work has been expended than on Cretaceous floras. Hence these genera did not come under the specific and detailed scrutiny of such workers as T. M. Harris and were not studied intensively until the 1960s, when an influx of young paleobotanists turned their attention to various Cretaceous floras. The type species of *Frenelopsis* was described from Carpathian material by Ettingshausen (1852) as *Thuites hoheneggeri*. Schenk (1869), describing different Carpathian material which he thought identical to Ettingshausen's, erected a new genus, *Frenelopsis*, that he considered to be closer to the living cupressaceous genus *Frenela* (now a synonym of *Callitris*) than to *Thuja*. All this material is lost, but Schenk undoubtedly had a mixture of at least two species with different leaf arrangements. Ettingshausen's diagnosis suggests that there are four leaves per whorl, whereas Schenk interpreted some of his specimens as opposite and decussate. It is almost certain that the species *Pseudofrenelopsis parceramosa* (Fontaine) Watson with spirally arranged leaves was also present in this material.

Thus the type-species *Frenelopsis hoheneggeri* (Ettingshausen) Schenk was established and subsequently used for specimens from many countries in its unsatisfactorily defined state, and the genus *Frenelopsis* Schenk was used for about a dozen new species. Fontaine, in using this genus for specimens in the Potomac flora from Virginia (1889) and the Glen Rose flora from Texas (1893), was quite clear about alternating whorls of three leaves in his new species *Frenelopsis ramosissima*. But in using the same genus for *Frenelopsis parceramosa* and *Frenelopsis varians*, he failed to recognize the spiral arrangement, thinking that some of the leaves were missing from the nodes. Nathorst (1893) recognized this spiral arrangement in Mexican material and established a new genus, *Pseudofrenelopsis*, at the same time drawing attention to the similarity of Fontaine's *Frenelopsis parceramosa*. Fontaine (1905) rejected Nathorst's observations, and the genus *Pseudofrenelopsis* fell into disuse as far as I know for over 70 years. The next significant event was the collecting of new frenelopsid material from Lower Cretaceous localities in England (Watson 1964) and Poland (Reymanówna 1965) followed by further discoveries in Czechoslovakia (Knobloch 1971, Hluštík 1974) and elsewhere. The renewed interest in these plants involved exchange and comparison of material from most of the Lower Cretaceous floras of Europe and the United States, and led to rapid progress in our understanding of the frenelopsid genera and species. At no time throughout this period of activity was their cupressaceous status even questioned, though I remember struggling to convince myself that they were really similar to *Callitris*, *Widdringtonia*, or *Tetraclinis*. However, no other family was available or even considered for comparison. It is interesting, in retrospect, to look at the stratigraphic ranges prepared at that time (Alvin et al. 1967). The Cheirolepidiaceae was interpreted as containing the single genus *Cheirolepidium*, restricted to around the Triassic–Jurassic boundary. The Cupressaceae, on the other hand, was indicated as stretching from the present back to the lowermost Jurassic, a range recently extended by the description of *Cupressinocladus ramonensis* Chaloner and Lorch (1960) from Israel.

This brings us back to the important discovery in 1975 that changed all that: male cones yielding *Classopollis* pollen were found by Alvin with *Pseudofrenelopsis parceramosa* from the Isle of Wight. At around the same time Hluštík and Konzalova discovered similar cones with *Frenelopsis alata* in Czechoslovakia. The news of this was spread in conversation at the 1975 International Botanical Congress in Leningrad, and mentioned by Reymanówna and Watson (1976) before formal descriptions were published. Subsequent progress in the light of this startling new information forms the following part of this review. It soon became quite clear that morphology of a kind hitherto regarded as distinctly cupressaceous is completely unreliable for family attribution in the Mesozoic. Indeed, I have already indicated elsewhere (Watson 1982) that I suspect that not a single well-authenticated member of the Cupressaceae remains before the Middle or Upper Cretaceous.

SHOOT MORPHOLOGY

Of the 22 foliage species presented here as fairly securely attributable to the family, 15 have actually been identified on the evidence of associated or even attached reproductive organs (table 9.1). These 22 species are currently distributed between the following seven genera, which are illustrated in figures 9.1–3 and considered individually in detail below:

Brachyphyllum Lindley and Hutton *ex* Brongniart (in part)
Cupressinocladus Seward (in part)
Frenelopsis Schenk
Hirmeriella Hörhammer (=*Cheirolepidium* Takhtajan)
Pagiophyllum Heer (in part)
Pseudofrenelopsis Nathorst
Tomaxellia Archangelsky

Other genera probably belonging in the family are the following:

Androvettia Hollick and Jeffrey
Geinitzia Endlicher (in part)
Glenrosa Watson and Fisher

Some of these are, of course, established artificial form-genera, and
to some extent their use for members of this family has been more by
accident than by considered application. In this context their usage
has probably already strayed farther from their original, intended, or
traditional use than is desirable, and a number of problems can be
foreseen that should be prevented before they become entrenched. For
instance, one grows increasingly uneasy about the continued use of
Brachyphyllum and *Pagiophyllum* (e.g., Harris 1979) once family membership seems established. These widely used genera show considerable variety of cuticle structure and must include members of a number of families—certainly the Araucariaceae, probably also the
Taxodiaceae, and perhaps the Podocarpaceae. There are difficulties
about removing such species to *Hirmeriella* Hörhammer because the
type-species is a whole plant with *Brachyphyllum*-type shoots but
with attached male and female cones. Harris (1979) has also used
Hirmeriella for two species of isolated female cones attributed to *Pagiophyllum* shoots in the Yorkshire Jurassic flora. In these two cases
there is not yet a problem because the genus *Hirmeriella sensu* Hörhammer can embrace the *Pagiophyllum* type of shoot. However, Harris is actually using *Hirmeriella* in this case in the old organ-genus
sense, and this is where a problem arises. *Hirmeriella*-type lobed female scales occur with *Pseudofrenelopsis parceramosa* (Fontaine) Watson and obviously cannot be called *Hirmeriella*. To my mind, Harris is
right that we actually need at this stage some clearly defined organ-
genera in the old sense. Whatever the objections to this now outlawed
genre, it cannot be denied that it was of great practical value and
would be a perfect answer to some of the problems outlined here. In
fact, *Classostrobus* Alvin, Spicer, and Watson for isolated cheirolepidiaceous male cones is nothing more than an organ-genus.

Additionally, as will become apparent, the diagnostic characters of
even the recently redefined genera are already becoming blurred as
our knowledge of the group advances.

Table 9.1 is a summary of details allowing comparison of characters
of all the species listed below that I regard as having little doubt
attached to their family membership. Precise details of cones given in
a previous table (Watson 1982) are not repeated here, as Van Konijnenburg-Van Cittert (1987) has published an updated version. Com-

Table 9.1. Well-Authenticated Members of the Cheirolepidiaceae

Vegetative Shoots	Locations	Ages	Male Cones	Female Cones	Stomatal Details	Other Details	References
Brachyphyllum crucis	England	Bajocian-Callovian	attached	attached	broad, prominent papillae around pit	cuticle 6 μm thick, rare decussate leaf pairs	Kendall 1947 Van Konijnenburg 1971, 1987 Harris, 1979
Brachyphyllum scottii	Scotland	Lower Lias	associated cone axis and pollen	—	thickened stomatal rim, non-papillate	cuticle 6 μm thick	Kendall 1949
Cupressinocladus pseudo-expansum	Iran	Bajocian-Bathonian	associated with Classostrobus rishra	—	papillate rim	cuticle 2–3 μm thick	Barnard 1968 Barnard & Miller, 1976
Cupressinocladus ramonensis	Israel	Lower Jurassic	associated with Masculostrobus harrisianus Classopollis not proved	—	broad papillae below thickened rim	cuticle about 6–8 μm thick	Chaloner & Lorch 1960, Lorch 1968.
Cupressinocladus valdensis	England	Purbeckian-Berriasian	associated with Classostrobus cone	—	rim variable: flush & stellate-polygonal or with thickened ring. Large papillae in pit	cuticle 15–20 μm thick, associated Protocupressinoxylon wood	Watson 1977, Francis 1983
Frenelopsis alata	Europe, U.S.A.	Albian-Senonian	attached	associated scales not distally lobed	stellate rim papillae inside pit	cuticle 30 μm thick	Alvin & Hluštik 1979 Pons 1979

Species	Location	Age			Rim	Cuticle	Reference
Frenelopsis harrisii	U.S.S.R.	Cenomanian	—	—	papillate rim papillae inside pit	cuticle up to 100 μm thick	Doludenko 1978, Doludenko & Reymanówna 1978
Frenelopsis hoheneggeri	Poland, Czechoslovakia	Hauterivian	—	—	lobed rim papillae inside pit	cuticle 40 μm thick	Reymanówna & Watson 1976
Frenelopsis occidentalis	Portugal	Aptian-Albian	—	—	rim polygonal, flush with surface, massive papillae inside pit	cuticle up to 60 μm thick. 3 leaves not confirmed.	Alvin 1977
Frenelopsis oligostomata	Portugal	Senonian	associated scales not distally lobed	attached	lobed ring forming rim, broad papillae inside pit	cuticle 30 μm thick, assoc. *Protopodocarpoxylon* wood	Alvin 1977, Pons & Broutin 1978
Frenelopsis ramosissima	U.S.A.	Barremian-Albian	—	—	papillate rim	cuticle about 30 μm plus hairs up to 120 μm	Watson 1977
Frenelopsis rubiesensis	Spain	? Berriasian	—	—	papillate subsidiary cells	cuticle 30 μm thick	Barale 1981
Frenelopsis silffloana	Sudan	Lower Cretaceous	—	—	papillae inside pit		Watson 1983
Frenelopsis teixeirae	Portugal	Hauterivian or Barremian	—	—	stellate or polygonal rim, large papillae inside pit	cuticle 50 μm thick, only opposite decussate leaves.	Alvin & Pais 1978
Hirmeriella airelensis	France	Rhaeto-Liassic	associated axis and sporophylls yielding *Classopollis harrisii*	associated	raised ring around rim sometimes papillate	cuticle usually 4–6 μm thick	Muir & Van Konijnenburg 1970

Table 9.1. (*Continued*)

Vegetative Shoots	Locations	Ages	Male Cones	Female Cones	Stomatal Details	Other Details	References
Hirmeriella muensteri	Germany, Poland, England, Wales.	Rhaeto-Liassic	attached	attached	raised ring around rim	cuticle about 12 μm, assoc. *Protocupressinoxylon* wood	Harris 1957 Jung 1968
Pagiophyllum araucarinum	England, France.	Upper Lias-Middle Deltaic	associated	*Hirmeriella estonensis*	long papillae around rim	cuticle 6 μm thick	Harris 1979 Barale 1981
Pagiophyllum maculosum	Yorkshire	Lower-Upper Deltaic	*Classostrobus cloughtonensis*	*Hirmeriella kendalliae*	non-papillate	cuticle up to 6 μm thick	Harris 1979 Van Konijnenburg 1987
Pseudofrenelopsis intermedia	China	Lower Cretaceous	*Classostrobus cathayanus*	—	? papillate rim (material eroded). Very deep pit	cuticle up to 250 μm thick, assoc. *Protopodocarpoxylon*-type wood.	Zhou 1983
Pseudofrenelopsis parceramosa	U.S.A., England, Europe, Ghana, Sudan.	Berriasian-early Cenomanian	*Classostrobus comptonensis*	associated scales lobed distally	papillate rim	cuticle about 30 μm thick assoc. *Protopodocarpoxylon*-type wood.	Watson 1977 Alvin, Spicer & Watson 1978
Pseudofrenelopsis varians	Mexico, U.S.A.	Aptian-Albian	attached	—	papillate rim, very deep pit.	cuticle up to 110 μm thick.	Watson, 1977.
Tomaxellia biforme	Argentina	Lower Cretaceous	attached	attached cones, lobed ovul. scales	non-papillate	cuticle 4 μm thick	Archangelsky 1968 Archangelsky & Gamerro 1967.

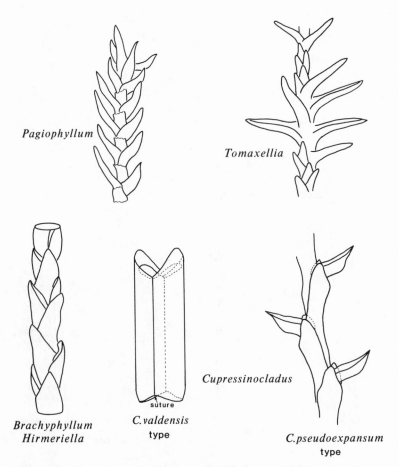

Pagiophyllum

Tomaxellia

Brachyphyllum
Hirmeriella

C. valdensis
type

suture

Cupressinocladus

C. pseudoexpansum
type

Figure 9.1. Shoot morphology of nonfrenelopsid members of the family. (Redrawn from Alvin 1982; Watson 1982.)

plete synonymies for individual species have not been given, but the bibliography in this review is as comprehensive as I am able to make it, with the exception of literature dealing with dispersed *Classopollis*, for which I have included only key references.

Family Members   Form-genus *Brachyphyllum* Lindley and Hutton *ex* Brongniart (Harris [1979] gives an emended diagnosis.)

Shoots have spirally arranged leaves with a basal cushion tapering to a short free part (figures 9.1; 9.4D). Some shoots have longer free

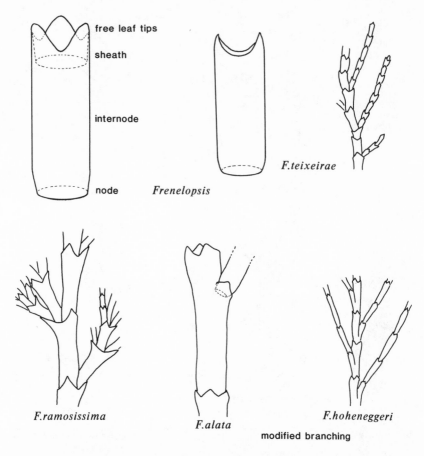

Figure 9.2. Shoot morphology of *Frenelopsis*. *F. teixeirae* and *F. ramosissima* show normal axillary branching. *F. alata* and *F. hoheneggeri* show displacement of the base of ultimate branchlets toward the node above the subtending leaf. *F. oligostomata* also has this habit. (Redrawn from Alvin 1982; Watson 1982.)

parts, thus overlapping the boundary of *Pagiophyllum* or *Geinitzia*—in effect, exactly like *Hirmeriella muensteri*, as shown in figure 9.5.

Brachyphyllum crucis Kendall 1947; Jurassic, England

See Kendall (1952—shoots, cuticle); Van Konijnenburg-Van Cittert (1971—male cones, pollen); Harris (1979—Kendall's 1952 figures repeated).

This species is known from large numbers of well-preserved shoots

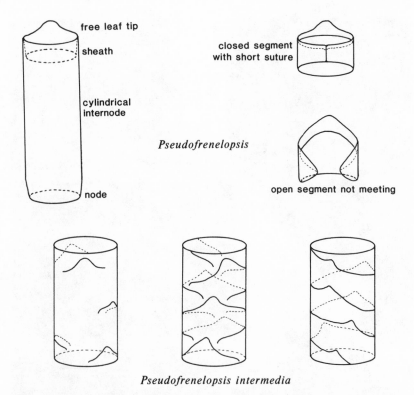

Pseudofrenelopsis intermedia

Figure 9.3. Shoot morphology of *Pseudofrenelopsis*. *P. intermedia* shoots often have leaves reduced to free tips only, widely spaced on a smooth stem (left) or closely spaced but with no overlap (center). The right-hand shoot shows a rare arrangement of free leaf tips in a continuous spiral. (Redrawn from Watson 1982; Zhou Zhiyan 1983.)

with the typical short leaves also and shoots with longer leaves of the *Geinitzia* type, which are probably juvenile foliage. A single specimen with opposite decussate leaf pairs on a side shoot is known. Harris (1979) discusses in some detail the correct genus for *B. crucis* and makes it quite clear that he regards it as a *Hirmeriella*.

Brachyphyllum scottii Kendall; Lias, Scotland
 See Kendall (1949—shoots, pollen, female cone scales); Harris (1979).
 Van Konijnenburg-Van Cittert (1971) confirmed that the pollen adhering to the associated male cone axis is indeed *Classopollis*. The female cone scale described is poorly defined, certainly not lobed, and

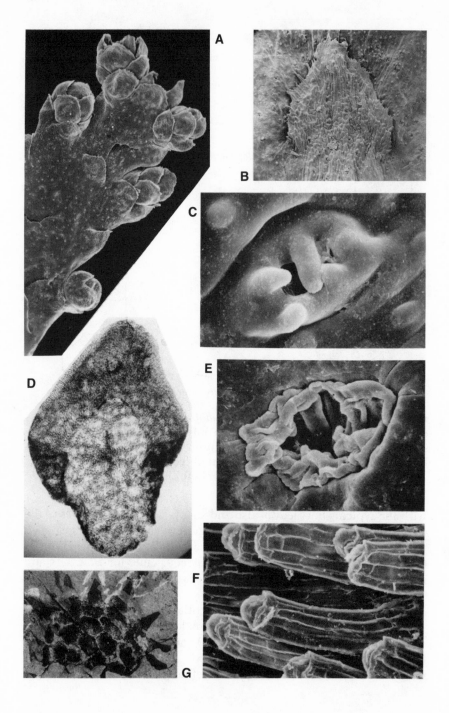

I can find no further reference to it in the literature. Harris (1979) merely states that he regards the plant as a *Hirmeriella*.

Genus *Cupressinocladus* Seward 1919:307 (Chaloner and Lorch [1960:237], generic diagnosis; Harris [1969:250], emended diagnosis; Barnard and Miller [1976:43] diagnosis refined; Watson [1977:742], combined Harris, Barnard, and Miller diagnosis repeated.)

Shoots have leaves in opposite decussate pairs or whorls of three, with the leaf bases separated by distinct sutures.

Cupressinocladus pseudoexpansum Barnard and Miller; Jurassic, Iran; figure 9.1

See Assereto, Barnard, and Fantini Sestini (1968—name in list only); Barnard (1968—male cones, pollen); Barnard and Miller (1976—comprehensive account, SEM details, male cone, pollen).

Cupressinocladus ramonensis Chaloner and Lorch 1960; Jurassic, Israel

Figures 9.1; 9.6D; 9.7E

See Lorch (1968—attributed male cones).

The pollen from these male cones is nonsaccate but has not been shown to be *Classopollis*. In new preparations I could not see any equatorial striations, but the pollen may not be mature.

Cupressinocladus valdensis (Seward) Seward; Purbeck-Wealden, England

Figures 9.1; 9.7F; 9.8A

See Seward (1895, 1919—removal to new genus); Watson (1977—

Figure 9.4. Two strong candidates for family membership that display particularly unusual features. **A–C** Scanning electron micrographs of *Androvettia*, kindly provided by Dr. F. M. Hueber; Upper Cretaceous, North Carolina. **A** Distal portion of phylloclade bearing tiny scale-like leaves, the lateral ones with bud-like shoots in their axils. × 20. **B** Fringed, papillate leaf on flat face of phylloclade. × 50. **C** Stoma with large papillae overhanging pit, × 1000. **D–G** *Brachyphyllum castatum*, widely dispersed in Wealden, Sussex. **D** Single leaf. × 25. **E** Outside view of stoma with convolutions of cuticle lining pit. × 750. **F** Longest stomatal tubes, showing guard cells at bottom. × 150. **G** *Conites berryi*, with long spines that have cuticle, closely matching leaves. × 1.

396   JOAN WATSON

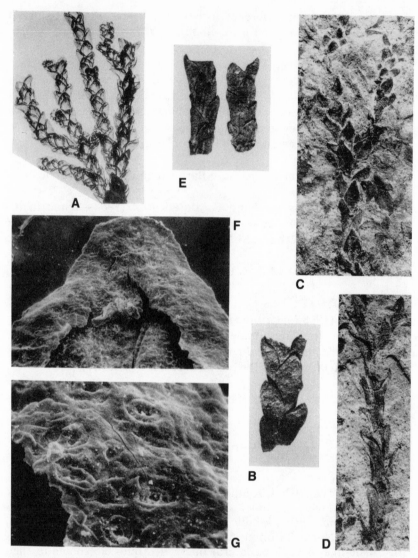

Figure 9.5. The Jurassic species *Hirmeriella muensteri*. **A** Shoot with closely adpressed leaves; from Hörhammer (1933). × 2. **B** Similar shoot. × 5. **C, D** Shoots with longer free parts of leaves. × 2. **A–D** from near Bayreuth, Germany. **E** Shoots from Odroważ, Poland. × 5. **F** Apex of Polish leaf viewed from adaxial side in SEM. × 50. **G** Same leaf, portion of adaxial cuticle showing stomata. × 250.

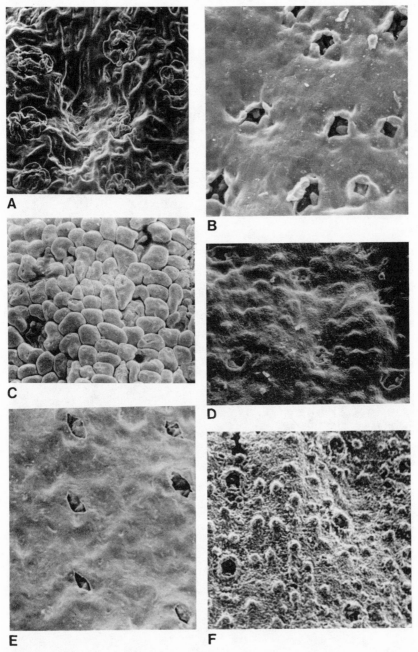

Figure 9.6. Outer surface of mid-internode showing details of stomatal distribution, papillation, and so on. All six species shown here are characterized by papillae within the stomatal pit. All scanning electron micrographs, × 250. **A** *Frenelopsis hoheneggeri*, Poland. **B** *Frenelopsis alata*, Texas. **C** *Frenelopsis harrisii*, U.S.S.R. **D** *Cupressinocladus ramonensis*, Israel. **E** *Frenelopsis occidentalis*, Portugal. **F** *Frenelopsis silfloana*, Sudan.

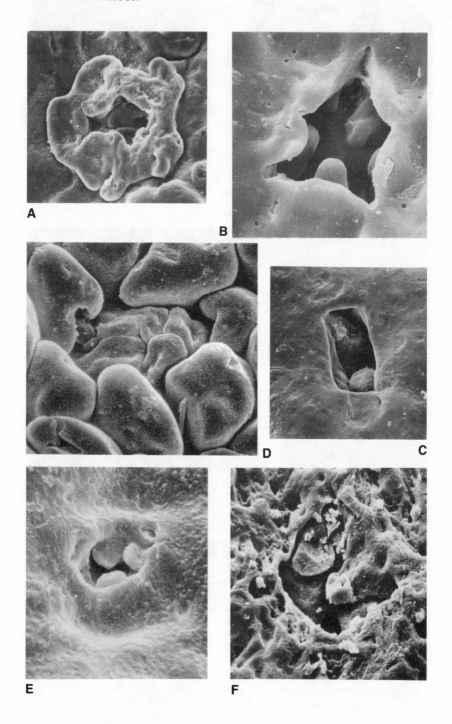

description of cuticle, attribution to Cheirolepidiaceae); Francis (1983 —shoots, wood, male cones, pollen, reconstruction).

Cupressinocladus Seward is one of the genera whose continued and increasing use, with attendant changes in definition, needs reconsideration. Seward established *Cupressinocladus* as a "non-committal generic term" for shoots resembling living Cupressaceae with opposite pairs of leaves decussately arranged. Although Seward does not give a formal diagnosis, he makes it quite clear that he would expect any species to be attributable to a cupressaceous genus if evidence from cones was available. The type species *Cupressinocladus salicornoides* (Unger) is from the Eocene of Europe and looks to me like a member of the Cupressaceae. Unfortunately, specimens of this species in the British Museum (Natural History) have no cuticle present, but they have the typical flattened appearance, with the facial and lateral pairs of leaves differing somewhat in shape and size. *Cupressinocladus cretaceus* (Heer) from the Upper Cretaceous of Sakhalin figured by Krassilov (1978: plate 106, figure 1) also shows this characteristic appearance. I have come to think that *Cupressinocladus* should be restricted as Seward originally intended for species like these, with the removal of those that prove to be cheirolepidiaceous. Francis (1983) has described the Purbeck material of *C. valdensis* as "arising in one plane to produce a flat, frond-like shoot," but those species of *Cupressinocladus* with decussate leaves that have proved to belong to the Cheirolepidiaceae seem to me to have their pairs of leaves absolutely equal and indistinguishable (figure 9.8A; Chaloner and Lorch 1960: plate 36, figure 3). The flattening may be preservational. These differences may be worth bearing in mind when appraising shoots with this leaf arrangement, although of course unflattened shoots are also known in the Cupressaceae.

Barnard and Miller (1976) further extended the use of *Cupressinocladus* to include *C. pseudoexpansum*, from the Jurassic of Iran, which has its leaves in whorls of three except on the ultimate branchlets. The shoot morphology of *C. pseudoexpansum* is shown in figure 9.1.

Figure 9.7. Scanning electron micrographs of stomata with papillae inside the pit. × 1000. **A** *Frenelopsis hoheneggeri*, which also has pouch-like papillae around the rim. **B** *Frenelopsis alata*, with lobed rim. **C** *Frenelopsis occidentalis* has small, simple pit openings flush with surface. **D** *Frenelopsis harrisii* has an outer set of papillae, which here fit together perfectly, closing the pit. For inner papillae, see figures 9.11C; 9.12F. **E** *Cupressinocladus ramonensis*. **F** *Cupressinocladus valdensis*.

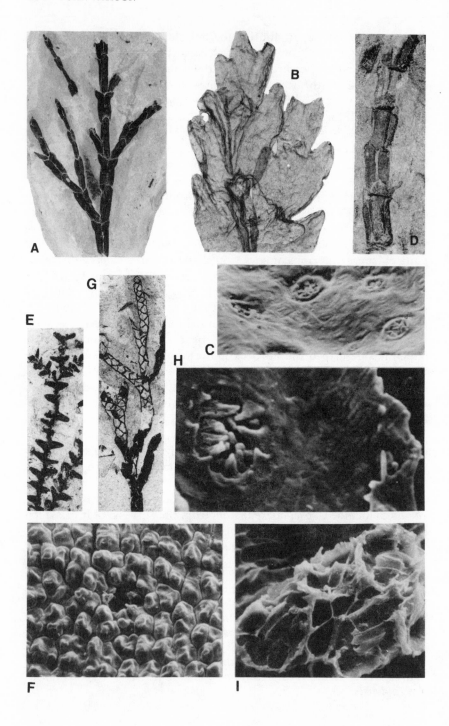

Barnard and Miller suspected dorsiventrality and attempted to prove it, but in the end decided that if present, it is rather weakly so. Figure 9.8D shows a shoot previously identified as *C. valdensis* (Daber 1960) that has now proved to have leaves in threes. The segment above the crack is now loose and certainly has a whorl of three leaves with long decurrent bases separated by sutures. The cuticle has yet to be studied in detail; the SEM reveals that it is not *C. valdensis* but certainly cheirolepidiaceous. In transferring *Cupressinocladus malaiana* from *Frenelopsis*, Barnard and Miller (1976) suggested that the specimens figured by Kon'no (1967, 1968) probably have three leaves per whorl. I agree. As Fontaine pointed out when describing *Frenelopsis ramosissima* (1889), it is very easy to mistake threes for decussate pairs when only one side of the specimen is visible. With the addition of these three-leaved species, the presence of a strong suture separating the basal cushions is probably now the one feature that distinguishes *Cupressinocladus* from *Frenelopsis*.

Genus *Frenelopsis* Schenk 1869 (Reymanówna and Watson [1976], emended diagnosis; Watson [1977], emendation to accommodate opposite decussate leaves.)

Shoots appear segmented, with leaves in whorls of three or opposite decussate pairs, the leaf bases of each segment combining to form a smooth cylindrical internode without sutures.

Frenelopsis alata (K. Feistmantel) Knobloch; Cretaceous, identified from Czechoslovakia, France, Portugal, United States
Figures 9.2; 9.6B; 9.7B; 9.9D, E; 9.10D; 9.11B

Figure 9.8. **A** *Cupressinocladus valdensis,* a well-established member of the family; Wealden, Sussex. Holotype showing opposite-decussate leaves with distinct sutures separating decurrent bases. × 1. For stoma, see figure 9.7F. **B–I** Various species strongly suspected of belonging to the family, see also figure 9.4. **B** Undescribed Wealden *Cupressinocladus* shoot. × 5. **C** Cuticle. × 250. Scanning electron micrograph. **D** Specimen of undetermined specific identity with three leaves per node and sutures between leaf bases; borehole, Wealden, German Democratic Republic. × 2. **E** *Glenrosa pagiophylloides* shoot. × 1. **F** *Glenrosa pagiophylloides* cuticle. × 200. **G** *Glenrosa texensis* shoot. × 1. **H** *Glenrosa texensis* cuticle. × 250. **F, H** show entrances to stomatal chambers. **I** *Glenrosa texensis,* inside view of cuticle showing four stomata of one chamber. × 250. **F, H, I,** scanning electron micrographs.

Synonyms: 1888 *Frenelopsis bohemica* Velenovsky
   1946 *Frenelopsis lusitanica* Romariz
See Feistmantel (1881); Knobloch (1971—name change only); Hluštík (1972, 1974, 1978, 1979); Hluštík and Konzalova (1976a—male
cones, pollen, 1976b); Alvin (1977—SEM details, Portugal); Watson
(1977—SEM details, Texas); Alvin and Hluštík (1979—unusual
branching pattern); Pons (1979—male cones, pollen, female cone scales,
France).

Frenelopsis harrisii Doludenko and Reymanówna 1978; Cretaceous,
Tajikistan, U.S.S.R.
   Figures 9.6C; 9.7D; 9.9B; 9.11C; 9.12F
   See Doludenko (1978).
This very sparsely branched species has one of the thickest and
most unusual cuticles seen in the family so far.

Frenelopsis hoheneggeri (Ettingshausen) Schenk; type-species, Cretaceous, Carpathians
   Figures 9.2; 9.6A; 9.7A; 9.10E, F
In the absence of the type material this species was revised using
dispersed segments from the type locality in the Carpathians (Reymanówna and Watson 1976). Hluštík (1979) has described the unusual
branching pattern in a Czechoslovakian specimen (figure 9.2).
   See Ettingshausen (1852), Schenk (1869), Reymanówna (1965), Purkyňová (1983).

Frenelopsis occidentalis Heer 1881; Cretaceous, Portugal
   Figures 9.6E; 9.7C
   See Alvin (1977—cuticle details).
The leaf arrangement of this species has not been determined, but
the isolated cylindrical segments and cuticle are distinctly frenelop-
sid.

Figure 9.9. *Frenelopsis* shoots. All × 1. **A** *Frenelopsis ramosissima*,
Lower Cretaceous, Virginia. **B** *Frenelopsis harrisii*, Cretaceous, Tajikistan, U.S.S.R. **C** *Frenelopsis silfloana*, Lower Cretaceous, Sudan.
Arrow indicates position of leaf whorl shown in Figure 9.13C. **D** *Frenelopsis alata*, Cretaceous, Czechoslovakia. **E** *Frenelopsis alata*, Lower
Cretaceous, Texas. Arrow indicates position of leaf whorl shown in
figure 9.10D.

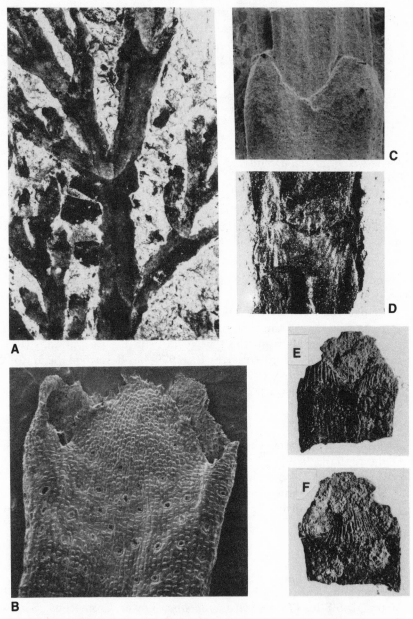

Figure 9.10. *Frenelopsis* leaf whorls. **A** *Frenelopsis ramosissima,* portion of shoot showing leaf whorls alternating and branches clearly arising in axils of leaves. × 5. Photographed under paraffin. **B** Portion of segment from fifth order branch on previous specimen, showing

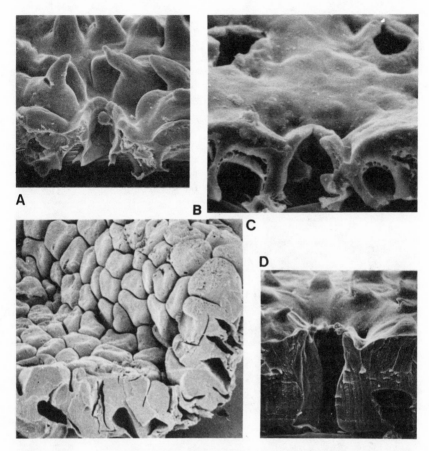

Figure 9.11. Vertical sections of stomata viewed at 60° tilt in scanning electron microscope. × 500. **A** *Frenelopsis ramosissima.* **B** *Frenelopsis alata.* **C** *Frenelopsis harrisii.* **D** *Pseudofrenelopsis varians.*

three free leaves, stomata, fringed margin, mildly papillate surface, and so on. × 50. Compare papillation in figure 9.13B. **C** SEM of silicone rubber cast of *Frenelopsis silfloana* showing leaf whorl indicated in figure 9.9C. × 20. **D** *Frenelopsis alata* showing leaf whorl indicated in figure 9.9E. × 6. **E, F** Both sides of leaf whorl of *Frenelopsis hoheneggeri*, Lower Cretaceous, Przenosza, Poland. × 7. Figures 9.10D–F show conspicuous converging ridges on the free leaf, a common feature that may be preservational.

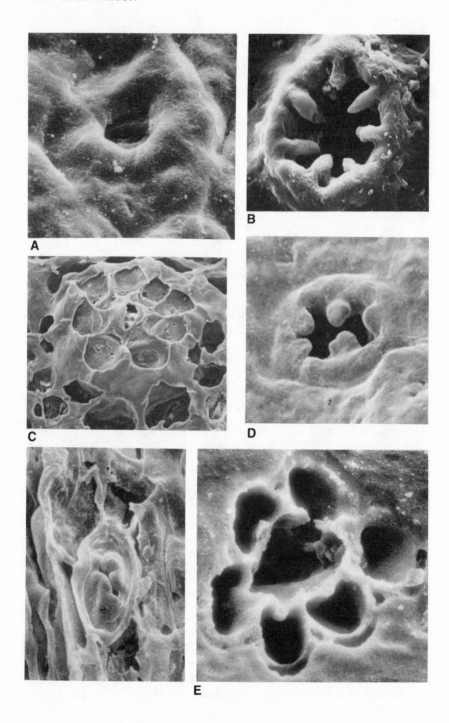

Frenelopsis oligostomata Romariz 1946; Cretaceous, Portugal
   Figure 9.13D
   See Broutin and Pons (1976—branched shoots, cuticle); Alvin (1977 —branched shoots, cuticle); Lauverjat and Pons (1978—Wood); Pons and Broutin (1978—cuticle, male cone, pollen, female cone scale).

Frenelopsis ramosissima Fontaine 1889; Lower Cretaceous, Maryland and Virginia
   Figures 9.2; 9.9A; 9.10A, B; 9.11A; 9.13B; 9.14
   See Berry (1910, 1911—cuticle drawings); Thompson (1912—cuticle drawing); Watson (1977—emended diagnosis, cuticle); Upchurch and Doyle (1981—paleoecology).

Frenelopsis rubiesensis Barale 1973; Lower Cretaceous, Spain
   The cuticle of this species is less well characterized than other species and needs SEM study for clear comparisons.

Frenelopsis silfloana Watson 1983. Cretaceous, Sudan
   Figures 9.6F; 9.9C; 9.10C
   Described from silicone rubber casts. See Watson and Alvin (1976); Edwards (1926).

Frenelopsis teixeirae Alvin and Pais 1978; Lower Cretaceous, Portugal
   Figure 9.2
   Based on a single specimen with opposite decussate leaves.

Genus *Hirmeriella* Hörhammer 1933
   Synonym: *Cheirolepidium* Takhtajan 1963
   Shoots with spirally arranged, adpressed scale leaves; *Classopollis*

Figure 9.12. All except **F** are stomata that do not have papillae within the pit. All scanning electron micrographs. All except **C** × 1000. **A** *Hirmeriella muensteri*, with guard cells visible. Lias, South Wales. **B** Outer surface of cuticle of *Pseudofrenelopsis varians*. **C** Inner surface. × 400. Cuticle is enormously thick; see figure 9.11D. Papillae around rim correspond to number of subsidiary cells. **D** *Pseudofrenelopsis parceramosa* from Virginia. English specimens usually have rounder and more prominent stomatal rim. **E** *Pseudofrenelopsis intermedia*, outer surface of hollow subsidiary cells eroded. **F** *Frenelopsis harrisii*, inner set of papillae more easily detected on inner surface when guard cells are missing.

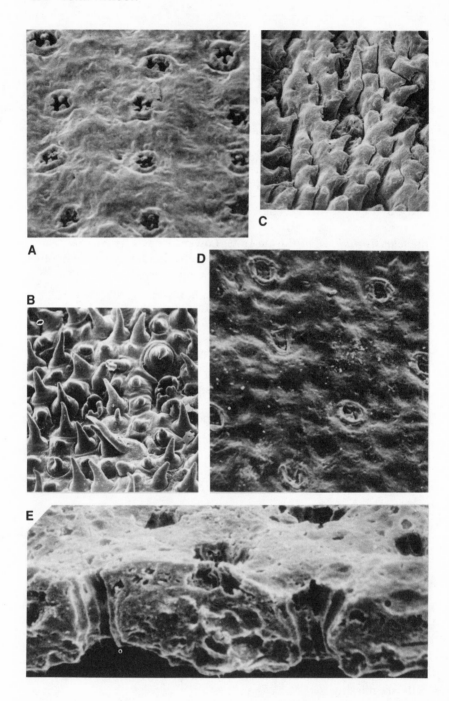

producing male cones; female cones with complex ovuliferous scale in axil of separate bract scale.

Hirmeriella airelensis Muir and Van Konijnenburg 1970; Rhaeto-Lias, France
This species should be reinvestigated in conjunction with new material of *H. muensteri*, which is currently under study from Poland (M. Reymanówna, personal communication) and south Wales (C. Hill, personal communication).

Hirmeriella muensteri (Schenk) Jung: Rhaeto-Liassic, Europe
Figures 9.1; 9.5A–G; 9.12A; 9.15; 9.16A–G
Synonym: 1933 *Hirmeriella Rhätoliassica* Hörhammer
See Hirmer and Hörhammer (1934), Harris (1957), Wood (1961), Chaloner (1962), Jung (1967, 1968), Krassilov (1982).
Figure 9.15 is redrawn from Jung's (1968) reconstruction of the shoots and cones of the type-species.

Genus *Pagiophyllum* Heer 1881 (Harris [1979], emended diagnosis)
As defined by Harris, the shoots have spirally arranged leaves that are longer than the width of their basal cushion, and the blade is broader than it is thick.

Pagiophyllum araucarinum (Pomel) Saporta; Jurassic, England, France, Germany.
Synonyms: *Pagiophyllum kurrii* (Schimper) Salfeld
Pagiophyllum connivens Kendall

Figure 9.13. All scanning electron micrographs, × 250. **A** *Pseudofrenelopsis parceramosa*, outer surface of mid-internode with smooth surface between rows of stomata, which have a ring of papillae overhanging the pit, one per subsidiary cell. **B** *Frenelopsis ramosissima*, internode surface from a first order branch of the specimen in figure 9.10A. **C** *Pseudofrenelopsis varians*, surface of extremely papillate open leaf; papillae joined in rows and pointing toward apex, at right; stomata somewhat cramped and distorted; compare figure 9.20E. **D** *Frenelopsis oligostomata*, outer surface of mid-internode with smooth surface between rather obscure rows of stomata, which have broad papillae within the pit. Cretaceous; Esgueira, Portugal. **E** *Pseudofrenelopsis intermedia*, vertical section through two stomata viewed at 60° tilt, showing hollow subsidiary cells and very thickly cutinized ordinary epidermal cells.

Figure 9.14. Reconstruction of a portion of *Frenelopsis ramosissima* showing succulent shoots with limited development of wood. Based on descriptions and dimensions given by Fontaine (1889) and on remaining specimens in the U.S.N.M. × 0.5 Outline of plant suggests a succulent, shrubby xerophyte.

See Kendall (1948—shoots, Yorkshire): Kendall (1952—male cones, pollen, female cone scales, Yorkshire); Harris (1979—male cones, pollen, female cone scales); Barale (1981—full synonymy, discussion, cuticle, and so on, France)

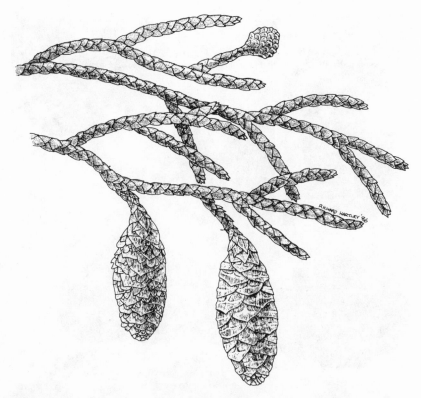

Figure 9.15. Reconstruction of *Hirmeriella muensteri*, with male and female cones. Female cones shown before and after shedding of lobed ovuliferous scales. Approx. × 1. (Modified and redrawn from Jung 1968.)

Pagiophyllum maculosum Kendall 1948; Jurassic, Yorkshire
   See Harris (1979 female cone, repeat of Kendall's drawings); Van Konijnenburg-Van Cittert (1987—male cones).

Genus *Pseudofrenelopsis* Nathorst 1893 (Watson [1977] emended diagnosis; Watson [1982], review)
   Synonyms: *Manica* Watson 1974
            Suturovagina Chow and Tsao 1977

Pseudofrenelopsis intermedia (Chow and Tsao) comb. nov.; Cretaceous, China
   Figures 9.3; 9.12E; 9.13E; 9.17J, K

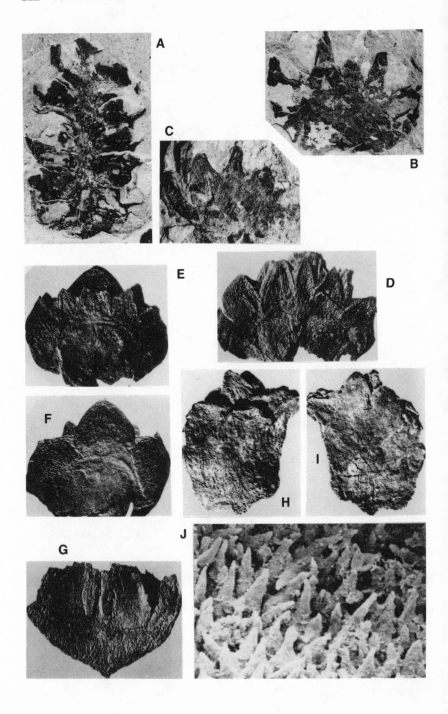

1977 *Suturovagina intermedia* Chow and Tsao, shoots, cuticle; Zhou
Zhiyan (1983), shoots, cuticle, male cone, pollen, wood
P. intermedia has interesting phyllotaxis. The shoots mostly have
open leaf sheaths that do not encircle the stem completely and only
occasionally have segments with closed cylinders. This led Chow and
Tsao to establish a new genus, but Alvin (1977, 1982) and Watson
(1977, 1982) have pointed out that both *Pseudofrenelopsis parceramosa*
and *Pseudofrenelopsis varians* also display this form of heterophylly,
though with a reversed preponderance of open and closed forms. *P.
intermedia* also has rare variants of the open type (figure 9.3).

Pseudofrenelopsis parceramosa (Fontaine) Watson; Lower Cretaceous,
recorded from United States, England, Europe, Sudan, Ghana, ?China,
?Japan.
  Figures 9.12D; 9.13A; 9.17A–C; 9.18; 9.19E; 9.20A–D
  See Fontaine (1889—shoots, Potomac); Watson (1974—generic diagnosis and name change only): Alvin (1977—shoots, cuticle, Portugal); Watson (1977—redescription of type material, full synonymy,
Potomac and southeast England); Alvin, Spicer, and Watson (1978—
male cones, pollen, Isle of Wight); Alvin, Fraser, and Spicer (1981—
wood anatomy, paleoecology, Isle of Wight); Upchurch and Doyle
(1981—paleoecology, Potomac); Alvin (1983—reconstruction).
  This is one of the best-known frenelopsids, with fairly voluminous
literature. The references quoted above are the major works, but the
following are also relevant: Oishi (1940); Teixeira (1948); Reymanówna (1965); Oldham (1976); Watson and Alvin (1976); Chow and
Tsao (1979); Zhou Zhiyan and Cao Zhengyao (1979); Hluštík (1979);
Alvin (1982); Watson (1982, 1983). The unpublished record from Ghana
has been established from a photograph provided by Prof. W. G.
Chaloner (personal communication).

Figure 9.16. *Hirmeriella* female cone scales. **A–G** *Hirmeriella muensteri* cone and scales, from the Rhaeto-Liassic, Bayreuth, Germany. ×
3. **A** Probable ripe cone of persistent bract scales. **B, C** Ovuliferous
scales from same British Museum (Natural History) collection. **D–F**
Ovuliferous scales from Hörhammer (1933); **E**, adaxial, **F**, abaxial
sides of same scale, the reverse of what was supposed by Hirmer and
Hörhammer. **G** From Hirmer and Hörhammer (1934); bract scale
with two scars where the ovuliferous scale was attached. **H, I** Lobed
scale found in association with *Pseudofrenelopsis parceramosa* [Fon-
taine] Isle of Wight. × 5. **J** Cuticle from similar scale. × 250.

Figure 9.17. *Pseudofrenelopsis* shoots. **A–C** *Pseudofrenelopsis parcera-mosa,* shoots showing cylindrical internodes of variable sizes, free leaf tips, stomata in rows. **A** Wealden, Isle of Wight. × 2. **B** Longest internodes so far encountered; borehole, Louisiana. × 2. **C** Portion of

Pseudofrenelopsis varians (Fontaine) Watson; Lower Cretaceous, Texas, Mexico.

Figures 9.11D; 9.12B; 9.13C; 9.17D–I; 9.19A–D; 9.20E–G; 9.21; 9.23C–H

Synonyms: 1893 *Pagiophyllum dubium* Fontaine: open leaf form, Texas

1893 *Pseudofrenelopsis felixi* Nathorst: open leaf form, Mexico

See Fontaine (1893—cylindrical segments, Texas); Watson (1977—redescription of type material); Daghlian and Person (1977—Texas).

Genus *Tomaxellia* Archangelsky 1963 Archangelsky (1966, 1968—emended diagnoses).

Tomaxellia includes shoots with long, spirally arranged decurrent leaves that are rhomboidal in cross section. They thus come within the range of Harris' diagnosis for *Geinitzia* Endlicher, but Archangelsky includes cuticle details in his diagnosis of *Tomaxellia*.

Tomaxellia biforme Archangelsky 1966; Cretaceous, Argentina

Figures 9.1, 9.22A–F

See Archangelsky and Gamerro (1967—male cones, pollen); Archangelsky (1968—female cones, cone scales, seed cuticles).

T. biforme is now known in some considerable detail, but the type species *T. degiustoi* remains less well defined and is known only from shoots. *T. biforme* is markedly heterophyllous (figure 9.22A–D), with short *Brachyphyllum*-type leaves occurring on the same shoot as the long leaves. The two leaves in figure 9.22C, D came from the specimen in figure 9.22A.

Other Members Omitted from the list above are several frenelopsid species that I cannot be certain are distinct. For instance, *Frenelopsis*

---

the lectotype. Trent's Reach, Virginia. × 2. **D–I** *Pseudofrenelopsis varians*. Glen Rose, Texas. **D–G** Shoots showing variable lengths and proportions of internode segments. **D, E, G,** closed cylinders; **F,** open type. × 2. **H** Top part of long segment broken to show very slender vascular cylinder. × 5. **I** Circular holes commonly found in segments (center) probably insect damage in life. × 4. **J, K** *Pseudofrenelopsis intermedia*. Cretaceous, near Nanjing, East China. × 2. **J** Typical shoots with most commonly found open-type leaves; compare **F**. **K** Two righthand shoots of type with only free leaf tips interrupting stem surface; see figure 9.3.

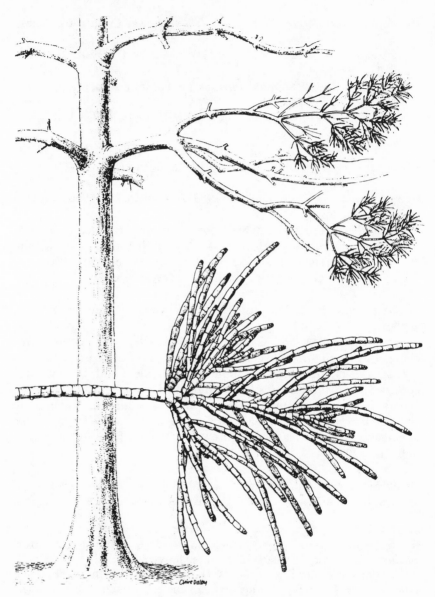

Figure 9.18. Reconstruction of *Pseudofrenelopsis parceramosa*, by Claire Dalby. (Reproduced by kind permission of Dr. K. L. Alvin.)

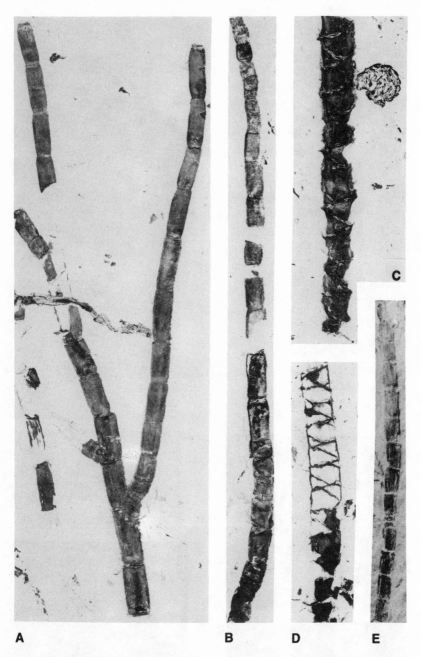

Figure 9.19. *Pseudofrenelopsis* shoots. All × 1. **A–D** *Pseudofrenelopsis varians*, Lower Cretaceous, Texas. **A** Lectotype showing long internodes and sparse branching. **B** Branch showing variable internode length. **C, D** Shoots with the short, open-type of leaf sheaths that do not completely encircle the stem. Cone in **C** is not attached. **E** *Pseudofrenelopsis parceramosa*, typical ultimate branchlet; see figure 9.18. Wealden, Isle of Wight.

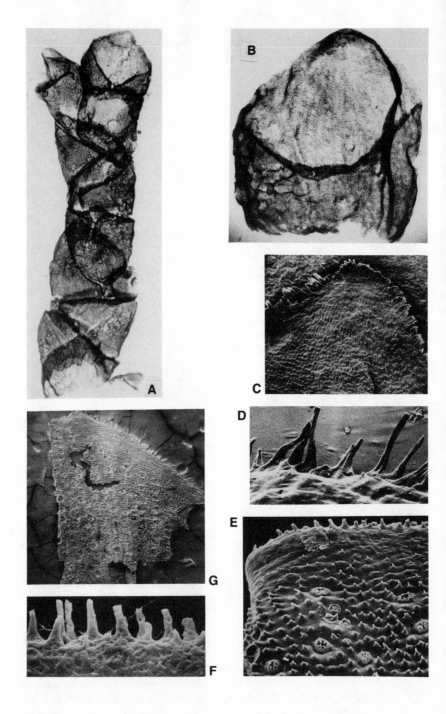

elegans Chow and Tsao (1977) and *Frenelopsis choshiensis* Kimura, Salki, and Arai (1985) look very similar to some of the European species. Apart from such frenelopsid species for which family membership must be fairly secure, there is a growing list of suspected members belonging to various other genera. These are listed in table 9.2. Some of them are extremely unusual in various ways and are considered individually in the section on cuticle characters. The following genera are probably wholly or in part members of the family.

Genus *Androvettia* Hollick and Jeffrey 1909
  Figure 9.4A–C
  Shoots of *Androvettia* superficially look like ferns but are flattened shoots forming phylloclades bearing tiny scale-like leaves (figure 9.4B). On some ultimate shoots the leaves have in their axils unexpanded buds (figure 9.4A), which have been mistaken in the past for male cones. Cones are unknown, but the cuticle has a strongly cheirolepidiaceous appearance (stoma, figure 9.4C). Four species have been described from the Upper Cretaceous of North America, and a revision of them is to be published elsewhere (Hueber, personal communication).

Genus *Geinitzia* Endlicher
  Used by Harris (1979) for shoots bearing long decurrent leaves, with the free part as thick as broad. It is like the type of *Tomaxellia* leaf shown in figure 9.22D).

Genus *Glenrosa* Watson and Fisher 1984
  Figure 9.8E–I
  Two species have been described from the Lower Cretaceous Glen Rose beds, Texas. *Glenrosa texensis* (Fontaine) looks like a *Brachyphyl-*

Figure 9.20. *Pseudofrenelopsis.* **C–G,** scanning electron micrographs. **A–C** *Pseudofrenelopsis parceramosa.* Hastings, Sussex. **A** Very small shoot with leaf segments of the type that meet along a short suture; see figure 9.3. × 10. **B** Leaf of similar dimensions with completely closed base. × 25. **C** Free leaf with fringe of hairs, papillate surface, and very few stomata. × 60. **D** U.S. specimens have longer, individual hairs around free leaf sheath. Trent's Reach, Virginia. × 250. **E–G** *Pseudofrenelopsis varians.* **E** Top part of long internode. × 100. **F** Hairs on leaf sheath margin; same specimen, × 500. **G** Open-type leaf from below a cone, surface extremely hairy. × 25.

Table 9.2. Suspected Members of the Cheirolepidiaceae

Species	Age	Localities	Features	References
Androvettia spp.	Upp. Cret.	U.S.A., Canada	phylloclades, papillate stomata	Hueber, personal communication.
Brachyphyllum ardenicum Harris	Jurassic	England	general similarity	Harris 1979.
Brachyphyllum carpentieri Fisher and Watson	Lr. Cret.	France	Lobed margin, large glands	Fisher and Watson 1983.
Brachyphyllum castatum Watson et al.	Wealden	England	long stomatal tubes, fluted stomatal rim	Watson, Fisher, and Hall 1987.
Brachyphyllum hegewaldia Ash	Upp. Trias	Arizona	cuticle strongly cheirolepidiaceous	Ash 1973.
Brachyphyllum ningshiaense Chow & Tsao	Cretaceous	E. China	like open type of *Pseudofrenelopsis*	Chow and Tsao 1977.
Brachyphyllum pulcher Lorch	Lr. Jurassic	Israel	strongly papillate stomatal rim	Lorch 1968; Watson, personal investigation.
Brachyphyllum squamosum (Velenovsky)	Mid-Upp. Cret.	Czechoslovakia	general similarity	Hluštík, personal communication.
35 CONIF BrB = *Brachyphyllum* sp.	Wealden	England	strongly papillate	Oldham 1976.

Species	Age	Location	Characteristic	Reference
Cupressinocladus malaiana (Kon'no)	Lr. Cret.	Malaya	whorls of 3 leaves	Barnard and Miller 1976.
33 CUPR CuA = *Cupressinocladus* sp.	Wealden	England	opp. decuss. leaves, papillate	Oldham 1976.
Geinitzia rigida (Phillips)	Jurassic	England	like *Tomaxellia*	Harris 1979.
Glenrosa pagiophylloides (Fontaine)	Lr. Cret.	Texas	highly papillate, unique stomatal pits	Watson and Fisher 1984.
Glenrosa texensis (Fontaine)	Lr. Cret.	Texas	unique stomatal pits	Watson and Fisher 1984.
4 spp. *Pagiophyllum*	Upp. Trias.	Arizona	papillate stomata look cheirolepidiaceous	Ash 1978.
Pagiophyllum insigne Kendall	Jurassic	England	general similarity	Harris 1979.
Pagiophyllum peregrinum (L. & H.)	Jurassic	England	deeply papillate stomata	Kendall 1948; Van Konijnenburg, personal communication.
Pagiophyllum ordinatum Kendall	Jurassic	England	general similarity	Harris 1979.
Pagiophyllum crassifolium Schenk	Wealden	England, Germany	papillate stomata of cheir. type	Fisher 1981.
Sphenolepis kurriana (Dunker)	Wealden	England, Germany	papillate stomata of cheir. type	Fisher 1981.

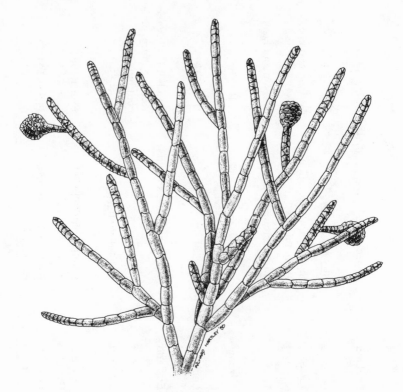

Figure 9.21. Reconstruction of *Pseudofrenelopsis varians*, based on specimens in the U.S.N.M., suggesting a fleshy, salt-marsh inhabitant. × 0.5.

lum and *G. pagiophylloides* (Fontaine) looks like a *Pagiophyllum*, but both have a stomatal arrangement unknown in any other conifer. The stomata are grouped together in chambers (figure 9.8I), thus sharing a single pit and pit opening with dense hairs around the rim (figure 9.8H), as in the angiosperm *Nerium oleander* L.

THE CUTICLE

Most of the cuticles of well-established members of the family and some of the suspected members are now known in considerable detail and have proved a most valuable means of recognizing or separating species. There is no doubt that the scanning electron microscope has played an enormous part in allowing us to decipher the fine details and subtleties of the structure of the stomata, which in some cases

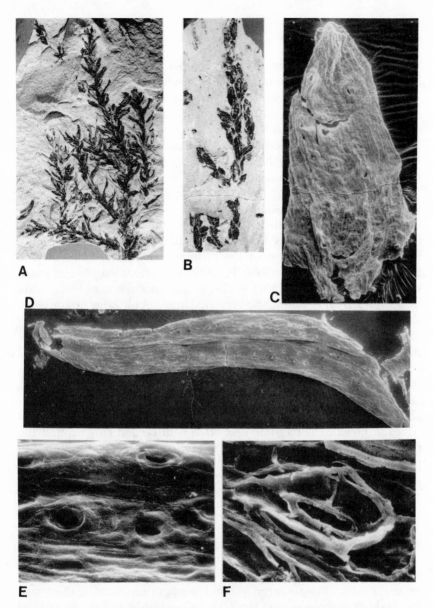

Figure 9.22. *Tomaxellia biforme,* from the Lower Cretaceous of Patagonia, Argentina. **B** through **F,** scanning electron micrographs. **A** Shoot clearly showing the long, falcate leaf form, × 2. Note especially that adpressed leaves are also present. **B** Shoot showing the short, adpressed leaf form. × 2. **C** Montage of short leaf from shoot in **A.** × 75. **D** Montage of falcate leaf from shoot in **A.** × 30. **E** Part of adaxial surface of leaf in **D** showing stomata; lateral keel below without stomata. × 250. **F** Inside view of stoma with five subsidiary cells. × 500.

verge on the bizarre. The position has perhaps been reached where lack of scanning electron micrographs seriously impedes comparison. The SEM has been so spectacularly useful with this particular group of plants because their cuticles present perfect subjects, thick and tough with a highly distinctive suite of characters found in different combinations. All the micrographs reproduced here are from the SEM, but light micrographs of the same cuticles can be found in the original papers (e.g., Reymanówna and Watson 1976; Alvin 1977; Watson 1977). The SEM photographs of *Tomaxellia* (figure 9.22), *Androvettia* (figure 9.4A–C), and *Cupressinocladus ramonensis* Chaloner and Lorch (figures 9.6D, 9.7E) are published for the first time.

The cuticular features in general include marginal hairs (figure 9.20C–G) and papillation of the ordinary epidermal cells varying from small bumps (figures 9.6D, F; 9.10B) via various lengths of pointed protuberances (figures 9.13C; 9.20E) to a dense covering of long hairs (figure 9.13B) and include the very striking "cobblestone" papillae of *Frenelopsis harrisii* (figures 9.6C; 9.7D). The stomata have randomly orientated apertures and arrangement is either scattered or in longitudinal rows, which in some species are irregular, in others well defined. The stomatal apparatus typically has a ring of four to six subsidiary cells, often six (figure 9.12E), but in *Pseudofrenelopsis varians* up to nine have been seen (figure 9.12C shows eight). The subsidiary cells form a deep pit associated with which may be one or two series of prominent papillae and often a thickened ring around the rim of the pit opening. The papillae are usually equal in number to the subsidiary cells and may form a ring extending over and constricting the mouth of the pit (figure 9.12B, D). In others, large papillae are present down inside the pit, often protruding at the mouth, and the rim may have papillae, lobes, or "pouches," thus forming a double structure. Such a construction gives a striking and, as far as I know, unique appearance to the stoma (figures 9.6; 9.7; 9.11B). This type of stoma is present in all the known species of *Frenelopsis* except *F. ramosissima* (figure 9.11A) and *F. rubiesensis* and has not actually been of practical application in family recognition because the *Frenelopsis* shoot morphology is so easily recognizable and is assumed to be a reliable indicator. *Cupressinocladus valdensis* (figures 9.7F; 9.8A) is the one example in which the stomatal structure was used to forecast cheirolepidiaceous affinity (Watson 1977). This species was later found to have *Classopollis* pollen (Francis 1983).

Of the stomata without this double structure, the type with a single set of papillae at the pit rim is found in a number of species of *Brachyphyllum* (table 9.1; Ash 1973), *Pagiophyllum* (Barale 1981: plate

50; Ash 1978), *Pseudofrenelopsis* (figure 9.12B, D) and *Cupressinocladus* (figure 9.8C; Barnard and Miller 1976), all extremely similar to each other. *Androvettia*, which is a strong contender for this family, also has this type of stoma (figure 9.4C).

Entirely nonpapillate stomata occur in several of the Jurassic *Brachyphyllum* and *Pagiophyllum* species, *Hirmeriella* (figures 9.5G; 9.12A) and *Tomaxellia* (figure 9.22D, E). Stomata of this kind seem to me very similar to species that have been reliably attributed to the Araucariaceae and Taxodiaceae (Harris 1979; Fisher 1981) and hence nonfrenelopsid members of the Cheirolepidiaceae need other supporting evidence, probably cones. However, I am convinced that, in the light of recent progress, routine reexamination of cuticles, particularly by SEM, would reap considerable rewards.

Figures 9.4 and 9.8 show a number of species with extremely unusual cuticle structure that are suspected members of the family; in particular, the two species of *Glenrosa* and *Brachyphyllum castatum* Watson, Fisher, and Hall (1987). *Brachyphyllum carpentieri* Fisher and Watson (1983) is also included in this increasing list of brachyphylls with extraordinary cuticular features. Their stomata do not accord with the types described above and suspicion of cheirolepidiaceous membership is really based on a hunch, but it is also difficult to imagine where else they might be placed.

INTERNAL ANATOMY

The wood is known for only five species within the family, all of which have been attributed (table 9.1) either to *Protocupressinoxylon* or *Protopodocarpoxylon* (Harris 1957; Lauverjat and Pons 1978; Alvin 1981; Francis 1983; Zhou Zhiyan 1983). These Mesozoic wood form-genera are characterized by semiaraucaroid or protopinaceous tracheid pitting, parenchymatous rays, and cross-field pitting, and are separated on details of the cross-field pitting. Thus the oldest member of the family, *Hirmeriella muensteri*, has very similar wood to the youngest member, *Frenelopsis oligostomata*. Alvin (1977, 1982) and Zhou Zhiyan (1983) have both summarized the known wood within the family, while describing in detail wood from species of *Pseudofrenelopsis*.

THE MALE CONE

Table 9.1 shows the 15 species for which male cones have been described, either attached or associated. Unattached cones yielding *Classopollis* pollen are now placed in the genus *Classostrobus* Alvin,

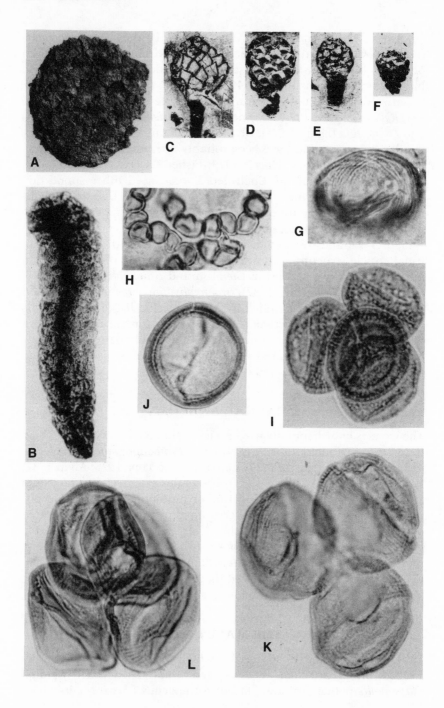

Spicer, and Watson (1978). The cones are all of typical coniferous structure (figures 9.23A, C–F), oval or round in shape with spirally arranged microsporophylls bearing pollen sacs on their lower surface. They vary in size from about 3 mm across to the largest recorded so far in *Pseudofrenelopsis intermedia*, which is 23 mm in diameter (Zhou Zhiyan 1983: plate 80, figure 1, text figure 4). The cones in figure 9.23C–F belong to *Pseudofrenelopsis varians*, some attached to short lengths of shoot. Although I formerly regarded these as female cones (Watson 1977), they are almost certainly all male cones. Only the smallest (figure 9.23F) has yielded *Classopollis* (figure 9.23G), however, and then not in pollen sacs. All the cones are heavily infested with fungal hyphae and spores (figure 9.23H). Heavy fungal infestations of frenelopsid remains have also been reported by Alvin and Muir (1970) and Pons and Boureau (1977). The cones of *P. varians* show the sporophylls with rhomboidal heads, as is found also in *Classotrobus comptonensis* (figure 9.23A) associated with *Pseudofrenelopsis parceramosa*. Male cones of *Hirmeriella* spp. have been interpreted as having almost peltate sporophylls (see Van Konijnenburg-Van Cittert 1987: table 1). The number of pollen sacs seems to vary according to the species, with numbers from 2 to 8 given. Harris (1957, 1979) has suggested that Hörhammer's description of *Hirmeriella muensteri* as having peltate *Taxus*-like sporophylls with 12 pollen sacs in a ring was a misinterpretation. The material from south Wales attributed to *H. muensteri* by Harris (1957) appears to have 2 pollen sacs per sporophyll. *Classostrobus comptonensis* has been interpreted

Figure 9.23. Male cones and *Classopollis* pollen. **A** *Classostrobus comptonensis*, the *Classopollis*-bearing male cone of *Pseudofrenelopsis parceramosa*. Isle of Wight, × 5. **B** Clump of *Classopollis* pollen from a similar cone, presumably the entire contents of a single cylindrical pollen sac. × 75. **C–F** Cones borne on tips of *Pseudofrenelopsis varians* shoots. × 1. **G** Oblique view of *Classopollis* grain yielded by cone in **F**. × 1000. **H** Preparation from the cones showing usual heavy fungal infestation with numerous spores. × 1000. **I–L** Dispersed *Classopollis* from various horizons showing a number of the distinctive features illustrated in figure 9.26. All × 1000. **I** Tetrad with thick equatorial band but not well-developed striae. Middle Jurassic, Brora, Scotland. (Kindly provided by Dr. R. Porter.) **J** Polar view of single grain in optical section. Wadhurst Clay, Wealden. **K** Triad with broad band of striae and thin subequatorial rimula. Purbeck Beds, southern England. **L** Typical tetrad, Ashdown Beds, Wealden. (**J–L** kindly provided by Dr. D. J. Batten.)

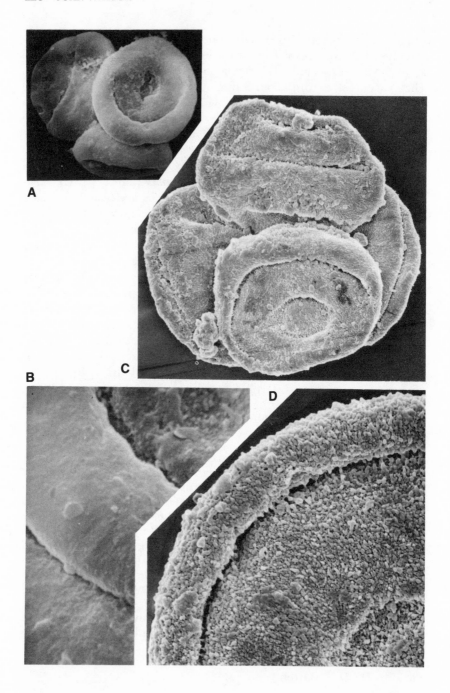

A

B

C

D

(Alvin et al. 1978) as having three cylindrical pollen sacs per sporophyll. The intact pollen mass from a single sac is shown in figure 9.23B. Harris (1957) isolated similar cylindrical pollen masses from *H. muensteri*. None of the other cones so far described are well enough preserved for pollen sac number and shape to be securely established.

CLASSOPOLLIS POLLEN

Considerably more is known about the morphology and stratigraphical and geographical distribution of *Classopollis* from palynological studies than from macrofossil remains. Tetrads of pollen from the male cone of the German *Hirmeriella muensteri* material were figured by Hörhammer (1933) and later single grains from the Welsh material that Harris (1957) identified as the same species. Chaloner (1962), investigating borehole material from southern England, also identified *H. muensteri* and recognized the associated pollen as the highly distinctive form known as *Classopollis* Pflug (1953). The use of the name *Classopollis* has been questioned by Cornet and Traverse (1975), who regard *Corollina* Malyavkina as the correct name for this pollen although most palynologists regard the original description and illustrations of the latter genus as so inadequate that its true identity cannot be ascertained. Srivastava (1976) has dealt in detail with the complex nomenclatural problems and, following Pocock and Jansonius (1961), recommends the continued use of *Classopollis* Pflug, which was more satisfactorily described. Figures 9.23–9.25 illustrate a variety of *Classopollis*, especially tetrads from various sources, whereas figure 9.26 is a diagrammatic representation of a typical single grain. It is approximately spherical with a trilete scar on the proximal pole and at the distal pole a circular cryptopore (figure 9.24A, C, D), which is an area where the exine is thinner and through which germination is thought to have occurred. Subequatorially toward the distal side is a furrow, the rimula (figure 9.23I, K) that forms a strong surface

Figure 9.24. *Classopollis* pollen. Scanning electron micrographs. (Photographs kindly provided by Dr. D. J. Batten.) **A** Tetrad from dispersed preparation, Hastings Beds, Lower Cretaceous, Sussex, England. × 1000. **B** Detail of front spore showing rimula (top right) and unsculptured exine. × 5000. **C** Dispersed tetrad from the Upper Cretaceous, Portugal. Front spore showing rimula and cryptopore on distal surface. × 1500. **D** Surface detail from same tetrad showing strongly sculptured exine. × 5000. Compare figures 9.25A, B.

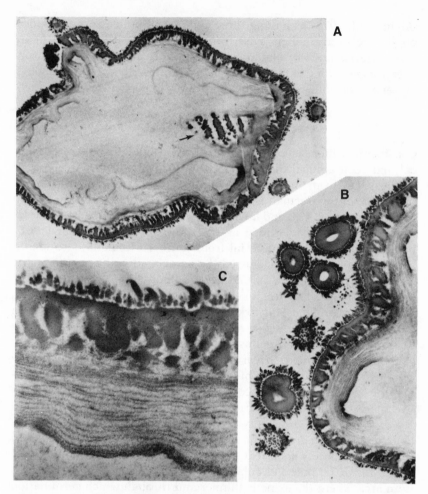

Figure 9.25. Transmission electron micrographs of *Classopollis* pollen from *Classostrobus comptonensis*, the male cone of *Pseudofrenelopsis parceramosa*. (Photographs kindly provided by Dr. T. N. Taylor.) **A** Ultrathin section of almost whole grain showing equatorial striations paradermally on right (arrow). × 3000. **B** Section of mature grain from same cone as **A**, showing layered wall and associated orbicules. × 7000. **C** Nonspecialized region of exine in ultrathin section, sculptured outer surface at top. × 20,000. See figure 9.26 for details of exine layers.

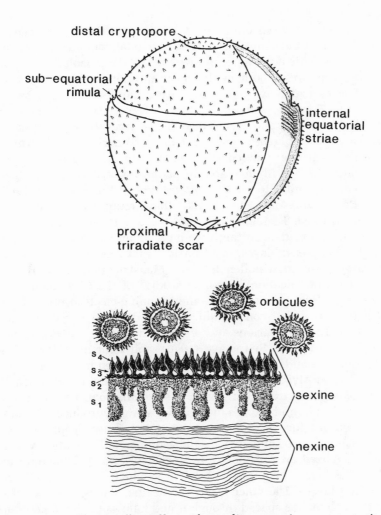

Figure 9.26. *Classopollis* pollen. *Above,* diagramatic representation of a single grain. Approximately × 1500. *Below,* section of mature pollen grain wall (redrawn from Taylor and Alvin 1984) showing lamellated nexine, the S_{1-4} layers of the sexine, and the identical S_4 ornamentation on the orbicules. Approx. × 15,000.

feature in the SEM (figure 9.24A–D). Equatorially the grain has a thickened band that is usually internally striated (figure 9.23G, K, L) and therefore cannot be seen in whole grains in the SEM (figure 9.24). The surface of the exine shows a variety of sculptural patterns (Reyre 1970) and is probably of taxonomic significance.

The earliest occurrence of *Classopollis* is from the Upper Triassic, not earlier than Carnian (D. J. Batten, personal communication), attaining worldwide occurrence at low and middle paleolatitudes. The earliest forms are smooth, with poorly developed equatorial striations. The typical *Classopollis* structure became established by the end of the Triassic, with increasingly diverse sculptural patterns through the Jurassic and Cretaceous. The youngest frenelopsid shoots known are those of *Frenelopsis oligostomata* Romariz, from the Turonian of Portugal, the male cones of which have yielded pollen (Pons and Broutin 1978) with an elaborate sculpture of anastomozing bacculae. This is similar to the *Classoidites* type of Van Amerom (1965), which Batten and MacClennan (1984) have found well preserved in younger strata in Portugal. Although there is doubt about the fine details of dating, it is certainly younger than Turonian, and, from overall pollen assemblages, they consider that this extends the range of *Classopollis*-related pollen into the Maastrichtian (D. J. Batten, personal communication). Detailed studies of the *Classopollis* wall structure by light, transmission- and scanning-electron microscopy, which have revealed considerable complexity and variation, have been carried out by various authors, including Pettitt and Chaloner (1964), Reyre (1970), Srivastava (1976), and Courtinat (1980). The latest contribution is a critical transmission electron microscope study by Taylor and Alvin (1984) of the ultrastructure and development of *Classopollis* grains extracted from *Classostrobus comptonensis* cones. Three of Taylor and Alvin's micrographs are reproduced by kind permission in figure 9.25, and figure 9.26 is redrawn from the same source using their terminology for the wall layers. The mature wall in a nonspecialized area of the grain consists of clearly defined nexine and sexine layers (figure 9.25C), each of which is subdivided into additional layers. The inner layer, the nexine, shows about 20 electron-dense lamellae spaced through it and follows the inner contours of the outer layer, the sexine. The sexine consists of four distinct layers that Taylor and Alvin have labeled S_{1-4}, with S_4 forming the surface sculpture of the grain and S_3 a uniform layer of small spaces where the tapering spicules of the sculpture are constricted at the base (figures 9.25, 9.26).

The S_1, the most conspicuous layer, is modified in the specialized parts of the pollen grain. The equatorial striations are seen at the ultrastructural level to be massive flanges of S_1 sporopollenin. At the rimula, cryptopore, and triradiate mark the S_1 layer is extremely reduced or absent. The S_2 layer is also thinner in these regions and often has perforations. Orbicules 1.5–3μm in diameter, which are often found adhering to pollen grains, are seen to be identical in their outer part to layers S_3 and S_4. Associated membranes often incorporating orbicules are interpreted as tapetal in origin, thus providing evidence that the outer sculptured layer is entirely tapetal in origin. Taylor and Alvin (1984) suggest that the enclosed layer of spaces (S_3) just below the surface may have been related functionally to pollination mechanisms involving the evolution of incompatibility systems.

Of course, we know nothing of the pollination mechanism associated with *Classopollis*, although it is notable that some forms commonly adhere in tetrads that survive even through laboratory preparation (figures 9.23I, K, L; 9.24A, C). Adhesion is thought to be by exinal threads arising from the triradiate scars (Courtinat 1980; Taylor and Alvin 1984) and is a rare occurrence in general. Hughes (1976) has pointed out that it may be regarded as a particularly advanced phenomenon in a gymnosperm, providing symmetrical distribution of the distal germinal apertures. Suggestions of animal vectors have arisen from the various unusual features involved but are purely speculative.

THE FEMALE CONE

Lack of information concerning the female cone (figures 9.15, 9.16) has been mentioned above. The cone of *Hirmeriella muensteri* (Schenk) Jung is the best known, with a reconstruction by Jung (1967, 1968) that shows the ovuliferous scale bearing a single seed covered by a flap-like structure. Harris (1979) studied similar cone scales of *H. kendalliae* from Yorkshire, though with very limited and difficult material. He found up to 10 layers of cuticle (including the bract), some of which were delicate double membranes he thought were sacs that contained the seeds and also had clumps of pollen trapped between the layers. He considered the seeds in question to be abortive. Krassilov (1982) macerated *Hirmeriella* scales from Poland and demonstrated cuticle lined locules within the scales. Other such scales studied by Archangelsky (1968) and Pons and Broutin (1978) also indicate a complex scale structure and add weight to the suggestion (Alvin 1982) that pollination was effected by some method other than pollen arriving in the micropyles in the usual gymnospermous manner. At present it is not clear how we can materially improve our understanding of the female cone at a level comparable to the information available for the pollen. Petrified cones of clearly cheirolepidiaceous affinity, which might seem our best hope, are yet to be identified, all such previous contenders having been eliminated (Alvin 1982). *Hirmeriella muensteri* is currently under reinvestigation in several laboratories (C. R. Hill, M. Reymanówna, J. H. Van Konijnenburg-Van Cittert, personal communication). A. Hluštík (personal communication) is investigating cone scales from Czechoslovakia attributed to *Frenelopsis alata* that indicate an enormous size for the cone of this species. Some of the scales are up to 6 cm wide, considerably larger than the scale figured by Pons and Broutin (1978). A cone of such size is comparable only to that of living members of the Araucariaceae.

Perhaps the large number of well-preserved scales of *H. muensteri* and *F. alata*, if sacrificed to maceration and subjected to the meticulous analysis of cuticles as attempted by Harris for *H. kendalliae*, will yield the information we await.

STRATIGRAPHIC OCCURRENCE

The stratigraphic range of *Classopollis* has been mentioned above, and one would suppose that this would be a most reliable reflection of the occurrence of macro-remains awaiting identification. Indeed, the degree of diversity of *Classopollis* indicates that, with regard to macro-remains, their recognition is probably in rather early stages and incomplete on a worldwide basis, so that only a broad picture emerges. The earliest recognized remains are of *Hirmeriella muensteri*, which is Lower Jurassic, and most of the later Jurassic members and suspected members are also of the *Brachyphyllum*, *Pagiophyllum*, and *Geinitzia* type. *Classopollis* is, of course, known from the Upper Triassic, and shoots of that age obviously await recognition. It is interesting that those of Ash (1973, 1978), from the Late Triassic of New Mexico and Arizona, which look very promising as members, are present in rocks from which *Classopollis* has not been identified (Stone 1978). These Triassic floras contain numerous other leafy shoots and cones of various sizes as well as several types of seeds and dispersed cone scales (S. R. Ash, personal communication), all of which look coniferous but are so far unstudied and undescribed.

The decussate leaf arrangement of *Cupressinocladus* is first seen in *C. ramonensis* from Israel and is thought to be Lower Jurassic. The presence of *Classopollis* has not been confirmed in this species. *Cupressinocladus pseudoexpansum* from Iran is Middle Jurassic in age. These leaf morphologies continue into the Cretaceous, with the heterophyllous *Tomaxellia* bearing both *Brachyphyllum*- and *Geinitzia*-type leaves, but from the lowest Cretaceous they are joined, and apparently overtaken, by the frenelopsids. It should be remembered, however, that members of this latter group are very easy to recognize, and the picture may still be considerably distorted. No stratigraphical pattern can be detected for *Frenelopsis* and *Pseudofrenelopsis*, which appear at about the same time and persist for about as long, the youngest being *F. oligostomata* from the Senonian of Portugal.

The various characteristics, like the different branching patterns, the double papillate stomata, and the incredibly thick cuticles, also seem distributed throughout the Cretaceous.

AFFINITIES

The phylogenetic relationships of this amazing family seem impossible to determine at this juncture based on what I would regard as solid factual evidence. Previous suppositions of derivation from or affinity with the Araucariaceae, Taxodiaceae, or Cupressaceae in many cases ensued merely from striking morphological similarity. In other cases the necessity to produce an evolutionary theory seems to have been overriding and the results speculative. It now seems more opportune to have a period of consolidation of facts rather than to add to the speculation.

The Voltziaceae has been suggested as a more attractive starting point for derivation of the cheirolepids (Archangelsky 1968; Němejc 1968; Jung 1968; A. Hluštík, personal communication); but until we understand the morphologies of the seeds and their scales, I am inclined to agree with Harris (personal communication) that, as in many past instances of phylogenetic conjecture, "the evidence for this is somewhere between precious little and damn all." Hughes (1976), writing before the "*Frenelopsis* connection," suggested for the early Cretaceous that family attributions based on a few individual characters should be abandoned in favor of a neutral term such as "early Cretaceous Brachyphylls or Linearphylls." Extending this suggestion to encompass the whole stratigraphic range of the Cheirolepidiaceae requires an enormous number of species to be reexamined openmindedly and may well provide some of the answers we seek.

DISTRIBUTION AND PALAEOECOLOGY

The geographical distribution of *Classopollis* producers was first established from palynological studies all over the world. Most particularly, data from the Soviet Union revealed their abundance at low latitudes and their rarity or absence toward the poles in the Mesozoic. Vakhrameev (1970) plotted the published *Classopollis* percentage data from numerous Russian palynologists' work both on maps and as graphs showing the changes of *Classopollis* content throughout the Jurassic and Cretaceous. When replotted on reconstructed paleogeographic maps (Hughes 1973), these data revealed a close approximation to latitudinal belts, with the *Classopollis* content increasing southward. For the Upper Jurassic, which has the highest abundance of this pollen, Vakhrameev (1978, 1981) demarcated three geographical zones. In the northern zone, above about 50°N paleolatitude, not more than 6–10 percent occurs and often only single grains. Lithological indicators and macro-plant remains support this Zone 3 as a belt of warm–temperate and humid climate. Southward the *Classopollis* content increases, to up to 50 percent in the central Zone 2, a warmer belt with decreased humidity. To the south of this, in Zone 1, the *Classopollis* concentration is over 50 percent, commonly 60–75 percent, and in the Oxfordian reaches 90 percent. These values, together with such indicators as red-beds and evaporites, point to an arid tropical belt dominated by the xeromorphic thermophilous Cheirolepidiaceae. *Classopollis* abundance in these belts also fluctuates considerably during the Mesozoic (Vakhrameev 1970, 1980, 1981), the maxima agreeing with lithological evidence of more widespread, warmer conditions. Similar evidence for increase in temperature and arid conditions corresponding with maximal abundance of *Classopollis* has also been noted for Australia, where *Classopollis* is prominent in the Jurassic, and then shows a considerable diminution until the Cenomanian, when there is a revival (Douglas 1985, and personal communication). The same indications have also been reported from Africa and northern South America (see Alvin 1982). Additionally it has been long known that *Classopollis* is especially abundant in transgressive marine deposits (Vakhrameev 1970).

Thus the distributional data clearly indicate the unusual but undoubted importance of conifers associated with sea margins at latitudes where they do not occur today. Until about a decade ago the known morphology was, of course, only of the *Hirmeriella* and *Tomaxellia* type, typically coniferous shoot genera, together with evidence that at least some were arborescent. As a result, the necessity to accommodate coniferous trees into a tropical flora led to a variety of suggestions, with most authors assuming ecological uniformity and with the emphasis on saline conditions. Vakhrameev, however (1970), envisaged low trees and shrubs forming unmixed thickets on highland slopes, whereas other suggestions included stilt-rooted mangrove species (Hughes and Moody-Stuart 1967) and lowland forests on seaward margins (Hughes 1976). Batten (1974), at that time trying to reconstruct the ecology of *Classopollis*-bearers in the nonmarine sequence of the Weald, suggested three possible habitats: sandy bars and coastal islands, mangrove communities, or upland slopes.

All these ecological models were suggested before the frenelopsids were linked with *Classopollis*, and of course this discovery, together with the increasing number of macro-fossils ascribed to the family, has broadened our thinking. The emerging picture is one of remarkable diversity with evidence of a wide range of both habit and habitat. A revised model for the Wealden of the Weald (Allen 1975), with braided alluvial sandplains and lagoonal alluvial mudplains, prompted Batten (1976) to a reappraisal of the possible habitats for the much expanded family, probably including small herbaceous forms. Other recent studies, including lithological and palynological evidence (Daghlian and Person 1977; Alvin, Fraser, and Spicer 1981; Upchurch and Doyle 1981; Francis 1983), have all added valuable insight to the paleoecology of the various morphological forms.

Most of the frenelopsids differ strikingly in the aspect of their shoots from any living conifer, being much more like some modern stem-succulent xeromorphic angiosperms. All frenelopsids to some extent share the same characters, having very reduced leaves, very thick cuticles, deeply sunken stomata, and photosynthetic stems, which, when the internodes are short, give a strongly jointed appearance (figures 9.17, 9.19). In *Pseudofrenelopsis varians*, which is one of the most strongly jointed and succulent looking, there was very restricted development of wood (figure 9.17H), and there is no evidence that it was anything other than of very small stature. *Pseudofrenelopsis parceramosa*, on the other hand, has ultimate shoots of similar succulent appearance and limited wood but is associated with wood that indicates it was a sizable tree.

Of the known arborescent forms three have been reconstructed as whole plants. Jung's (1968) reconstruction of the early Jurassic *Hirmeriella muensteri* shows a large, typically coniferous evergreen tree for which a limestone habitat is indicated by Harris' identification of the species in Carboniferous limestone fissures infilled by Liassic sediment. The Purbeck-Wealden *Cupressinocladus valdensis* has been reconstructed (Francis 1983) on the basis of *in situ* trunks, branches, foliage, and male cones as a large forest tree (figure 9.27). Variable growth rings in the wood indicate a strongly seasonal climate of alternating favorable and unfavorable conditions for growth. Associated evaporites suggest the close proximity of a hypersaline gulf. Irregular growth rings are also characteristic of the wood of *Pseudofrenelopsis parceramosa*, the reconstruction of which is reproduced in figure 9.18 (from Alvin 1982, 1983).

Figure 9.27. Reconstruction of *Cupressinocladus valdensis* as a tall forest tree; trunk approx. 1 meter diameter at the base. (Redrawn from Francis 1983.)

This species is found in large quantities in Lower Cretaceous Potomac and Wealden sediments, where it occurs with very few other plant megafossils. The Wealden sediment is interpreted as a fluviatile silt (Alvin, Fraser, and Spicer 1981), whereas the Potomac occurrence is suggestive of coastal, tidally influenced habitats (Upchurch and Doyle 1981). On the evidence of sedimentological differences at three Potomac localities, Upchurch and Doyle (1981) have discussed the possibility that *P. parceramosa* may have been adapted to a range of environments of different salinity or even to nonsaline habitats. Upchurch and Doyle (1981) have also made a paleoecological study of *Frenelopsis ramosissima* occurring in somewhat younger Potomac strata, where it belongs to a highly diverse flora including ferns, *Equisetum*, nonxeromorphic gymnosperms, and angiosperms. *F. ramosissima* has a peculiar morphology but is less xeromorphic than other frenelopsids, with a thinner cuticle (30 μm), relatively shallow stomatal pit (figure 9.11A), and lack of strong papillae overhanging the guard cells. The reconstruction presented in figure 9.14 is based on the measurements and detailed description given by Fontaine (1889). His graphic account of specimens seen in the field but now lost conjures up a short shrubby plant with a tiny core of wood in succulent stems up to 5 cm wide. Watson (1977) has demonstrated that a 2-cm-wide stem has a persistent unruptured cuticle, but Fontaine's description suggests that the larger stems may have eventually sloughed off the epidermis, perhaps with the formation of a periderm. The picture that emerges is something like a *Crassula* or *Kalanchoe*. Upchurch and Doyle (1981) in their nonhalophytic paleoecological model suggest growth on sandy or rocky places such as cacti inhabit.

Pseudofrenelopsis varians from Texas seems to have been truly halophytic. It occurs with an abundant marine fauna in the Glen Rose Limestone, where it forms thick, tangled mats (Daghlian and Person 1977; Watson 1977). The similarity of *F. varians* to *Salicornia* has already been noted (Watson 1977; Upchurch and Doyle 1981), and there is a strong geological evidence that it was a major inhabitant of salt marshes. The reconstruction in figure 9.21 is influenced by this similarity, although morphologically one could liken these frenelopsids to *Euphorbia, Ceropegia*, and so on. Other frenelopsids such as *F. alata* have longer internodes and narrower stems, with a tendency to unusual branching patterns (figure 9.2). This type is not unlike *Ephedra* in appearance but little has been written about their paleoecology, although Pons, Lauverjat, and Broutin (1980) have drawn a close parallel between the morphologies of *F. alata* and *Tetraclinis*. Hluštík (1984), studying *Tetraclinis* in North Africa, has concluded that the remains of a plant from such habitats would never be deposited in the sea and has reopened the discussion on possible mangrove forests.

However incomplete our knowledge of the full extent of the paleoecology of the cheirolepids, it has become clear that they enjoyed conspicuous success in the Mesozoic at least in tropical and subtropical climes, inhabiting many of the niches now dominated by angiosperms. It seems likely that all environments so far suggested were indeed represented, and more besides.

Dr. A. Hluštík (personal communication) has given considerable thought to the origins, relationships, and other details of this quite remarkable family and sees the Cheirolepidiaceae as a very large plastic group producing nearly all possible morphological variations of a coniferous habit, ecologically rather than phylogenetically controlled, with *Classopollis* as the sole phylogenetic fixed point. This of course takes us back to the pollination system and its likely importance. Alvin has written eloquently on the subject (1982) and should have the last word: "If a sporophytically controlled incompatibility system had been established in this conifer group at an early stage in its evolution, this might have contributed significantly to its phenomenal success in the Jurassic and Early Cretaceous."

ACKNOWLEDGMENTS

I am particularly indebted to those colleagues and friends who have most generously put their published work at my disposal as well as providing photographs and unpublished information and ideas, especially Drs. K. L. Alvin, S. R. Ash, D. J. Batten, W. A. Charlton, J. G. Douglas, A. Hluštík, F. M. Hueber, R. Porter, T. N. Taylor, and Han Van Konijnenburg-Van Cittert. I am also most grateful to Ian Miller, James P. Ferrigno, and Brian Atherton for photography; Richard Hartley for drafting reconstructions; and Catherine Hardy and Patricia Crook for typing the manuscript at short notice.

LITERATURE CITED

Allen, P. 1975. Wealden of the Weald: A new model. *Proc. Geol. Assoc.* 86:389–437.

Alvin, K. L. 1977. The conifers *Frenelopsis* and *Manica* in the Cretaceous of Portugal. *Palaeontology* 20:387–404.

—— 1982. Cheirolepidiaceae: Biology, structure and paleoecology, *Rev. Palaeobot. Palynol.* 37:71–98.

—— 1983. Reconstruction of a Lower Cretaceous conifer. *Bot. J. Linn. Soc.* 86:169–176.

Alvin, K. L., P. D. W. Barnard, T. M. Harris, N. F. Hughes, R. H. Wagner, and A. Wesley. 1967. Gymnospermophyta. In *The Fossil Record.* London: Geological Society of London.

Alvin, K. L., C. J. Fraser, and R. A. Spicer. 1981. Anatomy and paleoecology of *Pseudofrenelopsis* and associated conifers in the English Wealden. *Palaeontology* 24:759–778.

Alvin, K. L. and A. Hluštík. 1979. Modified axillary branching in species of the fossil genus *Frenelopsis:* A new phenomenon among conifers. *Bot. J. Linn. Soc.* 79:231–241.

Alvin, K. L. and M. D. Muir. 1970. An epiphyllous fungus from the Lower Cretaceous. *Biol. J. Linn. Soc.* 2:55–59.

Alvin, K. L. and J. J. C. Pais. 1978. A *Frenelopsis* with opposite decussate leaves from the Lower Cretaceous of Portugal. *Palaeontology* 21:873–879.

Alvin, K. L., R. A. Spicer, and J. Watson. 1978. A *Classopollis*-containing male cone associated with *Pseudofrenelopsis. Palaeontology* 21:847–856.

Archangelsky, S. 1963. A new Mesozoic flora from Ticó, Santa Cruz Province, Argentina. *Bull. British Museum (Natural History), Geol.* 8:45–92.

—— 1966. New gymnosperms from the Ticó flora, Santa Cruz Province, Argentina. *Bull. British Museum (Natural History), Geol.* 13:259–295.

—— 1968. On the genus *Tomaxellia* (Coniferae) from the Lower Cretaceous of Patagonia (Argentina) and its male and female cones. *J. Linn Soc. (Bot.)* 61:153–165.

Archangelsky, S. and J. C. Gamerro. 1967. Pollen grains found in coniferous cones from the Lower Cretaceous of Patagonia (Argentina). *Rev. Palaeobot. Palynol.* 5:179–182.

Ash, S. R. 1973. Two new Late Triassic plants from the Petrified Forest of Arizona. *J. Paleont.* 47:46–53.

—— 1978. Plant megafossils. In *Geology, Paleontology, and paleoecology of a Late Triassic Lake, Western New Mexico. Brigham Young Univ. Geol. Stud.* 25(2):23–43.

Assereto, R., P. D. W., Barnard, and N. Fantini Sestini. 1968. Jurassic stratigraphy of the Central Elburz (Iran). *Riv. Ital. Paleont.* 74:3–21.

Barale, G. 1973. Contribution à la connaissance de la flore des calcaires lithographiques de la province de Lerida (Espagne): *Frenelopsis rubiesensis* n. sp. *Rev. Palaeobot. Palynol.* 16:271–287.

—— 1981. La paléoflore jurassique du Jura francais. *Docum. Lab. Géol. Lyon.* no. 81:1–467.

Barnard, P. D. W. 1968. A new species of *Masculostrobus* Seward producing *Classopollis* pollen from the Jurassic of Iran. *J. Linn. Soc. (Bot.)* 61:167–176.

Barnard, P. D. W. and J. C. Miller. 1976. Flora of the Shemshak Formation (Elburz, Iran). Part 3. Middle Jurassic (Dogger) plants from Kutumbargah Vasek Gah and Imam Manak. *Palaeontographica* 155B:31–117.

Batten, D. J. 1974. Wealden palaeoecology from the distribution of plant fossils. *Proc. Geol. Assoc.* 85:433–458.

—— 1976. Wealden of the Weald—a new model: Written discussion of a paper previously published. *Proc. Geol. Assoc.* 87:431–433.

Batten, D. J. and MacLennan. 1984. The palaeoenvironmental significance of the conifer family Cheirolepidiaceae in the Cretaceous of Portugal. In W.-E. Reif and F. Westphal, eds., *3rd Symposium on Mesozoic Terrestrial Ecosystems, Short Papers:* 7:12. Tübingen.

Berry, E. W. 1910. The epidermal characters of *Frenelopsis ramosissima. Bot. Gaz.* 50:305–309.

—— 1911. Pteridophyta—Dicotyledonae. In *Lower Cretaceous, Maryland Geol. Surv.* Baltimore: Johns Hopkins Press.

Brenner, G. J. 1976. Middle Cretaceous floral provinces and early migrations of angiosperms. In C. B. Beck, ed., *Origin and Early Evolution of Angiosperms*, pp. 23–47. New York: Columbia University Press.

Broutin, J. and D. Pons. 1976. Novelles précisions sur la morphologie et la phytodermologie de quelques rameaux du genre *Frenelopsis* Schenk. *C. R. 100° Congr. Natl. Soc. Savantes (Paris)*, fasc. 2:29–46.

Carpentier, A. 1937. Remarques sur des empreintes de *Frenelopsis* trouvées dans le Campanien inférieur de la Sainte Baume. *Annls Mus. Hist. Nat. Marseille* 28:5–14.

Chaloner, W. G. 1962. Rhaeto-Liassic plants from the Henfield Borehole. *Bull. Geol. Surv. G.B.* 19:16–28.

—— 1976. The evolution of adaptive features in fossil exines. In I. K. Ferguson and J. Muller, eds., *The Evolutionary Significance of the Exine*, pp. 1–14. London: Academic Press.

Chaloner, W. G. and J. Lorch. 1960. An opposite-leaved conifer from the Jurassic of Israel. *Palaeontology* 2:236–242.

Chow Tseyan and Tsao Chengyao. 1977. On eight new species of conifers from the Cretaceous of East China with reference to their taxonomic position and phylogenetic relationship. *Acta Palaeont. Sin.* 16:165–181 (in Chinese, with English summary).

Cornet, B. and A. Traverse. 1975. Palynological contributions to the chronology and stratigraphy of the Hartford Basin in Connecticut and Massachusetts. *Geoscience & Man* 11:1–33.

Courtinat, B. 1980. Structure d'adhérence des grains de pollen en tétrade du genre *Classopollis* Pflug, 1953, de l'Hettangien de Saint-Fromont, Manche (France). *Géobios* 13:209–229.

Daber, R. 1960. Beitrag zur Wealden-Flora in Nordostdeutschland. *Geologie* 9:591–637.

Daghlian, C. and C. Person. 1977. The cuticular anatomy of *Frenelopsis varians* from the Lower Cretaceous of central Texas. *Am. J. Bot.* 64:564–569.

Doludenko, M. P. 1978. The genus *Frenelopsis* (Coniferales) and its occurrence in the Cretaceous of the U.S.S.R. *Paleontol. Zh.* 3:107–121 (in Russian).

Doludenko, M. P. and M. Reymanówna. 1978. *Frenelopsis harrisii* sp. nov. from the Cretaceous of Tajikistan, U.S.S.R. *Acta Palaeobot.* 19:3–12.

Douglas, J. G. 1985. The Albian extinctions in the great Aust-Antarctic Trough. *N.Z. Geol. Surv. Record* 9:38–40.

Douglas, J. G. and G. E. Williams. 1982. Southern polar forests: The early Cretaceous floras of Victoria and their palaeoclimatic significance. *Palaeogeogr., Palaeoclimatol., Palaeoecol.* 39:171–185.

Edwards, W. N. 1926. Fossil plants from the Nubian Sandstone of eastern Darfur. *Q. J. Geol. Soc. London* 82:94–100.

Ettingshausen, C. 1852. Beitrag zur näheren Kenntnis der Flora der Wealdenperiode. *Abh. geol. Bundesanst., Wien* 1(3/2):1–32.

Feistmantel, K. 1881. Der Hangendflötzung im Schlan-Rakonitzer Steinkohlenbecken. *Arch. naturw. Landdurchforsch. Böhm.* 4(6):96–98.

Fisher, H. L. 1981. A revision of some Lower Cretaceous conifer species. Ph.D. dissertation, University of Manchester.

Fisher, H. L. and J. Watson. 1983. A new conifer species from the Wealden beds of Féron-Glageon, Franc. *Bull. British Museum (Natural History) Geol.* 37:99–104.

Fontaine, W. M. 1889. The Potomac or younger Mesozoic flora. *Mon. U.S. Geol. Surv.* 15:1–377.

—— 1893. Notes on some fossil plants from the Trinity Division of the Comanche Series of Texas. *Proc. Nat. Mus.* 16:261–282.

—— 1905. In L. F. Ward, *Status of the Mesozoic Floras of the United States. Mon. U.S. Geol. Surv.* Vol. 48.

Francis, J. E. 1983. The dominant conifer of the Jurassic Purbeck Formation, England. *Palaeontology* 26:277–294.

Harris, T. M. 1957. A Liasso-Rhaetic flora in South Wales. *Proc. Roy. Soc.*, part B. 147:289–308.

—— 1969. Naming a fossil conifer. *J. Sen Memorial Volume*, pp. 243–252. Calcutta.

—— 1979. *The Yorkshire Jurassic Flora*, vol. 5. *Coniferales*. London: British Museum (Natural History).

Heer, O. 1881. Contributions à la flore fossile du Portugal. *Sect. Trav. Géol. Portugal*, Lisbon.

Heslop-Harrison, J. 1976. The adaptive significance of the exine. In I. K. Ferguson and J. Muller, eds., *The Evolutionary Significance of the Exine*. London: Academic Press.

Hirmer, M. and L. Hörhammer 1934. Zur weiteren Kenntnis von *Cheirolepis* Schimper und *Hirmeriella* Hörhammer mit Bemerkungen über deren systematische Stellung. *Palaeontographica*, 79B:67–84.

Hluštik, A. 1972. *Frenelopsis alata* (Cupress. fossil). *Taxon* 21:210.

—— 1974. New finds of *Frenelopsis* (Cupressaceae) from the Cretaceous of Czechoslovakia and their problems. *Cas. Miner. geol. Praha*, 19:263–268 (in Czech, with English summary).

—— 1978. Frenelopsid plants (Pinopsida) from the Cretaceous of Czechoslovakia. *Palaeontolgická Konf. 1977, Univ. Karlova, Praha*, pp. 129–141.

—— 1979. Fossil gymnosperms from the Planava Formation (Hauterivian), Stramberk (Moravia). *Acta Mus. Moraviae* 64:25–36.

—— 1984. Did fossil mangrove forests exist? *Int. Org. Palaeobot. Newsletter* 23:9–10.

Hluštík, A. and M. Konzalova. 1976a. Polliniferous cones of *Frenelopsis alata* (K. Feistm.) Knobloch from the Cenomanian of Czechoslovakia. *Vestn. Ustred. Ustavu. Geol.* 5:37–45.

—— 1976b. *Frenelopsis alata* (K. Feistm.) Knobloch (Cupressaceae) from the Cenomanian of Bohemia, a new plant producing *Classopollis* pollen. *Evol. Biol.*, Praha: 125–131.

Hollick, A. and E. C. Jeffrey. 1909. Studies of Cretaceous coniferous remains from Kreischerville, New York. *Mem. N.Y. Bot. Gardens* 3:1–76.

Hörhammer, L. 1933. Uber die Coniferen-Gattungen *Cheirolepis* Schimper und *Hirmeriella* nov. gen. aus dem Rhät-Lias Franken. *Bibl. Bot.* 107:1–34.

Hughes, N. F. 1973. Mesozoic and Tertiary distributions and problems of land-plant evolution. In *Organisms and Continents Through Time*. Special Papers in Palaeontology, 12:188–198. Palaeontological Association.

—— 1976. Palaeobiology of Angiosperm Origins. Cambridge: Cambridge University Press.

Hughes, N. F. and J. C. Moody-Stuart. 1967. Palynological facies and correlation in the English Wealden. *Rev. Palaeobot. Palynol.* 1:259–268.

Jung, W. 1967. Eine neue Rekonstruktion des Fruchtzapfens von *Cheirolepis münsteri* (Schenk) Schimper. *N. Jb. Geol. Paläont. Mh.* 1967:111–114.

—— 1968. *Hirmerella münsteri* (Schenk) Jung nov. comb., eine bedeutsame Konifere des Mesozoikums. *Palaeontographica* 122B:55–93.

Kendall, M. W. 1947. On five species of *Brachyphyllum* from the Jurassic of Yorkshire and Wiltshire. *Ann. Mag. Nat. Hist.*, ser. 11, 14:225–251.

—— 1948. On six species of *Pagiophyllum* from the Jurassic of Yorkshire and southern England. *Ann. Mag. Nat. Hist.*, ser. 12, 1:73–108.

—— 1949. On a new conifer from the Scottish Lias. *Ann. Mag. Nat. Hist.*, ser. 12, 2:299–307.

—— 1952. Some conifers from the Jurassic of England. *Ann. Mag. Nat. Hist.*, ser. 12, 5:583–594.

Kimura, T., K. Saiki, and T. Arai. 1985. *Frenelopsis choshiensis* sp. nov., a Cheirolepidiaceous conifer from the Lower Cretaceous Choshi Group in the Outer Zone of Japan. *Proc. Japan. Acad.* 61B:426–429.

Knobloch, E. 1971. Neue Pflanzenfunde aus dem böhmischen und Mährischen Cenoman. *N. Jb. Geol. Paläont. Abh.* 139:43–56.

Kon'no, E. 1967. Some younger Mesozoic plants from Malaya. *Geol. and Palaeont. of S.E. Asia* 3:135–164.

—— 1968. Additions to some younger Mesozoic plants from Malaya. *Geol. and Palaeont. of S.E. Asia* 4:139–155.

Krassilov, V. A. 1978. Late Cretaceous gymnosperms from Sakhalin, U.S.S.R., and terminal Cretaceous event. *Palaeontology* 21:893–905.

—— 1982. On the ovuliferous organ of *Hirmerella*. *Phyta, Studies on Living and Fossil Plants, Pant Comm. Vol.* 1982:141–144.

Lauverjat, J. and D. Pons. 1978. Le gisement Sénonien d'Esgueira (Portugal): Stratigraphie et flore fossile. *C. R. 103ᵉ Congr. Natl. Soc. Savantes (Nancy)*, fasc. 2:119–137.

Lecointre, G. and A. Carpentier. 1938. Sur des empreintes de *Frenelopsis* du Cénomanien provenant du forage de Monts-sur-Guesnes (Vienne). *Bull. Soc. Geol. Fr.*, ser. 5, 8:583–586.

Lorch, J. 1968. Some Jurassic conifers from Israel. *J. Linn. Soc. (Bot.)* 61:177–188.

Muir, M. D. and J. H. A. Van Konijnenburg-Van Cittert. 1970. A Rhaeto-Liassic flora from Airel, northern France. *Palaeontology* 13:433–442.

Nathorst, A. G. 1893. In J. Felix and H. Lenk, *Beiträge zur Geologie und Paläontologie der Republik Mexico. Part* 2:51–54. Leipzig.

Němejc, F. 1926. On the identity of *Sclerophyllum alatum* Feistm. and *Frenelopsis bohemica* Vel. *Sb. st. geol. Ust. csl. Repub.* 6:133–142 (in Czech, with English summary).

—— 1968. *Palaeobotanika*, vol. 3. Praha: Československé Akademie Věd.

Oishi, S. 1940. The Mesozoic floras of Japan. *J. Fac. Sci. Hokkaido Univ.* 5:123–480.

Oldham, T. C. B. 1976. Flora of the Wealden plant–debris beds of England. *Palaeontology* 19:437–502.

Pettitt, J. M. and W. G. Chaloner. 1964. The ultrastructure of the Mesozoic pollen *Classopollis*. *Pollen et Spores* 6:611–620.

Pflug, H. D. 1953. Zur Entstehung und Entwicklung des angiospermiden Pollens in der Erdgeschichte. *Palaeontographica* 95B:60–171.

Pocock, S. J. and J. Jansonius. 1961. The pollen genus *Classopollis* Pflug, 1953. *Micropaleontology* 7:439–449.

Pons, D. 1979. Les organes reproducteurs de *Frenelopsis alata* (K. Feistm.) Knobloch, Cheirolepidiaceae du Cénomanien de l'Anjou, France. *C. R. 104e Congr. Natl. Soc. Savantes (Bordeaux)*, fasc. 1:209–231.

Pons, D. and E. Boureau. 1977. Les champignons épiphylles d'un *Frenelopsis* du Cénomanien moyen de l'Anjou (France). *Rev. Mycol.* 41:349–361.

Pons, D. and J. Broutin. 1978. Les organes reproducteurs de *Frenelopsis oligostomata* (Crétacé, Portugal). *C. R. 103e Congr. Natl. Soc. Savantes (Nancy)*, Fasc. 2:139–159.

Pons, D., J. Lauverjat, and J. Broutin. 1980. Paléoclimatologie comparée de deux gisements du Crétacé supérieur d'Europe occidentale. *Mém. Soc. Géol. Fr.* 139:151–158.

Purkyňová, E. 1983. A new find of Lower Cretaceous plants in the Tesin–Hradisté Formation of the West Carpathians in northern Moravia (Czechoslovakia). *Cas. Slez. Muz. Opava.* 32:57–65 (in Czech, with German and Russian summaries).

Reymanówna, M. 1965. On *Weichselia reticulata* and *Frenelopsis hoheneggeri* from the western Carpathians. *Acta Palaeobot.* 6:15–26.

Reymanówna, M. and J. Watson. 1976. The genus *Frenelopsis* Schenk and the type species *Frenelopsis hoheneggeri* (Ettingshausen) Schenk. *Acta Palaeobot.* 17:17–26.

Reyre, Y. 1970. Stereoscan observations on the pollen genus *Classopollis* Pflug 1953. *Palaeontology* 13:303–322.

Romariz, C. 1946. Estudo e revisao das formas portuguesas de *Frenelopsis*. *Bolm. Mus. Lab. miner. geol. Univ. Lis.* (4)14:135–150.

Saporta, G. de. 1894. *Flore fossile du Portugal.* Lisbon.

Schenk, A. 1869. Beiträge zur Flora der Vorwelt. 3. Die fossilen Pflanzen der Wernsdorfer Schichten in den Nordkarpathen. *Palaeontographica* 19:1–34.

Seward, A. C. 1895. The Wealden flora, part 2. Gymnospermae. *Catalogue of Mesozoic Plants in the Department of Geology, British Museum (Natural History).*

—— 1919. *Fossil Plants: A Textbook for Students of Botany and Geology.* Cambridge: Cambridge University Press.

Srivastava, S. K. 1976. The fossil pollen genus *Classopollis. Lethaia* 9:437–457.

Stone, J. F. 1978. Pollen and spores. In *Geology, Palaeontology and Paleoecology of a Late Triassic Lake, Western New Mexico. Brigham Young Univ. Stud.* 25(2):45–59.

Takhtajan, A. L. 1963. Gymnosperms and angiosperms. In A. Orlov, ed., *Osnovy Paleontologii* 15:1–1743 (in Russian).

Taylor, T. N. and K. L. Alvin. 1984. Ultrastructure and development of Mesozoic pollen: *Classopollis. Am. J. Bot.* 71:575–587.

Teixeira, C. 1948. Flora Mesozoica Portuguesa, 1 parte. *Servicos Geologicos*, Lisbon.

Thompson, W. P. 1912. The structure of the stomata of certain Cretaceous conifers. *Bot. Gaz.* 54:63–67.

Upchurch, G. R. and J. A. Doyle. 1981. Paleoecology of the conifers *Frenelopsis* and *Pseudofrenelopsis* (Cheirolepidiaceae) from the Cretaceous Potomac Group of Maryland and Virginia. In R. C. Romans, ed., *Geobotany*, vol. 2. New York: Plenum Press.

Vakhrameev, V. A. 1970. Range and palaeoecology of Mesozoic conifers, the Cheirolepidiaceae. *Palaeontol. Zh.* 1:19–34.

—— 1978. The climates of the northern hemisphere in the Cretaceous in the light of paleobotanical data. *Paleontol. Zh.* 2:3–17.

—— 1980. *Classopollis* pollen as an indicator of Jurassic and Cretaceous climates. *Sov. Geol.* 8:48–56 (in Russian).

—— 1981. Pollen *Classopollis:* Indicator of Jurassic and Cretaceous climates. *Palaeobotanist* 28/29:301–307.

Van Amerom, H. W. J. 1965. Upper Cretaceous pollen and spore assemblages from the so-called "Wealden" of the Province of Léon (northern Spain). *Pollen et Spores* 7:93–133.

Van Konijnenburg-Van Cittert, J. H. A. 1971. *In situ* gymnosperm pollen from the Middle Jurassic of Yorkshire. *Acta Bot. Neerl.* 20:1–97.

—— 1987. New data on *Pagiophyllum maculosum* Kendall and its male cone from the Jurassic of North Yorkshire *Rev. Palaeobot. Palynol.* 51:95–105.

Velenovsky, J. 1888. Ueber einige neue Pflanzenformen der böhmischen Kreideformation. *Sber. K. böhm. Ges. Wiss.* 29:590–598.

Watson, J. 1964. Revision of the English Wealden fossil flora. Ph.D. dissertations, University of Reading, R. 1146.

—— 1974. *Manica:* A new fossil conifer genus. *Taxon* 23:428.

—— 1977. Some Lower Cretaceous conifers of the Cheirolepidiaceae from the U.S.A. and England. *Palaeontology* 20:715–749.

—— 1982. The Cheirolepidiaceae: A short review. *Phyta, Studies on Living and Fossil Plants, Pant Comm. Vol.* 1982:265–273.

—— 1983. A new species of the conifer *Frenelopsis* from the Cretaceous of Sudan. *Bot. J. Linn. Soc.* 86:161–167.

Watson, J. and K. L. Alvin. 1976. Silicone rubber casts of silicified plants from the Cretaceous of Sudan. *Palaeontology* 19:641–650.

Watson, J. and H. L. Fisher. 1984. A new conifer genus from the Lower Cretaceous Glen Rose Formation, Texas. *Palaeontology* 27:719–727.

Watson, J., H. L. Fisher, and N. A. Hall. 1987. A new species of *Brachyphyllum* from the English Wealden and its probable female cone. *Rev. Palaeobot. Palynol.* 51:169–187.

Wood, C. J. 1961. *Cheirolepsis muensteri* from the lower Lias of Dorset. *Ann. Mag. Nat. Hist.*, ser. 13, 14:505–510.

Yabe, H. 1922. Notes on some Mesozoic plants from Japan, Korea and China in the collection of the Institute of Geology and Paleontology of the Tohaku Imperial University. *Tohaku Imp. Univ. Sendai (Geol.)* 7:1–28.

Zeiller, M. R. 1882. Observations sur quelques cuticules fossiles. *Ann. Sci. Nat. Bot.*, Sér. 6, 13:217–238.

Zhou Zhiyan (Chow Tseyan). 1983. A heterophyllous cheirolepidiaceous conifer from the Cretaceous of East China. *Palaeontology* 26:789–811.

Zhou Zhiyan (Chow Tseyan) and Cao Zhengyao. 1979. Some Cretaceous conifers from southern China and their stratigraphical significance. in Inst. Verteb. Palaeont. and Palaeoanthrop. and Inst. Geol. and Palaeont., *Mesozoic and Cenozoic Red-beds of South China.* Beijing: Science Press (in Chinese).

10

The Origin of
Modern Conifer Families

CHARLES N. MILLER, JR.

▦

The object of this paper is to provide a synthesis of selected paleo-botanical evidence concerning the origin of modern conifer families. Determining the origin of these groups has been approached from several different lines of research. Praeger, Fowler, and Wilson (1976) and Price, Olsen-Stojkovich, and Lowenstein (1987) attempted to assess the degree of interrelationships of modern conifer families using similarities and differences in antigens of modern representatives. Schlarbaum and Tsuchiya (1984) looked at karyotypes. Florin (1938–1945, 1951) emphasized the construction of the ovulate cone; Ueno (1960) studied pollen grains; Greguss (1955) compared different types of secondary xylem; and Laubenfels (1953) surveyed foliage types. In addition, Namboodiri and Beck (1968a, 1968b) investigated primary vascular systems, and Suzuki (1979a, 1979b, and 1979c) studied the course of resin canals in stems. Many workers have contributed to our present understanding of embryology, and the topic is well summarized by Singh (1978). More recently, researchers have tried to use numerical methods to bring different types of information together in hopes of permitting more refined interpretations (Miller 1982; Hart 1987; R. A. Price, personal communication). While each of these approaches has contributed to a better understanding of conifer phylogeny, much remains to be done.

Assessment of the origin and relationships of the modern conifer families should be based on the possession of shared derived features. To do this, characters and their states must be defined, transitions must be determined, and their direction or polarity decided. In modern taxa this is done by comparing states of each character with those in an appropriate outgroup. The condition common to the taxon in question and to the outgroup is presumed to have occurred in the most immediate ancestor of the two and is determined to represent the ancestral state of the character for the group. Doing this for any modern conifer family requires some rationale for determining which one or more of the other fossil or modern groups represents an appropriate outgroup. Thus there needs to be some provisional view of the phylogenetic relationships of modern conifer families and fossil representatives. Such a preliminary determination is provided in this paper.

Some practitioners of phylogenetic inference contend that information from the fossil record is of little use in phylogenetic studies (Wiley 1981:148). For many groups that may well be true. The small number or poor preservation of fossils may not be sufficiently informative to be worth serious consideration. However, exceptions fall into three categories. First is a very important contribution to our understanding of homology. For example, prior to Florin's (1938–1945, 1951) studies on seed cone construction, the homologies of the bract and scale were explained by different theories that linked conifers to groups as divergent as lycopsids, sphensopids, and ferns (Florin 1954). None of these had more support than any of the others until Florin documented structural transitions from the fossil record showing the homology of the ovuliferous scale and its associated bract with the fertile dwarf-shoot and subtending bract in the Cordaitales. Thus conditions present in fossil material that are obscure or absent in modern plants may give evidence of homology.

A second important contribution of fossil record is the evidence it provides of combinations of character states (i.e. taxa) that are not represented by present-day forms. In the Pinaceae, for example, the 10 modern genera give us only a partial picture of the phylogeny of the family without taking into consideration an additional 22 genera represented by *Pseudoaraucaria* and the 21 species of *Pityostrobus*, each one of which probably represents a natural genus (Miller 1976, 1985a). Combinations of character states present in fossil forms may give evidence of the compatibility or incompatibility of character state transitions and thus provide an important source of information about homoplasy. In addition, alternative putative ancestors in the fossil record can be compared with one another and with hypothetical ancestors, and the most likely choice can be determined for a given set of assumptions about character state transitions, reversals, and parallelisms (Stein, Wight, and Beck 1984).

Third, the occurrence of character states in geologic time may be used partially independently of outgroup comparison to support hypotheses of transition and polarity (Stein 1987). Of course, a character state cannot be regarded as ancestral simply because it is ancient. Specialized conditions occurred in the past as well as the present, and ancestral conditions persist and may appear in living plants. Nonetheless, a hypothesis of polarity determined by outgroup comparison and supported by the occurrence of states in geologic time is stronger than one based on outgroup comparison alone.

Thus, fossils may hold great potential for studies of the phylogenetic relationships of taxa that have a useful fossil record. Evidence from the fossil record cannot be dismissed, but instead it must be evaluated and its usefulness determined as with any other evidence.

One principle of cladistic analysis, that relationships are inferred from the possession of shared derived features, applies broadly in studies of phylogeny. The evolution of such derived states must each be a unique event. Acquisition of a condition by parallel evolution or by reversal does not reflect phylogenetic relationship and is misleading. Thus a major thrust of this report is to consider various characters with regard to their bearing on the phylogenetic relationships of the modern families. While we recognize some of these families because of some feature that is unique to them (an apomorphy), such a feature does not reflect relationship but, rather, divergence of a group from its ancestor. It is more important to focus on shared derived features (synapomorphies) that may indicate relationship between groups having the feature.

One of the most serious deterrents to the inclusion of fossil forms in phylogenetic studies of modern taxa is the fragmentary nature of ancient plant remains. Only parts are preserved, and rarely are different organs of a given plant found attached. Particularly difficult to link to whole plants are organs that were dispersed naturally during the life cycle, such as pollen, pollen cones, seeds, and seed cones.

While many different species of coniferophytic secondary xylem have been described, often in detail that parallels that of modern plants, we remain ignorant of other organs of the same plants. Thus phylogenetic studies of different plant groups that have included fossils have tended to based on single organs or combinations of organs that remain attached to one another during the life of the plant. Examples are rhizomes and attached leaf bases (Miller 1971), seeds (Taylor 1965), pollen-bearing organs (Millay and Taylor 1979), and conifer cones (Florin 1951; Miller 1976, 1982).

In the case of conifers, basing phylogenetic studies on whole plants has proved most difficult. Studies based on leaves and leafy twigs, while blessed with an abundance of material, are difficult because relatively little information of phylogenetic meaning has been extracted from these organs. Laubenfels (1953) surveyed leaves of modern conifers and found that there were only four discrete types. To compound matters, a number of conifer species display a heteroblastic series from juvenile to mature foliage which includes as many as three of Laubenfels' types. Harris (1969) established eight form-genera for fossil foliage of unknown affinity; and some of these, even when known in microscopic detail, require knowledge of attachment of reproductive organs before a relationship with a modern family can be determined. Thus, while Crane (1985) regards the narrowly triangular leaf as a distinguishing feature of conifers (all other seed plants having large megaphylls), modifications of this structure within conifers have apparently included so many parallelisms and reversals that the organ is ambiguous in the phylogenetic information it carries. There are similar problems with pollen and secondary xylem.

Characters based on each of these, however, need to be considered and their usefulness in phylogenetic studies evaluated. The size of this task exceeds the space available in this paper. Therefore, only characters based on ovulate cones are treated. While this necessarily limits the nature of the results, it is a restriction imposed to some extent by the fossil record. Vegetative and pollen-producing organs are known for relatively few of the taxa described from seed cones. Furthermore, the fact that each modern family is distinguished by some unique feature of its seed cone gives added importance to characters of this organ. A hypothesis of phylogeny based on characters of one organ should be consistent with a hypothesis based on other organs if both hypotheses are correct. The limitation of using only cone features is not sufficiently serious to justify ignoring this important record of conifer fossils.

The distribution of character states in a large number of fossil and modern conifer taxa is presented. These data are analyzed for compatibility using the CLINCH program of Kent Fiala, the State University of New York at Stony Brook, and the resulting information is incorporated in the discussion of the potential for homoplasy of each character. The same data are analyzed using the PAUP program of David Swofford, Illinois Natural History Survey, which finds minimum-length trees. The results are summarized in a series of cladograms that form the basis for a discussion of the origin and evolution of modern conifer families as interpreted from the fossil record.

OUTGROUP

The Cordaitales, excluding the Vojnovskyaceae and Rufloriaceae, is regarded as the most appropriate outgroup for the Voltziales and Coniferales. Florin's (1938–1945, 1951) work provides good documentation that the fertile dwarf-shoot and subtending bract in *Cordaitanthus* is homologous with similar structures in the Walchiaceae and with the bract and ovuliferous scale in conifers. Information about structure and reproduction in the Cordaitales is relatively complete (Costanza 1985; Rothwell 1982b; Rothwell and Warner 1984; Stewart 1983; Taylor 1981). In fact, except for certain details of embryogeny, our knowledge of the cordaite plant is on a par with that of many conifers and better than some of them. Florin (1950, 1951) thought that conifers evolved from the Cordaitales and viewed character states present in the latter group as ancestral conditions. Current authorities (Clement-Westerhof 1984; Mapes and Rothwell 1984; Meyen 1984; Miller 1982; Rothwell 1982a) no longer believe that the conifers evolved directly from the Cordaitales but that the two more likely diverged from a common ancestor. The Cordaitales are somewhat older than the earliest conifers, but the two groups coexisted for over 50 million years before the former group died out (Rothwell 1987).

Other possible outgroups include the cycads, ginkgos, Gnetales, and the putative cordaite relatives, the Vojnovskyaceae and the Rufloriaceae (Crane 1985; Doyle and Donoghue 1986; Hart 1987). Cycads and ginkgos lack homologues of the conifer bract and ovuliferous scale. The ovuliferous structure in *Ginkgo* occurs axillary to a vegetative leaf and is analogous to the ovuliferous scale and subtending bract of conifers, but homology remains to be documented and even analogy of groups of *Ginkgo* ovuliferous structures with the compound strobili of conifer seed cones is vague (Rothwell 1987). While Crane (1985), Doyle and Donoghue (1986) and Hart (1987) treat *Ginkgo* as sharing a common ancestor with the conifer-cordaite clade, Meyen (1984) regards *Ginkgo* as having evolved from pteridosperm ancestors different from conifers and cordaites. Similarly, although the ovulate strobili in the Gnetales appear to be compound, as in conifers, homology between comparable parts remains somewhat conjectural, and modifications in the gnetalean ovulate strobili make direct comparison with conifers difficult. The Vojnovskyaceae and Rufloriaceae share some vegetative features with the cordaites and are treated within the Cordaitales by Meyen (1984). However, it is not clear that the fertile dwarf-shoot is an axillary structure in the Vojnovskyaceae or if it occurs in the same ontogenetic helix as the leaves (Meyen 1982, 1984; Miller 1985). Similarly, strobili of the Rufloriaceae have not been found attached, so homology with the fertile dwarf-shoot of *Cordaitanthus* cannot be demonstrated. Given these problems, *Cordaitanthus* is the best choice of an outgroup for polarizing character state transitions in the ovulate structures of conifers.

CHARACTERS

This section provides a discussion of the usefulness of certain seed cone characters for making phylogenetic comparisons. For each character a proposal is offered that defines the states and indicates their polarity. The latter is assessed from the standpoint of outgroup comparison and the occurrence of the states in geologic time. In addition, the possibility that the character is homoplasous is discussed, based on the occurrence of the states in the various fossil and modern taxa, the compatibility of the character with others, and the relative ease of structural changes leading to derived states.

1. *Relative fusion of the bract and the fertile dwarf-shoot complex*

   A. Proposal: Freedom of the fertile dwarf-shoot complex or its homologues from the subtending bract is ancestral; fusion of the two is derived.
   B. Outgroup comparison: In the Cordaitales the fertile dwarf-shoot is free from the bract (Costanza 1985: plate 1, figure 6).
   C. Fossil record: The fertile dwarf-shoot is free from the bract in the earliest conifer fossils that have sufficient preservation to show the condition clearly (*Lebachia lockardii, Ortiseia, Moyliostrobus*). There are no examples of Pennsylvanian or early Permian conifers in which the two appendages are fused. However, fusion is evident in the late Permian *Pseudovoltzia* (Schweitzer 1963), and there are many examples which freedom or fusion cannot be determined because of the

nature of preservation. It is clear that with some exceptions, many species in younger sediments show fusion between the bract and fertile dwarf-shoot or its homologues (Florin 1951). Exceptions are the Triassic *Compsostrobus* (Delevoryas and Hope 1973), the Cheirolepidiaceae (Alvin 1982; Watson 1982), and the Pinaceae. In these conifers the bract and ovuliferous scale remain free from one another.

Certain early conifers exhibit substantial fusion. In the Late Permian *Pseudovoltzia*, the bract and ovuliferous scale are fused for about one-half their length, and the evidence comes from permineralized material that leaves no doubt (Schweitzer 1963). Similarly, bract and scale fusion is apparent in the Late Triassic conifers *Aethophyllum* and *Swedenborgia* (Grauvogel-Stamm 1978) and may also be present in *Tricranolepis* (Roselt 1958). Many modern forms and their close fossil counterparts exhibit nearly complete fusion of bract and ovuliferous scale. The Jurassic *Araucaria mirabilis* (Stockey 1978) and the early Cretaceous *Elatides* and *Sphenolepidium* (Harris 1953) provide examples in the Araucariaceae and Taxodiaceae, respectively. Thus the transition from freedom to fusion of the fertile dwarf-shoot and its homologue the ovuliferous scale and the subtending bract is supported by the occurrence of these states in geologic time.

D. Discussion: A critical question in phylogenetic studies is whether or not the fusion indicated in this character was a unique event. Alternatively, did fusion of the bract and axillary fertile dwarf-shoot occur at different rates and at different times in different clades? While it is difficult to find evidence to support either alternative, two arguments in particular favor the latter view over the former.

The first argument is intuitive. Fusion of the fertile dwarf shoot or ovuliferous scale with the subtending bract is a structurally simple modification. Furthermore, we find in modern cones many different degrees of fusion, and the degree of fusion often does not correlate directly with relationship. Thus, similar degrees of fusion in families thought to be unrelated or only distantly related on other grounds probably result from homoplasy.

Similarly, the transition from freedom to fusion in this character is compatible with one-half of the other characters considered in this paper (table 10.3). Lack of compatibility with other characters suggests that this transition is homoplasous. Admittedly, the transition from free to fused would be expected to be a gradual one with a theoretically infinite number of states between complete independence and complete fusion. Trying to define discrete states and imposing a somewhat artificial evaluation of particular taxa may have led to some of the incompatibility, and the incompatibility of this character with certain others could disappear with restructuring of the other character in these pairs. While it denotes an evident trend in the evolution of conifer seed cones, the trend occurred widely and in different conifer lineages.

2. Symmetry of the fertile dwarf-shoot

A. Proposal: Radial symmetry of the fertile dwarf-shoot complex and its homologues is regarded as ancestral; bilateral symmetry is derived.

B. Outgroup comparison: In the Cordaitales the fertile dwarf-shoot is essentially radially symmetrical (Costanza 1985; Florin 1951: figures 16, 17, 18; Rothwell 1982b).

C. Fossil record: Structurally preserved fossils of early conifers that show the three-dimensional arrangement of appendages *(Lebachia lockardii, Ortiseia, Moyliostrobus)* have nearly radial fertile dwarf-shoots. The number of species exemplifying this condition is even greater if one accepts Florin's (1938–1945) interpretation of the arrangement of parts in fossils such as *Walchiostrobus* and *Ernestiodendron*. The transition toward bilateral symmetry is already apparent in the late Permian *Pseudovoltzia* (Schweitzer 1963), *Ullmannia*, and *Glyptolepis* and continues into the early Mesozoic with the appearance of essentially modern forms (Miller 1977). Concurrently, conifers with radially symmetrical fertile dwarf-shoots drop out.

D. Discussion: Some of the change in symmetry correlates with transitions in other characters. The preferential reduction or loss of parts from one side of the fertile dwarf-shoot, such as that facing the bract, the reduction in number of parts, and fusion of the fertile dwarf-shoot with the bract all correlate with a change in symmetry. Selection favoring a more compact cone during seed development seems to lead logically to these kinds of changes. Important selective pressures may have been dessication and predation.

The change from radial to bilateral symmetry was probably not a unique event but instead one that occurred in different lineages at different times in different ways. Support for this contention comes mainly from a consideration of the variety of ways asymmetry can be achieved.

It seems unlikely that reversal in this character played an important role in the evolution of various conifer lineages. Once radial symmetry was lost, it would be structurally difficult to regain it. This depends, of course, on how it was lost. Changes in symmetry due to reduction or loss of parts would be difficult to regain because this

456   CHARLES N. MILLER, JR.

would entail reversal of the original loss or reduction or repositioning
of the remaining structures in the ontogenetic helix to compensate for
a loss. Flattening due to compression of parts from restriction of space
for growth between tightly imbricated bracts, for example, could be
reversed if subsequent changes resulted in more space for dwarf-shoot
enlargement. Such reversals would probably be seen in lax cones or
ones with few bract scale complexes. There are no examples of lax
cones with radial fertile dwarf-shoots that would document such a
reversal. Both the Podocarpaceae and the Taxaceae have solitary bract
scale complexes. They are bilateral in the Podocarpaceae and radial
in the Taxaceae. However, there are no transitional forms on record
in the latter family to suggest reversal, and the condition is best
interpreted as representing the ancestral state.

Three states of this character are recognized. Radial symmetry is
the ancestral condition; bilateral symmetry without much flattening
of the fertile dwarf-shoot represents an intermediate condition; and
evident flattening of the fertile dwarf-shoot is the advanced state.
Assessing fertile dwarf-shoot symmetry from fossil compressions and
impressions presents some problems in scoring individual taxa.

This character shows compatibility with more than one-half of the
other characters (table 10.3). This may reflect the evaluation of the
states in various taxa because all but four were scored with the derived condition. A scoring regime that reflects the reality of more
transitional states may result in less compatibility.

3. *Reduction of the axis of the fertile dwarf-shoot*

A. Proposal: Possession of an evident dwarf-shoot axis is ancestral;
reduction of the axis is derived.

B. Outgroup comparison: The dwarf-shoot axis is readily apparent
in *Cordaitanthus* (Costanza 1985; Florin 1951; figures 17f, 20d; Rothwell 1982b).

C. Fossil record: An axis to which fertile and sterile appendages are
attached occurs only in ancient cones and not even in all of them. It is
clearly present in *Lebachia lockardii* and *Moyliostrobus*. Its presence
in *Lebachia piniformis* and *Ortiseia* is inferred from the mode of attachment of appendages.

D. Discussion: Loss of an evident dwarf-shoot axis is another manifestation of reduction of the axillary complex possibly in keeping
with selective pressures favoring more compact cones. As such, reduction of the dwarf-shoot axis was probably gradational and correlated
with reduction and loss of appendages on the axis and fusion of those

that remained. The three states recognized for purposes of this analysis (table 10.1) represent an oversimplification; but since there is only one actual example of an intermediate state, the present scheme seems justified. There are no examples on which to base the definition of other intermediate states that might represent additional stages in the transition from the ancestral to the derived condition. The compatibility of this character with more than one-half of the other characters (table 10.3) supports its usefulness.

While reduction of the dwarf-shoot axis could have occurred at different times in different lineages, the distribution of the states (table 10.2) suggests that such parallelism was in fact minimal; and this is supported by the wide compatibility of this character (table 10.3). Those that display ancestral and intermediate conditions are all in late Pennsylvanian and early Permian age taxa. Thus the dwarf-shoot axis was lost early in the evolution of conifers. It is unlikely that it was regained, since there is no evidence of reversal in younger fossil cones.

It is interesting to speculate about possible vestiges of the axis in modern cones. I was unable to find in modern pinaceous cones any single vascular strand in the ovuliferous scale that runs from the scale base to the apex and might thus be construed as a vestige of the vascular supply of an ancestral axis. The ring of vascular tissue that supplies the ovuliferous scale in pinaceous, taxodiaceous, and cupressaceous cones may, as it enters the base of the scale unbranched, represent a remnant of a former axis. Its radial symmetry argues in favor of this point of view. In cupressaceous and taxodiaceous cones this ring of vascular tissue divides, forming many vascular strands that continue outward into the bract scale complex arranged more or less in a ring, consistent with the arrangement of strands around a vestigial axis. The lack of structurally preserved forms that permits tracing these patterns back to ancestral species in the fossil record leaves only the opportunity to speculate about these points.

4. *Fusion of appendages of the fertile dwarf-shoot*

A. Proposal: Freedom of the appendages of the fertile dwarf-shoot from one another is ancestral; fusion of appendages is derived.

B. Outgroup comparison: Appendages of the fertile dwarf-shoot are free from one another in *Cordaitanthus* (Costanza 1985; Florin 1951; Rothwell 1982b).

C. Fossil record: A trend toward fusion of the appendages of the fertile dwarf-shoot is apparent in the fossil record. In late Pennsylvan-

Table 10.1. Scoring of character states in the data matrix.

Character 1
  A. Axillary complex free from bract.
  B. Axillary complex partially fused to bract.
  C. Axillary complex fused to bract obscuring identity.
Character 2
  A. Dwarf-shoot complex radially symmetrical.
  B. Dwarf-shoot complex somewhat flattened.
  C. Dwarf-shoot complex flattened.
Character 3
  A. Dwarf-shoot axis evident.
  B. Dwarf-shoot axis not evident; appendages arising from basal attachment.
Character 4
  A. Appendages of dwarf-shoot free from one another.
  B. Appendages partly fused; tips free.
  C. Appendages fused, obscuring identity.
  D. Some appendages fused; some free (derived from A).
Character 5
  A. Ovules erect.
  B. Ovules recurved.
Character 6
  A. Bract not reduced; as large as fertile dwarf-shoot or nearly so.
  B. Bract reduced.
Character 7
  A. Ovules per dwarf-shoot 6 or more.
  B. Ovules per dwarf-shoot 3 to 5.
  C. Ovules per dwarf-shoot 1 or 2.
Character 8
  A. Ovules naked.
  B. Ovules with an epimatium.
Character 9
  A. Sterile appendages per dwarf-shoot more than 10.
  B. Sterile appendages per dwarf-shoot 6 to 10.
  C. Sterile appendages per dwarf-shoot less than 6.
Character 10
  A. Bract with one vascular strand.
  B. Bract with more than one vascular strand.
Character 11
  A. Bract apex acute.
  B. Bract apex bifurcate.
Character 12
  A. Seeds large.
  B. Seeds small.

Character 13
  A. Seed wing from lateral extension of integument
  B. Seed wing absent
  C. Seed wing from scale epidermis
  D. Seed wing from bract and scale (derived from B).
Character 14
  A. Many bract scale complexes per cone.
  B. Few bract scale complexes per cone.
  C. One bract scale complex per cone.
Character 15
  A. Aril absent.
  B. Aril present.
Character 16
  A. Ovule lateral.
  B. Ovule terminal.

---

ian and early Permian Walchiaceae, parts are essentially free (Mapes and Rothwell 1984). In late Permian cones fusion is already apparent —e.g., *Pseudovoltzia, Ullmannia, Glyptolepis*. A trend toward greater fusion of parts continues in younger forms.

   D. Discussion: Fusion of parts is difficult to characterize independently of loss and reduction of parts. Furthermore, no distinction is made in this character between fertile and sterile elements; they are considered homologous. Clement-Westerhof (1984) takes a different view of the broad megasporophyll with an attached recurved seed in *Ortiseia* while admitting that stalked ovules may be homologous with sterile elements. The changes that resulted in the broad megasporophyll remain unknown. One explanation is that the stalk became flattened and the ovule recurved. Under these circumstances I would consider a megasporophyll and attached ovule homologous with a simple stalked ovule. If the megasporophyll resulted from the fusion of several elements, it would not be homologous with a simple stalked ovule; but I see no evidence for this kind of fusion.

   Because many parts are present on the ancestral dwarf-shoot, fusion may have taken different routes that would be difficult to distinguish from one another in the end result. How many elements have fused may be reflected in the ovuliferous scale by the presence of unfused tips of elements. Similarly, the number of seeds may reflect the number of ancestral fertile elements. Thus, the cone scale of *Cryptomeria* can be interpreted as resulting from the fusion of five sterile elements and three to six fertile elements (Bierhorst 1971: figure 25-5).

Table 10.2. Character states of the OTU's used in the CLINCH and PAUP analyses.

OTU	Character Number															
	1	2	3	4	5	6	7	8	9	10	11	12	13	14	15	16
HYP-ANCESTOR (!)	A	A	A	A	A	A	A	A	A	A	A	A	A	A	A	A
WAL-ERN-WLK (!)	A	B	A	A	B	A	C	A	A	A	B	A	A	A	A	A
MOYLI-ORT (!)	A	B	A	A	C	A	C	A	A	A	A	A	A	A	A	A
GLYPTOLEPIS	A	C	C	B	B	A	C	A	B	A	A	A	A	A	A	A
DREPANOLEPIS	A	C	C	A	C	A	C	A	C	A	A	A	A	A	A	A
HIRMERIELLA	A	C	C	B	A	A	C	A	C	A	A	A	B	A	A	A
VOLTZIOPSIS	A	C	B	B	C	A	B	A	A	A	C	A	A	A	A	A
VOLTZIA	A	C	C	B	C	A	B	A	B	A	A	A	A	A	A	A
PSEUDOVOLTZIA	B	C	C	B	C	A	B	A	B	A	A	A	A	A	A	A
ULLMANNIA	A	C	C	C	C	A	C	A	C	A	A	A	A	A	A	A
AETHOPHYLLUM	B	C	C	B	C	A	B	A	B	A	A	A	C	A	A	A
PINACEAE	A	C	C	C	A	A	C	A	A	A	A	A	A	B	A	A
CUPRESSUS	C	C	C	C	A	A	A	A	A	A	A	C	A	B	A	A
CHAMAECYPARIS	C	C	C	C	C	A	B	A	A	A	A	C	A	A	A	A
SCIADOPITYS	B	C	C	C	A	A	A	A	A	A	A	C	B	B	A	A
JUNIPERUS	C	C	C	C	A	A	C	A	B	A	A	C	A	B	A	A
CALOCEDRUS	C	C	C	B	C	A	C	A	B	A	A	B	B	A	A	A
SWEDENBORGIA	B	C	C	B	C	A	B	A	B	A	A	B	B	B	A	A
TRICRANOLEPIS FRI.	B	C	C	B	A	A	B	A	B	A	A	B	B	A	A	A
SCHIZOLEPIS	B	C	C	B	A	A	C	A	C	A	A	B	B	A	A	A
PACHYLEPIS	B	C	C	B	A	A	B	A	B	A	A	B	B	A	A	A
CEPHALOTAXUS	A	C	C	C	A	A	C	A	C	A	A	A	B	B	A	A

Taxon																	
PHYLLOCLADUS	A	C	C	A	A	A	C	B	A	A	C	A	B	A	B	A	B
CRYPTOMERIA	B	C	B	A	B	A	B	A	C	B	B	A	A	A	A	A	A
THUJA	C	C	C	A	A	A	B	A	C	A	B	A	B	A	B	A	A
TRICRANOLEPIS MON.	B	C	B	B	C	A	C	A	C	A	A	A	A	A	A	A	A
CYCADOCARPIDIUM ER.	B	C	B	B	C	A	C	A	C	A	A	A	A	A	B	A	A
DACRYDIUM	A	C	C	A	A	A	B	A	C	A	A	A	B	A	A	A	B
MICRO-SAXEGOTHAEA	A	C	B	A	A	A	C	B	B	B	B	A	A	A	A	A	B
CYCADOCARPIDIUM PI.	B	C	C	A	A	A	C	B	A	A	A	A	A	A	A	A	B
ARAUCARIA	C	C	C	A	A	A	A	C	A	A	A	A	D	A	A	A	A
PODOCARPUS	A	C	C	A	A	A	C	B	A	A	B	A	B	A	B	A	B
TAXODIUM	C	C	C	A	A	A	C	A	A	A	B	A	B	A	A	A	A
METASEQUOIA	C	C	C	A	A	A	A	A	A	A	A	A	A	A	A	A	A
SEQUOIADENDRON	C	C	C	C	A	A	A	B	A	A	A	A	A	A	A	A	A
CUNNINGHAMIA	B	C	C	B	A	A	B	B	B	A	A	A	A	A	A	A	A
ATHROTAXIS	C	C	C	B	A	A	B	A	A	A	A	A	A	A	A	A	A
SEQUOIA	C	C	C	B	A	A	B	A	A	A	A	A	A	A	A	A	A
AGATHIS	C	C	C	C	A	A	C	A	A	A	A	A	D	A	A	A	A
GLYPTOSTROBUS	C	C	C	C	A	A	C	A	A	A	A	A	A	A	A	A	A
TAIWANIA	C	C	C	C	A	A	C	A	A	A	A	A	A	A	A	A	A
CALLITRIS	C	C	C	A	A	A	C	B	B	A	A	A	B	A	B	A	B
TAXUS	A	A	A	A	A	A	A	A	A	A	A	A	B	A	B	A	B
AM-NO-TORREYA (!)	A	A	A	A	A	A	A	A	A	A	A	A	B	A	B	A	B
THUJOPSIS	C	C	C	A	A	A	B	B	A	B	A	A	A	A	B	A	A
TETRACLINIS	C	C	C	A	A	A	C	A	A	A	A	A	B	A	A	A	A
PARARAUCARIA	A	C	C	A	A	A	C	C	B	C	C	B	C	A	A	A	A

Notes: (!) = HYP-ANCESTOR, Hypothetical ancestor; WAL-ERN-WLK = *Walchia, Ernestiodendron,* and *Walchia lockardii;* MOYLI-ORT = *Moyliostrobus* and *Ortiseia;* AM-NO-TORREYA = *Amentotaxus, Nothotaxus,* and *Torreya;* MICRO-SAXEGOTHAEA = *Microcachrys* and *Saxegothaea.*

Table 10.3. Compatibility matrix for data in table 10.2 (0 = incompatible; 1 = compatible.)

	1	2	3	4	5	6	7	8	9	10	11	12	13	14	15
2	1														
3	1	1													
4	0	1	0												
5	0	0	0	0											
6	0	1	1	0	1										
7	0	0	0	0	0	0									
8	1	1	1	1	0	1	1								
9	0	1	1	0	0	0	0	1							
10	0	1	1	0	0	1	0	1	0						
11	1	0	0	0	0	1	0	1	1	1					
12	0	1	1	0	0	0	0	1	0	0	1				
13	0	0	0	0	0	0	0	1	0	0	1	0			
14	0	0	0	0	0	1	0	0	0	1	1	0	0		
15	1	1	1	1	1	1	1	1	1	1	1	1	1	1	
16	1	0	0	0	0	1	1	1	0	1	1	1	1	0	1

One possible reflection of the number of original parts is the number of vascular strands in the ovuliferous scale. The 20 or so strands in pinaceous cones may reflect derivation from an ancestor, with little or no reduction in the number of elements. The occurrence of the strands in a circle with their phloem to the outside in cross sections of the ovuliferous scale in certain cupressaceous and taxodiaceous cones may similarly reflect derivation from ancestral sources that had radially symmetrical dwarf-shoots. Whether those strands represent vestiges of original elements or the remnants of the vascular system of the dwarf-shoot axis is difficult to determine.

Fusion was certainly not a unique event, and homoplasy is indicated by the incompatibility of this character in the data set (table 10.3). Reversal seems less likely, but the possibility cannot be ruled out.

5. *Recurvation of the ovule*

A. Proposal: The possession of erect ovules is ancestral; the possession of recurved ovules is derived.

B. Outgroup comparison: Ovules seem to be typically erect in *Cordaitanthus*. Florin (1951: figure 16) illustrates C. pseudofluitans with ovules that appear somewhat recurved. However, the degree of recurvation is less than that of *Lebachia lockardii* and *Ortiseia* and may represent a natural drooping of an erect ovule or a bending of the peduncle under growth pressure from an adjacent ovule. Thus the greatest amount of evidence supports the erect state as being typical of *Cordaitanthus*.

C. Fossil record: Some modern families such as the Taxaceae exhibit no ovule recurvation. Its occurrence in the fossil cones is difficult to assess. Florin (1938–1945, 1951) described most ancient cones with erect ovules, and the ovule was described as erect in *Moyliostrobus* (Miller and Brown 1973). However, the occurrence of recurved ovules in *Ortiseia* and *Lebachia lockardii* and the strong possibility that the "seed" of *L. piniformis* is merely an *Ortiseia*-like megasporophyll give reason to suspect that not all Walchiaceae had erect seeds. Reinvestigation of this feature in *Moyliostrobus* indicates that it has *Ortiseia*-like seeds (Mapes 1987). The occurrence of erect seeds diminishes in cones from younger sediments.

D. Discussion: It is difficult to envision seed recurvation as a unique event in the evolution of conifers. Instead, it probably occurred at different times in different lineages. The occurrence of an ovule on the abaxial side of the large megasporophyll in *Ortiseia*, however, probably represents a unique event. If such an ovule and its megasporophyll were reduced and fused to other dwarf-shoot elements, it would be difficult to distinguish from ones at the tips of more slender stalks that had become recurved and fused. Thus there are different routes that could have led to recurved ovules in conifers.

Once compact cones with tightly imbricated ovuliferous scales had evolved in any linage, erect or recurved seeds might merely reflect the position of the ovule primordium at the time adjacent parts came too close together to allow further ovule movement. Thus reversal and parallelism are likely, and this is further indicated by the incompatibility of this character (table 10.3). Perhaps the best guide is to note the condition in modern representatives. In the Pinaceae and Araucariaceae there is no variation, and the recurved condition probably resulted from a single transition. In the Cupressaceae, Podocarpaceae, and Taxodiaceae some genera have erect seeds and others recurved. Thus homoplasy is likely.

6. Reduction of the bract

A. Proposal: Possession of a large bract is ancestral; significant reduction of the bract is derived.

B. Outgroup comparison: A large bract subtends each fertile dwarf-shoot in *Cordaitanthus* (Costanza 1985; Florin 1951; Rothwell 1928b).

C. Fossil record: Bracts are large and evident in Paleozoic conifer cones and smaller in certain Mesozoic cones (Miller 1977). The widespread occurrence of the putative ancestral state in the more ancient cones supports the above hypothesis of polarity.

D. Discussion: In cones of certain groups, bracts are much smaller than the scale they subtend. Examples occur in several families that are not thought to be closely related on other grounds. This suggests that bract reduction occurred independently in different lineages.

Determining which taxa display a reduced bract presents some problems because bract size is usually determined relative to scale size. In many groups the bract is much larger than the scale in the younger cone, but the scale enlarges more than the bract and is much longer at cone maturity (e.g., *Pinus*). Furthermore, in other groups the seed becomes very large, obscuring both the bract and the scale. In scoring states of this character, seed size has been disregarded; only the relative size of the bract and scale at cone maturity is considered.

7. *Reduction in number of ovules*

A. Proposal: A large number of ovules per fertile dwarf-shoot or its homologues is ancestral; a low number is derived.

B. Outgroup comparison: In *Cordaitanthus* there is a relatively large number of immature ovules, each at the tip of a fertile element or sporophyll, although many abort (Rothwell 1982b).

C. Fossil record: Certain of the late Pennsylvanian and early Permian conifers described by Florin (1938–1945, 1951), such as *Walchiostrobus*, have a large number of ovules, while others have few or only one (Clement-Westerhof 1984). In *Lebachia lockardii* only one ovule matures per fertile dwarf-shoot, but there is evidence of aborted ovules on other elements of the dwarf-shoot (Mapes and Rothwell 1984). Cones of the Voltziaceae typically have five or fewer seeds per scale (Florin 1951; Miller 1977).

D. Discussion: Reduction in seed number probably took place a number of times. It occurred at least once early in the history of conifers, resulting in the single seed per fertile dwarf-shoot of the late Paleozoic genera *Ernestiodendron*, *Lebachia*, *Moyliostrobus*, *Ortiseia*, and *Ullmannia*. It probably occurred again within the Podocarpaceae.

While reduction may have occurred more than once, reversal seems less likely to have taken place. Once an appendage of the fertile dwarf-shoot lost the capacity to produce an ovule, it would probably be difficult to regain it. However, this might vary from group to group.

In the Pinaceae, where two seeds per scale appear well fixed in the lineage, addition of ovules would seem unlikely. By comparison, in the Taxodiaceae and certain Cupressaceae, where seed number is highly variable, gain or loss of a single ovule might be developmentally simple. This would probably apply to phylogenetic transition as well.

Evaluating seed number by category allows for some homoplasous change within a given increment without material effect on the resulting analysis. Even so, the low level of compatibility (table 10.3) of this character supports the existence of homoplasy.

8. *Evolution of an epimatium*

A. Proposal: Absence of an epimatium is ancestral; the presence of one is derived.

B. Outgroup comparison: Nothing like an epimatium occurs in the Cordaitales.

C. Fossil record: The earliest conifer known to have an epimatium is *Stalagma* from the late Triassic (Zhou 1983). None of the early Walchiaceae or Voltziaceae shows an epimatium. The existence of an epimatium is not apparent in early cones assigned to the Podocarpaceae by Townrow (1967a, 1967b, 1969). Thus the condition may have evolved after the podocarp lineage diverged from a common ancestor with other conifers. It is clear, however, that the epimatium is a feature unique to the Podocarpaceae, and this further supports the above hypothesis of polarity.

D. Discussion: The evolution of an epimatium is viewed as a unique event, a conclusion supported by the high degree of compatibility of this character with others (table 10.3). The epimatium is interpreted as being derived from the fusion of sterile elements of the fertile dwarf-shoot around the base of a terminal fertile element. In some forms the number of vascular strands in the epimatium many refect the number of elements involved in the fusion. If the structure evolved after the lineage had diverged, it may not be present in all ancient forms—e.g., *Mataia* and *Rissikia*.

9. *Reduction in the number of sterile elements on the dwarf shoot*

A. Proposal: A large number of sterile elements on the dwarf-shoot represents the ancestral condition; a low number is derived.

B. Outgroup comparison: The dwarf-shoot of *Cordaitanthus* is characterized by a large number of sterile elements (Florin 1951; Rothwell 1982b).

C. Fossil record: Ancient cones such as *Lebachia*, *Ortiseia*, and *Moyliostrobus* have a large number of sterile elements. Others that are contemporaneous have relatively few. In general, however, those in younger sediments have relatively few elements.

D. Discussion: Reduction in the number of sterile elements is another manifestation of the production of a more compact seed cone. To a limited extent this character duplicates character 7, reduction in number of ovules, in that there were probably factors that operated to favor reduction in the total number of appendages of the fertile dwarf-shoot. On the other hand, there were probably factors that affected reduction in fertile elements independently of sterile ones, and vice versa. Thus character 9 may be distinct from character 7.

There is also a problem in determining the number of elements in ovuliferous scales that are completely fused and have no free tips to serve as indicators. It is tempting to use the number of vascular strands; however, some of these may represent vestiges of the dwarf-shoot axis rather than its appendages. Furthermore, after fusion to form a scale, its enlargement or reduction may affect the number of strands therein, and therefore not reflect the original number of sterile elements.

The low level of compatibility of this character (table 10.3) indicates homoplasy. While reduction in the number of sterile elements denotes a broad evolutionary trend, it apparently occurred repeatedly in different lineages of conifers.

10. *Number of vascular strands in the bract*

A. Proposal: The presence of single vascular strand in the bract is ancestral; the presence of more than one is derived.

B. Outgroup comparison: In *Cordaitanthus* the bract has a single vascular strand.

C. Fossil record: All of the Paleozoic conifers known have bracts supplied by a single vascular strand. The Triassic *Cycadocarpidium* is the oldest example of a cone in which the bracts have more than one vascular strand. The condition is also seen in the Araucariaceae and certain Taxodiaceae *(Cunninghamia, Athrotaxis,* and *Taiwania)*. The occurrence of the states in geologic time thus supports the above hypothesis of polarity.

D. Discussion: There is no reason to regard the development of multiple strands in the bract as a unique event. The condition could have evolved more than once and in different ways. Specifically, multiple strands might have resulted from fusion of parts of the fertile dwarf-shoot with the bract, or they could have resulted from a developmental broadening of the bract. In modern cones with multiple bract strands, all of the latter divide from a single strand that enters the bract. This plus the few occurrences of multiple bract strands suggests that parallelism was not common.

11. *Bifurcate apex on bract*

A. Proposal: The presence of a pointed, undivided bract apex is ancestral; the presence of a bifurcate apex is derived.

B. Outgroup comparison: In *Cordaitanthus* the bract apex is undivided.

C. Fossil record: Both states occur in late Pennsylvanian and early Permian conifer cones. Those of the Mesozoic except *Voltziopsis* all have undivided bract apices.

D. Discussion: Florin (1938–1945), 1951) contended that the bifurcate bract apex was ancestral and corresponded to the *Gomphostrobus* leaves on penultimate vegetative branches of certain early conifers. Among the Mesozoic cones known to him, only *Voltziopsis* has a bifurcate bract. Thus the fossil record supported his idea.

We now know that both states occur in ancient conifers; *Moyliostrobus* and *Ortiseia* lack bifurcate bracts. Thus outgroup comparison provides the main basis for interpreting polarity.

Trifurcate bract apices occur in certain Pinaceae and are regarded as derived. *Rissikia*, an early podocarp (Townrow 1967a), also has a trifurcate bract.

This character is compatible with more than one-half of the other characters in the data set (table 10.3).

12. *Reduction in seed size*

A. Proposal: The presence of large seeds is ancestral; the presence of small seeds is derived.

B. Outgroup comparison: Seeds of *Cordaitanthus* tend to be large —e.g., *Cardiocarpon, Mitrospermum.*

C. Fossil record: Many Walchiaceae have large seeds; while some modern groups have large seeds, most taxa have small ones.

D. Discussion: The production of small seeds was probably favored by more efficient utilization of limited resources because it led to greater dispersal. This character is incompatible with many other characters, indicating homoplasy. Small seeds probably evolved more than once, and reversal is also possible.

A problem with this character is to determine definable states— i.e., what constitutes large seeds versus small ones. Seed size in modern and fossil conifers appears to display continuous variation between the extremes, making definitions either subjective or arbitrary. Seed size is unknown for many fossil conifers.

13. *Seed wing*

A. Proposal: The formation of a wing by lateral extension of the integument is ancestral; several derived conditions are: state "B," no wing; state "C," a terminal wing from the scale epidermis; and state "D," the formation of a wing by the fusion of the bract and scale to the seed.

B. Outgroup comparison: Seeds of the Cordaitales have a wing formed by the lateral extension of the integument.

C. Fossil record: Seeds of the putative ancestral type occur commonly in ancient cones (Clement-Westerhof 1984; Florin 1951; Mapes and Rothwell 1984; Miller and Brown 1973).

D. Discussion: While there is good support for the putative ancestral state, the manner by which the various derived states evolved is unknown. States "C" and "D" seem to be typical of the Pinaceae and Araucariaceae, respectively, but it is not known whether they evolved from state "B" or directly from the ancestral state. In the Araucariaceae the seed itself lacks integumental extensions (Eames 1913; Stockey 1978; Wilde and Eames 1948), so this condition may have been derived directly from the ancestral state or the wingless state. Similarly, pinaceous seeds tend to be flattened slightly in cross section, and some have evident lateral ridges that could represent vestiges of ancestral wings. Without intermediate forms, the determination remains speculative.

14. *Reduction in the number of bract scale complexes per cone*

A. Proposal: A large number of bract scale complexes per cone is ancestral; a small number is derived.

B. Outgroup comparison: In *Cordaitanthus* there is a large number of fertile dwarf-shoots and subtending bracts in each cone (Costanza 1985; Florin 1951; Rothwell 1982b).

C. Fossil record: Ovulate cones of the Walchiaceae have a large number of bracts and fertile dwarf-shoots. Similarly, in the Permian Voltziaceae, there are a large number of bract scale complexes per cone. An ancient exception is the late Permian *Buriadia*, in which the fertile dwarf-shoot is conjecturally a solitary structure borne on the stem among vegetative leaves (Pant and Nautiyal 1967). In general, however, cones with relatively few bract scale complexes occur in younger sediments, and many belong to modern families.

D. Discussion: The derived state is seen most frequently in modern conifers of different families—e.g., Cupressaceae, Taxaceae, certain Podocarpaceae, and certain Taxodiaceae. Rather than these families sharing an ancestor that has a reduced number of bract scale complexes, the feature more likely evolved independently in the different lineages. Homoplasy is indicated by the low level of compatibility of the character in the data set (table 10.3).

15. *Evolution of an aril*

A. Proposal: The absence of an aril is ancestral; its presence is derived.

B. Outgroup comparison: There is nothing comparable to an aril in the Cordaitales.

C. Fossil record: Early conifers lack arils. The earliest evidence of such a structure is in the late Triassic *Paleotaxus* (Florin 1951) and in the early Jurassic *Marskea* (Harris 1976, 1979). Members of the Triassic Palissyaceae are described as having arillate seeds (Florin 1951), but the construction of seed cones in this group is too poorly understood to determine homologies with cones of either ancient or modern conifers, and the Palissyaceae have not been included in the analysis.

D. Discussion: With the Palissyaceae excluded, the aril is a synapomorphy of the genera of the Taxaceae. The complete compatibility (table 10.3) with other characters indicates that its evolution was a unique event within this monophyletic group. While the homologies of the aril are not understood, it is interpreted conjecturally as the result of fusion of elements on a fertile dwarf shoot. This parallels the interpretation of the epimatium in the Podocarpaceae. While the aril and epimatium are not regarded as homologous, the similarity gives some basis for the interpretation of the homologies of the aril. Without such an interpretation, the seed-bearing structure in the Taxaceae cannot be regarded as fundamentally homologous with that of other conifers, and they cannot be part of the same monophyletic group.

16. *Position of the ovule on the fertile dwarf shoot*

A. Proposal: The lateral position of the ovule is ancestral; the terminal position is derived.

B. Outgroup comparison: Ovules are lateral on the fertile dwarf-shoot in *Cordaitanthus* (Costanza 1985; Florin 1951; Rothwell 1982b).

C. Fossil record: The oldest coniferophytic plants known have ovules placed laterally on the fertile dwarf-shoot (Florin 1951; Mapes and Rothwell 1984). Terminal ovules occur in the late Triassic *Paleotaxus* (Florin 1951) and in the early Jurassic *Marskea* (Harris 1976, 1979). The late Permian *Buriadia* (Pant and Nautiyal 1967) also has ovules, each terminating a stalk that is borne on stems among vegetative leaves. Whether the stalk is a sporophyll or a stem—i.e., a reduced fertile dwarf-shoot—is unknown. The fossil record, based on conclusions about what is reasonably well known, supports the hypothesis that the terminal position of the ovule is derived.

D. Discussion: The distinction between terminal and lateral ovules was the basis for Florin's (1954) segregation of the Taxaceae from the conifers and establishment of the class Taxopsida. More recently, investigators have found reason to doubt that this group should be separated from other families. Keng (1969) finds sufficient similarity between *Amentotaxus* and *Phyllocladus* to suggest a close relationship between the Taxaceae and Podocarpaceae. Wilde (1975) interpreted the peltate microsporophyll in the Taxaceae as the fusion product of ancestral leafy sporophylls and thereby further reduced the distinction between taxads and conifers. Harris (1976) proposed a hypothetical sequence by which a lateral ovule could become terminal. Ovules are nearly terminal in many Podocarpaceae (Aase 1915; Sinnott 1913; Wilde 1944). Thus the transition from lateral to terminal ovules may be just one more modification of conifer seed cones.

MODERN CONIFER FAMILIES

Since this work focuses on the phylogenetic relationships of modern conifer families, it is desirable to characterize the latter with regard to the seed cone features each displays and to describe briefly the fossil record of each family. Eckenwalder (1976) noted that each modern conifer family except the Cupressaceae appears to be distinguished by some unique feature of its ovulate cone. While such unique features (autapomorphies) fail to tell us about relationships with other families, the observation suggests that the ovulate cone may contain other features of importance. Those that are derived and shared with other modern families or with fossil taxa will help us understand relationships. The essential features of each modern family are evaluated from this perspective.

Pinaceae Seed cones of the Pinaceae consist of numerous ovuliferous scales, each subtended by a bract, arranged in a compact helix around the cone axis. The ovuliferous scale is flattened and bears two recurved seeds at its base. The bract is flattened, tongue-shaped, and free from the scale. It is supplied by a single vascular strand. In *Pseudotsuga* and in some species of *Abies* and *Larix*, the bract is longer than the scale and has a trifurcate apex. This is viewed as a specialization of the bract with the unbranched apex. The ovuliferous scale is supplied by a vascular strand that is semicircular in cross section. It divides in the scale to produce about 20 strands. In each of these the phloem is oriented adaxially. The constancy of this large number of strands suggests that they may represent vestiges of elements that fused to form the scale. There is no basis for distinguishing strands that might have supplied ovules from those representing sterile appendages. Typically, the seed is winged, with the wing derived from scale tissue rather than from the seed. The wing is a specialization. Whether it was derived directly from seeds with lateral wings or from seeds without wings is unknown.

If these assumptions of homologies are correct, the ancestor of the Pinaceae must have had two or more ovules on a fertile dwarf-shoot with a large number of appendages. These requirements suggest a prototype with a level of specialization comparable to that of the Walchiaceae rather than a more advanced group.

The oldest members of the Pinaceae are early Cretaceous in age. Certain older fossils may belong to the family, but their affinities remain uncertain. These are isolated cones scales of *Schizolepis* (Harris 1979), needle leaves attached to dwarf-shoots treated as *Pityocladus* (Harris 1979), and imprints of cones called *Compsostrobus* (Delevoryas and Hope 1973). More than two dozen species of Cretaceous seed cones are known. Some of these belong to *Pinus*, but others must be classified in fossil genera because they lack the necessary combination of states to allow inclusion in one of the remaining modern genera. Six species belong to *Pseudoaraucaria*, which is viewed as a natural group. Over 20 species are classified in *Pityostrobus*, and each of these species may represent a natural genus. Modern genera other than *Pinus* appear at the onset of the Cenozoic, but some, such as *Larix*, are unknown until late Tertiary.

Araucariaceae   Seed cones are compact, with many bracts and scales arranged in a helix around the axis. The ovuliferous scale is highly reduced and is fused almost completely to the bract. The scale bears a single recurved ovule, although occasional cones are found with seed scales having two or three ovules. Some view the latter as reflecting the ancestral condition (Mitra 1927; Wilde and Eames 1948, 1955). The bract receives a single trace that divides to form numerous strands. In *Araucaria bidwillii* a strand from each side of the gap above the divergent bract trace branches toward the scale. These fuse with one another while still in the cortex of the cone axis. The scale trace divides, forming vascular bundles that supply the scale and ovule (Eames 1913: figure 1-15). Vascularization in other *Araucaria* species differs in lacking a scale trace. Only the bract trace enters the bract scale complex. In some species adaxial strands with adaxial phloem originate from the bract trace and also provide a source of strands supplying the larger bract. In *A. cooki* no upper series is produced; two strands supply the ovule but none extends into the scale.

In *Agathis* there is essentially no ovuliferous scale, the only remnant being the recurved ovule and its point of attachment to the bract. A single bract trace enters the bract and branches to form numerous strands. The central one branches to supply the ovule (Eames 1913). *Agathis vitiensis* is an exception. A raised pad of tissue on the adaxial side of the bract represents the scale, and several of the central bract strands branch to supply it and the ovule.

Imprints isolated cone scales with attached seeds are known in late Triassic sediments, and seed cones of modern *Araucaria* occur in the Jurassic (Stockey 1978, 1980a, 1980b, 1982). The similarity of these ancient cones to modern counterparts helps us understand the history and relationships of modern species groups, but questions remain about the origin and early evolution of the family.

Taxodiaceae   Cones of the Taxodiaceae are compact structures with numerous bract scale complexes arranged in a helix except in *Metasequoia*, in which they are decussate. Cones of *Athrotaxis*, *Cunninghamia*, and *Taiwania* are unusual because the bract is more prominent than the ovuliferous scale. The bract is supplied by a single vascular strand that ramifies within the bract to form many. The scale is highly reduced and fused to the bract, and it is supplied by strands that branch from the bract trace. There are three recurved seeds per scale in *Cunninghamia*, two in *Athrotaxis*, and two in *Taiwania* (Bierhorst 1971).

In *Cryptomeria* and *Taxodium*, seeds are erect. The bract is slightly reduced, as is the bract vascular supply. The single bract trace is unbranched in *Cryptomeria*, and the nature of the vascular supply is not clear in *Taxodium*.

In *Sequoia*, *Sequoiadendron*, and *Metasequoia* the bract and scale contribute about equally to the complex, and the scale and bract vasculatures are about equal. The vascular strands are arranged in an open pattern. Seeds are recurved in the former two genera and erect in the latter.

In *Sciadopitys* the bract scale complex consists of a bract and scale fused about two-thirds of their length and free at the tips. The bract is supplied by a single vascular strand that extends from the base of a gap in the cone axis to the bract apex. The scale is supplied by a single strand that branches and diverges from the top of the gap. There are five to nine recurved ovules.

The family has an impressive fossil record. Noteworthy are ovulate cones from the Jurassic and Cretaceous that show features of different present-day genera in combinations precluding assignment to any of them. Examples are *Athrotaxites* (Miller and LaPasha 1983), *Austrosequoia* (Peters and Christophel 1978), *Cunninghamiostrobus* (Miller 1975; Ogura 1930), *Elatides* (Harris 1943, 1953), *Nephrostrobus* (LaPasha and Miller 1981), *Parataxodium* (Arnold and Lowther 1955), *Rhombostrobus* (LaPasha and Miller 1981), and *Sphenolepidium* (Harris 1953). Other Mesozoic conifers, such as *Swendenborgia* and *Cycadocarpidium* (Grauvogel-Stamm 1978), are presently included in the Voltziaceae but show similarities to the Taxodiaceae.

Cupressaceae Seed cone construction in the many genera of the Cupressaceae is poorly known. There was sufficient information to include only 7 of the 20 modern genera in the cladistic analysis. In cones of these genera, bracts and scales are fused, with only the tips free in some instances. Bracts tend to be larger than scales. Bract scale complexes are reduced in number and are borne in opposite pairs on the cone axis. There are up to eight complexes in *Cupressus* and as few as 2 in *Calocedrus* and *Juniperus*. The bract is supplied by a single vascular strand, whereas there are about 15 in the scale. The erect ovules mature into relatively small winged seeds. There are many seeds per complex in *Callitris* and only one or two in other genera (Bierhorst 1971). More work needs to be done on seed cone construction in this family.

The fossil record of this family is obscure. Nearly all Mesozoic fossils assigned to the Cupressaceae have proved to belong to the extinct family Cheirolepidiaceae (Alvin 1982; Watson 1982). Vaudois and Prive (1971) recognize two species of wood belonging to the family from the Jurassic and six species from the Cretaceous. Certain structurally preserved seed cones described by Penny (1947) from the late Cretaceous of New Jersey are probably valid cupressaceous remains, and fertile branches of *Juniperus* occur in sediments of this age from Alaska (C. J. Smiley, personal communication). Thus the family was in existence by the late Mesozoic, but its time of origin is not clear.

Podocarpaceae The ovulate cone in this family consists of one or more bracts on an axis, each subtending an ovule with an epimatium or covering (Aase 1915; Sinnott 1913; Wilde 1944). The bract is considered to be homologous with that of *Cordaitanthus* and other conifers, while the stalked ovule and epimatium are homologues of the fertile dwarf-shoot. Reduction of the latter leaves the ovule in an apparent terminal position subtended by remnants that are represented by the epimatium (Sinnott 1913).

In all modern species of the family, the bract subtends a single ovule that is recurved in all *Podocarpus* and some *Dacrydium* species. The ovule is erect or nearly so in all other genera. The epimatium may entirely envelop the seed or may be reduced to a small basal sheath.

In the genera *Podocarpus* and *Dacrydium* the cone axis is reduced to a short, thickened stalk bearing a few sterile, bract-like scales at its base and one or two ovuliferous appendages above. In the genera *Saxegothaea, Microcachrys, Pherosphaera,* and *Phyllocladus* fertile scales are more numerous and are arranged in more typical cones.

In *Podocarpus* subgenus *Stachycarpus,* ovules are in spikes or borne singly or in pairs at the end of a branchlet; no receptacle is formed at maturity. In the subgenus *Protopodocarpus,* ovules are generally borne on a fleshy receptacle formed by fusion of bract bases. In the sections *Nageia, Eupodocarpus,* and *Microcarpus* of this subgenus, the bract is not equal in length to the epimatium, and the latter is entirely free from the former. In the section *Dacrycarpus* the bract and epimatium are about equal in length and are fused together (Wilde 1944).

Several members of the Podocarpaceae have been described from vegetative and reproductive structures from Mesozoic sediments (Archangelsky 1966; Townrow 1967a, 1967b, 1969; Vishnu-Mittre 1958; Zhou 1983). *Rissikia,* from the basal to late Triassic, indicates that the family had evolved by the onset of the Mesozoic. The organization of *Rissikia* seed cones (Townrow 1967a) suggests origin of the family from ancestors that would be classified either as specialized members of the Walchiaceae or unspecialized members of the Voltziaceae.

Cephalotaxaceae *Cephalotaxus* seed cones have a series of decussate bracts on an axis. Except for the basal pair, each bract has an axillary seed scale complex consisting of a pair of erect ovules and a ridge-like outgrowth of the cone axis between them. The ovules and ridge are viewed as representing a fertile dwarf-shoot (Singh 1961). Three strands go into each fertile bract. The anterior bundle supplies the bract and is unbranched. Each lateral bundle bifurcates. Two daughter bundles orient themselves phloem to phloem and pass into the integument this way, each giving off a horizontal branch midway toward the inner part of the ovule (Singh 1961).

The fossil record of this family is a modest one based mainly on compressed foliage in which the epidermal structure is similar to that of modern species. The oldest fossil genus is *Thomasiocladus*, from the Jurassic of England (Harris 1979). While such fossils show that the family had evolved by the Jurassic, they tell us little about its ancestry.

Taxaceae This family includes the genera *Taxus* (seven species), *Torreya* (five), *Pseudotaxus* (one), *Austrotaxus* (one), and *Amentotaxus* (four). In the latter four of these genera, ovulate axes occur directly on long shoots in the axils of leaves. The ovulate axis bears a varying number of bracts, usually in decussate arrangement, and terminates in an erect ovule. In *Taxus* secondary shoots, each of which bears an ovule on an axis having helically arranged scales, occur on primary shoots in the axil of a leaf.

The homologies of these structures with counterparts in other conifers are obscure. Florin (1951, 1954) thought that the terminal ovule in the Taxaceae is not homologous with the lateral ovules in other conifers. He believed that while the other conifer families evolved from the Walchiaceae, the Taxaceae did not. If this is the case, the lack of homology would extend as well to the Cordaitales and would imply that all other coniferophytic features of taxads evolved by convergence from some other ancestral source.

A simpler explanation is that Taxaceae share a more immediate ancestor with other conifer families and that the ovulate structure of taxads is somehow homologous with all or part of the coniferophytic compound strobilus. Harris (1976) proposed that the ovule may have become terminal simply by the loss of all parts more distal to it. A similar situation occurs in certain Podocarpaceae (Sinnott 1913), placing the ovule in near-terminal position.

Determining a relationship between a bract and a modified fertile dwarf-shoot remains a problem. Possibly the entire axis with its bracts and terminal ovule represents a fertile dwarf-shoot, and the vegetative leaf subtending it is the bract homologue. A problem with this interpretation is that the bract is a modified leaf in all conifers and in the outgroup. A major reversal would have to have occurred, and there are no intermediate forms or other evidence of such a major change.

An interpretation better supported by evidence is that the axis represents a cone axis and the bracts are homologous with those of other conifers. The ovule is the remnant of a fertile dwarf-shoot that was once axillary to one of the bracts; those axillary to the other bracts have been lost. This explanation is consistent with the trends of fertile dwarf-shoot compaction by reduction and loss of its appendages. In addition, there are numerous examples in other modern conifers—e.g., the Cephalotaxaceae, Cupressaceae, Podocarpaceae—in which usually the basal bracts of the seed cone are sterile. There is no anatomical evidence for such a loss in the Taxaceae; but only *Taxus canandensis* has been examined in microscopic detail (Dupler 1920), and studies of other genera might be revealing. This interpretation, while admittedly conjectural, is followed here.

The fossil record of the Taxaceae shows that the family had evolved by the Middle Jurassic, but there is no clue about possible ancestors. *Marskea* (Harris 1976, 1979) is interesting in combining features of different modern genera in a unique way.

DISCUSSION

Over 100 equally parsimonious trees resulted from the parsimony analysis of data in table 10.2 by PAUP. Despite evidence of homoplasy in many of the characters, there is only one basic form to the trees (figure 10.1). Variations in the trees are due to differences in character state transitions and rearrangements of taxa within the Cupressaceae-Taxodiaceae clade (figures 10.1, 10.3, 10.4). Only two instances of character state reversal or parallelism occur in the equally parsimonious trees leading to the main clades (figure 10.1), and only three instances of reversal or parallelism occur within the Araucariaceae–Cephalotaxaceae–Pinaceae–Podocarpaceae clade (figure 10.2). Thus the broad trends indicated by the characters have phylogenetic significance at the family level (figures 10.1, 10.2) but not at the genus level

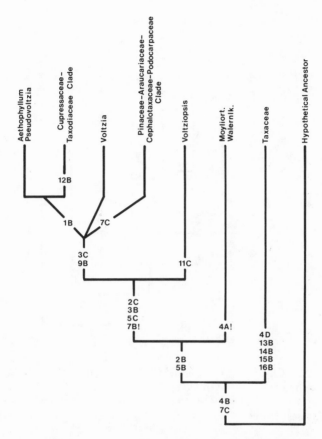

Figure 10.1. Cladogram showing general relationships of the modern families and fossil groups. Numbers and letters represent character states as in table 10.1 and show the transitions that result in the cladogram. ! = reversal; "Moyliort." = *Moyliostrobus* and *Ortiseia;* "Walernlk." = *Walchia, Ernestiodendron,* and *Lebachia lockardii.*

within the Cupressaceae–Taxodiaceae clade (figures 10.3, 10.4). For this reason, character state transitions are given for the family level clades (figures 10.1, 10.2) only.

Consistent in all trees is the division of most of the modern families into three main groups (figure 10.1). One group includes the modern genera of the Cupressaceae and Taxodiaceae along with several fossil genera generally regarded as "transition conifers" (Florin 1951) or Voltziaceae (Miller 1977, 1982). A second group includes the Araucariaceae, Cephalotaxaceae, Pinaceae, and Podocarpaceae as well as cer-

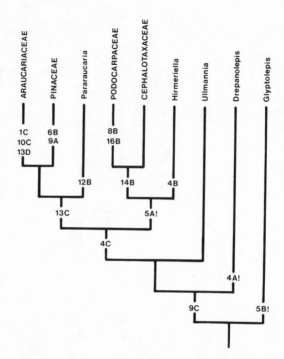

Figure 10.2. Cladogram showing relationships within the Araucari-aceae–Cephalotaxaceae–Pinaceae–Podocarpaceae clade. Numbers and letters represent character states as in table 10.1 and show the transitions that result in the cladogram. ! = reversal.

tain fossil genera of the Voltziaceae (figure 10.2). The genera of the Taxaceae constitute a third group that appears in the cladogram (figure 10.1) as a clade distinct from the other conifers included in the analysis. This arrangement recalls Florin's (1951, 1954) removal of the Taxaceae from the Coniferopsida into its own class, the Taxopsida.

Within the Cupressaceae–Taxodiaceae clade the equally parsimonious trees show many instances of reversal and parallelism in the character state transitions, even though there are only a few differences in the arrangement of the taxa (figures 10.3, 10.4). The latter consistency supports the proposal by Eckenwalder (1976) that the two families are so closely related that they should be merged into a single family. Hart's (1987) study, based on a broad array of characters from modern species, also shows a close relationship between these families. However, Hart (1987) and R. A. Price (personal communication)

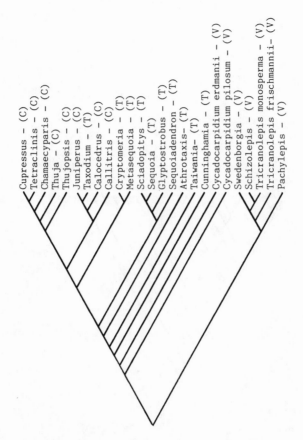

Figure 10.3. Cladogram showing one extreme in the suite of equally parsimonious trees of relationships within the Cupressaceae–Taxodiaceae clade.

have found *Sciadopitys* to be sufficiently distinct to warrant treatment in its own family. Perhaps because of its limited scope, the present work fails to support the latter relationship. The numerous instances of reversal and parallelism within this clade suggest that the number and nature of the characters selected for this study are inadequate to resolve generic level relationships accurately. It is clear, however, that the fossil genera *Aethophyllum*, *Pachylepis*, *Pseudovoltzia*, *Schizolepis*, and *Swendenborgia* are closely related to the Cupressaceae and Taxodiaceae and may represent an ancestral complex. Furthermore, *Aethophyllum* and *Pseudovoltzia* are sufficiently well known (Grauvogel-Stamm 1978; Schweitzer 1963) to be useful as outgroups when

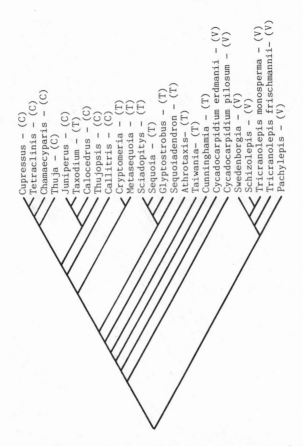

Figure 10.4. Cladogram showing the other extreme in the suite of equally parsimonious trees of relationships within the Cupressaceae–Taxodiaceae clade.

polarizing morphological and anatomical characters in cladistic studies of the modern Cupressaceae and Taxodiaceae. Indeed, they would make a better choice than any of the other modern families (figure 10.1).

The clade that includes the Cephalotaxaceae, Podocarpaceae, Araucariaceae, and Pinaceae is consistent in character state transitions in all the equally parsimonious trees. The resulting relationships determined for the taxa (figure 10.2) are well founded within the context of the characters used. In all trees the Podocarpaceae and Cephalotaxaceae occur close to one another, as do the Pinaceae and Araucariaceae. This arrangement recalls the scheme proposed by Sinnott (1913) to account for the phylogeny of the Podocarpaceae. The position of the fossil genus *Hirmeriella* close to the Cephalotaxaceae and Podocarpaceae probably reflects our poor understanding of the construction of the seed cone in this genus rather than a true close relationship. The placement of *Pararaucaria* near the Araucariaceae and Pinaceae agrees with Stockey's (1977) recognition of the many pinaceous features of these fossil cones. However, Stockey's suggestion that *Pararaucaria* represents a link between the Pinaceae and Taxodiaceae is not supported by the analysis. If more were known about the morphology and anatomy of *Drepanolepis* and *Glyptolepis*, they might serve as good outgroups in cladistic studies of these modern families. However, until we have more information about these fossil genera, sister groups are better selected from closely related modern families. This study indicates that the Cephalotaxaceae and Podocarpaceae are sister groups, as are the Araucariaceae and Pinaceae (figures 10.1, 10.2).

The exact role of the Voltziales in the evolution of modern conifer families remains elusive. Florin (1938–1945, 1951) believed that the modern families evolved from the Lebachiaceae via "transition conifers" of the Voltziaceae, and the present work provides no reason to reject that general hypothesis. Genera of the Voltziaceae occur in the cladogram both in the Cupressaceae–Taxodiaceae clade and in the Araucariaceae–Cephalotaxaceae–Pinaceae–Podocarpaceae clade, and they diverge from positions basal to the modern families (figures 10.1, 10.2, 10.3, 10.4). Still, it is impossible to identify specific ancestor candidates for any of the modern families. In addition, our present understanding of seed cone construction in the earliest conifers differs from that of Florin. Recent work (Clement-Westerhof 1984; Visscher et al. 1986), based on study of the type material and new specimens, led to renaming the Lebachiaceae the Walchiaceae and to a redefinition of the latter to exclude cones having more than one ovule on each fertile dwarf-shoot. The construction of cones described by Florin as having more than one ovule per fertile dwarf-shoot is poorly understood, and these cones have not been included in the present study. This may explain why the Walchiaceae (figure 10.1, "Walernlk.–Moyliort.") appears to be such a divergent group.

The apparent divergence of the Taxaceae may also reflect the use of the more restricted Walchiaceae (Clement-Westerhof 1984; Visscher et al. 1986). If ancient conifer cones having more than one ovule per fertile dwarf-shoot were included in the analysis, they would probably have formed a clade diverging basal to the Taxaceae and bringing the latter group into greater unity with other conifers. Even so, the Taxaceae would appear distinct from the other modern families, which is surprising in view of the interpretation of the seed-bearing structure of the family like that of the Podocarpaceae. Another explanation is simply that evolution did not follow the most parsimonious path.

The previous discussion assumed that conifers are monophyletic, and current phylogenetic reviews (Crane 1984; Doyle and Donoghue 1986; Hart 1987) support that view. However, such late Paleozoic plants as *Buriadia, Walkomiella*, and the Ferugliocladaceae Archangelsky and Cuneo (1987) exhibit coniferophytic vegetative organization, but their ovules were not produced on scales in compound strobili. Rothwell (1982a) states that the Callistophytales, Cordaitales, and Lyginopteridales all have aspects of coniferophytic reproduction and vegetative construction. Furthermore, plants in each of these groups produced small scale leaves in heteroblastic series with their more typical foliage. Rothwell hypothesizes that each of these had the potential to give rise to conifer-like plants by heterochrony, specifically the production of the scale leaves of the heteroblastic series all along the vegetative shoot. The potential for at least three possible ancestral sources for conifers, our inability to accommodate certain late Paleozoic conifer-like plants in our present scheme, and the wide gaps apparent in the present study between certain modern families and between them and the Walchiaceae lend plausibility to a hypothesis of polyphyly in conifers.

Our knowledge of fossil and modern conifers and their possible ancestors is still too imperfect to allow solutions to these problems, and there is good reason to keep an open mind.

LITERATURE CITED

Aase, H. C. 1915. Vascular anatomy of the megasporophylls of conifers. *Bot. Gaz.* 60:277–313.

Alvin, K. L. 1982. Cheirolepidiaceae: Biology, structure, and paleoecology. *Rev. Palaeobot. Palynol.* 37:71–98.

Archangelsky, S. 1966. New gymnosperms from the Ticó flora, Santa Cruz Province, Argentia. *Bull. British Museum (Natural History), Geol.* 13:259–295.

Archangelsky, S. and R. Cuneo. 1987. Ferugliocladaceae, a new conifer family from the Permian of Gondwana. *Rev. Palaeobot. Palynol.* 51:3–30.

Arnold, C. A. and J. S. Lowther. 1955. A new Cretaceous conifer from Alaska. *Amer. J. Bot.* 42:522–528.

Bierhorst, D. W. 1971. *Morphology of Vascular Plants.* New York: Macmillan.

Clement-Westerhof, J. A. 1984. Aspects of Permian paleobotany and palynology. 4. The conifer *Ortiseia* Florin from the Val Gardena Formation of the Dolomites and Vicentinian Alps (Italy) with special reference to a revised concept of the Walchiaceae (Goppert) Schimper. *Rev. Palaeobot. Palynol.* 41:51–166.

Costanza, S. H. 1985. *Pennsylvanioxylon* of Middle and Upper Pennsylvanian coals from the Illinois Basin and its comparison with *Mesoxylon. Palaeontographica* 197B:81–121.

Crane, P. R. 1985. Phylogenetic analysis of seed plants and the origin of angiosperms. *Ann. Missouri Bot. Gard.* 72:716–793.

Delevoryas, T. and R. C. Hope. 1973. Fertile coniferophyte remains from the Late Triassic Deep River Basin, North Carolina. *Amer. J. Bot.* 60:810–818.

Doyle, J. A. and M. J. Donoghue. 1986. Seed plant phylogeny and the origin of angiosperms: An experimental cladistic approach. *Bot. Rev.* 52:321–431.

Dupler, A. W. 1920. Ovuliferous structures of *Taxus canadensis. Bot. Gaz.* 69:492–520.

Eames, A. J. 1913. The morphology of *Agathis australis. Ann. Bot.* 27:1–38.

Eckenwalder, J. E. 1976. Re-evaluation of Cupressaceae and Taxodiaceae: A proposed merger. *Madrono* 23:237–256.

Florin, R. 1938–1945. Die Koniferen des Oberkarbons und des Unteren Perms. 2. Palaeontographica, Abt. B, 85:1–729.

—— 1950. Upper Carboniferous and Lower Permian conifers. *Bot. Rev.* 16:258–388.

—— 1951. Evolution in cordaites and conifers. *Acta Horti Bergiani* 15:285–388.

—— 1954. The female reproductive organs of conifers and taxads. *Biol. Rev.* 29:367–389.

Grauvogel-Stamm, L. 1978. La flore du Grès à Voltzia (Bundsandstein supérieur) des Vosges du Nord (France): Morphologie, anatomie, interprétations phylogénetique, et paléogéographique. *Univ. Louis Pasteur de Strasbourg, Inst. Géol., Mém.* 50:1–225.

Greguss, P. 1972. *Xylotomy of Living Conifers.* Budapest: Akadémiai Kiadó.

Harris, T. M. 1943. The fossil conifer *Elatides williamsoni. Ann. Bot.,* 7:325–339.

—— 1953. Conifers of the Taxodiaceae from the Wealden of Belgium. *Inst. Roy. Sci. Nat. Belg. Mem.* 126:1–43.

—— 1969. Naming a fossil conifer. *Bot. Soc. Bengal, J. Sen Mem. Vol.,* 243–252.

—— 1976. The Mesozoic gymnosperms. *Rev. Palaeobot. Palynol* 21:119–134.

—— 1979. The Yorkshire Jurassic flora. 5. Coniferales. London: British Museum (Natural History).

Hart, J. 1987. A cladistic analysis of conifers: Preliminary results. *J. Arnold Arboretum* 68:269–307.

Keng, H. 1969. Aspects of the morphology of *Amentotaxus formosana* with a note on the taxonomic position of the genus. *J. Arnold Arboretum* 50:432–446.

—— 1977. *Phyllocladus* and its bearing on the systematics of conifers. In K. Kubitzki, ed., *Flowering Plants, Evolution and Classification of Higher Categories,* pp. 235–252. Vienna: Springer-Verlag.

LaPasha, C. A. and C. N. Miller. 1981. New taxodiaceous seed cones from the Upper Cretaceous of New Jersey. *Amer. J. Bot.* 68:1374–1382.

Laubenfels, D. J. de. 1953. The external morphology of coniferous leaves. *Phytomorphology* 3:1–20.

Mapes, G. 1987. Ovule inversion in the earlies conifers. *Amer. J. Bot.* 74:1205–1210.

Mapes, G. and G. W. Rothwell. 1984. Permineralized ovulate ones of *Lebachia* from the late Paleozoic limestones of Kansas. *Palaeontology* 27:69–94.

Meyen, S. V. 1982. The Carboniferous and Permian floras of Angaraland (synthesis). *Biol. Mem.* 7:1–109.

—— 1984. Basic features of gymnosperm systematics and phylogeny as evidenced by the fossil record. *Bot. Rev.* 50:1–111.

Millay, M. A. and T. N. Taylor. 1979. Paleozoic seed fern pollen organs. *Bot. Rev.* 45:301–375.

Miller, C. N. 1971. Evolution of the fern family Osmundaceae based on anatomical features. *Contrib. Univ. Michigan Mus.Paleo.* 23:105–169.

—— 1975. Petrified cones and needle bearing twigs of a new taxodiaceous conifer from the early cretaceous of California. *Amer. J. Bot.* 62:706–713.

—— 1976. Early evolution in the Pinaceae. *Rev. Palaeobot. Palynol.* 21:101–117.

—— 1977. Mesozoic conifers. *Bot. Rev.* 43:218–280.

—— 1982. Current status of Paleozoic and Mesozoic conifers. *Rev. Palaeobot. Palynol.* 37:99–144.

—— 1985a. *Pityostrobus pubescens*, a new species of pinaceous cones from the late Cretaceous of New Jersey. *Amer. J. Bot.* 72:520–529.

—— 1985b. A critical review of S. V. Meyen's "Basic features of gymnosperm systematics and phylogeny as evidenced by the fossil record." *Bot. Rev.* 51:295–318.

Miller, C. N. and J. T. Brown. 1973. A new voltzialean cone bearing seeds with embryos from the Permian of Texas. *Amer. J. Bot.* 60:561–569.

Miller, C. N. and C. A. LaPasha. 1983. Structure and affinities of *Athrotaxites berryi* Bell, an early Cretaceous conifer. *Amer. J. Bot.* 70:772–779.

Mitra, A. K. 1927. On the occurrence of two ovules on araucarian cone-scales. *Ann. Bot.* 41:461–471.

Namboodiri, K. K. and C. B. Beck. 1968a. A comparative study of the primary vascular system of conifers. 1. Genera with helical phyllotaxis. *Amer. J. Bot.* 55:447–457.

—— 1968b. A comparative study of the primary vascular system of conifers. 2. Genera with opposite and whorled phyllotaxis. *Amer. J. Bot.* 55:458–463.

Ogura, Y. 1930. On the structure and affinities of some Cretaceous plants from Hokkaido. *Tokyo Univ. Fac. Sci. Jour.* sec. 3 Botany 2:381–412.

Pant, D. D. and D. D. Nautiyal. 1967. On the structure of *Buriadia heterophylla* (Feistmantel) Seward & Sahni and its fructifications. *Philos. Trans. Roy. Soc. London* 252B:27–48.

Penny, J. S. 1947. Studies on conifers of the Magothy flora. *Amer. J. Bot.* 34:281–296.

Peters, M. D. and D. C. Christophel. 1978. *Austrosequoia winttonensis*, a new taxodiaceous cone from Queensland, Australia. *Can. J. Bot.* 56:3119–3128.

Praeger, E. M., D. Fowler, and A. C. Wilson. 1976. Rates of evolution in conifers (Pinaceae). *Evol.* 30:637–649.

Price, R. A., J. Olsen-Stojkovich, and J. M. Lowenstein. 1987. Relationships among genera of Pinaceae: An immunological comparison. *Syst. Bot.* 12:91–97.

Roselt, G. 1958. Neue Koniferen aus dem Unteren Keuper und ihre Beziehungen zu verwandten fossilen und rezenten. *Friedrich-Schiller-Univ. Wiss. Zeitschr.*, Jahrg. 5:75–118.

Rothwell, G. W. 1982a. New interpretations of the earliest conifers. *Rev. Palaeobot. Palynol.* 37:7–28.

—— 1982b. *Cordaianthus duquesnensis* sp. nov., anatomically preserved ovulate cones from the Upper Pennsylvanian of Ohio. *Amer. J. Bot.* 69:239–243.

—— 1987. The role of development in plant phylogeny: A paleobotanical perspective. *Rev. Palaeobot. Palynol. 50:97–114.*

Rothwell, G. W. and S. Warner. 1984. *Cordaixylon dumusum*, n. sp. (Cordaitales). 1. Vegetative structures. *Bot. Gaz.* 145:275–291.

Schlarbaum, S. E. and T. Tsuchiya. 1984. Cytotxonomy and phylogeny in certain species of Taxodiaceae. *Pl. Syst. Evol.* 174:29–54.

Schweitzer, H. J. 1963. Der weibliche Zapfen von *Pseudovoltzia liebeanna* und seine Bedeutung für die Phylogenie der Koniferen. *Palaeontographica* 113B:1–29.

Singh, H. 1961. The life history and systematic position of *Cephalotaxus drupacea* Sieb. et Zucc. *Phytomorphology* 11:153–197.

—— 1978. *Embryology of Gymnosperms: Encyclopedia of Plant Anatomy.* Berlin: Gebruder Borntraeger.

Sinnott, E. W. 1913. The morphology of the reproductive structures in the Podocarpineae. *Ann. Bot.* 27:39–82.

Stein, W. E. 1987. Phylogenetic analysis and fossil plants. *Rev. Palaeobot. Palynol.* 50:31–61.

Stein, W. E., D. Wight, and C. Beck. 1984. Possible alternatives for the origin of the Sphenopsida. *Sys. Bot.* 9:102–118.

Stockey, R. A. 1977. Reproductive biology of the Cerro Cuadrado (Jurassic) fossil conifers: *Pararaucaria patagonica. Amer. J. Bot.* 64:733–744.

—— 1978. Reproductive biology of Cerro Cuadrado fossil conifers: Ontogeny and reproductive strategies in *Araucaria mirabilis* (Spegazzini) Windhausen. *Palaeontographica* 166B:1–15.

—— 1980a. Anatomy and morphology of *Araucaria sphaerocarpa* Carruthers from the Jurassic Inferior Oolite of Bruton, Somerset. *Bot. Gaz.* 141:116–124.

—— 1980b. Jurassic araucarian cone from southern England. *Palaeontology* 23:657–666.

—— 1982. The Araucariaceae: An evolutionary perspective. *Rev. Palaeobot. Palynol* 37:133–154.

Suzuki, M. 1979a. The course of resin canals in the shoots of conifers. 1.

Taxaceae, Cephalotaxaceae and Podocarpaceae. *Bot. Mag. Tokyo* 92:235–251.

—— 1979b. The course of resin canals in the shoots of conifers. 2. Araucariaceae, Cupressaceae and Taxodiaceae. *Bot. Mag. Tokyo.* 92:253–274.

—— 1979c. The course of resin canals in the shoots of conifers. 3. Pinaceae and summary. *Bot. Mag. Tokyo* 92:333–353.

Taylor, T. N. 1965. Paleozoic seed studies: A monograph of the American species of *Pachytesta*. *Palaeontographica,* 117B;1–46.

—— 1981. *Paleobotany: An Introduction to Fossil Plant Biology.* New York: McGraw-Hill.

Townrow, J. A. 1967a. On *Rissikia* and *Mataia*, podocarpaceous conifers from the Lower Mesozoic of southern lands. *Pap. Proc. Roy. Soc. Tasmania* 101:103–136.

—— 1967b. On a conifer from the Jurassic of east Antarctica. *Pap. Proc. Roy. Soc. Tasmania* 101:137–147.

—— 1969. Some Lower Mesozoic Podocarpaceae and Araucariaceae. in *Gondwana Stratigraphy*, pp. 159–184. UNESCO. Gap, France: Louis-Jean.

Ueno, J. 1960. Studies on pollen grains of Gymnospermae, concluding remarks to the relationships between Coniferae. *J. Inst. Polyt. Osaka* 11:109–136.

Vaudois, N. and C. Prive. 1971. Révision des bois fossiles de Cupressaceae. *Palaeontographica* 134B:61–86.

Vishnu-Mittre. 1958. Studies on the fossil flora of Nipania (Rajmahal series), Bidar—Coniferales. *Paleobotanist* 6:82–122.

Visscher, H., J. H. F. Kerp, and J. A. Clement-Westerhof. 1986. Aspects of Permian paleobotany and palynology. 4. Towards a flexible system of naming Paleozoic conifers. *Acta Bot. Neerl.* 35:87–99.

Watson, J. 1982. The Cheirolepidiaceae: A short review. *Phyta, Pant Comm. Vol.,* pp. 265–273.

Wilde, M. 1975. A new interpretation of microsporangiate cones in the Cephalotaxaceae and Taxaceae. *Phytomorphology* 25:434–450.

Wilde, M. and A. J. Eames. 1948. The ovule and "seed" of *Araucaria bidwilli* with discussion of the taxonomy of the genus. 1. Morphology. *Ann. Bot.,* 12:311–326.

—— 1955. The ovule and "seed" of *Araucaria bidwilli* with discussion of the taxonomy of the genus. 3. Anatomy of multiovulate cone scales. *Ann. Bot.,* 19:343–349.

Wiley, E. O. W. 1981. *Phylogenetics: The Theory and Practice of Phylogenetic Systematics.* New York: Wiley.

Zhou, Z. 1983. *Stalagma samara*, a new podocarpaceous conifer with monocolpate pollen from the Upper Triassic of Hunan, China. *Paleontographica* 185B:56–78.

Index

498 INDEX